MECHANICAL ENGINEERING THEORY AND APPLICATIONS

FLUID POWER, MATHEMATICAL DESIGN OF SEVERAL COMPONENTS

MECHANICAL ENGINEERING THEORY AND APPLICATIONS

Additional books in this series can be found on Nova's website under the Series tab.

Additional e-books in this series can be found on Nova's website under the e-book tab.

MECHANICAL ENGINEERING THEORY AND APPLICATIONS

FLUID POWER, MATHEMATICAL DESIGN OF SEVERAL COMPONENTS

JOSEP M. BERGADA
AND
SUSHIL KUMAR

nova publishers
New York

For permission to use material from this book please contact us:
Telephone 631-231-7269; Fax 631-231-8175
Web Site: http://www.novapublishers.com

Library of Congress Cataloging-in-Publication Data

ISBN: 978-1-62948-316-0

Published by Nova Science Publishers, Inc. † New York

CONTENTS

Preface xiii

About the Authors xv

Chapter 1 Introduction 1

Josep M. Bergada

1.1. Fluid, A Molecular Point of View 1

1.2. Fluid, A Thermodynamic Point of View 2

1.3. Fluid, A Mechanical Point of View 4

1.4. Continuum Theory 4

1.5. Local Thermodynamic Equilibrium 5

1.6. Fluid Properties 5

1.6.1. Bulk Modulus of a Fluid 5

1.6.2. Thermal Expansion Coefficient 6

1.6.3. Relation between Fluid Volume, Bulk Modulus and Thermal Expansion Coefficient 7

1.6.4. Effective Bulk Modulus 8

1.6.5. Surface Tension 11

1.6.6. Definition of Viscosity 12

1.7. Fluid Kinematics 15

1.7.1. Concept of Material or Total Derivative 15

1.7.2. Concept of Convective Flow 16

1.7.3. Circulation 17

1.7.4. Streamlines, Path Lines and Streaklines 17

1.7.4.1. Pathlines 18

1.7.4.2. Streaklines 19

1.7.4.3. Streamlines 19

1.7.5. Concept of Vorticity and Non Rotational Flow 21

1.7.6. Kinematic Study of a Fluid Particle 23

1.8. Nomenclature 26

1.9. References 28

Chapter 2 Main Fluid Mechanics Equations **29**
Josep M. Bergada
2.1. Introduction. Reynolds Transport Equation *29*
2.2. Continuity Equation, Integral Form *34*
2.2.1. Continuity Equation, Differential Form *34*
2.3. Momentum Equation, Integral Form *36*
2.3.1. Momentum Equation, Differential Form *37*
2.4. Momentum Equations for a Non Inertial Coordinate System, Integral Form *42*
2.4.1. Momentum Equations for a Non Inertial Coordinate System, Differential Form *45*
2.5. Equation of Angular Momentum for an Inertial Coordinate System. Integral Form *46*
2.5.1. Application of the Angular Momentum Equation to Turbomachinery *48*
2.5.2. Equation of Angular Momentum for Non Inertial Coordinate Systems *50*
2.6. Energy Equation. Integral Form *51*
2.6.1. Composition of the Mechanical Work *52*
2.6.2. Energy Equation Applied to Turbomachinery, Case Thermal and Hydraulic Machines *54*
2.6.3. Energy Equation. Differential Form *57*
2.7. Application of Differential Equations: Flow under Dominant Viscosity *59*
2.7.1. Flow between Two Parallel Plates *59*
2.7.1.1. Plane Couette - Poiseulle Flow 62
2.7.1.1.1. Couette Flow 63
2.7.1.1.2. Hagen-Poiseulle or Plane Poiseulle Flow 63
2.7.2. Time Dependent Flow, Rayleich Flow *64*
2.7.3. Stationary Flow inside Circular Ducts *69*
2.7.3.1. Poiseulle Flow 69
2.7.4. Flow between Annular Tubes *72*
2.7.4.1. Example 1. Flow between Two Concentric Pipes. Boundary conditions 1a 74
2.7.4.2. Example 2. Flow between Two Concentric Pipes. Boundary conditions 4 75
2.7.4.3. Example 3. Flow between Two Concentric Pipes. Boundary conditions 2b 78
2.7.5. Flow between Concentric Rotating Tubes *80*
2.7.5.1. Example 1. Case Boundary Conditions 1 83
2.7.5.2. Example 2. Case Boundary Conditions 2 84
2.7.5.3. Example 3. Case Boundary Conditions 3 86
2.7.5.4. Example 4. Case Boundary Conditions 1 (Modified) 88
2.8. Introduction to Flow with Negligible Acceleration *91*
2.8.1. Introduction *91*
2.8.2. Reynolds Lubrication Theory. Hydrodynamic Plane Journal Bearings *94*

2.8.3. Reynolds Lubrication Equation in Cartesian Coordinates, Case Two Dimensional Flow *100*
2.8.4. Reynolds Lubrication Equation in Cartesian Coordinates and for Two Directional Three Dimensional Time Independent Flow *101*
2.8.5. Reynolds Lubrication Equation in Cartesian Coordinates and for Two Directional Three Dimensional Time Dependent Flow *102*
2.8.6. Flow with Negligible Acceleration, Case Cylindrical Journal Bearings Statically Loaded *104*
2.8.7. Reynolds Equation of Lubrication in Cylindrical Coordinates *111*
2.9. Nomenclature *117*
2.10. References *119*

Chapter 3 Introduction to Computer Fluid Dynamics (CFD) **121**
Sushil Kumar
3.1. Step by Step Numerical Formulation *121*
3.1.1. Selecting an Appropriate Grid and Integration Formulation *121*
3.1.2. Selection of an Appropriate Reference of Frame for the Problem *122*
3.1.3. Selecting Appropriate Boundary Conditions for the Problem *123*
3.2. Basic Fluid Dynamic Equations and Their Physical Interpretation *124*
3.2.1. Understanding Momentum Equation as Flux Equation *125*
3.3. Discretization of Momentum Equation *126*
3.3.1. Temporal Discretization of Generalized Momentum Equation *126*
3.3.2. Spatial Discretization of Generalized Momentum Equation Using Finite Volume Method *127*
3.3.2.1. Source Term Linearization 130
3.3.3. Spatial Discretization of Generalized Momentum Equation Using Finite Difference Method *131*
3.3.4. Pressure and Velocity Coupling for Finite Volume and Finite Difference Method *132*
3.3.5. Spatial Discretization of Generalized Momentum Equation Using Finite Element Method *134*
3.3.5.1. Weak form of NVS 135
3.3.5.2. Galerkin Finite Element Approximation 137
3.4. Solving a Finite Element Problem via Finite Volume Through Coordinate Transformation *138*
3.4.1. Source Term Linearization for Transformed NVS equations *139*
3.4.2. Spatial Discretization of Generalized Transformed Momentum Equation *140*

3.4.3. Different Transformed Momentum Equations in Discrete Form *142*
3.4.4. Pressure and Velocity Coupling for Transformed Equation *143*
3.5. Convergence Criteria *146*
3.5.1. Grid Independency Test *146*
3.6. Closing Remarks *147*
3.6.1. Solving a Steady and Transient Flow Problem *147*
3.6.2. Mesh Topology *147*
3.6.3. Mesh less Method *148*
3.7. Nomenclature *148*
3.8. References *149*

Chapter 4 Valves **151**
Josep M. Bergada
4.1. Introduction *151*
4.2. Conical Seat Relief Valves *154*
4.2.1. Previous Research on Conical Seat Relief Valves *154*
4.2.2. Mathematical Development Based on Laminar Flow across a Conical Valve Seat *158*
4.2.2.1. Theoretical Background 158
4.2.2.2. Force on a Conical Poppet Assuming Laminar Flow 162
4.2.3. CFD modelling *167*
4.2.4. Experimental Results *169*
4.2.5. Conclusion *172*
4.2.6. References *173*
4.3. Some Measured Steady State Characteristics on Proportional Directional Control Valves *174*
4.4. Servovalve Performance *177*
4.4.1. Introduction to the Four Nozzle Two Flapper Single Stage Servovalve *177*
4.4.2. Directional Control Four Nozzle Two Flapper First Stage Servovalve *179*
4.4.2.1. Forces Acting onto the Flappers 179
4.4.2.2. Servovalve Discharge Coefficients 188
4.4.2.3. Flow Instability 190
4.4.2.4. Servovalve Erratic Performance 192
4.4.2.5. Conclusion 204
4.4.2.6. Servovalve Static Performance Curves 205
4.4.2.7. References 206
4.5. Nomenclature *209*

Chapter 5 Pumps and Motors **211**
Josep M. Bergada, Sushil Kumar and John Watton
5.1. Introduction *211*
5.1.1. General Classification of Pumps and Motors *212*
5.1.2. Axial Piston Pump under Research *213*

5.2. Effect of Piston-Barrel Clearance and Grooves 214
5.2.1. Previous Research 214
5.2.2. Mathematical Analysis 216
5.2.3. 2-D CFD Approach 221
5.2.4. Piston-Cylinder Numerical Model under Tilt Conditions 221
5.2.5. Results. Piston without Grooves 226
5.2.5.1. Results. The Effect of Grooves 226
5.2.5.2. Effect of Grooves on Piston-Barrel Pressure Distribution 227
5.2.5.3. Effect of Grooves on Piston-Barrel Leakage 229
5.2.5.4. Effect of the Grooves on Piston-Barrel Cavitation 232
5.2.5.5. Effect of the Grooves on Total Piston Force and Y-Directional Torques 233
5.2.5.6. Effect of the Piston Diameter on Leakage and Torque 234
5.2.6. Conclusion 237
5.2.7. References 238
5.3. Slipper Performance, Effect of Grooves on Slipper Surface 239
5.3.1. Previous Research on Slippers 239
5.3.2. Flat Slipper with Grooves, Static Equations 244
5.3.3. Tilted Slipper with Grooves, Static Analytical Equations 249
5.3.4. CFD Model of Flat Slipper under Static Conditions 258
5.3.5. Flat Slipper with Grooves, Static and Dynamic Numerical Model 260
5.3.6. Numerical Solution Technique 261
5.3.7. Experimental Test Rigs 264
5.3.8. Results 268
5.3.8.1. Leakage and Pressure Distribution for a Non Tilted Static Slipper 268
5.3.8.2. Influence of Groove Position on Non Tilted Static Slipper Leakage and Force 271
5.3.8.3. Non Tilted Dynamic Slipper 273
5.3.8.3.1. Non Tilted Dynamic Slipper, Pressure, Force and Torque, Experimental and Numerical Results 274
5.3.8.3.2. Non Tilted Dynamic Slipper, Effect on Slipper/Swash Plate Leakage, Experimental and Numerical Results 276
5.3.8.3.3. Non Tilted Slipper, Vorticity Inside the Groove Under Static and Dynamic Conditions 278
5.3.8.4. Tilt Slipper, Static Performance 282
5.3.8.4.1. Tilt Static Slipper Leakage 282
5.3.8.4.2. Tilt Static Slipper. Pressure Distribution 285
5.3.8.4.3. Tilt Static Slipper, Vorticity Inside the Groove 287
5.3.8.5. Tilt Slipper, Dynamic Performance 290
5.3.9. Conclusion 292
5.3.10. References 294

5.4. Barrel-Port Plate Performance 297
5.4.1. Previous Research 297
5.4.2. Mathematical Analysis 300
5.4.2.1. Pressure Distribution and Leakage between Barrel and Port Plate. Main Groove Effect 302
5.4.2.2. Barrel/Port Plate, Pressure Distribution and Leakage. Effect of the Entrance Timing Groove 306
5.4.2.3. Force and Torque on the Barrel Due to the Pressure Distribution. Main Groove Effect 308
5.4.2.4. Force and Torque Caused by the Action of the Timing Groove 311
5.4.2.5. The Effect of Cylinder Pressure 312
5.4.3. Barrel Port Plate, Numerical Simulation 313
5.4.4. Experimental Test Rig and Measuring Procedure 314
5.4.5. Results 317
5.4.5.1. Numerical and Analytical Results 317
5.4.5.1.1. Pressure Distribution. 317
5.4.5.1.2. Leakage in the Main Groove and the Timing Groove 318
5.4.5.1.3. Force Acting on the Barrel 319
5.4.5.1.4. Mean Torques about the XX and YY Axes 320
5.4.5.2. Experimental Results 323
5.4.5.2.1. Position Transducers Direct Measurements 323
5.4.5.2.2. Average Distance between Port Plate and Barrel Aluminium Disc 324
5.4.5.2.3. Barrel Dynamics, Fluctuation Wave 329
5.4.5.2.4. Barrel Dynamics, Simulated Results 331
5.4.6. Conclusion 334
5.4.7. References 336
5.5. Spherical Journal Bearing 339
5.5.1. Introduction 339
5.5.2. Mathematical Analysis 340
5.5.3. Results 345
5.5.4. Conclusion 346
5.5.5. References 346
5.6. Piston Pump Full Dynamic Model 347
5.6.1. Introduction 347
5.6.2. Leakage Equation between Piston and Barrel 348
5.6.3. Leakage Equation in the Clearance Tilt Slipper and Swash Plate 349
5.6.4. Leakage Equations in the Clearance Barrel Port Plate 349
5.6.5. Leakage Equation in the Piston Slipper Spherical Journal Bearing 350
5.6.6. Flow Leaving Each Piston-Barrel Chamber 350
5.6.7. Temporal Piston Cylinder Differential Equation 351
5.6.8. Temporal Outflow Ripple, Combination of Nine Pistons 352

5.6.9. Computational Technique *353*
5.6.10. Experimental Test Rig *353*
5.6.11. Results *355*
5.6.11.1. Experimental Results 355
5.6.11.2. Numerical Results 357
5.6.12. Conclusion *358*
5.6.13. References *359*
5.7. Some New Trends on Piston Pumps *360*
5.8. Nomenclature *360*

Chapter 6 Accumulators **367**
Josep M. Bergada
6.1. Introduction to Accumulators *367*
6.2. Types of Accumulators *367*
6.3. Accumulators Design *369*
6.3.1. Accumulator Used As Volume Accumulator/ Energy Storage *369*
6.3.2. Accumulators Used As Pulsation Compensator *373*
6.3.3. Accumulator Used As a Shock Damper *373*
6.4. Accumulators Application *374*
6.4.1. Examples Accumulator Used as Energy Storage *374*
6.4.2. Example Accumulator Used as Pulsation Compensator *383*
6.4.3. Example Accumulator Used as a Shock Damper *384*
6.5. Nomenclature *385*
6.6. References *385*

Chapter 7 Contamination Control in Fluid Power Systems **389**
Josep M. Bergada
7.1. Introduction *389*
7.2. Sorts of Contamination *389*
7.2.1. Definitions Regarding Filtration *391*
7.2.2. Sort of Wear and Erosion in Hydraulic Systems *392*
7.3. Hydraulic Filters *393*
7.3.1. Filtering Elements *395*
7.3.2. Pressure Losses in Filters *395*
7.4. Strategies for Contamination Control *397*
7.5. Nomenclature *399*
7.6. References *399*

Chapter 8 Introduction to Cartridge Valves **401**
Josep M. Bergada
8.1. Introduction *401*
8.2. Cartridge Valves, Main Parts and Classification *402*
8.3. Main Cartridge Valve Configurations *405*
8.3.1. Main Configurations of Pressure Control Cartridge *405*
8.3.1.1. Pressure Relief Cartridge Valves 405
8.3.1.2. Pressure Reducing Cartridge Valves 411

8.3.2. Directional Cartridge Valves *415*
8.3.3. Flow Control Cartridge Valves *419*
8.4. Example of Application *425*
8.5. References *426*

Index **427**

PREFACE

Fluid Power merges the knowledge of three different basic fields, Control, Fluid Mechanics and Materials Technology. Control is needed to characterize the dynamics of each component, to control the dynamic performance, position, pressure etc. of a given component or circuit, electronics play a decisive role in this field. Fluid mechanics provides the tools to understand fluid behavior, static and dynamic fluid forces on components, phenomena like water hammer and cavitation among others, need to be understood thanks to a proper fluid mechanics background. Choosing materials with appropriate elasticity, hardness, friction properties etc. is also a decisive factor in the design of fluid power components. Nevertheless as these components are usually being provided by the manufacturers, the user cannot in general play with this parameter.

The present book focuses on the fluid mechanics understanding of several components. The book first three chapters are designed to give a proper background to the reader regarding the main fluid characteristics, chapter 1, the main fluid mechanics equations, chapter 2 and a strategic background of the Computer Fluid Dynamics (CFD) techniques, chapter 3. It must be kept in mind that nowadays, conventional mechanics as well as fluid mechanics, are fully immersed in the CFD era, therefore the components design desperately needs the use of this relatively new tool.

Chapter 4 introduces original research based on fluid mechanics understanding of relief valves and servovalves, dynamic and stability considerations are being given in both cases, hints to solve stability problems are provided. Chapter 5 also provides original research on, very likely, the most complex machines in the fluid power field; these are piston pumps and motors. In fact, chapter 5 focuses on axial piston pumps, although the information gathered in this chapter can be directly extrapolated to other piston pumps and motors configurations. In Chapter 5 the reader will find a thorough mathematical description of how slippers with non vented grooves can be designed, the effect of grooves on pistons is also thoroughly analyzed; the barrel dynamic movements are also being introduced, piston pump pressure dynamics under different operating conditions is among the information to be found in this chapter. In all cases, the reader will be able to extract ideas of how a proper design shall be obtained. It is important o highlight that all experiments presented in chapter 5 were done by the book first author in the Professor John Watton fluid Power laboratory at Cardiff University UK, this is why Professor John Watton has to be seen as a co-author of this particular chapter.

Chapters 6, 7 and 8 are designed to introduce some details which are often forgotten in many publications, this is the use of accumulators, the importance of proper filtration and the use of cartridge valves whenever fluid pressure and flow are overcoming a certain value. It is crucial to realize that accumulators can vastly improve a given circuit efficiency, often saving large amounts of energy. A proper filtration is crucial to increase components life and prevent system failures.

It is our wish to help Manufacturers, Engineers and Scientist to gather the appropriate knowledge in order to be able to thoroughly design the few fluid power components presented here, may this book serve this purpose.

About the Authors

Josep M. Bergada - Dr. Eng. J. M. Bergada received his PhD from University Politecnica of Catalunya (UPC) (1996), Barcelona, Spain. His dissertation involved CFD flow simulation inside a servovalve and acoustic servovalve vibrations linked with flow instabilities. From 1996 to 2001 he developed several research projects at the Textile Research Institute (UPC). During the period 2000-2010 his research focused on CFD simulations and mathematical development of flow in relieve valves and axial piston pumps, this research being developed in collaboration with Prof. John Watton at Cardiff University UK. Several measurement test rigs were co-designed and build at Cardiff University to validate the theoretical results. From 2011 until the present, his research is based in collaboration with TU-Berlin, and so far the research focused on performance evaluation of fluidic amplifiers, mathematical study of vortexes generated below airplane wings nearby the ground, and dynamic frequency and amplitude variations inside small pipes used in turbulent flow measurements.

From January 1990 until the present, he has always been working in the Fluid Mechanics department at ETSEIAT-UPC. During this period he has been in charge of three main subjects, Fluid Mechanics, Fluid Power and Hydraulic Machinery. He has over 60 papers published in international Journals and national and International conferences. He has written four books on Fluid Mechanics and two solved problems books with other co-authors, related with hydraulic machinery and Fluid Power.

Contact information: Dr. Eng. Josep M. Bergada
Reader in Fluid Mechanics / Assistant Professor
ETSEIAT-UPC
Fluid Mechanics Department
Colon 7-11 08222 Terrassa, Spain
Tel.: 0034-937398771
Fax: 0034-937398101
bergada@mf.upc.edu

Sushil Kumar - Dr. S. Kumar received his PhD from University Politecnica of Catalunya (UPC) (2010), Barcelona, Spain. His dissertation involved research related algorithm development for solving Partial Deferential Equations (PDE) by numerical simulation. A practical case of axial piston pump machine was chosen and complete analyses were performed by doing Numerical Simulations of coupled PDE equations to optimize pump performance.

Prior to Olx, India, he worked as a Research Scientist at CEMEF, France. His work there involved development of a novel numerical technique to transfer data between two meshes for Forging ALE simulations. He also worked as a Data scientist at Essex and Lake Group, India where he mostly focused on predictive modeling. His other educational background includes a B. Tech. (chemical Engineering) which he gained in 2006 from the IIT (Indian Institute of Technology), Guwahati, India.

His current focus involves developing advance analytical techniques and machine learning techniques for modelling and information extraction from structured and un-structured data.

Contact Information: Dr. Sushil Kumar
Senior Research Data Scientist
OLX India
DLF Corporate Park
Ground Floor, Tower – III
M.G. Road
Gurgaon – 12202
Tel. 0091-7838338571
kumarsush@gmail.com

Chapter 5 is written by three authors; J. M. Bergada, S. Kumar and J. Watton.

Professor John Watton
Fluid Power Emeritus Professor
Cardiff University, UK
wattonj@cardiff.ac.uk

Chapter 1

INTRODUCTION

1.1. FLUID, A MOLECULAR POINT OF VIEW

Fluid Mechanics is a branch of physics which focuses in studding the static and dynamic equilibrium of fluids. When studding the fluid under the molecular point of view, it can be established that when fluid is to be found in gaseous phase, it means that intermolecular forces are weak, explaining why molecules separation distance is certainly big.

For fluid in liquid phase, and in order to study the fluid under a molecular point of view, the concept of radial distribution function g (r) shall be employed. Such function is the quotient between the average density of the fluid gathered inside an sphere of generic radius "r", divided by the average density ρ (R_0) of the fluid located inside a sphere of radius R(0), understanding that millions of molecules fit in the mentioned space. It is to be noticed that the radial distribution function, g (r) can reach a value smaller or bigger than one.

$$g(r) = \frac{\rho(r)}{\rho(R_0)} \begin{cases} <1 \\ =1 \\ >1 \end{cases} \tag{1.1}$$

The radial distribution function can be presented as a function of the radius r sphere, see figure 1.1 Notice that when the sphere has a very small generic radius. Which means, very few molecules are inside the generic sphere, the radial distribution function tend to zero, as the number of molecules inside the generic radius sphere increases, the value of the radial distribution function tends to one.

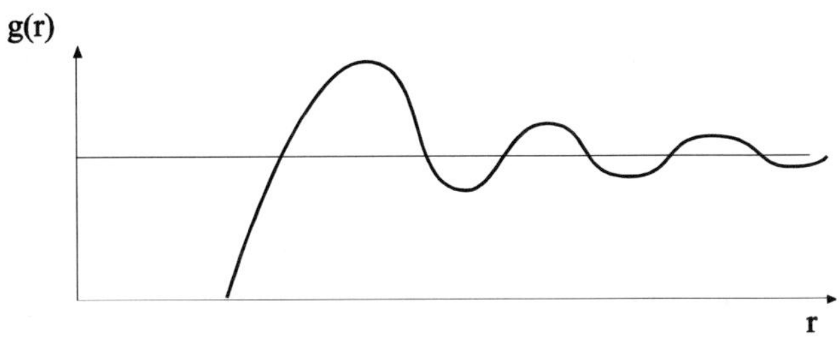

Figure 1.1. Radial distribution function.

To understand why density suffers such variations, it is necessary to study the molecules attraction and repulsion forces. Figure 1.2 presents the attraction and repulsion forces as a function of the distance between two molecules, Notice that at very tiny distances, the repulsion forces are much bigger than the attraction ones, but once the distance r_0 is overcome both forces are nearly under equilibrium, just the attraction forces are slightly bigger than the repulsion forces.

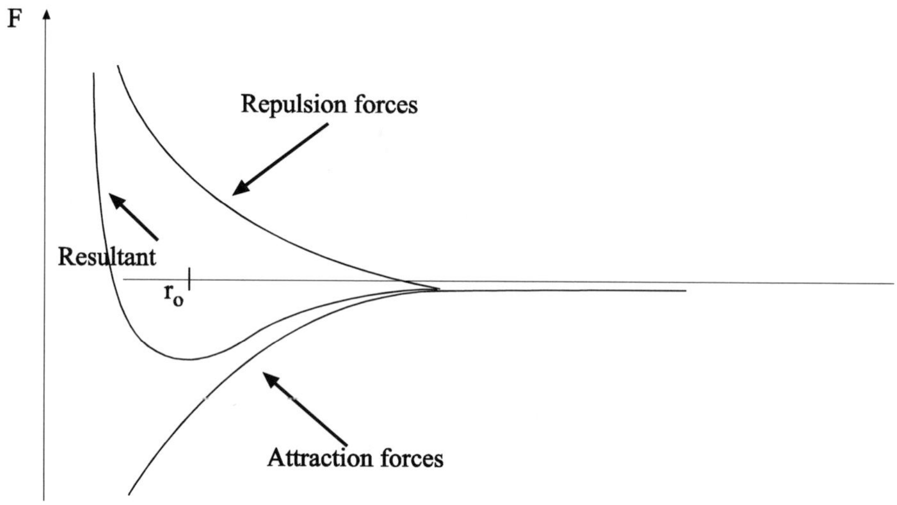

Figure 1.2. Attraction and repulsion molecular forces.

The energy necessary to displace an atom a distance dr against a force F(r), it is being calculated as: $du = -F(r)\,dr$, the sign (-) establishes that as the radius r increases, the force F(r) decreases. The total energy required to bring an atom from the infinite to a distance "r" can be defined:

$$u = -\int_{\infty}^{r} F(r)dr \tag{1.2}$$

This equation it is called Lennard-Jones potential.

1.2. Fluid, a Thermodynamic Point of View

From the thermodynamic point of view, the matter can be taken three different states, liquid solid and gaseous. Therefore, whenever fluid is being under consideration, it is necessary to consider its thermodynamic status.

Figure 1.3 presents the diagram pressure-temperature P-T of water. Notice that the three states are clearly seen. The critical point and triple point are clearly defined.

The critical parameters, critical pressure, critical temperature, critical volume and critical enthalpy, are defined in this point. Under these conditions, the fluid is able to change phase, from liquid to vapour or vice versa without any head addition or subtraction, (vaporization heat is null). In a homologous way, in the triple point, sublimation heat is also zero.

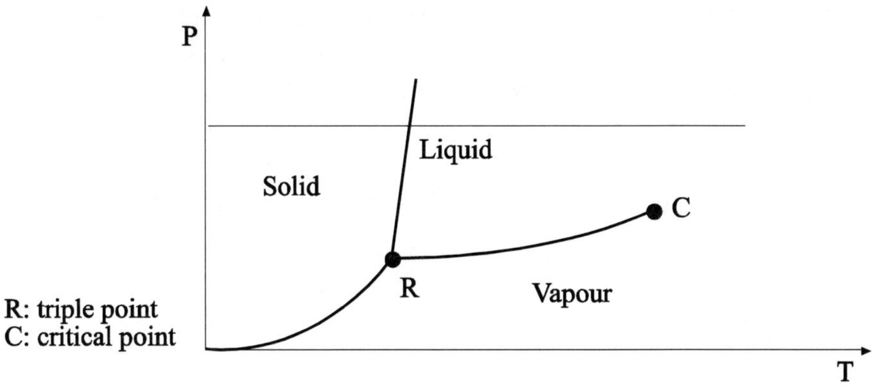

Figure 1.3. Pressure temperature diagram for water.

Figures 1.4 and 1.5, show the diagrams P-V (pressure-volume) and T-S (temperature-entropy) for water. Again can be clearly seen the three matter states as well as the critical and triple points.

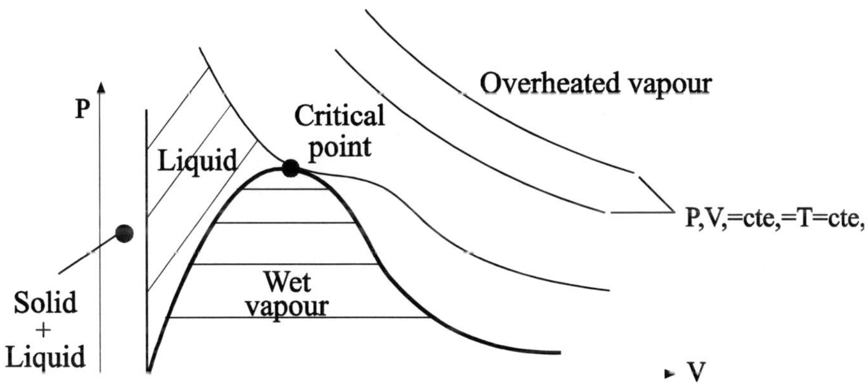

Figure 1.4. Pressure volume diagram for water.

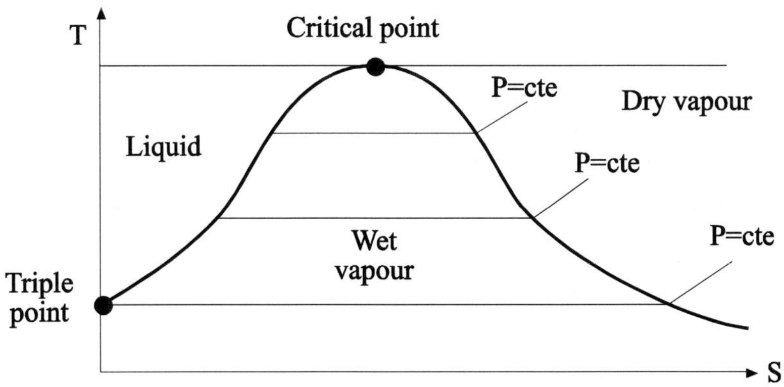

Figure 1.5. Temperature entrophy diagram for water.

Under the thermodynamic point of view, it can be concluded that whenever talking about fluid, it needs to be clarified which is the fluid thermodynamic state, then fluid properties very much depend on the thermodynamic conditions the fluid is subjected to.

1.3. FLUID, A MECHANICAL POINT OF VIEW

Matter is to be seen as fluid if experiences a continuum deformation while subjected to a tangential tension. Liquids and gasses cannot hold tangential tensions without appearing a velocity gradient. A solid matter on the other hand, requires a finite tension before deformation appears.

The non dimensional number called Deborah (De) number, allows, from a mechanical point of view and based on experimental measurements, to determine if the matter under study is a fluid or a solid. The Deborah number definition is the quotient between the time during which a tangential tension is applied to the body under study and the time needed to evaluate the deformation appearing into the body.

$$D_e = \frac{\tau}{t_0}; \frac{\text{relaxation time}}{\text{observation time}}; \frac{\text{Time during which a tension is applied}}{\text{time needed to evaluate the deformation velocity}} \tag{1.3}$$

If the observation time is longer than the time the tangential tension is applied, the matter under study has to be seen as a fluid, for a solid, the observation time will always be smaller than the relaxation time.

Therefore:

De << the substance is a fluid

De >> the substance is a solid

1.4. CONTINUUM THEORY

In theory, it is possible to describe the behavior of a substance in any state via studding the dynamics of the molecules. In reality, this is impossible, due to the huge number of molecules a given substance is having. Nearly in all cases it is possible to ignore the molecular nature of the matter and therefore can be seen as continuum. As a result, the physical and chemical phenomena can be usually studied in a macroscopic scale, the molecular structure of any substance can be generally ignored, fluid will be considered as isotropic. As a consequence of the previous statement, properties defining a substance represent in reality the average characteristics of its molecular structure. Properties will be described as continuum functions along time and space.

In continuum mechanics, it is sufficient to study the density, velocity and internal energy as a function of position and time.

1.5. LOCAL THERMODYNAMIC EQUILIBRIUM

Whenever a fluid is to be studied, it will be considered that each fluid differential is in mechanical and thermal equilibrium with the surrounding differentials of fluid.

Thermodynamics show that the macroscopical state of a fluid under equilibrium can be defined via employing some state variables, like: pressure, density, temperature, entropy, internal energy etc. Thermodynamics also clarify that if fluid is homogeneous, it is sufficient to know two state variables to be able to find out the rest as a function of these two. State equations link the different state variables.

Fluid Mechanics it is characterized for the existence of non uniformity in the mechanical and thermal properties of a fluid. Nevertheless, (at least for gases), the kinetic theory shows that, whenever the average molecular distance can be regarded as small when compared with the characteristic length of the macroscopical no uniformities, and the time between molecular collisions is also small when compared with the time needed for a macroscopical variable to experience a local change, exist local thermodynamic equilibrium.

Such hypothesis can be justified for the fact that a molecule is having a great number of collisions with its neighbors before reaching regions where macroscopic magnitudes are different, therefore the fluid particle adapts its movement and energy to the ones existing locally and keeps on loosing memory of its previous states.

Knudsen number measures the relation between the average molecular distance λ and the characteristic macroscopic length L in which fluid properties change, $Kn = \lambda/L$. Whenever $Kn \ll 1$, it will be considered that local thermodynamic equilibrium exists. Although in reality it is also necessary to fulfill that the time between molecular collisions is small versus the time needed for the macroscopic variables to experience appreciable local changes.

1.6. FLUID PROPERTIES

Fluid properties can be subdivided into mechanical and thermal. Some mechanical properties of fluids are, pressure, bulk modulus, density, surface tension and fluid viscosity.

Some of the thermal properties are, temperature, internal energy, enthalpy, entropy, specific heat at constant at pressure and at constant volume and thermal expansion coefficient.

In what follows it is presented a brief description of few of these properties.

1.6.1. Bulk Modulus of a Fluid

The bulk modulus of a fluid β is a property which indicates how easy a fluid can be compressed when submitted to a pressure differential. It is defined as:

$$\beta = -\forall \frac{dP}{d\forall} = -\upsilon \frac{dP}{d\upsilon} = \rho \frac{dP}{d\rho} \tag{1.4}$$

The bulk modulus is equivalent to Young modulus in solid mechanics.
For a gas it is stated $P = P(\rho, T)$

Therefore it can be defined: $dP = \left(\frac{\partial P}{\partial \rho}\right)_T d\rho + \left(\frac{\partial P}{\partial T}\right)_\rho dT$ (1.5)

If the process is at constant temperature, the isothermal bulk modulus is given as:

$$\beta_{v,t} = \rho\left(\frac{dP}{d\rho}\right) = \rho\frac{\left.\frac{\partial P}{\partial \rho}\right|_T d\rho}{d\rho} = \rho\left(\frac{\partial P}{\partial \rho}\right)_T \quad (1.6)$$

And if the gas can be considered as ideal, the following expression is relevant $P = \rho RT$, where:

$\beta_{v,t} = \rho\left(\frac{\partial(\rho RT)}{\partial \rho}\right)_T = \rho RT = P$ The bulk modulus is in reality the pressure of the gas.

In general, for any fluid, the bulk modulus generic equation takes the form $\frac{d\rho}{\rho} = \frac{dP}{\beta}$. This equation can be used to mathematically define a fluid as compressible or incompressible. If it is for example accepted that a density change of 1% is non significant, a fluid shall be regarded as incompressible whenever:

$$\frac{dP}{\beta} < 0,01\,; \quad (1.7)$$

Relating the pressure differential with the kinetic energy associated to the fluid, it is established:

$$\Delta P = \frac{1}{2}\rho\, v^2 \quad (1.8)$$

Then, the consideration of incompressible fluid can be given as $\frac{\rho\, v^2}{\beta} < 0,02$.

For air under standard conditions, this condition is equivalent to a velocity around 40 m/s.

1.6.2. Thermal Expansion Coefficient

A property equivalent to bulk modulus β, but from the thermal point of view is the thermal expansion coefficient, α_T. This property measures the fluid expansion effect as a function of temperature variation, its mathematical formulation is:

$$\alpha_T = \frac{1}{\forall}\frac{d\forall}{dT} = \frac{1}{\upsilon}\frac{d\upsilon}{dT} = -\frac{1}{\rho}\frac{d\rho}{dT} \quad (1.9)$$

All fluids have a relationship between density, pressure and temperature, $\rho = (P,T)$, then, the density variation as a function of fluid pressure and temperature is mathematically expressed as:

$$d\rho = \left(\frac{\partial \rho}{\partial T}\right)_P dT + \left(\frac{\partial \rho}{\partial P}\right)_T dP \tag{1.10}$$

For a thermal expansion process at constant pressure, the previous equation reads:

$$d\rho = \left(\frac{\partial \rho}{\partial T}\right)_P dT\ ; \frac{d\rho}{dT} = \left(\frac{\partial \rho}{\partial T}\right)_P \tag{1.11}$$

Substituting equation (1.11) in (1.9) the thermal expansion coefficient can be given as:

$$\alpha_{T,P} = -\left(\frac{1}{\rho}\frac{\partial \rho}{\partial T}\right)_P , \tag{1.12}$$

Equation (1.12) characterizes the thermal expansion coefficient for a constant pressure process.

For an ideal gas it can be established:

$$\rho = \frac{P}{RT} \tag{1.13}$$

Therefore for a constant pressure process, it can be said:

$$\alpha_{T,P} = -\left(\frac{1}{\rho}\frac{\partial}{\partial T}\left(\frac{P}{RT}\right)\right)_P = -\frac{1}{\rho}\left[\frac{1}{RT}\left(\frac{\partial P}{\partial T}\right)_P + P\left(\frac{\partial}{\partial T}\left(\frac{1}{RT}\right)\right)_P\right] = -\frac{1}{\rho}\frac{P}{R}\left(-\frac{1}{T^2}\right) = \frac{1}{T} \tag{1.14}$$

The conclusion is that the thermal expansion coefficient of a gas is equal to the inverse of its temperature.

1.6.3. Relation between Fluid Volume, Bulk Modulus and Thermal Expansion Coefficient

Consider a given volume of fluid, $\forall$. If it is required to study the variation of volume as a function of both pressure and temperature variation, it can be established:

$$d\forall = \left(\frac{\partial \forall}{\partial T}\right)_p dT + \left(\frac{\partial \forall}{\partial P}\right)_T dP \tag{1.15}$$

$$\frac{d\forall}{\forall} = \frac{1}{\forall}\left(\frac{\partial\forall}{\partial T}\right)_P dT + \frac{1}{\forall}\left(\frac{\partial\forall}{\partial P}\right)_T dP \tag{1.16}$$

Substituting equations (1.4) and (1.9) in (1.15) it is obtained:

$$\frac{d\forall}{\forall} = \alpha_{T,P} dT - \frac{1}{\beta_{V,T}} dP \tag{1.17}$$

Equation (1.17) characterizes the volume variation of a fluid as a function of pressure and temperature variation, the bulk modulus and the thermal expansion coefficient are the known fluid properties.

The relation between the fluid volume variation with the fluid density variation can be expressed as:

$$\frac{d\forall}{\forall} = \frac{\forall - \forall_0}{\forall} = \frac{\frac{\forall}{m} - \frac{\forall_0}{m}}{\frac{\forall}{m}} = \frac{\upsilon - \upsilon_0}{\upsilon} = \frac{\frac{1}{\rho} - \frac{1}{\rho_0}}{\frac{1}{\rho}} = \frac{\rho_0 - \rho}{\rho_0} = -\frac{\rho - \rho_0}{\rho_0} \tag{1.18}$$

Substituting equation (1.18) in (1.17) the fluid density variation can be expressed as a function of pressure and temperature variations.

$$\frac{\rho - \rho_0}{\rho_0} = \frac{1}{\beta_{V,T}}(P - P_0) - \alpha_{T,P}(T - T_0) \tag{1.19}$$

As an example, for water under thermodynamic conditions $P = 10^5$ Pa and $T = 277$ K, the bulk modulus and thermal expansion factor have a value of: $\beta_{V,T} = 1{,}96.10^9 \frac{N}{m^2}$; $\alpha_{T,P} = 1{,}53.10^{-4} K^{-1}$. It must be kept in mind that all fluid properties depend on the thermodynamic conditions, being the dependence on temperature especially relevant.

1.6.4. Effective Bulk Modulus

When considering fluid elasticity/compressibility, it must be considered the possibility of having a fluid mixture, liquid and gas for example; it is also interesting to consider that the fluid container may expand when submitted to a pressure increase. Effective bulk modulus will consider not only the compressibility of a fluid or mixture of fluids but also the mechanical characteristics of the container surrounding the fluid.

Equation 1.4 defined the concept of bulk modulus of a generic fluid, in what follows this concept will be used to obtain the relation between the effective bulk modulus o a system composed of several fluids and a container. In fluid power systems, the effect of fluid compressibility may induce noise and even vibration in several components, since high frequency resonance is associated to fluid elastic behavior.

Let's assume there is a volume of fluid, liquid and gas which is submitted to a pressure increase.

Understanding that the container is also elastic, the final volume occupied by the liquid and gas, shall be given as the initial volume plus the volume increase due to the container expansion minus the volume decrease due to the fluid mixture contraction.

$$\forall_{\text{final}} = \forall_{\text{liquid}} + \forall_{\text{gas}} = \forall_{\text{initial}} + \forall_{\text{increase container}} - \forall_{\text{decrease fluid}} \quad (1.20)$$

Defining the effective volume $\forall_{\text{effective}}$ as:

$$\forall_{\text{effective}} = \forall_{\text{liquid}} + \forall_{\text{gas}} - \forall_{\text{increase container}} = \forall_{\text{initial}} - \forall_{\text{decrease fluid}} \quad (1.21)$$

Effective volume is mend to be determined easily since is not necessarily involving the knowledge of the container deformation characteristics.

Differentiating the previous equation it is obtained:

$$d\forall_{\text{effective}} = d\forall_{\text{liquit}} + d\forall_{\text{gas}} - d\forall_{\text{increase container}} \quad (1.22)$$

The four terms defined in the previous equation, evaluate each elemental volume variation when submitted to a differential pressure increase. Remembering the generic equation 1.4 it can be established:

$$\beta_{\text{liquid}} = -\forall_{\text{liquid}} \frac{dP}{d\forall_{\text{liquid}}} \quad (1.23)$$

$\forall_{\text{liquid}}$ being the initial volume of liquid contained in the container.

$$\beta_{\text{gas}} = -\forall_{\text{gas}} \frac{dP}{d\forall_{\text{gas}}} \quad (1.24)$$

$\forall_{\text{gas}}$ being the initial volume of gas contained in the container.

$$\beta_{\text{effective}} = -\forall_{\text{effective}} \frac{dP}{d\forall_{\text{effective}}} = -\forall_{\text{initial}} \frac{dP}{d\forall_{\text{effective}}} \quad (1.25)$$

$\forall_{\text{effective}} = \forall_{\text{initial}}$ is the initial volume of liquid and gas contained in the container. Notice that the effective bulk modulus is the parameter needed and represents the compressibility effect of fluid and container.

Using as well the concept of bulk modulus for the container it can be stated:

$$\beta_{container} = \forall_{container} \frac{dP}{d\forall_{increase\ container}} = \forall_{initial} \frac{dP}{d\forall_{increase\ container}} \tag{1.26}$$

Notice that the initial volume of the container is the same as the initial liquid and gas volume.

$\forall_{container} = \forall_{effective} = \forall_{initial}$. It is also important to realize that as fluid pressure increases the volume of the container also increases, this is why equation (1.26) is having a positive sign associated.

Substituting equations (1.23), (1.24), (1.25) and (1.26) into (1.22) it is obtained:

$$\forall_{initial} \frac{dP}{\beta_{effective}} = \forall_{liquid} \frac{dP}{\beta_{liquid}} + \forall_{gas} \frac{dP}{\beta_{gas}} + \forall_{initial} \frac{dP}{\beta_{container}} \tag{1.27}$$

Substituting equation (1.21) into (1.27) it is obtained:

$$\frac{1}{\beta_{effective}} = \left(\forall_{initial} - \forall_{gas} + \forall_{increase\ container}\right) \frac{1}{\forall_{initial}\ \beta_{liquid}} + \frac{\forall_{gas}}{\forall_{initial}} \frac{1}{\beta_{gas}} + \frac{1}{\beta_{container}} \tag{1.28}$$

and after further arrangement it is reached:

$$\frac{1}{\beta_{effective}} = \left(1 + \frac{\forall_{increase\ container}}{\forall_{initial}}\right) \frac{1}{\beta_{liquid}} + \left(1 - \frac{\beta_{gas}}{\beta_{liquid}}\right) \frac{\forall_{gas}}{\forall_{initial}} \frac{1}{\beta_{gas}} + \frac{1}{\beta_{container}} \tag{1.29}$$

And considering: $\forall_{initial} >> \forall_{increase\ container}$; and $\beta_{liquid} >> \beta_{gas}$ from equation (1.29) it is reached:

$$\frac{1}{\beta_{effective}} = \frac{1}{\beta_{liquid}} + \frac{\forall_{gas}}{\forall_{initial}} \frac{1}{\beta_{gas}} + \frac{1}{\beta_{container}} \tag{1.30}$$

Equation (1.30) is just an approximation, nevertheless is has to be seen as a rather accurate equation. Notice that the effective bulk modulus not only depends on the liquid, gas and container bulk modulus, but also on the gas volume, usually air, mixed with the fluid. This particular parameter is very difficult to evaluate although ideally there should be no air mixed with the hydraulic oil. Chapter 7 of the present book clarifies that proper filtering removes air from hydraulic systems.

Therefore assuming: $\forall_{initial} >> \forall_{gas}$, equation (1:30) can be reduced to:

$$\frac{1}{\beta_{effective}} = \frac{1}{\beta_{liquid}} + \frac{1}{\beta_{container}} \tag{1.31}$$

Equation (1.31) is the simplest one to be obtained and evaluates the combined effect of liquid and container (usually a pipe) compressibility. Despite its simplicity, equation (1.31) is to be seen as precise enough in many real applications. It is interesting to realize that the effective bulk modulus will always be smaller than the liquid or pipe bulk modulus, notice as well that whenever the pipe is not metallic effective bulk modulus might be very much affected by it.

1.6.5. Surface Tension

Surface tension, usually being given by the Greek letter σ, $\sigma = \left[\frac{N}{m}\right]$, describe the forces appearing in the interface between a liquid and a gas, or between two liquids or between a liquid a solid and a gas. These forces are weak because its origin is the non equilibrium molecules at the liquid surface.

Surface tension plays a decisive role in problems involving liquid droplets, jet sprays and liquids under no gravity conditions among others.

In order to demonstrate the weak character of the forces due to surface tension, the experiment depicted in figure 1.6 is presented. A metal wire having a U form has another straight piece of metal wire which can slide along the previous one. In the closed surface formed between the sliding wire and the U form one it is placed a soap film, there is no need to say that the soap film thickness is very small.

The soap film generates a force which tends to decrease the soap film area. The external opposite force needed to maintain the sliding wire in position can be calculated as $F = 2\ \sigma\ l$. The number two is due to the fact that the soap film has two surfaces in contact with the air, the upper surface and the lower one, in both surfaces exists molecules in non equilibrium state, the ones responsible of the force generated. If the wire is allowed to move a distance Δx, the work produced will be: $W = F(\Delta x) = 2\ \sigma\ l\ (\Delta x) = 2\ \sigma\ \Delta S$, being ΔS the surface differential swept by the sliding wire.

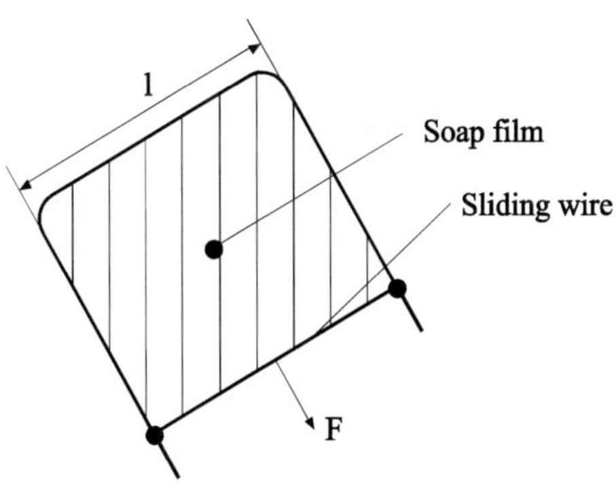

Figure 1.6. Experimental evaluation of surface tension.

1.6.6. Definition of Viscosity

Probably the wider known fluid property is the viscosity; this might be because in fact the reologic equation of the fluids depends on it. Viscosity was initially defined by Sir. Isaac Newton, and he demonstrated that viscosity related the shear stresses applied to a fluid layer with the velocity of the deformation the fluid was experiencing.

To clarify this concept, the following experiment is being presented. Figure 1.7 presents a two dimensional differential of a solid, which it is initially under static conditions, and to which shear stresses on its upper and lower sides are applied. A solid, when submitted to shear stresses it deforms an angle differential $\delta\theta$, which is not depending on the time shear stresses are being applied. Whenever shear stress sees to be applied, the surface differential turns to its original position, (providing the elastic limit has not been overcome). If the same experiment is being undertaken with a differential of fluid, it is observed that while shear stresses are being applied, the fluid keeps on deforming, which means the deformation angle $\delta\theta$ is for a fluid time dependent.

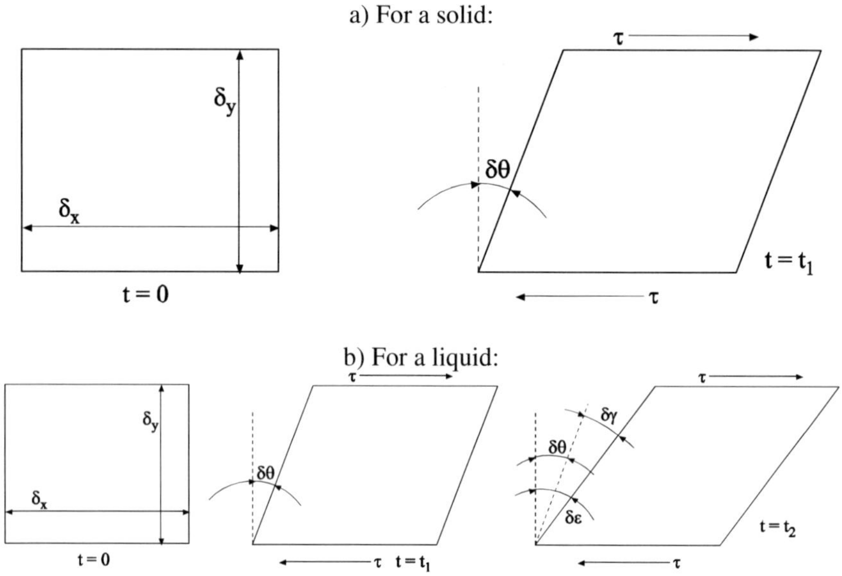

Figure 1.7. Deformation of the solid and liquid differential elements.

Let's suppose now there is a layer of moving fluid, figure 1.8 left hand side, from this layer of fluid it is extracted a surface differential, figure 1.8 right hand side, notice that the fluid velocity onto the upper side of the surface differential is different than the one acting onto the lower side.

The velocity difference between the upper and lower part is $\frac{du}{dy}\delta y$. The distance δe is the distance between the upper and lower part of the surface differential, due to the different velocity associated. This distance δe can be mathematically defined as:

$$\delta e = \frac{du}{dy}\delta y \delta t \cdot \qquad (1.32)$$

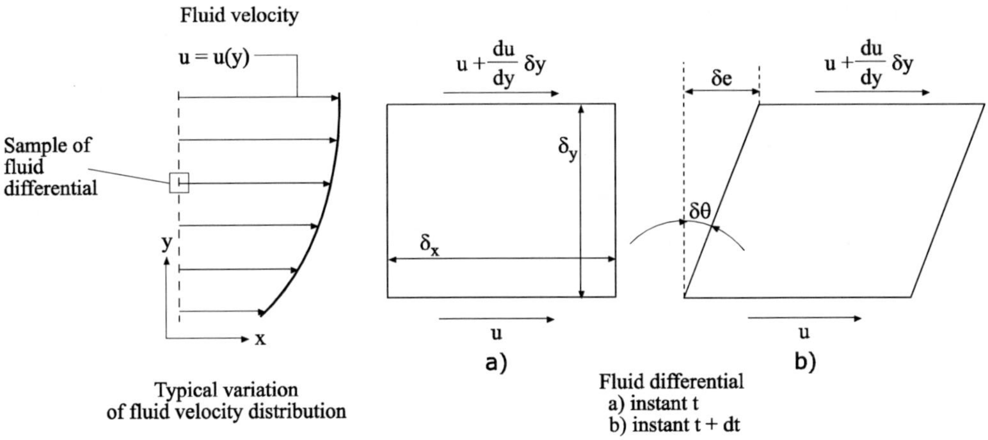

Figure 1.8. Surface differential of a moving fluid.

The angle displaced due to the velocity differential will be very small for a small time differential, calling $\delta\theta$ to the differential angle displaced due to the difference in fluid velocity, it can be calculated as:

$$\delta\theta \approx \tan(\delta\theta) = \frac{\delta e}{\delta y} = \frac{du}{dy}\delta t \tag{1.33}$$

The angular velocity deformation is defined as the temporal variation of angle differential, and according to equation (1.33) can be expressed as:

$$\frac{\delta\theta}{\delta t} = \frac{du}{dy} \tag{1.34}$$

In reality, if a velocity gradient between the upper and lower part of the surface differential exists, it must exist shear stresses between the two layers of fluid. Considering that shear stresses τ are proportional to the angular velocity deformation, it can be stated that $\tau \; \alpha \; \frac{\delta\theta}{\delta t}$, where the parameter α is the constant of proportionality. Such constant of proportionality it has to be a parameter characteristic of the fluid, this parameter is being called absolute viscosity. As a result it is obtained the Newton law of viscosity, also called the reologic equation of a Newtonian fluid, and it is expressed as:

$$\tau = \mu \frac{du}{dy} \tag{1.35}$$

Equation (1.35) can also be given as:

$$\tau = \mu.\dot{\gamma}^{n} \;;\; \text{being } \dot{\gamma} = \frac{\partial u}{\partial y}. \tag{1.36}$$

In reality, the exponent n can take the following values:

For n = 1; the fluid is Newtonian

For n < 1 or n > 1; the fluid is non Newtonian

As a conclusion, Newtonian fluids are the ones having a linear relation between the shear stresses and the angular deformation velocity of a fluid. Figure 1.9 presents the reologic characteristics of several conventional fluids.

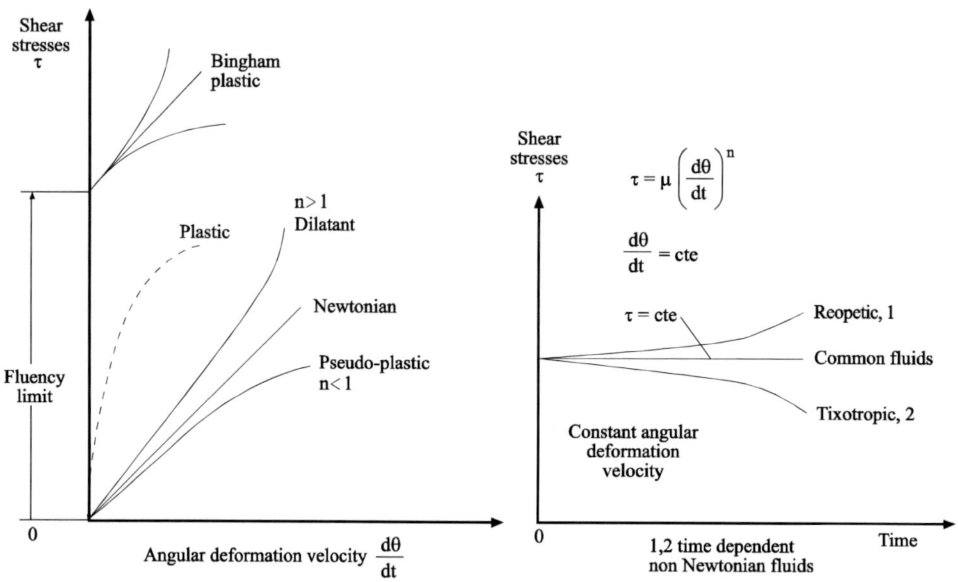

Figure 1.9. Reologic equations of conventional fluid.

A fluid called dilatant, it is characterized by an increase of the resistance to deformation when shear stresses increase.

The fluid called pseudo-plastic has a decrease of the resistance to deformation when shear stresses increase. Whenever this effect is particularly relevant, the fluid is called plastic.

A fluid requiring a minimum shear stress to start flowing, it is called Bingam' plastic, its reologic equation would be:

$$\tau = \tau_0 + \mu . \dot{\gamma}^n ; \qquad 1 \leq n \geq 1 \tag{1.37}$$

Fluids needing a progressive increase of shear stresses to maintain constant the deformation velocity are being called reopectics, and the ones needing a decrease on the shear stresses to maintain constant the deformation velocity are called tixotropic.

Relative or kinematic viscosity is the relation between the absolute or dynamic viscosity and the density.

$$\text{Kinematic viscosity } \nu = \frac{\mu}{\rho} \tag{1.38}$$

1.7. FLUID KINEMATICS

1.7.1. Concept of Material or Total Derivative

Let's define a generic intensive fluid property ϕ (density, temperature, etc.) associated to any fluid particle. Following the movement of a fluid particle, the magnitude ϕ associated to the particle, may change with the position $\vec{x}$ and time t. If the temporal variation of the magnitude ϕ associated to a given fluid particle and versus an observer which is moving with the fluid, is to be determined, it is obtained the material derivative. Developing the generic property ϕ in series, it can be said:

$$\delta\phi = \phi(\vec{x}+\delta\vec{x};\, t+\delta t) - \phi(\vec{x};t) \approx \delta\vec{x}\,\nabla\phi. + \frac{\partial\phi}{\partial t}\delta t + ... \tag{1.39}$$

Dividing the previous equation by the time differential and ignoring the terms associated to high order derivatives, it is reached:

$$\frac{\delta\phi}{\delta t} = \frac{\partial\phi}{\partial t} + \frac{\delta\vec{x}}{\delta t}\nabla\phi \tag{1.40}$$

Expression which can be given as:

$$\frac{d\phi}{dt} = \frac{D\phi}{Dt} = \frac{\partial\phi}{\partial t} + \vec{v}\,\nabla\,\phi \tag{1.41}$$

The term $\frac{D\phi}{Dt}$ is defined as the material derivative of the property ϕ associated to a fluid particle.

The term $\frac{\partial\phi}{\partial t}$ it is called the local derivative, and the term $\vec{v}\nabla\phi$ it is the convective derivative.

In fluid Mechanics, the acceleration a particle is experiencing, it is defined as the material derivative of the particle velocity, and can be expressed as:

$$\vec{a} = \frac{D\vec{v}}{Dt} = \frac{\partial\vec{v}}{\partial t} + (\vec{v}\nabla)\vec{v} = \left(\frac{\partial}{\partial t} + \vec{v}\nabla\right)\vec{v} \tag{1.42}$$

The first term it is called local acceleration and the second one it is called convective acceleration.

Please notice that according to vector algebra, the term, $(\vec{v}.\nabla)\vec{v}$ can be presented as:

$\nabla\left(\frac{v^2}{2}\right)-\vec{v}\wedge(\nabla\wedge\vec{v})$, and it is interesting to notice that the term $\vec{v}\wedge(\nabla\wedge\vec{v})$ explain why a particle tends to change its position.

In Cartesian coordinates and using sub index notation, the particle acceleration can be given as:

$$a_i=\frac{\partial v_i}{\partial t}+v_j\frac{\partial v_i}{\partial x_j} \tag{1.43}$$

If the reference system is non inertial, the absolute acceleration it is obtained via finding out the relative acceleration and adding to it all accelerations associated to the moving system of reference, thus obtaining:

$$-\vec{a}_r=\underset{1}{\vec{a}_0}+\underset{2}{\frac{d\vec{\Omega}}{dt}\wedge\vec{r}}+\underset{3}{\vec{\Omega}\wedge(\vec{\Omega}\wedge\vec{r})}+\underset{4}{2\vec{\Omega}\wedge\vec{v}} \tag{1.44}$$

1. Linear acceleration
2. Tangential acceleration
3. Centripetal acceleration
4. Coriolis acceleration

$\vec{a}_0$ = Is the linear acceleration the moving reference frame is having versus the fixed one.
Ω = Angular velocity associated to the moving reference frame.

1.7.2. Concept of Convective Flow

Let's consider a surface fixed to a reference system, from this surface, a surface differential ds is chosen. All fluid particles able to reach the surface differential ds in a time differential dt, will be the ones located in a distance equal or inferior to $\vec{v}\,dt$. Among them, just the particles having the appropriate orientation $\vec{v}\,d\vec{s}\,dt$ will reach the surface differential ds. The magnitude of ϕ associated to the fluid and crossing the surface differential in a unit of time will be $\phi\,\vec{v}\,\vec{n}\,ds$.

The total flow of property ϕ crossing the entire fixed surface will be given as $\int_S \phi\,\vec{v}\,\vec{n}\,ds$, where the term $\phi\vec{v}\,\vec{n}$ is the flow of property ϕ crossing the surface differential having an orientation $\vec{n}$.

- If ϕ is a scalar, $\phi\bar{v}$ will be a vector denominated vector flow of property ϕ, (if ϕ is for example, the density ρ, the term $\rho\bar{v}$ will be the vector mass flow per unit surface).
- If ϕ is a vector, the term $\phi\bar{v}$ is a tensor called tensor flow of the magnitude ϕ, (if ϕ is the fluid momentum $\rho\bar{v}$, the term $\rho\bar{v}\bar{v}$ will be the momentum flow per unit surface).

If the surface is closed, the term $\phi\bar{v}$ has to be seen as a continuous function, and the Gauss-Ostrogradsky theorem transforming the surface integral to a volume one, can be applied, resulting

$\oint_S \phi \vec{v}\, \hat{n}\, ds = \int_{\forall} \nabla (\phi \vec{v}) d\forall$, where $\nabla(\phi\bar{v})$ is the flow of magnitude ϕ per unit volume.

The concept of volumetric flow is given as

$$\dot{\forall} = Q = \oint_S \vec{v}\, d\vec{s} = \int_{\forall} \nabla\vec{v}\, d\forall = \int_{\forall} \operatorname{div}\vec{v}\, d\forall \tag{1.45}$$

And the mass flow will be given as $\dot{m} = \oint_S \rho \vec{v}\, d\vec{s} = \int_{\forall} \nabla(\rho \vec{v})\, d\forall$ (1.46)

1.7.3. Circulation

The circulation Γ along a given streamline L, it is defined as $\Gamma = \oint_L \vec{v}\, dl$, and it is representing the work developed by the velocity vector $\bar{v}$ along the line L. If the given line L is closed, the stokes theorem says that the circulation equals to the curl vector flux $\nabla \wedge \bar{v}$ across a surface delimitated by the closed line L.

$$\Gamma = \oint_L \vec{v}\, dl = \int_S (\nabla\wedge\vec{v})\, d\vec{s} = \int_S \operatorname{curl}\vec{v}\, d\vec{s} \tag{1.47}$$

If the circulation along a given closed line is null, the curl vector $\nabla \wedge \bar{v}$ will also be zero in all fluid field. $\Gamma = 0 \Rightarrow \nabla \wedge \bar{v} = 0$; the flow will be called non rotational, the velocity $\bar{v}$ derives from a scalar potential. $\bar{v} = \nabla\vartheta$.

The reciprocal is in general not true, although curl along all fluid field is equal to zero, ($\nabla \wedge \bar{v} = 0$), it might be possible to find some closed lines inside the fluid where $\Gamma \neq 0$.

1.7.4. Streamlines, Path Lines and Streaklines

Given a vectorial velocity field which characterizes the moving fluid, fluid particles spatial movement can be described once known the position of the particles at initial time. To do so, the concepts of streamlines, streaklines and pathlines will be used.

Streamlines are defined as a family of curves that are instantaneity tangent to the velocity vectors associated to the flow.

Steaklines are the locus of points of all fluid particles that have passed continuously through a particular spatial point in the past. The dye steadily injected into the fluid at a fixed point, extends along a streakline.

Pathlines are described as the trajectories followed by individual fluid particles. It can be seen as the time recording of the path of a fluid element.

Each of these three lines is characterized by a differential equation; its form will depend on the coordinate system used. It is also interesting to point out that these three lines fall into a single one for a time independent fluid flow.

The definition of a fluid line is of a set of particles which at a given time follow a straight line. No mathematical description is associated to this line.

The differential equations describing these lines are presented next.

1.7.4.1. Pathlines

Pathlines are given as a function of a velocity field according to the following differential equation.

$$\frac{d\vec{x}}{dt} = \vec{v}(\vec{x}, t) \tag{1.48}$$

The boundary conditions to solve the differential equations are: at initial time, it is known the position of the fluid particle. $t = 0; \quad \vec{x}(t_0) = \vec{x}_0$.

Extending equation (1.48) for the different coordinate systems, it is obtained:

For a Cartesian coordinate system:

$$\vec{x} = (x, y, z); \quad \vec{v} = (v_X, v_Y, v_Z)$$

$$\frac{dx}{dt} = v_X(\vec{x}, t); \quad \frac{dy}{dt} = v_Y(\vec{x}, t); \quad \frac{dz}{dt} = v_Z(\vec{x}, t) \tag{1.49}$$

In cylindrical coordinates it is given as:

$$\vec{x} = (r, \theta, z); \quad \vec{v} = (v_r, v_\theta, v_Z)$$

$$\frac{dr}{dt} = v_r(\vec{x}, t); \quad r\frac{d\theta}{dt} = v_\theta(\vec{x}, t); \quad \frac{dz}{dt} = v_Z(\vec{x}, t) \tag{1.50}$$

Pathlines differential equations in spherical coordinates are:

$$\vec{x} = (r, \theta, \phi); \quad \vec{v} = (v_r, v_\theta, v_\phi)$$

$$\frac{dr}{dt} = v_r(\vec{x}, t); \quad r\,\mathrm{sen}\phi\frac{d\theta}{dt} = v_\theta(\vec{x}, t); \quad r\frac{d\phi}{dt} = v_\phi(\vec{x}, t) \tag{1.51}$$

The boundary conditions needed to solve the differential equations, and for each coordinate system are:

Cartesian coordinates: $t=0;\ x=x_0;\ y=y_0;\ z=z_0$

Cylindrical coordinates: $t=0;\ r=r_0;\ \theta=\theta_0;\ z=z_0$

Spherical coordinates: $t=0;\ r=r_0;\ \theta=\theta_0;\ \phi=\phi_0$

1.7.4.2. Streaklines

These lines study a set of particles passing through a given point at different times, $\vec{x}=\vec{x}(\vec{x}_0,t_0,t)$. Differential equations defining Streaklines are the same as the ones defining Pathlines. The difference between them is to be seen in the boundary conditions. The boundary conditions for Streaklines in different coordinate systems are:

Cartesian coordinates: $t=\varepsilon;\ x=x_0;\ y=y_0;\ z=z_0$

Cylindrical coordinates: $t=\varepsilon;\ r=r_0;\ \theta=\theta_0;\ z=z_0$

Spherical coordinates: $t=\varepsilon;\ r=r_0;\ \theta=\theta_0;\ \phi=\phi_0$

Notice that boundary conditions used to calculate Streaklines indicate that it is necessary to determine the equation of all trajectories passing to the reference point (x_0, y_0, z_0) in different instant times $t_0=\varepsilon$ and eliminate t_0 from the resulting equations.

1.7.4.3. Streamlines

To determine flow streamlines, it is necessary to study a set of particles which in a given instant and at all particle locations are tangent to the velocity vector. In Cartesian coordinates, streamlines differential equation it is obtained from the relation presented in figure 1.10. Notice that the proportionality between the tangent vector and the velocity vector is what evaluates the streamlines differential equation.

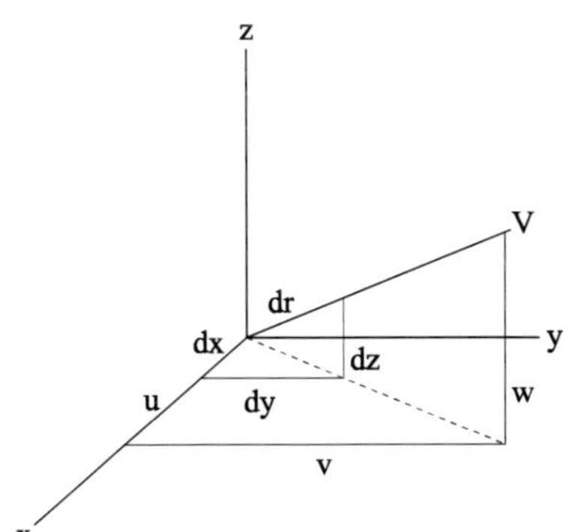

Figure 1.10. Geometric relation used to determine the streamlines differential equation.

Streamlines differential equations for the different coordinate systems are:

Cartesian coordinates:

$$\vec{x}=(x,y,z);\quad \vec{v}=(v_X,v_Y,v_Z)$$

$$\frac{dx}{v_X(\vec{x},t)} = \frac{dy}{v_Y(\vec{x},t)} = \frac{dz}{v_Z(\vec{x},t)} \tag{1.52}$$

Cylindrical coordinates:

$$\vec{x} = (r,\theta,z); \quad \vec{v} = (v_r, v_\theta, v_Z)$$

$$\frac{dr}{v_r(\vec{x},t)} = \frac{r\,d\theta}{v_\theta(\vec{x},t)} = \frac{dz}{v_Z(\vec{x},t)} \tag{1.53}$$

Spherical coordinates:

$$\vec{x} = (r,\theta,\phi); \quad \vec{v} = (v_r, v_\theta, v_\phi)$$

$$\frac{dr}{v_r(\vec{x},t)} = \frac{r\,\text{sen}\phi\, d\theta}{v_\theta(\vec{x},t)} = \frac{r\,d\phi}{v_\phi(\vec{x},t)} \tag{1.54}$$

The boundary conditions needed to solve streamlines differential equations are the same used for the Pathlines case and they were:

Cartesian coordinates: $t = 0; x = x_0; y = y_0; z = z_0$

Cylindrical coordinates: $t = 0; r = r_0; \theta = \theta_0; z = z_0$

Spherical coordinates: $t = 0; r = r_0; \theta = \theta_0; \phi = \phi_0$

In order to easily solve the Streamlines differential equation integration for complex vectorial fields, it is convenient to use the Streamlines differential equation in parametric form. The generic parameter S is chosen for the parameterization, notice that the final equation must not depend on the parameter S, it therefore has to be extracted.

In Cartesian coordinates, streamlines differential equations in parametric form, are

$$\frac{dx}{dS} = v_X(\vec{x},t); \; \frac{dy}{dS} = v_Y(\vec{x},t); \; \frac{dz}{dS} = v_Z(\vec{x},t) \tag{1.55}$$

In cylindrical coordinates, take the form:

$$\frac{dr}{dS} = v_r(\vec{x},t); \; \frac{r\,d\theta}{dS} = v_\theta(\vec{x},t); \; \frac{dz}{dS} = v_Z(\vec{x},t) \tag{1.56}$$

In spherical coordinates:

$$\frac{dr}{dS} = v_r(\vec{x},t); \; \frac{r\,\text{sen}\phi\, d\theta}{dS} = v_\theta(\vec{x},t); \; \frac{r\,d\phi}{dS} = v_\phi(\vec{x},t) \tag{1.57}$$

The boundary conditions needed to solve this second group of differential equations are:
Cartesian coordinates: $t = 0; x = x_0; y = y_0; z = z_0; S = 0$
Cylindrical coordinates: $t = 0; r = r_0; \theta = \theta_0; z = z_0; S = 0$
Spherical coordinates: $t = 0; r = r_0; \theta = \theta_0; \phi = \phi_0; S = 0$

It is important to remember that Streamlines, Pathlines and Streaklines, are in reality the same line whenever the vectorial field characterizes a time independent fluid movement.

1.7.5. Concept of Vorticity and Non Rotational Flow

A given fluid it is called non rotational when the angular velocity versus any coordinate axis is zero. Figure 1.11 shows two fluid lines of infinitesimal length and initially forming an angle of 90°. The fluid, which is seen as bi-dimensional, is characterized by a velocity field of components v_x and v_y. After a time differential, the initial two fluid lines which were forming initially an angle of 90° will have now moved and its angle will now be a different one, the lines have tuned an angle differential. Regarding the deformation of the fluid element, the following relations can be obtained.

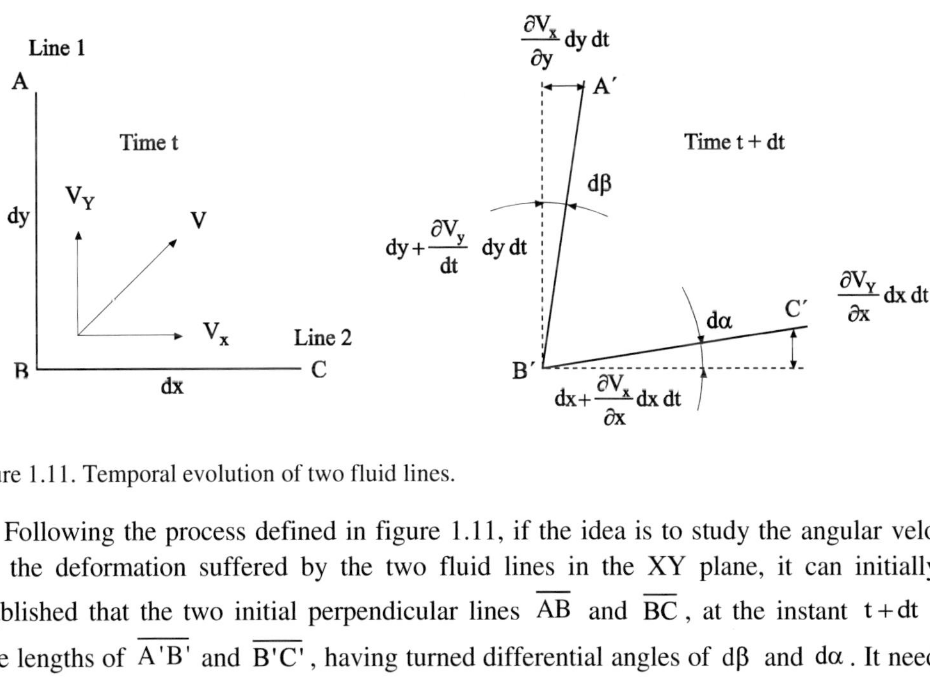

Figure 1.11. Temporal evolution of two fluid lines.

Following the process defined in figure 1.11, if the idea is to study the angular velocity and the deformation suffered by the two fluid lines in the XY plane, it can initially be established that the two initial perpendicular lines $\overline{AB}$ and $\overline{BC}$, at the instant $t + dt$ will have lengths of $\overline{A'B'}$ and $\overline{B'C'}$, having turned differential angles of $d\beta$ and $d\alpha$. It needs to be understood that this deformation appears when considering that the velocity field is not spatially uniform.

Defining the angular velocity ω_z versus the XY plane perpendicular axis, Z, as the temporal average angle turned by each fluid line, and considering anticlockwise direction as positive, it is obtained:

$$\omega_z = \frac{1}{2}\left(\frac{d\alpha}{dt} - \frac{d\beta}{dt}\right) \tag{1.58}$$

Figure 1.11 shows the relation between $d\alpha$, $d\beta$, and the x and y components of the two fluid lines. Mathematically it can be established:

$$d\alpha = \lim_{dt\to 0}\left(\text{arctg}\frac{\frac{\partial v}{\partial x}dxdt}{dx + \frac{\partial u}{\partial x}dxdt}\right) = \frac{\partial v}{\partial x}dt \tag{1.59}$$

$$d\beta = \lim_{dt\to 0}\left(\text{arctg}\frac{\frac{\partial u}{\partial y}dydt}{dy + \frac{\partial v}{\partial y}dydt}\right) = \frac{\partial u}{\partial y}dt \tag{1.60}$$

Substituting equations (1.59) and (1.60) in (1.58) it is obtained:

$$\omega_z = \frac{1}{2}\left(\frac{\partial v}{\partial x} - \frac{\partial u}{\partial y}\right) \tag{1.61}$$

A similar process could be done for angular displacements on the planes YZ and XZ, the average angular velocities will be:

$$\omega_x = \frac{1}{2}\left(\frac{\partial w}{\partial y} - \frac{\partial v}{\partial z}\right) \quad ; \quad \omega_y = \frac{1}{2}\left(\frac{\partial u}{\partial z} - \frac{\partial w}{\partial x}\right) \tag{1.62}$$

Notice that the angular velocity vector of a three dimensional moving fluid will be given as: $\omega = i\omega_x + j\omega_y + k\omega_z$, in reality, the fluid angular velocity can also be obtained via solving the following matrix, associated with the definition of curl vector.

$$\omega = \frac{1}{2}\text{curl}\vec{v} = \frac{1}{2}\begin{vmatrix} i & j & k \\ \frac{\partial}{\partial x} & \frac{\partial}{\partial y} & \frac{\partial}{\partial z} \\ u & v & w \end{vmatrix} \tag{1.63}$$

In order to avoid using $\frac{1}{2}$; it is often used the vector two times the angular velocity, also called fluid vorticity, its mathematical expression is:. $\xi = 2\omega = \text{curl}\,\vec{v} = \nabla \wedge \vec{v}$

Non rotational fluids are the ones characterized by $\text{curl}\,\vec{v} = 0$.

Another parameter which can be obtained from figure 1.11 is the deformation velocity of the two fluid lines. This velocity can be defined as the speed the two initial lines separate from their initial position. And it is mathematically expressed as:

$$\varepsilon_{xy} = \frac{d\beta}{dt} + \frac{d\alpha}{dt} = \frac{\partial u}{\partial y} + \frac{\partial v}{\partial x} \tag{1.64}$$

The average angular deformation velocity it is defined as: $\phi_{xy} = \frac{1}{2}\varepsilon_{xy}$, and therefore can be said:

$$\phi_{XY} = \frac{1}{2}\left(\frac{d\beta}{dt} + \frac{d\alpha}{dt}\right) = \frac{1}{2}\left(\frac{\partial u}{\partial y} + \frac{\partial v}{\partial x}\right) \tag{1.65}$$

For the case of three dimensional fluid, there exist nine average angular deformation velocities, taking the form:

$$\begin{aligned}
&\phi_{XX} = \frac{1}{2}\left(\frac{\partial u}{\partial x} + \frac{\partial u}{\partial x}\right) = \frac{\partial u}{\partial x}; \quad \phi_{XY} = \frac{1}{2}\left(\frac{\partial u}{\partial y} + \frac{\partial v}{\partial x}\right); \quad \phi_{XZ} = \frac{1}{2}\left(\frac{\partial u}{\partial z} + \frac{\partial w}{\partial x}\right);\\
&\phi_{YX} = \frac{1}{2}\left(\frac{\partial v}{\partial x} + \frac{\partial u}{\partial y}\right); \quad \phi_{YY} = \frac{1}{2}\left(\frac{\partial v}{\partial y} + \frac{\partial v}{\partial y}\right) = \frac{\partial v}{\partial y}; \quad \phi_{YZ} = \frac{1}{2}\left(\frac{\partial v}{\partial z} + \frac{\partial w}{\partial y}\right);\\
&\phi_{ZX} = \frac{1}{2}\left(\frac{\partial w}{\partial x} + \frac{\partial u}{\partial z}\right); \quad \phi_{ZY} = \frac{1}{2}\left(\frac{\partial w}{\partial y} + \frac{\partial v}{\partial z}\right); \quad \phi_{ZZ} = \frac{1}{2}\left(\frac{\partial w}{\partial z} + \frac{\partial w}{\partial z}\right) = \frac{\partial w}{\partial z};
\end{aligned} \tag{1.66}$$

It is interesting to point out that shear stresses defined for a two dimensional flow in section 1.6.6, can now be defined in a generic three dimensional flow via using the concept of angular deformation velocity. shear stresses for a generic direction ij are obtained via multiplying the kinematics viscosity with the angular deformation velocity in direction ij. $\tau_{ij} = \mu\, \varepsilon_{ij} = 2\, \mu\, \phi_{ij}$

1.7.6. Kinematic Study of a Fluid Particle

To perform the kinematics study of a fluid particle movement, it needs to be considered the superposition of four independent movements, translation, rotation, linear deformation and angular deformation. The four independent movements are presented in figure 1.12. Its mathematical description is defined next.

Mathematical description of the independent movements.

– -Displacement. Given by the vector velocity or acceleration. $\vec{V};\ \vec{a}$;

$$\vec{a} = \frac{D\vec{v}}{Dt} = \frac{\partial \vec{v}}{\partial t} + (\vec{v}\nabla)\vec{v}$$

- -Rotation. Defined by the angular velocity. $\vec{\omega} = \frac{1}{2}\text{curl}\vec{v} = \frac{1}{2}(\nabla \wedge \vec{v})$
- -Linear deformation. Defined by the velocity divergence. $\frac{1}{\delta\forall}\frac{d(\delta\forall)}{dt} = \nabla\vec{v} = \frac{\partial u}{\partial x} + \frac{\partial v}{\partial y} + \frac{\partial w}{\partial z}$
- -Angular deformation. Defined by the average angular deformation velocity. ϕ_{ij}; $\phi_{XY} = \frac{1}{2}\left(\frac{\partial u}{\partial y} + \frac{\partial v}{\partial x}\right)$

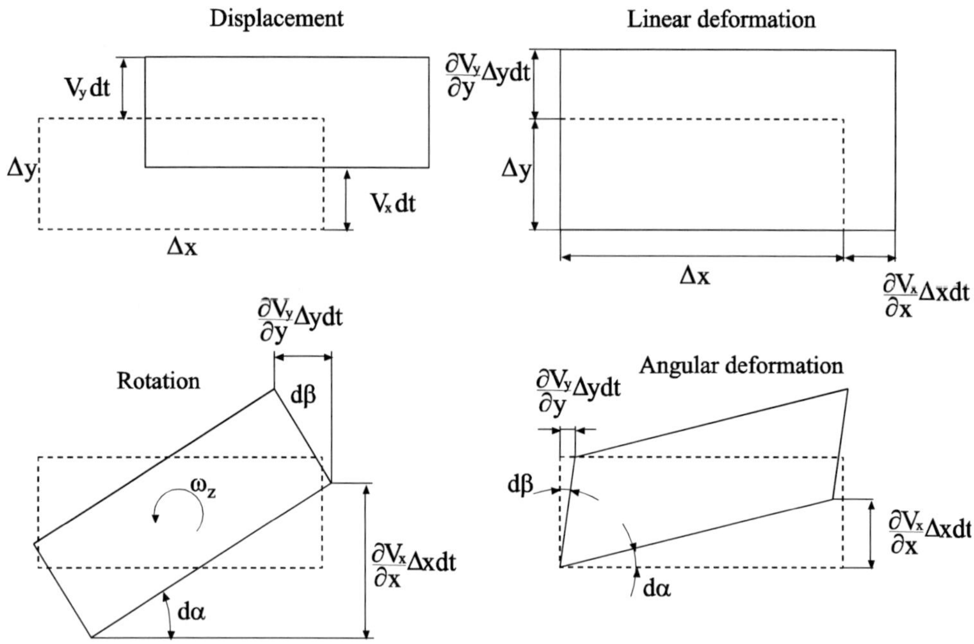

Figure 1.12. The four independent movements associated to the kinematical study of a fluid particle.

The concepts of acceleration, angular velocity and angular deformation velocity have already been explained, in what follows it will be clarified the concept of linear deformation, see figure 1.13.

Let's assume we have a generic fluid volume $\forall$, on the fluid surface we take a surface differential ds, and this generic surface differential is crossed by a fluid particle of velocity V. After a time differential dt, the differential volume increase due to the flow crossing the surface differential can be defined as

$$(\Delta\forall) = (\vec{V}\,dt)\,d\vec{s}\,. \tag{1.67}$$

On the other hand, the overall volume increase per unit time, it has to be the addition of all differential volume increases associated to each surface differential, mathematically can be expressed as:

$$\frac{d\forall}{dt} = \oint_S \vec{V}\, d\vec{s} \qquad (1.68)$$

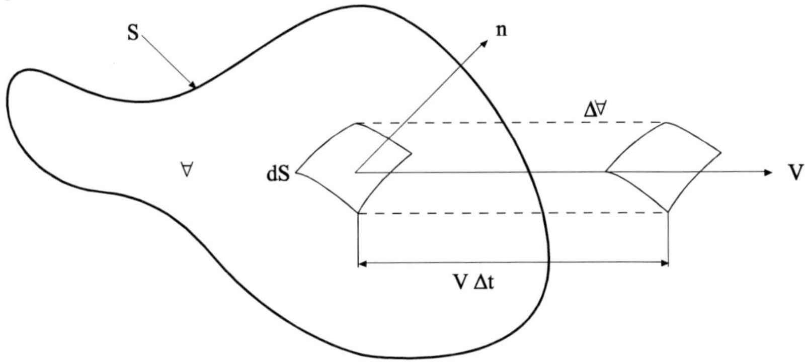

Figure 1.13. Fluid volume used to study the concept of linear deformation.

Supposing, that the initial volume of fluid $\forall$ it is now reduced to a point, in other words, it is reduced to a volume differential and applying the divergence theorem it is obtained:

$$\frac{d(\delta\forall)}{dt} = \oint_S \vec{V}\, d\vec{s} = \int_\forall \nabla\vec{V}\, d(\delta\forall) = \int_\forall \operatorname{div}\vec{V}\, d(\delta\forall) \qquad (1.69)$$

Assuming that for the elemental fluid volume chosen $\delta\forall$, the divergence of the velocity vector is constant across the entire volume, it can be defined:

$$\frac{d(\delta\forall)}{dt} = \int_\forall \operatorname{div} V\, d(\delta\forall) = \operatorname{div}\vec{V}\,\delta\forall \qquad (1.70)$$

Equation which can be presented as:

$$\operatorname{div}\vec{V} = \frac{1}{\delta\forall}\frac{d(\delta\forall)}{dt} \qquad (1.71)$$

As a conclusion it can be said that the linear deformation generated onto a differential volume, it is defined by the velocity divergence.

Once the four independent movements have been clarified, it is important to group them in what is called the kinematics study of a fluid particle, which is defined by the velocity gradient tensor $\bar{\bar{G}}_V$, this tensor is the addition of other two tensors, the deformation tensor which is a symmetrical one, and the vorticity tensor.

$$\bar{\bar{G}}_v = \phi_{ij} + \Omega_{ij} \qquad (1.72)$$

ϕ_{ij} = Deformation tensor.

Ω_{ij} = Vorticity tensor.

Using the sub index notation, the velocity gradient tensor can be expressed as:

$$\bar{\bar{G}}_v = \frac{1}{2}(T_{ij} + T_{ji}) + \frac{1}{2}(T_{ij} - T_{ji}) = \phi_{ij} + \Omega_{ij} \tag{1.73}$$

$$\bar{\bar{G}}_v = \frac{\partial v_i}{\partial r_j} = \frac{1}{2}\left(\frac{\partial v_i}{\partial r_j} + \frac{\partial v_j}{\partial r_i}\right) + \frac{1}{2}\left(\frac{\partial v_i}{\partial r_j} - \frac{\partial v_j}{\partial r_i}\right) = \phi_{ij} + \Omega_{ij} \tag{1.74}$$

The velocity gradient tensor as well as the deformation and vorticity tensors for Cartesian coordinates, are presented explicitly next.

Notice that the main diagonal of the deformation tensor, defines the linear deformation, which is characterized by the divergence of the velocity vector. For example, the term $\frac{\partial u}{\partial x}$ represents the dilatation speed of a fluid line, proportional to its length and towards the x direction. The terms outside the main diagonal, represent the average speed the two initial fluid lines deform.

The vorticity tensor represents the average angular velocity the three dimensional fluid is having versus any coordinate axis.

$$\bar{\bar{G}}_v = \left\| \frac{\partial V_i}{\partial r_j} \right\| = \begin{vmatrix} \frac{\partial u}{\partial x} & \frac{\partial u}{\partial y} & \frac{\partial u}{\partial z} \\ \frac{\partial v}{\partial x} & \frac{\partial v}{\partial y} & \frac{\partial v}{\partial z} \\ \frac{\partial w}{\partial x} & \frac{\partial w}{\partial y} & \frac{\partial w}{\partial z} \end{vmatrix} = \begin{vmatrix} \frac{\partial u}{\partial x} & \frac{1}{2}\left(\frac{\partial u}{\partial y} + \frac{\partial v}{\partial x}\right) & \frac{1}{2}\left(\frac{\partial u}{\partial z} + \frac{\partial w}{\partial x}\right) \\ \frac{1}{2}\left(\frac{\partial v}{\partial x} + \frac{\partial u}{\partial y}\right) & \frac{\partial v}{\partial y} & \frac{1}{2}\left(\frac{\partial v}{\partial z} + \frac{\partial w}{\partial y}\right) \\ \frac{1}{2}\left(\frac{\partial w}{\partial x} + \frac{\partial u}{\partial z}\right) & \frac{1}{2}\left(\frac{\partial w}{\partial y} + \frac{\partial v}{\partial z}\right) & \frac{\partial w}{\partial z} \end{vmatrix} +$$

$$\begin{vmatrix} 0 & \frac{1}{2}\left(\frac{\partial u}{\partial y} - \frac{\partial v}{\partial x}\right) & \frac{1}{2}\left(\frac{\partial u}{\partial z} - \frac{\partial w}{\partial x}\right) \\ \frac{1}{2}\left(\frac{\partial v}{\partial x} - \frac{\partial u}{\partial y}\right) & 0 & \frac{1}{2}\left(\frac{\partial v}{\partial z} - \frac{\partial w}{\partial y}\right) \\ \frac{1}{2}\left(\frac{\partial w}{\partial x} - \frac{\partial u}{\partial z}\right) & \frac{1}{2}\left(\frac{\partial w}{\partial y} - \frac{\partial v}{\partial z}\right) & 0 \end{vmatrix} = \phi_{ij} + \Omega_{ij} \tag{1.75}$$

1.8. NOMENCLATURE

$\vec{a}$	Fluid acceleration. (m/s^2)
dr	Radius differential. (m).
D_e	Deborah number.
F	Force. (N)
$g_{(r)}$	Radial distribution function.

$\bar{\bar{G}}_V$	Velocity gradient tensor. (s^{-1}).
m	Mass. (Kg).
P	Pressure. (Pa)
r	Generic radius. (m).
R	Radius. (m).
R	Gas constant. (J/(Kg K))
T	Temperature. (K).
T_{ij}	Generic tensor. (s^{-1})
t	Time. (s).
u	Velocity. (m/s).
u	Lennard-Jones potential. (N m)
$V = v$	Velocity. (m/s).
$V_x = u$	Velocity X direction component. (m/s).
$V_y = v$	Velocity Y direction component. (m/s).
$V_z = w$	Velocity Z direction component. (m/s).
W	Work. (N m).
$\forall$	Volume. (m^3).
α	Generic angle. (rad).
α	Thermal expansion coefficient. (K^{-1}).
β	Generic angle. (rad).
β	Bulk modulus. (Pa).
$\dot{\gamma}$	Angular velocity. (rad/s).
$\delta e = \delta\vec{x}$	Length differential. (m).
$\delta\theta$	Angle differential. (rad)
$\delta\gamma$	Angle differential. (rad).
$\delta\varepsilon$	Angle differential. (rad).
$\varepsilon = S$	Generic parameter.
θ	Generic angle. (rad).
μ	Dynamic viscosity. (Kg((m s)).
ν	Kinematic viscosity. (m^2/s).
ρ	Density. (Kg/m^3).
σ	Surface tension. (N/m).
τ	Shear stresses. (N/m^2).
$\upsilon = \frac{1}{\rho}$	Specific volume. (m^3/Kg).
ϕ	Fluid generic property.
ϕ	Generic angle. (rad).
ϕ_{ij}	Deformation tensor. (s^{-1}).
ϕ_{ij}	Angular deformation velocities. (rad/s).
$\vec{\omega} = \vec{\Omega}$	Angular velocity. (rad/s).
Ω_{ij}	Vorticity tensor. (s^{-1}).

1.9. REFERENCES

[1] Bergadà Josep M. (2012). *Mecánica de Fluidos*. Breve introducción teórica con problemas resueltos. Barcelona. Iniciativa digital politécnica UPC.

[2] Douglas JF, Gasiorek JM, Swaffield JA. (1998). *Fluid Mechanics* 3rd edition. Singapore. Longman.

[3] Manring Noah D. (2005). *Hydraulic Control Systems*. Hoboken, New Jersey. John Willey & Sons.

[4] Watton John. (1989). *Fluid power Systems*. Singapore. Prentice Hall.

[5] White Frank M. (2004). *Mecánica de Fluidos*. 5th edition. Madrid. McGraw Hill.

Chapter 2

MAIN FLUID MECHANICS EQUATIONS

2.1. INTRODUCTION. REYNOLDS TRANSPORT EQUATION

The estate and movement of a given fluid can be determinate from the following principles and laws.

Mass conservation principle. In a given finite volume, mass remains constant, even if its position and form is modified.

Momentum balance. The fluid momentum variation per unit time in a given control volume, is equal to the resultant forces acting on the control volume surface (pressure and shear stresses), and the forces acting onto the full volume, volumetric forces.

Energy conservation principle. The variation of energy in a fluid control volume, is equal to the work due to the external forces acting onto the volume, plus the external head transferred via conduction and radiation.

Local thermodynamic equilibrium. Local thermodynamic equilibrium will exist, if the macroscopic status of the fluid in each point and instant can be characterized by its velocity $\vec{V}$ and the thermodynamic state variables $\rho, T, P, u, s, etc.$ measured by an observer moving with the fluid particle.

In Fluid Mechanics, the equations can be applied to a finite control volume or to a differential one, some advantages and disadvantages of using a finite or differential control volume are resumed in the following table.

The determination of the basic fluid mechanics equations in integral form shall be obtained from the Reynolds transport theorem. Such theorem provides a way to identify a finite system and evaluates the velocity of change of any fluid property or characteristic in this system, flow is examined via using a control volume. In what follows the concept of control system and control volume is defined.

A control system it is defined as a given fluid mass of fixed identity and identified by its spatial position. The fluid mass can be finite or elemental.

A control volume is stated as a finite region with open boundaries through which mass flow, momentum flow and energy flow can be defined. Between the control volume frontiers a balance of incoming, outgoing and remaining flow is established.

In order to determine the velocity of change of a generic property associated to the fluid located inside a control system, it is conceived a way to identify a specific volume of fluid which can move and deform, the fluid mechanic fundamental laws shall be applied to this system, the second Newton law which studies the velocity of change of fluid momentum inside the system and the first law of thermodynamics which evaluates the velocity of change of the energy associated to the system.

	Advantages	Disadvantages
Differential Formulation	1. It reveals with high precision all the flow details. 2. Fluid is forced to obey the fundamental laws at all points. 3. To solve the problem, boundary conditions are needed. 4. Very accurate and precise information about the fluid it is obtained.	1. The resulting differential equations are difficult or impossible to solve analytically. 2. The resulting system of equations is usually solved via a computer program.
Integral Formulation	1. The mathematical equations employed are relatively simple. 2. Information obtained is approximate although very useful. The hypotheses established are simple. 3. The time needed to obtain the required information is short.	1. Fluid is not forced to obey the fluid mechanics fundamental laws at each point. 2. It is common that results obtained are just approximate. 3. It requires more initial information, like velocity distribution at the control volume frontiers. 4. The information obtained is in many cases fewer than what it would be desirable.

To calculate the velocity of change of a generic property associated to the fluid, a control system which moves with the fluid is defined in figure 2.1. A control volume which shall be regarded as static is also defined in the same figure; initially the control system and the control volume are positioned to one another.

As the control system is moving with the fluid, after a time differential, the volume of fluid inside the control system will not be the same remaining inside the control volume, figure 2.2 presents the spatial evolution of the control system versus the control volume at different temporal differentials.

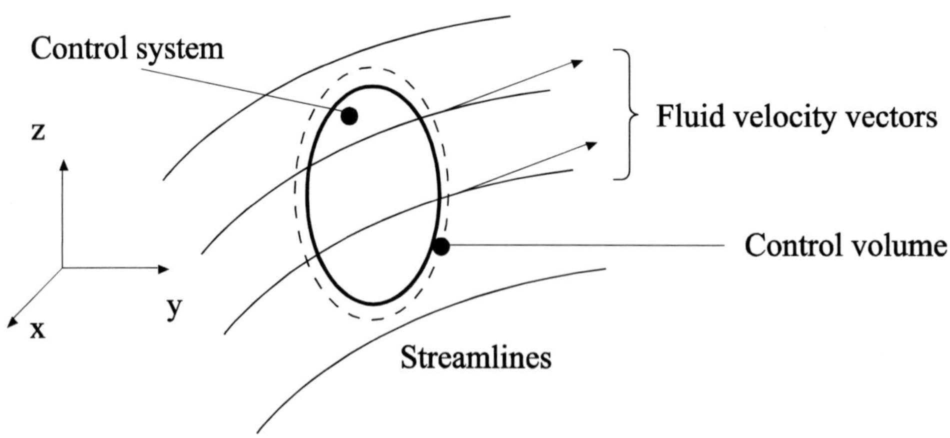

Figure 2.1. A control system and a control volume are defined and located inside a moving fluid.

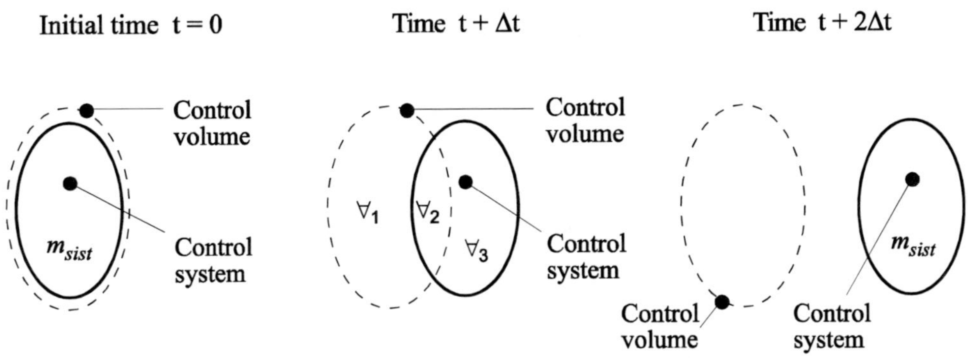

Figure 2.2. Spatial evolution of the control system versus the control volume at different time steps.

It is to be noticed that the mass of fluid inside the control volume at initial time m_{sist} , it is exactly the same as the one inside the control system.

As established previously the problem consists of evaluating the velocity of change of a generic fluid property, called B, associated to the control system.

The property b is defined as a generic property per unit mass associated to the fluid.

$$b = \frac{B}{m}$$

The generic property B associated to the fluid inside the control system is defined as.

$$B_{sist} = \int_{msist} b dm = \int_{\forall sist} b \rho d\forall \tag{2.1}$$

$\forall$ = volume of the control system

The velocity of change of the generic fluid property B associated to fluid inside the control system shall be defined as:

$$\frac{dB_{sist}}{dt} = \lim_{\delta t \to 0} \frac{B_{sist}\big|_{t+\delta t} - B_{sist}\big|_t}{\delta t} \tag{2.2}$$

where $B_{sist}\big|_{t+\delta t}$ is the generic fluid property B associated to the fluid inside the control system at time t+δt.

Substituting equation (2.1) into (2.2) it is obtained:

$$\frac{dB_{sist}}{dt} = \lim_{\delta t \to 0} \frac{\int_{\forall sist} \rho b d\forall \Big|_{t+\delta t} - \int_{\forall sist} \rho b d\forall \Big|_t}{\delta t} \tag{2.3}$$

Remembering that, at initial time, the volume of fluid inside the control system and the control volume is the same.

$$\forall_{sist}\big|_t = \forall_{control\ volume} \tag{2.4}$$

On the other hand, the volume of fluid inside the control system at time $t+\delta t$, can be evaluated according to figure 2.2 as:

$$\forall_{sist}\Big|_{t+\delta t} = \forall_{II} + \forall_{III} = \forall_{VC} - \forall_{I} + \forall_{III} \tag{2.5}$$

Considering equation (2.5), equation (2.3) takes the form:

$$\frac{dB_{sist}}{dt} = \lim_{\delta t\to 0} \frac{\int_{\forall_{VC}} \rho b d\forall\Big|_{t+\delta t} - \int_{\forall_I} \rho b d\forall\Big|_{t+\delta t} + \int_{\forall_{III}} \rho b d\forall\Big|_{t+\delta t} - \int_{\forall_{VC}} \rho b d\forall\Big|_{t}}{\delta t} \tag{2.6}$$

Equation (2.6) can be also presented as:

$$\frac{dB_{sist}}{dt} = \underbrace{\lim_{\delta t\to 0} \frac{\int_{\forall_{VC}} \rho b d\forall\Big|_{t+\delta t} - \int_{\forall_{VC}} \rho b d\forall\Big|_{t}}{\delta t}}_{\text{Term I}} + \underbrace{\lim_{\delta t\to 0} \frac{\int_{\forall_{III}} \rho b d\forall\Big|_{t+\delta t}}{\delta t}}_{\text{Term II}} - \underbrace{\lim_{\delta t\to 0} \frac{\int_{\forall_{I}} \rho b d\forall\Big|_{t+\delta t}}{\delta t}}_{\text{Term III}} \tag{2.7}$$

The term I, represents the temporal variation of the generic fluid property B contained inside the control volume, (velocity of change of the property B associated to the fluid inside the control volume), and can also be presented as:

$$I = \frac{d}{dt}\int_{\forall_{VC}} \rho b d\forall = \frac{d}{dt}(B_{VC}) \tag{2.8}$$

The term II, represents the amount of property B associated to the fluid leaving the control volume between the instants t and $t+dt$.

The term II, can also be presented as $\dot{B}_{out}$, which evaluates the temporal variation of property B associated to the mass flow leaving the control volume per unit time.

The term III, represents the property B associated to the mass flow entering the control volume between the instants t and $t+dt$, and can be presented as $\dot{B}_{ent}$.

Therefore, equation (2.7) can also be presented as:

$$\frac{dB_{sist}}{dt} = \frac{d}{dt}\int_{VC} \rho b d\forall + \dot{B}_{out} - \dot{B}_{ent} \tag{2.9}$$

Notice that the negative sign (-) clarifies that the flow enters the control volume.

Equation (2.9) establishes that the velocity of change of a generic property associated to the fluid located inside the volume control system, is equal to the accumulation velocity of this property inside the control volume, plus the difference between the property associated to the instant mass flows leaving and entering the control volume.

The mass flows leaving and entering the control volume can be given as:

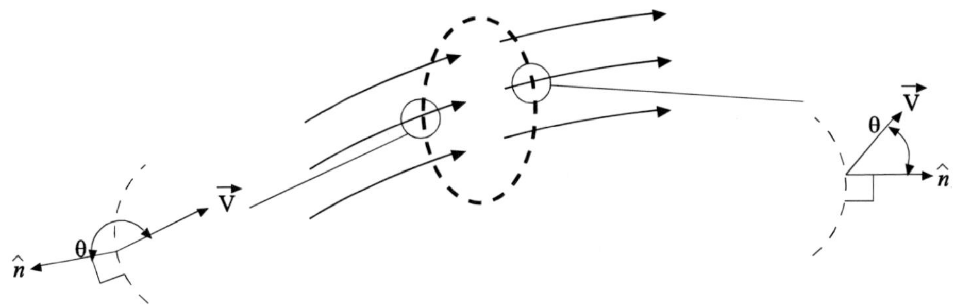

Figure 2.3. Velocity vectors and surface differentials at the inlet and outlet regions of the control volume.

a) The mass flow leaving the control volume is defined:

$$\dot{m}_{out} = \int_{s-out} \rho \vec{V} \vec{ds} = \int_{s-out} \rho V \cos\theta \, ds = \int_{s-out} \rho \vec{V}\hat{n} \, ds \tag{2.10}$$

Remembering that B is a generic property associated to the fluid mass. The temporal variation of property B associated to the fluid mass flow leaving the control volume can be expressed:

$$\begin{aligned} d\dot{B}_{out} &= b.(d\dot{m})_{out} \\ \dot{B}_{out} &= \int d\dot{B}_{out} = \int b.(d\dot{m})_{out} = \int_{s-out} b\rho \vec{V}\hat{n} \, ds = \int_{s-out} b\rho V \cos\theta \, ds = \int_{s-out} b\rho \vec{V} d\vec{s} \end{aligned} \tag{2.11}$$

Notice that for the flow leaving the control volume, the value of $\cos\theta$ will always be positive, then the angle θ can have values between 0 and 90 degrees, see figure 2.3.

b) For the mass flow entering the control volume it can be established:

$$\dot{B}_{ent} = \int_{s-ent} b\rho \vec{V}\hat{n} ds = \int_{s-ent} b\rho V \cos\theta \, ds = \int_{s-ent} b\rho \vec{V} d\vec{s} \tag{2.12}$$

Notice that for the flow entering the control volume, the sign associated to $\cos\theta$ will be negative, since the angle θ will be having values between 90 and 180 degrees.

Substituting equations (2.11) and (2.12) in (2.9) it is obtained:

$$\begin{aligned} \frac{dB_{sist}}{dt} &= \frac{d}{dt}\int_{VC} \rho b \, d\forall + \int_{s-out} \rho b \vec{V}\hat{n} ds + \int_{s-ent} \rho b \vec{V}\hat{n} \, ds = \\ &\frac{d}{dt}\int_{VC} \rho b \, d\forall + \int_{s-out} \rho b V \cos\theta \, ds - \int_{s-ent} \rho b V \cos\theta \, ds \end{aligned} \tag{2.13}$$

Equation (2.13) can be expressed in a compact form as:

$$\frac{dB_{sist}}{dt} = \frac{d}{dt}\int_{VC} \rho\, b\, d\forall + \oint_{SC} \rho\, b\, \vec{v}\, d\vec{s} \tag{2.14}$$

The term $\oint_{SC} \rho\, b\, \vec{v}\, d\vec{s}$ represents the net flow of property B across the control volume frontiers. Equation (2.14) is defined as the Reynolds transport theorem.

2.2. Continuity Equation, Integral Form

The Reynolds transport theorem shall be now used to determine the integral form of the continuity equation. The fluid will be assumed to be tri-dimensional, tri-directional and temporal dependent.

The generic property B to be evaluated is the fluid mass, B = mass. It is at this point important to realize that the control system was defined in the previous section as having a fix identity mass, which means the mass inside the control system is temporal independent. As a result, the left hand side of equation (2.14) can be expressed:

$$\frac{dB_{sis}}{dt} = \frac{dm_{sis}}{dt} = 0 \tag{2.15}$$

The property B per unit mass will now be: $b = \frac{B_{sist}}{m_{sist}} = \frac{m_{sist}}{m_{sist}} = 1$ (2.16)

Substituting equations (2.15) and (2.16) in (2.14) it is found:

$$0 = \frac{d}{dt}\int_{\forall_{vc}} \rho\, d\forall + \oint_{SC} \rho\, \vec{v}\, d\vec{s} \tag{2.17}$$

Equation (2.17) is the integral form of the continuity equation. The first term on the right hand side, represents the velocity the mass associated to the fluid increases or decreases inside the control volume. The second term on the right hand side represents the net mass flow across the control volume surfaces.

2.2.1. Continuity Equation, Differential Form

The differential form of the continuity equation can be obtained from the integral definition of this equation, via applying the divergence theorem.

Notice that the integral across the control volume surface, second term of the right hand side of equation (2.17), can be transformed to an integral across the control volume when applying to it the divergence theorem, obtaining:

$$\oint_{SC} \rho\, \vec{V}\, d\vec{s} = \int_{VC} \nabla \left(\rho\, \vec{V}\right) d\forall \tag{2.18}$$

Substituting (2.18) into (2.17), it is reached:

$$\frac{\partial}{\partial t}\int_{VC} \rho\, d\forall + \int_{VC} \nabla\left(\rho \vec{V}\right) d\forall = 0 \tag{2.19}$$

Both integrals in equation (2.19) are defined across the control volume, both terms could be written inside a single integral which is equal to zero, since the volume differential cannot be zero the terms inside the integral have to be zero, obtaining:

$$\frac{\partial \rho}{\partial t} + \nabla\left(\rho \vec{V}\right) = 0 \tag{2.20}$$

Equation (2.20) represents the differential form of the continuity equation, the first term characterizes the temporal variation of density inside the control volume and the second term evaluates the spatial variation of fluid density and velocity. Equation (2.20) can also be presented as:

$$\frac{\partial \rho}{\partial t} + \rho \nabla \vec{V} + \vec{V} \nabla \rho = 0 \tag{2.21}$$

And using the concept of substantial derivative, it is obtained:

$$\frac{D\rho}{Dt} + \rho \nabla \vec{V} = 0 \tag{2.22}$$

Equations (2.20) to (2.22) are different forms of the continuity equation in differential form. The representation presented here is valid for any coordinate system; it is just required to use the differential operator ∇ in the desired coordinate.

In cylindrical and spherical coordinates, the differential form of the continuity equation is represented as:

Cylindrical coordinates:

$$\frac{\partial \rho}{\partial t} + \frac{1}{r}\frac{\partial}{\partial r}(\rho\, r\, v_r) + \frac{1}{r}\frac{\partial}{\partial \theta}(\rho\, v_\theta) + \frac{\partial}{\partial x}(\rho\, v_x) = 0 \tag{2.23}$$

Spherical coordinates.

$$\frac{\partial \rho}{\partial t} + \frac{1}{r^2}\frac{\partial}{\partial r}(\rho\, r^2\, v_r) + \frac{1}{r \sin\theta}\frac{\partial}{\partial \theta}(\rho\, v_\theta \sin\theta) + \frac{1}{r \sin\theta}\frac{\partial}{\partial \phi}(\rho\, v_\phi) = 0 \tag{2.24}$$

2.3. Momentum Equation, Integral Form

As previously done in the continuity equation, the momentum equation will also be obtained from the Reynolds transport theorem, for the present case Newton second law will also be employed. For the momentum equation, the generic property B associated to the fluid will be the product of the fluid particle mass by its associated velocity, also called the fluid particle momentum: $B = \vec{M} = m\vec{v}$.

Substituting the generic property B into the left hand side of equation (2.14) and remembering the second Newton equation, it is obtained:

$$\frac{dB_{sist}}{dt} = \frac{d\vec{M}_{sist}}{dt} = \frac{d(m\vec{v})}{dt} = m\frac{d\vec{v}}{dt} = m\vec{a} = \sum F_{sist} \tag{2.25}$$

Notice that the temporal variation of the momentum associated to the fluid inside the control system is equal to the forces acting onto this system.

$\vec{M}_{sist}$ is the momentum associated to the fluid inside the control system.

$\sum F_{sist}$ defines de overall forces acting onto the control system, it is important to realize that at the initial time, the control system and the control volume are exactly the same.

The generic property per unit mass b, is equal by definition to the generic property B divided by the mass, and for the present case the generic property per unit mass b will be the fluid velocity $b = \vec{V}$.

Substituting the value of b and equation (2.25) into equation (2.14) it is obtained:

$$\Sigma \vec{F}_{vc} = \frac{d}{dt}\int_{vc} \rho\, \vec{V}_i\, d\forall + \oint_{SC} \rho\, \vec{V}_i\, \vec{V}\, d\vec{s} \tag{2.26}$$

Equation (2.26) is the integral form of the momentum equation, which can be defined as: The addition of all the forces acting onto the control volume is equal to the momentum accumulation velocity in the control volume plus the net momentum flux across the boundaries of the control volume.

Notice that two velocities are being distinguished, the velocity $\vec{V}$ represents the fluid velocity at the surface boundary, measured versus the surface, the term $\vec{V}\, d\vec{s}$ is conceptually the same as the one found in the continuity equation and has a sign associated to the scalar product. The velocity $\vec{V}_i$ is the component of the fluid velocity crossing the control volume surface but measured versus the axis coordinate reference system, the sign associated will be given by the velocity component direction.

$\sum \vec{F}$ represents the addition of all vector forces acting onto fluid inside the control volume. In general these forces will act onto the fluid volume or onto the fluid surface.

For an inertial coordinate system the forces acting onto the fluid control volume will be due to the gravitational acceleration. The forces acting onto the fluid surface are the ones due to pressure and the ones due to shear stresses.

$$\underbrace{\sum \vec{F} = \sum \vec{F}_{gravity}}_{F_{fluid\ volume}} + \underbrace{\sum \vec{F}_{pressure} + \sum \vec{F}_{shear\ stresses}}_{F_{fluid\ surface}} \tag{2.27}$$

Each term of equation (2.27) can be defined as:

$\sum \vec{F}_{gravity} = \int_{vc} \rho \vec{g} \, d\forall =$ represents a force per unit mass of fluid.

$\sum \vec{F}_{pressure} + \sum \vec{F}_{shear\ stresses}$ these two terms represent the forces acting onto the control volume fluid surface. The first term, the pressure forces are given as:

$$\Sigma \vec{F}_{pressure} = -\oint_{SC} P \, d\vec{s} \tag{2.28}$$

The direction of the forces due to the pressure acting onto the control volume surface, will always be opposite to the surface differential normal vector, this is why the negative sign is stated in equation (2.28).

The forces due to shear stresses will be represented as:

$$\Sigma \vec{F}_{shear\ stresses} = \oint_{SC} \bar{\bar{\tau}}_{ij} \, ds \tag{2.29}$$

The sign associated to this force shall be given by the direction of the shear stresses versus a coordinate system.

When introducing equations (2.27) to (2.29) in (2.26) the momentum equation for an inertial coordinate system is obtained, equation (2.30). Notice that each term of the momentum equation has a vector form, therefore, equation (2.30), needs to be applied to each coordinate direction. In fact, the sub-index i attached to one of the velocities, represents the velocity component in the direction the momentum equation is applied.

$$\int_{VC} \rho \vec{g} \, d\forall - \oint_{SC} P \, d\vec{s} + \oint_{SC} \bar{\bar{\tau}}_{ij} \, ds = \frac{\partial}{\partial t} \int_{VC} \rho \vec{V}_i \, d\forall + \oint_{SC} \rho \vec{V}_i \vec{V} \, d\vec{s} \tag{2.30}$$

2.3.1. Momentum Equation, Differential Form

Probably the quickest way to determine the momentum equation in differential form is via applying the divergence theorem to the momentum equation in integral form, the idea is to transform the integrals across the surface to integrals across the entire volume. From equation (2.30) it is seen that just three terms are defined across the control volume surface, applying the Gauss Ostrogradsky or divergence theorem to these three terms it is obtained:

$$-\oint_{SC} P \, d\vec{s} = -\int_{VC} \nabla P \, d\forall \tag{2.31}$$

$$\oint_{SC} \bar{\bar{\tau}}\, ds = \int_{VC} \nabla \bar{\bar{\tau}}\, d\forall \tag{2.32}$$

$$\oint_{SC} \rho\, \vec{V}_i\, \vec{V}\, d\vec{s} = \int_{VC} \nabla(\rho\, \vec{V}_i\, \vec{V})\, d\forall \tag{2.33}$$

Substituting equations (2.31) to (2.33) into (2.30), it is obtained:

$$\frac{\partial}{\partial t}\int_{VC} \rho\, \vec{V}_i\, d\forall + \int_{VC} \nabla(\rho \vec{V}_i\, \vec{V})\, d\forall = -\int_{VC} \nabla P\, d\forall + \int_{VC} \rho\, \vec{g}\, d\forall + \int_{VC} \nabla \bar{\bar{\tau}}\, d\forall \tag{2.34}$$

Equation (2.34) can be written inside a single integral across the control volume, integral which will be equal to zero, and therefore obtaining:

$$\frac{\partial\left(\rho\, \vec{V}_i\right)}{\partial t} + \nabla(\rho \vec{V}_i \vec{V}) = -\nabla P + \rho\, \vec{g} + \nabla \bar{\bar{\tau}} \tag{2.35}$$

Equation (2.35) is the differential form of the momentum equation, nevertheless some further development can be made, the left hand side of the equation, can be given as:

$$\rho \frac{\partial \vec{V}_i}{\partial t} + \vec{V}_i \frac{\partial \rho}{\partial t} + \vec{V}_i \nabla(\rho \vec{V}) + \rho \vec{V} \nabla \vec{V}_i = \rho\left[\frac{\partial \vec{V}_i}{\partial t} + \vec{V} \nabla \vec{V}_i\right] + \vec{V}_i\left[\frac{\partial \rho}{\partial t} + \nabla(\rho \vec{V})\right] \tag{2.36}$$

According to the continuity equation: $\dfrac{\partial \rho}{\partial t} + \nabla(\rho \vec{V}) = 0$

and remembering at this point the concept of material derivative, equation (2.36) shall be presented as:

$$\rho \frac{\partial \vec{V}_i}{\partial t} + \rho \vec{V} \nabla \vec{V}_i = \rho\left[\frac{\partial \vec{V}_i}{\partial t} + \vec{V} \nabla \vec{V}_i\right] = \rho \frac{D\vec{V}}{Dt} \tag{2.37}$$

Therefore, equation (2.35) can also be expressed as:

$$\rho \frac{D\vec{V}}{Dt} = -\nabla P + \rho\, \vec{g} + \nabla \bar{\bar{\tau}} \tag{2.38}$$

Equation (2.38) is to be seen as a compact presentation of the momentum equation in differential form, also called Cauchy equation. This equation is generic and can be applied to any coordinate system. The left hand side represents the fluid inertial forces per unit volume, and on the right hand side are represented the pressure, gravitational and viscous forces per unit volume.

For an ideal fluid, fluid without viscosity, equation (2.38) will take the form:

$$\rho\frac{D\vec{V}}{Dt}=-\nabla P+\rho\vec{g} \tag{2.39}$$

Equation (2.39) is called the Euler equation and can be used for compressible or incompressible fluid.

It is interesting to point out that Cauchy equation, as presented in (2.38) has no solution, since the shear stresses are at this point unknown. In order to solve this problem, it is necessary to relate the viscous forces with the fluid deformation velocity, notice that the fluid viscosity is the proportionality constant between these two terms.

For a multidimensional and multidirectional fluid, shear stresses are proportional to the angular deformation velocity; normal viscous stresses are proportional to the normal deformation velocity. Therefore, and recalling the deformation tensor in kinematics, shear and normal stresses shall be presented as:

$$\tau_{xy}=\tau_{yx}=2\mu\phi_{xy}=\mu\left(\frac{\partial u}{\partial y}+\frac{\partial v}{\partial x}\right);\phi=\text{angular deformation velocity};\ \phi_{xy}=\frac{1}{2}\left(\frac{\partial u}{\partial y}+\frac{\partial v}{\partial x}\right);(2.40)$$

$$\tau_{xz}=\tau_{zx}=2\mu\phi_{xz}=\mu\left(\frac{\partial u}{\partial z}+\frac{\partial w}{\partial x}\right);\phi_{xz}=\frac{1}{2}\left(\frac{\partial u}{\partial z}+\frac{\partial w}{\partial x}\right) \tag{2.41}$$

$$\tau_{yz}=\tau_{zy}=2\mu\phi_{yz}=\mu\left(\frac{\partial v}{\partial z}+\frac{\partial w}{\partial y}\right);\phi_{yz}=\frac{1}{2}\left(\frac{\partial v}{\partial z}+\frac{\partial w}{\partial y}\right) \tag{2.42}$$

$$\sigma_x=2\mu\frac{\partial u}{\partial x};\quad \sigma_y=2\mu\frac{\partial v}{\partial y};\quad \sigma_z=2\mu\frac{\partial w}{\partial z}; \tag{2.43}$$

In fact, normal stresses consider two main terms, one characterizes the linear deformation of the fluid volume differential, and the second evaluates the volumetric deformation defined as the addition of the velocity gradients along the three coordinate axes. As a result, equations (2.43) will take the form:

$$\sigma_x=2\mu\frac{\partial u}{\partial x}+\lambda\left(\frac{\partial u}{\partial x}+\frac{\partial v}{\partial y}+\frac{\partial w}{\partial z}\right) \tag{2.44}$$

$$\sigma_y=2\mu\frac{\partial v}{\partial y}+\lambda\left(\frac{\partial u}{\partial x}+\frac{\partial v}{\partial y}+\frac{\partial w}{\partial z}\right) \tag{2.45}$$

$$\sigma_z=2\mu\frac{\partial w}{\partial z}+\lambda\left(\frac{\partial u}{\partial x}+\frac{\partial v}{\partial y}+\frac{\partial w}{\partial z}\right) \tag{2.46}$$

The parameter λ is being called the second viscosity coefficient, its value is small, and a good approximation is the one given by Stokes:

$$\lambda = -\frac{2}{3}\mu \tag{2.47}$$

Substituting the normal and shear stresses on the right hand side of Cauchy's equation, term $\nabla\overline{\overline{\tau}}$, and considering Cartesian coordinates, x direction, it is obtained:

$$\frac{\partial\sigma_x}{\partial x}+\frac{\partial\tau_{yx}}{\partial y}+\frac{\partial\tau_{zx}}{\partial z}=\frac{\partial}{\partial x}\left(2\mu\frac{\partial u}{\partial x}\right)-\frac{2}{3}\mu\frac{\partial}{\partial x}\left(\frac{\partial u}{\partial x}+\frac{\partial v}{\partial y}+\frac{\partial w}{\partial z}\right)+\frac{\partial}{\partial y}\left[\mu\left(\frac{\partial u}{\partial y}+\frac{\partial v}{\partial x}\right)\right]+\frac{\partial}{\partial z}\left[\mu\left(\frac{\partial u}{\partial z}+\frac{\partial w}{\partial x}\right)\right] \tag{2.48}$$

After some arrangement it is reached, x direction:

$$\frac{\partial\sigma_x}{\partial x}+\frac{\partial\tau_{yx}}{\partial y}+\frac{\partial\tau_{zx}}{\partial z}=\mu\left(\frac{\partial^2 u}{\partial x^2}+\frac{\partial^2 u}{\partial y^2}+\frac{\partial^2 u}{\partial z^2}\right)+\mu\frac{1}{3}\frac{\partial}{\partial x}\left[\frac{\partial u}{\partial x}+\frac{\partial v}{\partial y}+\frac{\partial w}{\partial z}\right] \tag{2.49}$$

Once the equations equivalent to (2.49) in y and z directions are being generated, and when substituting them in equation (2.38) it is reached:

$$\rho\left(\frac{\partial u}{\partial t}+u\frac{\partial u}{\partial x}+v\frac{\partial u}{\partial y}+w\frac{\partial u}{\partial z}\right)=-\frac{\partial p}{\partial x}-\rho g_x+\mu\left(\frac{\partial^2 u}{\partial x^2}+\frac{\partial^2 u}{\partial y^2}+\frac{\partial^2 u}{\partial z^2}\right)+\mu\frac{1}{3}\frac{\partial}{\partial x}\left[\frac{\partial u}{\partial x}+\frac{\partial v}{\partial y}+\frac{\partial w}{\partial z}\right] \tag{2.50}$$

$$\rho\left(\frac{\partial v}{\partial t}+u\frac{\partial v}{\partial x}+v\frac{\partial v}{\partial y}+w\frac{\partial v}{\partial z}\right)=-\frac{\partial p}{\partial y}-\rho g_y+\mu\left(\frac{\partial^2 v}{\partial x^2}+\frac{\partial^2 v}{\partial y^2}+\frac{\partial^2 v}{\partial z^2}\right)+\mu\frac{1}{3}\frac{\partial}{\partial y}\left[\frac{\partial u}{\partial x}+\frac{\partial v}{\partial y}+\frac{\partial w}{\partial z}\right] \tag{2.51}$$

$$\rho\left(\frac{\partial w}{\partial t}+u\frac{\partial w}{\partial x}+v\frac{\partial w}{\partial y}+w\frac{\partial w}{\partial z}\right)=-\frac{\partial p}{\partial z}-\rho g_z+\mu\left(\frac{\partial^2 w}{\partial x^2}+\frac{\partial^2 w}{\partial y^2}+\frac{\partial^2 w}{\partial z^2}\right)+\mu\frac{1}{3}\frac{\partial}{\partial z}\left[\frac{\partial u}{\partial x}+\frac{\partial v}{\partial y}+\frac{\partial w}{\partial z}\right] \tag{2.52}$$

Equations (2.50) to (2.52) are being called the Navier-Stokes equations, applicable to compressible and incompressible flow, laminar conditions. These three equations can be expressed in vector form; therefore the Navier-Stokes equations can be represented as:

$$\rho\frac{D\vec{V}}{Dt}=-\nabla p+\rho\vec{g}+\mu\nabla^2\vec{V}+\frac{1}{3}\mu\nabla\left(\nabla\vec{V}\right) \tag{2.53}$$

Notice that each term of equation (2.53) represents forces per unit volume.

It is also interesting to realize that the term $\left[\frac{\partial u}{\partial x}+\frac{\partial v}{\partial y}+\frac{\partial w}{\partial z}\right]$ characterizes the velocity divergence; as a result, the Navier-Stokes equations for laminar flow conditions and incompressible flow, Cartesian coordinates are given as:

$$\rho\left(\frac{\partial u}{\partial t}+u\frac{\partial u}{\partial x}+v\frac{\partial u}{\partial y}+w\frac{\partial u}{\partial z}\right)=-\frac{\partial p}{\partial x}-\rho g_x+\mu\left(\frac{\partial^2 u}{\partial x^2}+\frac{\partial^2 u}{\partial y^2}+\frac{\partial^2 u}{\partial z^2}\right) \tag{2.54}$$

$$\rho\left(\frac{\partial v}{\partial t}+u\frac{\partial v}{\partial x}+v\frac{\partial v}{\partial y}+w\frac{\partial v}{\partial z}\right)=-\frac{\partial p}{\partial y}-\rho g_y+\mu\left(\frac{\partial^2 v}{\partial x^2}+\frac{\partial^2 v}{\partial y^2}+\frac{\partial^2 v}{\partial z^2}\right) \tag{2.55}$$

$$\rho\left(\frac{\partial w}{\partial t}+u\frac{\partial w}{\partial x}+v\frac{\partial w}{\partial y}+w\frac{\partial w}{\partial z}\right)=-\frac{\partial p}{\partial z}-\rho g_z+\mu\left(\frac{\partial^2 w}{\partial x^2}+\frac{\partial^2 w}{\partial y^2}+\frac{\partial^2 w}{\partial z^2}\right) \tag{2.56}$$

These three equations plus the continuity equation, form a system of four equations with four unknowns (u,v,w,p), Cartesian coordinates.

The vector form of Navier-Stokes equations under laminar and incompressible flow conditions is represented as:

$$\rho\frac{D\vec{V}}{Dt}=-\nabla p+\rho\vec{g}+\mu\nabla^2\vec{V} \tag{2.57}$$

Please notice that equations (2.53) and (2.57) are applicable to any coordinate system, Cartesian, Cylindrical and Spherical. In what follows and for completeness, continuity and Navier-Stokes equations in cylindrical and Spherical coordinates are being presented.

Continuity equation in Cylindrical coordinates.

$$\frac{\partial\rho}{\partial t}+\frac{1}{r}\frac{\partial}{\partial r}(\rho r V_r)+\frac{1}{r}\frac{\partial}{\partial\theta}(\rho V_\theta)+\frac{\partial}{\partial x}(\rho V_x)=0 \tag{2.58}$$

Navier-Stokes equations in Cylindrical coordinates.

$$\begin{aligned}&\rho\left(\frac{\partial V_r}{\partial t}+V_r\frac{\partial V_r}{\partial r}+\frac{V_\theta}{r}\frac{\partial V_r}{\partial\theta}+V_x\frac{\partial V_r}{\partial x}-\frac{V_\theta^2}{r}\right)=\\&\rho\vec{g}_r-\frac{\partial P}{\partial r}+\mu\left[\frac{\partial}{\partial r}\left(\frac{1}{r}\frac{\partial}{\partial r}(rV_r)\right)+\frac{1}{r^2}\frac{\partial^2 V_r}{\partial\theta^2}+\frac{\partial^2 V_r}{\partial x^2}-\frac{2}{r^2}\frac{\partial V_\theta}{\partial\theta}\right]\end{aligned} \tag{2.59}$$

$$\begin{aligned}&\rho\left(\frac{\partial V_\theta}{\partial t}+V_r\frac{\partial V_\theta}{\partial r}+\frac{V_\theta}{r}\frac{\partial V_\theta}{\partial V_\theta}+V_x\frac{\partial V_\theta}{\partial x}+\frac{V_r V_\theta}{r}\right)=\\&\rho\vec{g}_\theta-\frac{1}{r}\frac{\partial P}{\partial\theta}+\mu\left[\frac{\partial}{\partial r}\left(\frac{1}{r}\frac{\partial}{\partial r}(rV_\theta)\right)+\frac{1}{r^2}\frac{\partial^2 V_\theta}{\partial\theta^2}+\frac{\partial^2 V_\theta}{\partial x^2}+\frac{2}{r^2}\frac{\partial V_r}{\partial\theta}\right]\end{aligned} \tag{2.60}$$

$$\rho\left(\frac{\partial V_x}{\partial t}+V_r\frac{\partial V_x}{\partial r}+V_\theta\frac{\partial V_x}{r\partial\theta}+V_x\frac{\partial V_x}{\partial z}\right)=\rho\vec{g}_x-\frac{\partial P}{\partial x}+\mu\left[\frac{1}{r}\frac{\partial}{\partial r}\left(r\frac{\partial V_x}{\partial r}\right)+\frac{1}{r^2}\frac{\partial^2 V_x}{\partial\theta^2}+\frac{\partial^2 V_x}{\partial x^2}\right] \tag{2.61}$$

Continuity equation in Spherical coordinates.

$$\frac{\partial \rho}{\partial t}+\frac{1}{r^2}\frac{\partial}{\partial r}(\rho\, r^2\, v_r)+\frac{1}{r\sin\theta}\frac{\partial}{\partial \theta}(\rho\, v_\theta \sin\theta)+\frac{1}{r\sin\theta}\frac{\partial}{\partial \phi}(\rho\, v_\phi)=0 \tag{2.62}$$

Navier-Stokes equations in Spherical coordinates.

$$\rho\left(\frac{\partial V_R}{\partial t}+V_R\frac{\partial V_R}{\partial R}+\frac{V_\theta}{R}\frac{\partial V_R}{\partial \theta}+\frac{V_\phi}{R\sin\theta}\frac{\partial V_R}{\partial \phi}-\frac{V_\theta^2+V_\phi^2}{R}\right)$$
$$=\rho g_R-\frac{\partial p}{\partial R}+\mu\left[\frac{1}{R^2}\frac{\partial}{\partial R}\left(R^2\frac{\partial V_R}{\partial R}\right)+\frac{1}{R^2\sin\theta}\frac{\partial}{\partial \theta}\left(\sin\theta\frac{\partial V_R}{\partial \theta}\right)\right]+ \tag{2.63}$$
$$\mu\left[\frac{1}{R^2\sin^2\theta}\frac{\partial V_R}{\partial \phi^2}-\frac{2V_R}{R^2}-\frac{2}{R^2}\frac{\partial V_\theta}{\partial \theta}-\frac{2V_\theta\cot\theta}{R^2}-\frac{2}{R^2\sin\theta}\frac{\partial V_\phi}{\partial \phi}\right]$$

$$\rho\left(\frac{\partial V_\theta}{\partial t}+V_R\frac{\partial V_\theta}{\partial R}+\frac{V_\theta}{R}\frac{\partial V_\theta}{\partial \theta}+\frac{V_\phi}{R\sin\theta}\frac{\partial V_\theta}{\partial \phi}-\frac{V_R V_\theta}{R}-\frac{V_{\phi^2}\cot\theta}{R}\right)$$
$$=\rho g_\theta-\frac{1}{R}\frac{\partial p}{\partial \theta}+\mu\left[\frac{1}{R^2}\frac{\partial}{\partial R}\left(R^2\frac{\partial V_\theta}{\partial R}\right)+\frac{1}{R^2\sin\theta}\frac{\partial}{\partial \theta}\left(\sin\theta\frac{\partial V_\theta}{\partial \theta}\right)\right]+ \tag{2.64}$$
$$\mu\left[\frac{1}{R^2\sin^2\theta}\frac{\partial^2 V_\theta}{\partial \phi^2}+\frac{2}{R^2}\frac{\partial V_R}{\partial \theta}-\frac{V_\theta}{R^2\sin^2\theta}-\frac{2\cos\theta}{R^2\sin^2\theta}\frac{\partial V_\phi}{\partial \phi}\right]$$

$$\rho\left(\frac{\partial V_\phi}{\partial t}+V_R\frac{\partial V_\phi}{\partial R}+\frac{V_\theta}{R}\frac{\partial V_\phi}{\partial \theta}+\frac{V_\phi}{R\sin\theta}\frac{\partial V_\phi}{\partial \phi}+\frac{V_\phi V_R}{R}+\frac{V_\theta V_\phi\cot\theta}{R}\right)$$
$$=\rho g_\phi-\frac{1}{R^2\sin\theta}\frac{\partial p}{\partial \phi}+\mu\left[\frac{1}{R^2}\frac{\partial}{\partial R}\left(R^2\frac{\partial V_\phi}{\partial R}\right)+\frac{1}{R^2\sin\theta}\frac{\partial}{\partial \theta}\left(\sin\theta\frac{\partial V_\phi}{\partial \theta}\right)\right]+ \tag{2.65}$$
$$\mu\left[\frac{1}{R^2\sin^2\theta}\frac{\partial^2 V_\phi}{\partial \phi^2}-\frac{V_\phi}{R^2\sin^2\theta}+\frac{2}{R^2\sin^2\theta}\frac{\partial V_R}{\partial \phi}+\frac{2\cos\theta}{R^2\sin^2\theta}\frac{\partial V_\phi}{\partial \phi}\right]$$

2.4. Momentum Equations for a Non Inertial Coordinate System, Integral Form

When performing the analysis of atmospheric flows for weather forecast, it is necessary to consider that a control volume attached to the earth surface, has, due to the earth rotation, a small acceleration versus the stars, which for this purpose can be considered static.

When analyzing the momentum associated to a fluid located inside an accelerating control volume, it is necessary to consider that such a control volume will be regarded as non inertial when studied versus a fixed coordinate system.

Figure 2.4 presents a control volume attached to a non inertial reference system x', y', z'. The non inertial reference system has a relative movement versus an inertial reference system x, y, z. It is at this point important to remember that the second Newton equation is only applicable in an inertial reference system x, y, z.

For the control system located at a given time t inside the control volume, it can be established:

$$\frac{d}{dt}\left(\vec{M}_{SIST}\right)_{xyz} = \sum \vec{F} \tag{2.66}$$

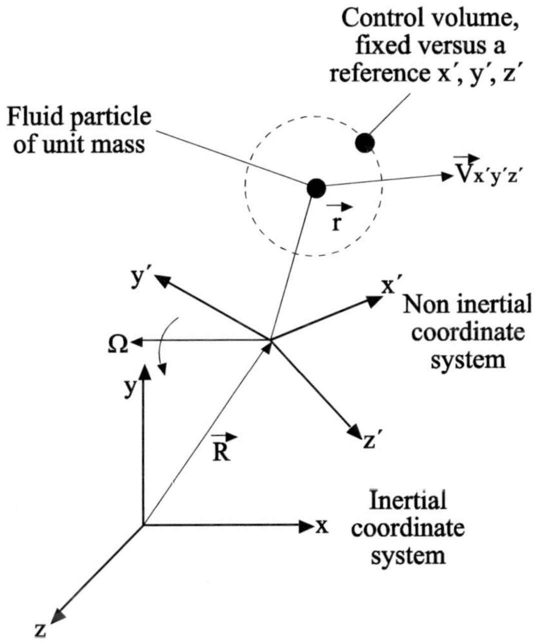

Figure 2.4. Inertial and non inertial reference frames.

In order to develop an expression able to evaluate the temporal momentum variation inside the control volume, it will initially be considered an elemental fluid particle located inside the control volume.

The momentum associated to this particle, $\vec{M}_{XYZ}$ is to be defined as the product of its elemental mass by its velocity.

$$\delta\vec{M}_{xyz} = (\delta_m)\cdot\vec{V}_{xyz} \tag{2.67}$$

From the Reynolds transport theorem equation (2.14), left hand side, it can be established, see equation (2.25):

$$\sum \vec{F}_{XYZ} = \frac{dB_{sc\,XYZ}}{dt} = m\frac{d\vec{V}_{XYZ}}{dt} = \int_{VC} \rho \frac{d\vec{V}_{XYZ}}{dt}\, d\forall \tag{2.68}$$

The temporal variation of a fluid particle velocity versus an inertial coordinate system, given as a function of the particle velocity referenced to a non inertial coordinate system, can be presented as: See figure 2.4.

$$\frac{d\vec{V}_{xyz}}{dt}=\frac{d\vec{V}_{x'y'z'}}{dt}+\frac{d^2\vec{R}}{dt^2}+\frac{d\vec{\Omega}}{dt}\wedge\vec{r}+2\vec{\Omega}\wedge\vec{V}_{x'y'z'}+\vec{\Omega}\wedge\vec{\Omega}\wedge\vec{r} \tag{2.69}$$

where Ω represents the generic angular velocity associated to the non inertial coordinate system.

Therefore, the forces acting versus an inertial coordinate system when considering the different accelerations a non inertial coordinate system can be subjected to, is given by:

$$\sum\vec{F}_{XYZ}=\int_{VC}\rho\left\{\frac{d\vec{V}_{x'y'z'}}{dt}+\frac{d^2\vec{R}}{dt^2}+\frac{d\vec{\Omega}}{dt}\wedge\vec{r}+\vec{\Omega}\wedge\vec{\Omega}\wedge\vec{r}+2\vec{\Omega}\wedge\vec{V}_{x'y'z'}\right\}d\forall \tag{2.70}$$

$$\sum\vec{F}_{XYZ}=\int_{VC}\rho\left\{\frac{d\vec{V}_{x'y'z'}}{dt}\right\}d\forall+\int_{VC}\rho\left\{\frac{d^2\vec{R}}{dt^2}+\frac{d\vec{\Omega}}{dt}\wedge\vec{r}+\vec{\Omega}\wedge\vec{\Omega}\wedge\vec{r}+2\vec{\Omega}\wedge\vec{V}_{x'y'z'}\right\}d\forall \tag{2.71}$$

Notice, that the first term of the right hand side from equation (2.71), represents the fluid forces versus a moving, non inertial, reference system, equation (2.71) can therefore be presented as:

$$\sum\vec{F}_{XYZ}=\sum\vec{F}_{x'y'z'}+\int_{VC}\rho\left\{\frac{d^2\vec{R}}{dt^2}+\frac{d\vec{\Omega}}{dt}\wedge\vec{r}+\vec{\Omega}\wedge\vec{\Omega}\wedge\vec{r}+2\vec{\Omega}\wedge\vec{V}_{x'y'z'}\right\}d\forall \tag{2.72}$$

And recalling at this point equation (2.26) it can be established:

$$\sum\vec{F}_{XYZ}=\frac{\partial}{\partial t}\int_{VC}\rho\,\vec{V}_{(i)x'y'z'}\,d\forall+\oint_{SC}\rho\,\vec{V}_{(i)x'y'z'}\,\vec{V}_{x'y'z'}\,d\vec{s}$$
$$+\int_{VC}\rho\left\{\frac{d^2\vec{R}}{dt^2}+\frac{d\vec{\Omega}}{dt}\wedge\vec{r}+\vec{\Omega}\wedge\vec{\Omega}\wedge\vec{r}+2\vec{\Omega}\wedge\vec{V}_{x'y'z'}\right\}d\forall \tag{2.73}$$

Extending the left hand side of equation (2.73) in the same form as the one already presented in equation (2.30), it is reached:

$$-\oint_{SC}P\,d\vec{s}+\oint_{SC}\bar{\bar{\tau}}\,ds+\int_{VC}\rho\,\vec{g}\,d\forall=\frac{\partial}{\partial t}\int_{VC}\rho\,\vec{V}_{(i)x'y'z'}\,d\forall+\oint_{SC}\rho\,\vec{V}_{(i)x'y'z'}\,\vec{V}_{x'y'z'}\,d\vec{s}$$
$$+\int_{VC}\rho\left\{\frac{d^2\vec{R}}{dt^2}+\frac{d\vec{\Omega}}{dt}\wedge\vec{r}+\vec{\Omega}\wedge\vec{\Omega}\wedge\vec{r}+2\vec{\Omega}\wedge\vec{V}_{x'y'z'}\right\}d\forall \tag{2.74}$$

Equation (2.74) is the momentum equation to be used when a non inertial coordinate system is to be considered. Notice that the terms.

$\frac{d^2\vec{R}}{dt^2}$ represents the linear acceleration of the non inertial coordinate system is having versus an inertial coordinate system.

$\frac{d\vec{\Omega}}{dt}\wedge\vec{r}$ is the acceleration due to variable angular velocity, tangential acceleration.

$\vec{\Omega}\wedge\vec{\Omega}\wedge\vec{r}$ represents the centripetal acceleration.

$2\vec{\Omega}\wedge\vec{V}_{x'y'z'}$ considers the Coriolis acceleration.

2.4.1. Momentum Equations for a Non Inertial Coordinate System, Differential Form

To determine the momentum equation applicable for non inertial coordinates, differential form, it will be used the same methodology as the one used for the conventional momentum equation, section 2.3.1. This is, the Gauss Ostrogradsky theorem will be applied to all surface integrals found in the integral momentum equation (2.74), as a result, all terms will be evaluated across volume integrals, grouping all terms inside a single volume integral and equaling to zero, it is obtained:

$$-\nabla P+\nabla\overline{\overline{\tau}}+\rho\vec{g}=\frac{\partial}{\partial t}\left(\rho\vec{V}_i\right)+\nabla\left(\rho\vec{V}_i\ \vec{V}\right)+\rho\left\{\frac{d^2\vec{R}}{dt^2}+\frac{d\vec{\Omega}}{dt}\wedge\vec{r}+\vec{\Omega}\wedge\vec{\Omega}\wedge\vec{r}+2\vec{\Omega}\wedge\vec{V}_{x'y'z'}\right\} \quad (2.75)$$

The first two terms on the right hand side of the equation (2.75) can be expressed as:

$$\frac{\partial}{\partial t}\left(\rho\vec{V}_i\right)+\nabla\left(\rho\vec{V}_i\ \vec{V}\right)=\rho\frac{\partial\vec{V}_i}{\partial t}+\vec{V}_i\frac{\partial\rho}{\partial t}+\vec{V}_i\ \nabla\left(\rho\vec{V}\right)+\rho\left(\vec{V}\nabla\right)\vec{V}_i \quad (2.76)$$

$$\frac{\partial}{\partial t}\left(\rho\vec{V}_i\right)+\nabla\left(\rho\vec{V}_i\ \vec{V}\right)=\rho\left[\frac{\partial\vec{V}_i}{\partial t}+\left(\vec{V}\nabla\right)\vec{V}_i\right]+\vec{V}_i\left[\frac{\partial\rho}{\partial t}+\nabla\left(\rho\vec{V}\right)\right] \quad (2.77)$$

Remembering the continuity equation in differential form and the concept of material derivative, equation (2.77) can be expressed:

$$\frac{\partial}{\partial t}\left(\rho\vec{V}_i\right)+\nabla\left(\rho\vec{V}_i\ \vec{V}\right)=\rho\left[\frac{D\vec{V}_i}{Dt}\right] \quad (2.78)$$

The velocity $\vec{V}_i$ is the velocity component versus a generic direction (i), then, for a generic tri dimensional flow, it can be established:

$$\rho\left[\frac{D\vec{V}_i}{Dt}\right]=\rho\left[\frac{D\vec{V}}{Dt}\right] \quad (2.79)$$

As a result, the momentum equation in differential form applicable to non inertial coordinates shall take the final form:

$$-\nabla P+\nabla\bar{\bar{\tau}}+\rho\vec{g}=\rho\frac{D\vec{V}}{Dt}+\rho\left\{\frac{d^2\vec{R}}{dt^2}+\frac{d\vec{\Omega}}{dt}\wedge\vec{r}+\vec{\Omega}\wedge\vec{\Omega}\wedge\vec{r}+2\vec{\Omega}\wedge\vec{V}_{x'y'z'}\right\} \tag{2.80}$$

Comparing equations (2.80) with (2.38) it can be realized that, for inertial and non inertial coordinate systems the equations look very much alike, but for a non inertial reference frame an extra tem is needed, term which evaluates the forces per unit volume, due to the different possible accelerations the non inertial reference system might be having versus an inertial reference system.

2.5. Equation of Angular Momentum for an Inertial Coordinate System. Integral Form

As in the previous equations, to determine the angular momentum one, the Reynolds transport theorem, equation (2.14) will be used. For the present case the generic property associated to the fluid, B, shall be defined as:

$$B=\int_{VC}\rho\left(\vec{r}\wedge\vec{v}\right)d\forall \tag{2.81}$$

The generic property per unit mass "b" will therefore be: $b=\vec{r}\wedge\vec{v}$; (2.82)
Substituting equation (2.81) into the left hand side of equation (2.14) it is obtained:

$$\begin{aligned}\frac{dB_{SC}}{dt}&=\frac{d}{dt}\int_{VC}\rho\left(\vec{r}\wedge\vec{v}\right)d\forall=\frac{d}{dt}\left(\vec{r}\wedge m\vec{v}\right)=\frac{d\vec{r}}{dt}\wedge m\,\vec{v}+\vec{r}\wedge m\frac{d\vec{v}}{dt}=\\ &m\left(\vec{v}\wedge\vec{v}\right)+\vec{r}\wedge m\frac{d\vec{v}}{dt}=\vec{r}\wedge m\frac{d\vec{v}}{dt}=\vec{r}\wedge m\vec{a}=\vec{r}\wedge\vec{F}=\vec{M}_0\end{aligned} \tag{2.83}$$

Equation which clarifies that, the temporal variation of a generic property B associated to the fluid inside the control system, is equivalent for the present case to the torque generated by the fluid versus a fixed coordinate system.

A conceptual way to understand the torque generated by a fluid particle moving with a velocity $\vec{V}$ versus an inertial coordinate system, is gathered via using the second Newton equation, which defines the concept of force as: $\vec{F}=\frac{d}{dt}(m\,\vec{v})$

Let's assume a given fluid particle of mass m, is moving with a velocity $\vec{V}_1$ versus an inertial coordinate system, see figure 2.5, the net force acting over the particle is defined as $\vec{F}$
.

The torque generated by the fluid particle versus a fixed coordinate system shall be represented as:

$$\vec{M}_0 = \vec{r}_0 \times \vec{F} = \vec{r}_0 \times \frac{d}{dt}\left(m_p \cdot \vec{V}\right) \tag{2.84}$$

Notice that equations (2.83) and (2.84) represent the same concept.

Substituting equations (2.82) and (2.83) into (2.14) it is obtained:

$$\sum \vec{M}_0 = \frac{\partial}{\partial t}\int_{VC} \rho(\vec{r} \wedge \vec{v})d\forall + \oint_{SC} \rho(\vec{r} \wedge \vec{v})\, \vec{v}\, d\vec{s} \tag{2.85}$$

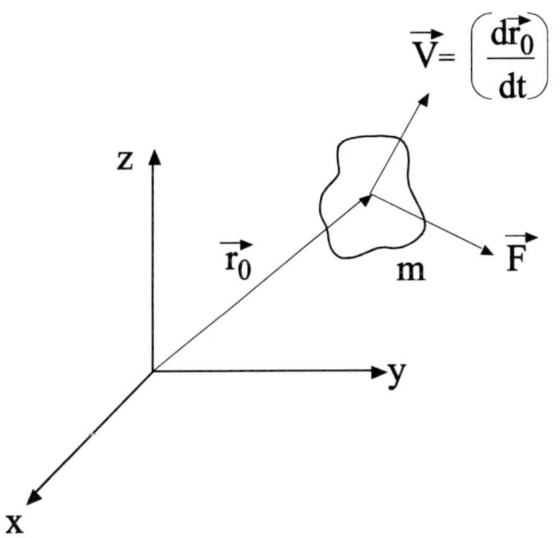

Figure 2.5. Fluid particle moving versus an inertial coordinate system.

Expression (2.85) is the integral form of the angular momentum equation for inertial coordinate systems.

The first term on the right hand side of the equation, represents the temporal variation of the angular momentum inside the control volume, the second term on the right hand evaluates the angular momentum flow across the control volume surfaces. As in the previous defined equations, this equation has a vector form, which enables to determine the angular momentum versus any coordinate axis.

Remembering what was established in the momentum equation, see equations (2.27)-(2.30), the forces acting onto a given control volume can be due to shear stresses, pressure or gravitational acceleration. As in the present equation the momentum generated by these forces is evaluated, the left hand side of equation (2.85) can be extended to:

$$\vec{r}\wedge \int_{VC} \rho\, \vec{g} d\forall + \vec{r}\wedge - \oint_{SC} P\, d\vec{s} + \vec{r}\wedge \oint_{SC} \bar{\bar{\tau}}\, ds = \frac{\partial}{\partial t}\int_{VC} \rho(\vec{r}\wedge\vec{v})d\forall + \oint_{SC} \rho(\vec{r}\wedge\vec{v})\vec{v}\, d\vec{s} \tag{2.86}$$

Equation (2.86) is the explicit integral form of the angular momentum equation for inertial coordinate systems. Notice that the generic distance $\vec{r}$, which on the left hand side of the equation (2.86) is located outside the integrals, needs to be placed inside the integrals whenever the terms inside the integrals are function of such distance.

2.5.1. Application of the Angular Momentum Equation to Turbomachinery

Figure 2.6 presents the rotor of a centrifugal pump; the dotted lines represent the control volume chosen. The angular momentum equation will be applied to the control volume understanding the flow as two dimensional. Angular momentum will be evaluated versus the z direction.

Assuming the flow as two dimensional and acting on the XY plane, the angular momentum equation applied to the control volume presented in figure 2.6 will take the form:

$$\sum M_Z = \left[\frac{d}{dt}\int_{\forall_{VC}} \rho(\vec{r} \times \vec{V}) d\forall\right]\vec{k} + \left[\oint_{SC} \rho(\vec{r} \times \vec{V})(\vec{V}.\hat{n}) ds\right]\vec{k} \tag{2.87}$$

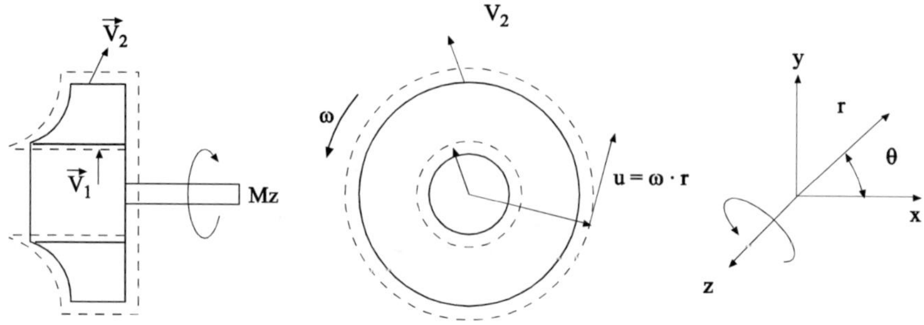

Figure 2.6. Rotor of a turbomachine, centrifugal pump.

The following hypotheses are being considered.

- The only torque considered will be versus the z axis.
- There is no temporal variation of the angular momentum inside the control volume.
- Flow at the control volume inlet and outlet is considered uniform.
- Gravitational forces are neglected.

Considering and Eulerian reference system, at the control volume inlet and outlet the velocity triangles presented in figure 2.7 can be defined. Notice that in figures 2.6 and 2.7 the blades inside the rotor are for clarity, not presented. Velocity triangles are formed when adding the tangential velocity U due to the rotor rotation and the relative velocity W, which direction will depend on the blades inclination angle at the inlet and outlet. The resultant vector is the absolute velocity V, which can be decomposed in a tangential and a radial component.

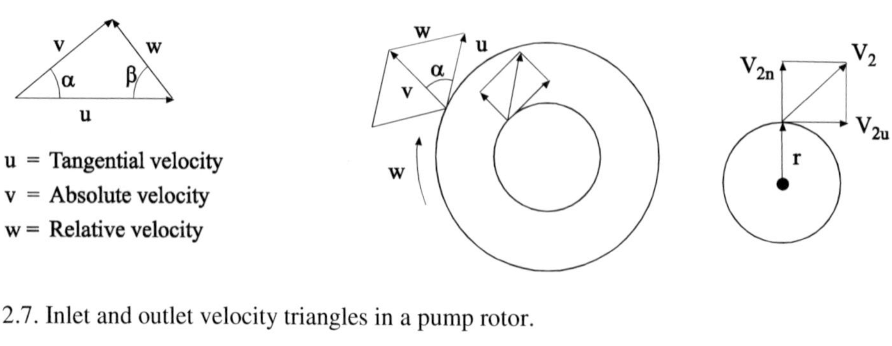

Figure 2.7. Inlet and outlet velocity triangles in a pump rotor.

According to the hypotheses assumed, the angular momentum equation will be reduced to:

$$\Sigma M_Z = \oint_{SC} \rho(\vec{r}_0 \times \vec{V})(\vec{V}_r . \hat{n}) ds \tag{2.88}$$

Extending equation (2.88) to the inlet and outlet surfaces it is obtained:

$$\sum M_z = \int_{SO} \rho(\vec{r}_2 \times \vec{V}_2)(\vec{V} \cdot \hat{n}) ds \, \hat{k} + \int_{SI} \rho(\vec{r}_1 \times \vec{V}_1)(\vec{V} \cdot \hat{n}) ds \, \hat{k} \tag{2.89}$$

The vector products defined in equation (2.89) can be transformed into:

$$\vec{r}_2 \times \vec{V}_2 = \vec{r}_2 \times \vec{V}_{2n} + \vec{r}_2 \times \vec{V}_{2u} = r_2 \, V_{2n} \sin(r_2 V_{2n}) + r_2 \, V_{2u} \sin(r_2 V_{2u}) = r_2 \, V_{2u} \sin(r_2 V_{2u}) \tag{2.90}$$

Substituting (2.90) into (2.89) and remembering that sine of 90 degrees is equal to one, it results:

$$\sum M_z = \int_{SO} \rho(r_2 V_{2u})(\vec{V}_r \cdot \hat{n}) ds + \int_{SI} \rho(r_1 V_{1u})(\vec{V}_r \cdot \hat{n}) ds \tag{2.91}$$

The remaining scalar product can be evaluated as:

$$\vec{V}_r \cdot \hat{n} = V_{ru} \cdot n \cos\left(\vec{V}_{ru} \hat{n}\right) + V_{rn} \cdot n \cos\left(\vec{V}_{rn} \hat{n}\right) = V_{rn} \cdot n \cos\left(\vec{V}_{rn} \hat{n}\right) \tag{2.92}$$

It must be noticed that the angle formed by the velocity radial component and the normal to the surface is, of zero degrees at the outlet and 180 degrees at the inlet, this is why the cosine at the inlet will bring a negative sign. Substituting equation (2.92) into (2.91) it is obtained:

$$\sum M_z = \int_{SO} \rho(r_2 V_{2u})(V_{rn} \, n) ds + \int_{SI} -\rho(r_1 V_{1u})(V_{rn} n) ds \tag{2.93}$$

Integrating equation (2.93) and remembering the concept of mass flow it is reached:

$$\sum M_z = \left| r_2 \, V_{2u} \, \dot{m} \right|_{SS} - \left| r_1 \, V_{1u} \, \dot{m} \right|_{SE} = \dot{m} \, (r_2 \, V_{2u} - r_1 \, V_{1u}) \tag{2.94}$$

The torque defined by equation (2.94) needs to be transferred to the pump by the electrical motor. Multiplying both terms of equation (2.94) by the angular turning speed and remembering that the power transferred to the pump is the product of the torque by the angular speed, it is obtained:

$$\dot{W}_{axis} = M_z \cdot \omega = \dot{m} \cdot \omega (r_2 V_{2u} - r_1 V_{1u}) \tag{2.95}$$

$$\dot{W}_{axis} = \dot{m}(u_2 V_{2u} - u_1 V_{1u}) \tag{2.96}$$

$$\frac{\dot{W}_{axis}}{\dot{m}} = Y = (u_2 V_{2u} - u_1 V_{1u}) \tag{2.97}$$

Equation (2.97) is the Euler equation for turbomachinery, it gives the energy per unit mass transferred to the fluid, this equation can be applied to pumps and turbines.

For pumps: $u_2 V_{2u} > u_1 V_{1u} \rightarrow \dot{W}_{axis} > 0$

For turbines: $u_2 V_{2u} < u_1 V_{1u} \rightarrow \dot{W}_{axis} < 0$

2.5.2. Equation of Angular Momentum for Non Inertial Coordinate Systems

The angular momentum equation for an inertial coordinate system was established as, see equation (2.83).

$$\sum \vec{M}_0 = \frac{d\vec{B}_{SC}}{dt} = \vec{r} \wedge m \frac{d\vec{V}_{XYZ}}{dt} = \vec{r} \wedge \int_{VC} \rho \frac{d\vec{V}_{XYZ}}{dt} d\forall \tag{2.98}$$

The acceleration of a fluid particle versus an inertial coordinate system, given as a function of the particle acceleration referenced to a non inertial coordinate system, can be presented as: See figure 2.4.

$$\frac{d\vec{V}_{XYZ}}{dt} = \frac{d\vec{V}_{x'y'z'}}{dt} + \frac{d^2\vec{R}}{dt^2} + \frac{d\vec{\Omega}}{dt} \wedge \vec{r} + \vec{\Omega} \wedge \vec{\Omega} \wedge \vec{r} + 2\vec{\Omega} \wedge \vec{V}_{x'y'z'} \tag{2.99}$$

Substituting (2.99) into (2.98) it is obtained:

$$\vec{M}_0 = \vec{r} \wedge \int_{VC} \rho \left\{ \frac{d\vec{V}_{x'y'z'}}{dt} + \frac{d^2\vec{R}}{dt^2} + \frac{d\vec{\Omega}}{dt} \wedge \vec{r} + \vec{\Omega} \wedge \vec{\Omega} \wedge \vec{r} + 2\vec{\Omega} \wedge \vec{V}_{x'y'z'} \right\} d\forall \tag{2.100}$$

$$\vec{M}_0 = \vec{r} \wedge m \frac{d\vec{V}_{x'y'z'}}{dt} + \vec{r} \wedge \int_{VC} \rho \left\{ \frac{d^2\vec{R}}{dt^2} + \frac{d\vec{\Omega}}{dt} \wedge \vec{r} + \vec{\Omega} \wedge \vec{\Omega} \wedge \vec{r} + 2\vec{\Omega} \wedge \vec{V}_{x'y'z'} \right\} d\forall \tag{2.101}$$

Notice that the first term of the right hand side of the equation (2.101), characterizes the angular momentum of a fluid particle versus a non inertial coordinate system.

$$\vec{M}_{0\,x'y'z'} = \vec{r} \wedge m \frac{d\vec{V}_{x'y'z'}}{dt} = \frac{\partial}{\partial t} \int_{VC} \rho (\vec{r} \wedge \vec{V}_{x'y'z'}) d\forall + \oint_{SC} \rho (\vec{r} \wedge \vec{V}_{x'y'z'}) \vec{V}_{x'y'z'} \, d\vec{s} \tag{2.102}$$

And substituting equation (2.102) in (2.101) it is reached:

$$\vec{M}_0 = \frac{\partial}{\partial t}\int_{\forall C} \rho(\vec{r}\wedge\vec{V}_{x'y'z'})d\forall + \oint_{SC} \rho(\vec{r}\wedge\vec{V}_{x'y'z'})\vec{V}_{x'y'z'}\, d\vec{s} + \vec{r}\wedge \int_{\forall C} \rho \left\{\frac{d^2\vec{R}}{dt^2} + \frac{d\vec{\Omega}}{dt}\wedge\vec{r} + \vec{\Omega}\wedge\vec{\Omega}\wedge\vec{r} + 2\vec{\Omega}\wedge\vec{V}_{x'y'z'}\right\} d\forall \quad (2.103)$$

Equation (2.103) represents the angular momentum equation applicable to non inertial coordinate systems.

2.6. Energy Equation. Integral Form

In a generic control volume, the five possible different forms of energy associated to the fluid located inside the control system which at time t=0 occupies the same volume as the control volume, are:

- Kinetic energy. Associated to the fluid particle movement. $E_c = \frac{V^2}{2}$
- Potential energy. Associated to the particle position. $E_p = g\,z$
- Internal energy. Associated to the molecules structure and movement.
- Chemical energy. Associated to the possible occurring chemical reactions and to the atoms disposition inside molecules.
- Nuclear energy. Associated to the internal atomic structure, this energy is liberated in a nuclear fusion or fission.

In most of the conventional processes and or when working with a single fluid, Nuclear and Chemical energies are not to be considered, in what follows just the three conventional forms of energy associated to a fluid particle, Kinetic, potential and Internal will be considered.

Energy can also be classified as thermal or mechanical. Thermal energy is associated to the fluid temperature, molecular structure and heat transfer, mechanical energy is associated to particle movement and the forces acting on it. A simple classification of mechanical and thermal energies would be:

- Mechanical energy: Kinetic, potential and work.
- Thermal energy: Internal energy and heat.

Given a generic control system, the energy balance can be expressed via evaluating the temporal velocity change of the intrinsic energy associated to the fluid, and equaling it to the velocity of heat transferred to the fluid, plus the velocity of work applied onto the system.

Using the thermodynamic sign convention, which considers the incoming heat flow as positive and the work given to the control system as negative, it can be stated:

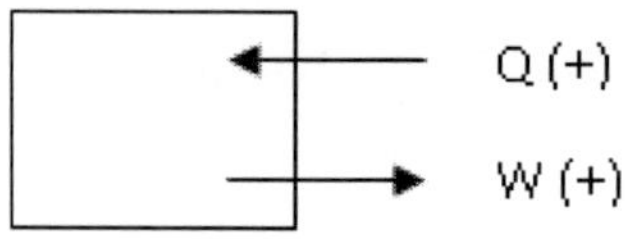

Figure 2.8. Sign convention adopted.

$$\frac{dQ}{dt}-\frac{dW}{dt}=\dot{Q}-\dot{W}=\frac{dE_{sis}}{dt}=\frac{dB_{sis}}{dt} \tag{2.104}$$

Remembering at this point the Reynolds transport equation (2.14), where the balance of a generic property B associated to the fluid was evaluated, and defining in the present case this property as the energy associated to the system, it can be said:

$$B_{sis}=E_{sis}=\int_{\forall_{sis}}\rho\ (u+\frac{V^2}{2}+gz)d\forall \tag{2.105}$$

The generic property b, defined for the present case as the energy per unit mass, will be:

$$b=\frac{B}{m}=u+\frac{V^2}{2}+gz \tag{2.106}$$

Substituting equations (2.104) and (2.106) into the Reynolds transport equation (2.14), it is obtained:

$$\dot{Q}-\dot{W}=\frac{dQ}{dt}-\frac{dW}{dt}=\frac{\partial}{\partial t}\int_{\forall C}\rho\left(u+\frac{V^2}{2}+gz\right)d\forall+\oint_{SC}\rho\left(u+\frac{V^2}{2}+gz\right)\vec{V}_r\,d\vec{s} \tag{2.107}$$

It is important to remember:

$\dot{W}=\frac{dW}{dt}$ represents the mechanical power transferred to or obtained from the control volume considered.

$\dot{Q}=\frac{dQ}{dt}$ represents the heat per unit time transferred or obtained from the control volume chosen.

2.6.1. Composition of the Mechanical Work

Three different types of work acting onto a given control system can be differentiated:

$$W=W_{axis}+W_{\tau}+W_P \tag{2.108}$$

The work given or taken by an axis (W_{axis}).

The work done by shear stresses (W_τ), due to the shear stresses acting onto the control volume frontiers.

The work done by the fluid pressure (W_P), due to the fluid pressure acting onto the control volume frontiers.

The power associated to each term can be mathematically presented as:

a) Power associated to the shear stresses.

$$\dot{W}_\tau = \oint_{Avc} \vec{V}\ d\vec{F}_\tau = \oint_{Avc} \vec{V}\ (\bar{\bar{\tau}}\ ds) \tag{2.109}$$

F_τ = is the force associated to the shear stresses.

This power is to be understood as the power dissipated by the viscosity forces.

b) Power associated to a mechanical axis.

$$\dot{W}_{axis} = M \cdot \omega \tag{2.110}$$

The power associated to a mechanical axis is the torque M multiplied by the angular velocity ω.

c) Power associated to pressure forces.

In reality the power associated to pressure forces, can be divided into power due to pressure forces associated to the fluid flow and the one associated to the deformation of the control volume. The second one is just possible if the control volume allows deformation when pressure acts onto its surface.

The sign associated to the power due to the fluid flow pressure, is given by the scalar product of the fluid velocity crossing the control volume surfaces and the vector surface differential.

$$d\dot{W}_{P\ flow} = PV_n d\vec{s} = P.\vec{V}.\hat{n}\, ds = P.V \cos\theta\, ds = P\vec{V} d\vec{s} \tag{2.111}$$

V is the fluid velocity relative to the control volume surface.

Integrating equation (2.111) along the control volume surface it is obtained: $\dot{W}_{P\ flow} = \oint_{SC} P\,\vec{V}\ d\vec{s}$

The power obtained by the possible deformation of the control volume and due to pressure forces $\dot{W}_{P\ D}$, is represented as the product of the pressure acting onto the control volume surface and the velocity the surface is deforming. Whenever the control volume is not deformable, this term will not exist.

$$\dot{W}_{P\ D} = \oint_{SC} \vec{V}_d\ d\vec{F}_P = \oint_{SC} P\,\vec{V}_d\ \hat{n}\ ds = \oint_{SC} P\,\vec{V}_d\ \ d\vec{s} \tag{2.112}$$

$\vec{V}_d$ represents the control surface velocity of deformation.

The power associated to the pressure forces can be given as:

$$\dot{W}_P = \dot{W}_{P\,flow} + \dot{W}_{P\,D} = \oint_{SC} P\,\vec{V}\,\hat{n}\,ds = \oint_{SC} P\,\vec{V}_d\,\hat{n}\,ds \tag{2.113}$$

The overall mechanical power will be: $\dot{W} = \dot{W}_{axis} + \dot{W}_{\tau} + \dot{W}_{P\,flow} + \dot{W}_{P\,D}$ (2.114)

Substituting equations (2.114) and (2.113) into (2.107) it is obtained:

$$\dot{Q} - \dot{W}_{axis} - \dot{W}_{\tau} - \oint_{SC} P\,\vec{V}\,\hat{n}\,ds - \oint_{SC} P\,\vec{V}_d\,\hat{n}\,ds = \frac{\partial}{\partial t}\int_{\forall c} \rho\left(u + \frac{V^2}{2} + gz\right) d\forall + \oint_{SC} \rho\left(u + \frac{V^2}{2} + gz\right)\vec{V}\,\hat{n}\,ds \tag{2.115}$$

Equation (2.115) is usually presented as:

$$\dot{Q} - \dot{W}_{axis} - \dot{W}_{\tau} - \oint_{SC} P\,\vec{V}_d\,\hat{n}\,ds = \frac{\partial}{\partial t}\int_{\forall c} \rho\left(u + \frac{V^2}{2} + gz\right) d\forall + \oint_{SC} \rho\left(u + \frac{P}{\rho} + \frac{V^2}{2} + gz\right)\vec{V}\,\hat{n}\,ds \tag{2.116}$$

It is important to remember that fluid enthalpy is defined as: $u + \frac{P}{\rho} = h$.

Notice from equation (2.116) that for a static and rigid control volume, under permanent conditions, for incompressible flow and with no heat transfer or mechanical work, equation (2.116) gives birth to:

$$\frac{P_1}{\rho_1} + \frac{V_1^2}{2} + g z_1 = \frac{P_2}{\rho_2} + \frac{V_2^2}{2} + g z_2 \tag{2.117}$$

Equation (2.117) is called the Bernoulli equation, being equivalent to the first thermodynamic principle.

2.6.2. Energy Equation Applied to Turbomachinery, Case Thermal and Hydraulic Machines

Figure 2.9 represents a schematic representation of a turbomachine, fluid enters the machine through the suction inlet and leaves the machine through the outlet. Flow properties at the inlet and outlet are known.

The machine has an axis though which power is transferred or extracted from the control volume.

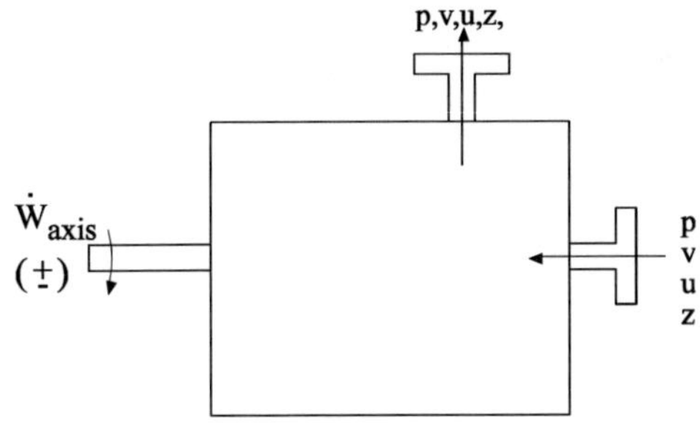

Figure 2.9. Schematic representation of a turbomachine.

Choosing as a control volume the square box presented in figure 2.9, and the control volume static and rigid, the energy equation will take the form:

$$\dot{Q} - \dot{W}_{axis} - \dot{W}_{\tau} = \frac{\partial}{\partial t}\int_{\forall c} \rho\left(u + \frac{V^2}{2} + gz\right) d\forall + \oint_{SC} \rho\left(u + \frac{P}{\rho} + \frac{V^2}{2} + gz\right)\vec{V} d\vec{s} \tag{2.118}$$

Considering the regime as permanent and extending the energy flow terms at the inlet and outlet it is obtained:

$$\dot{Q} - \dot{W}_{axis} - \dot{W}_{\tau} = \int_{SI} \rho\left(h + gz + \frac{V^2}{2}\right)\vec{V} d\vec{s} + \int_{SO} \rho\left(h + gz + \frac{V^2}{2}\right)\vec{V} d\vec{s} \tag{2.119}$$

This equation is used in thermal machinery, where the power generated mostly depends on the enthalpy rate between inlet and outlet.

For hydraulic machinery, the appropriate equation would be:

$$-\dot{W}_{axis} - \dot{W}_{\tau} = \int_{SI} \rho\left(\frac{P}{\rho} + gz + \frac{V^2}{2}\right)\vec{V} d\vec{s} + \int_{SO} \rho\left(\frac{P}{\rho} + gz + \frac{V^2}{2}\right)\vec{V} d\vec{s} \tag{2.120}$$

If the fluid density is considered as constant and the fluid velocity is uniform:

$$-\dot{W}_{axis} - \dot{W}_{\tau} = \dot{m}\left(\frac{P_O}{\rho} - \frac{P_I}{\rho} + \frac{V_O^2}{2} - \frac{V_I^2}{2} + g(z_O - z_I)\right) \tag{2.121}$$

where the terms:

$\frac{\dot{W}_{axis}}{\dot{m}} = Y$; represents the energy per unit mass transferred to the fluid or taken from the fluid.

$\frac{\dot{W}_\tau}{\dot{m}} = \Delta y$; represents the energy per unit mass lost inside the machine and due to shear stresses.

It is important to notice that the energy lost inside the machine due to shear stresses is being transformed into heat; this heat is being absorbed by the fluid and the solid boundaries around it. According to the sign convention adopted this energy shall be seen as positive, since the fluid is losing it, therefore:

$$-Y + \Delta y = \left(\frac{P_S}{\rho} - \frac{P_E}{\rho} + \frac{V_S^2}{2} - \frac{V_E^2}{2} + g(z_S - z_E) \right) \tag{2.122}$$

The net energy transferred to the fluid, case of a centrifugal pump, will be:

$$Y_{real} = Y - \Delta y \tag{2.123}$$

$$-Y_{real} = \left(\frac{P_S}{\rho} - \frac{P_E}{\rho} + \frac{V_S^2}{2} - \frac{V_E^2}{2} + g(z_S - z_E) \right) \tag{2.124}$$

The overall efficiency of a pump, considering the mechanical efficiency as 100% shall be defined as:

$$\eta_T = \frac{\rho Y_{real} Q_{real}}{\dot{W}_{axis}} \tag{2.125}$$

The hydraulic efficiency for pumps is defined as:

$$\eta_H = \frac{Y_{real}}{Y} = \frac{Y - \Delta y}{Y} = \frac{\dot{W}_{axis} - \dot{W}_\tau}{\dot{W}_{axis}} = \frac{\dot{W}_{axis} - \sum \text{losses}}{\dot{W}_{axis}} \tag{2.126}$$

The relation between the overall and the hydraulic efficiency is:

$$\eta_T = \frac{\rho(Y - \Delta y)Q_{real}}{\dot{W}_{axis}} = \frac{\rho(Y - \Delta y)Q_{real}}{\rho Y Q} = \eta_H \frac{Q_{real}}{Q} = \eta_H \, \eta_V \tag{2.127}$$

Notice that the relation $\eta_V = \frac{Q_{real}}{Q}$ is defined as the pump volumetric efficiency.

For hydraulic turbines, the overall efficiency is given as:

$$\eta_T = \frac{\dot{W}_{axis}}{\rho Y_{real} Q_{real}} \tag{2.128}$$

And the hydraulic efficiency for turbines is defined as:

$$\eta_H = \frac{Y}{Y_{real}} = \frac{Y}{Y+\Delta y} = \frac{\dot{W}_{axis}}{\dot{W}_{axis} + \dot{W}_{\tau}} = \frac{\dot{W}_{axis}}{\dot{W}_{axis} + \sum \text{losses}} \quad (2.129)$$

Notice that for turbines:

$$Y_{real} = Y + \Delta y \quad (2.130)$$

Which means that according to the sign convention established, the power transferred to the axis as well as the energy lost due to shear stresses is energy extracted from the fluid. The energy equation to be used for hydraulic turbines will be:

$$Y_{real} = \left(\frac{P_O}{\rho} - \frac{P_I}{\rho} + \frac{V_O^2}{2} - \frac{V_I^2}{2} + g(z_O - z_I) \right) \quad (2.131)$$

And the overall, hydraulic and volumetric efficiencies for hydraulic turbines are defined as:

$$\eta_T = \frac{\dot{W}_{axis}}{\rho\, Y_{real} Q_{real}} = \frac{\rho\, Y\, Q}{\rho\,(Y+\Delta y) Q_{real}} = \eta_H \frac{Q}{Q_{real}} = \eta_H\, \eta_V \quad (2.132)$$

The term $\eta_V = \frac{Q}{Q_{real}}$ defines the volumetric efficiency of a turbine.

Notice that Q_{real} represents the flow entering the turbine, and Q represents the flow used to transfer the power to the axis. For turbines it is understood that $Q_{real} > Q$.

2.6.3. Energy Equation. Differential Form

Following the same procedure used to determine the differential form of the previous equations, the differential form of the energy equation will be obtained from the integral form of the energy equation and via transforming the surface integrals into volume integrals.

The integral form of the energy equation was taking the form, equation (2.116).

$$-(\dot{W}_{axis} + \dot{W}_{\tau}) + \dot{Q} = \frac{\partial}{\partial t}\int_{VC} \rho \left(u + \frac{V^2}{2} + gz\right) d\forall + \oint_{SC} \left(h + \frac{V^2}{2} + gz\right) \rho\, \vec{V}\, d\vec{s} \quad (2.133)$$

Applying the Gauss-Ostrogradsky theorem to the surface integral it is obtained:

$$-(\dot{W}_{axis} + \dot{W}_{\tau}) + \dot{Q} = \frac{\partial}{\partial t}\int_{VC} \rho(u + \frac{V^2}{2} + gz) d\forall + \int_{VC} \nabla \left[(h + \frac{V^2}{2} + gz)\rho \vec{V} \right] d\forall \quad (2.134)$$

Differentiating equation (2.134) versus the volume it is reached:

$$\frac{d}{d\forall}\left[-\left(\dot{W}_{axis}+\dot{W}_{\tau}\right)+\dot{Q}\right]=\frac{\partial}{\partial t}\left[\rho(u+\frac{V^2}{2}+gz)\right]+\nabla\left[(h+\frac{V^2}{2}+gz)\rho\vec{V}\right] \tag{2.135}$$

This equation can be given as:

$$-\left(\delta\dot{W}_{axis}+\delta\dot{W}_{\tau}\right)+\delta\dot{Q}=\frac{\partial}{\partial t}\left[\rho(u+\frac{V^2}{2}+gz)\right]+\nabla\left[(h+\frac{V^2}{2}+gz)\rho\vec{V}\right] \tag{2.136}$$

Each term of equation (2.136) represents the power per unit volume. The first term of the right hand side of the previous equation can be expressed as:

$$\frac{\partial}{\partial t}\left[\rho(u+\frac{V^2}{2}+gz)\right]=\frac{\partial}{\partial t}\left[\rho(u+\frac{V^2}{2}+gz)\right]+\frac{\partial}{\partial t}\left[\rho(P\cdot\upsilon)\right]-\frac{\partial}{\partial t}\left[\rho(P\cdot\upsilon)\right] \tag{2.137}$$

$$\frac{\partial}{\partial t}\left[\rho(u+\frac{V^2}{2}+gz)\right]=\frac{\partial}{\partial t}\left[\rho(u+P\cdot\upsilon+\frac{V^2}{2}+gz)\right]-\frac{\partial}{\partial t}\left[\rho(P\cdot\upsilon)\right] \tag{2.138}$$

$$\frac{\partial}{\partial t}\left[\rho(u+\frac{V^2}{2}+gz)\right]=\frac{\partial}{\partial t}\left[\rho(h+\frac{V^2}{2}+gz)\right]-\frac{\partial}{\partial t}P \tag{2.139}$$

Substituting equation (2.139) into (2.136) it is obtained:

$$-\left(\delta\dot{W}_{axis}+\delta\dot{W}_{\tau}\right)+\delta\dot{Q}=\frac{\partial}{\partial t}\left[\rho(h+\frac{V^2}{2}+gz)\right]+\nabla\left[(h+\frac{V^2}{2}+gz)\rho\vec{V}\right]-\frac{\partial}{\partial t}P \tag{2.140}$$

The right hand side of equation (2.140) can be expressed as:

$$\begin{aligned}&\frac{\partial\rho}{\partial t}(h+\frac{V^2}{2}+gz)+\rho\frac{\partial}{\partial t}(h+\frac{V^2}{2}+gz)+(h+\frac{V^2}{2}+gz)\cdot\nabla(\rho\vec{V})\\&+\rho(\vec{V}\nabla)(h+\frac{V^2}{2}+gz)-\frac{\partial}{\partial t}P\end{aligned} \tag{2.141}$$

Grouping terms it is obtained:

$$(h+\frac{V^2}{2}+gz)\left(\frac{\partial\rho}{\partial t}+\nabla(\rho\vec{V})\right)+\rho\frac{\partial}{\partial t}(h+\frac{V^2}{2}+gz)+\rho(\vec{V}\nabla)(h+\frac{V^2}{2}+gz)-\frac{\partial}{\partial t}P \tag{2.142}$$

Remembering the continuity equation in differential form, the first term of equation (2.142) is zero, therefore:

$$\rho\frac{\partial}{\partial t}(h+\frac{V^2}{2}+gz)+\rho(\vec{V}\cdot\nabla)(h+\frac{V^2}{2}+gz)-\frac{\partial P}{\partial t} \tag{2.143}$$

Via using the concept of material derivative, the two first terms of equation (2.142) can be grouped to:

$$\rho\frac{D}{Dt}(h+\frac{V^2}{2}+gz)-\frac{\partial P}{\partial t} \tag{2.144}$$

Substituting equation (2.144) into (2.140), the energy equation in differential form will take the form:

$$-\left(\delta\dot{W}_{eje}+\delta\dot{W}_{\tau}\right)+\delta\dot{Q}=\rho\frac{D}{Dt}(h+\frac{V^2}{2}+gz)-\frac{\partial P}{\partial t} \tag{2.145}$$

Equation (2.145) is one of the representations of the energy equation in differential form. Notice that each term has units of power per unit volume $\frac{J}{m^3 s}$.

2.7. Application of Differential Equations: Flow under Dominant Viscosity

Under viscosity dominant, it is understood that the inertial forces associated to a fluid particle play an irrelevant role, the fluid movement is dominated by the viscosity forces, the Reynolds number is small. In what follows several typical cases where viscosity forces are relevant are presented, notice that flow in narrow gaps are a typical example where viscosity forces become dominant.

2.7.1. Flow between Two Parallel Plates

Figure 2.10 presents two infinite parallel plates, the lower plate remains static while the upper one moves towards the x direction with a time dependant velocity U(t). The distance between the two plates d is very small and this narrow gap between the two plates is filled with fluid. Initially it is interesting to find out the fluid velocity distribution between the two plates.

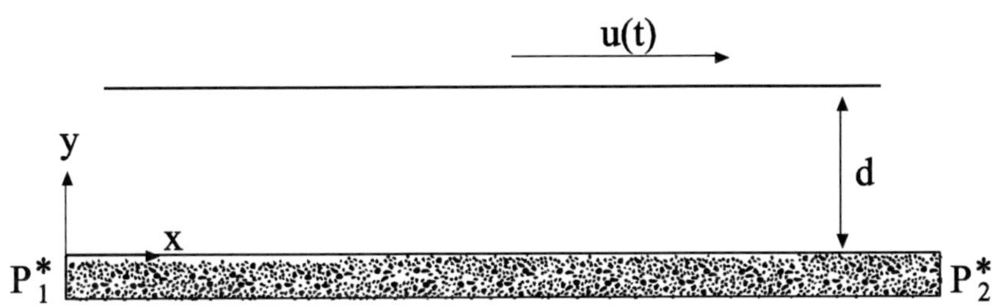

Figure 2.10. Flow between two parallel plates.

The working hypothesis shall be:

- Fluid is considered as incompressible.
- Both plates are considered as infinite.
- The upper plate moves towards the x direction with a velocity U, which might be time dependant.
- Flow movement is considered unidirectional and two-dimensional.
- A pressure gradient might exist between plate's borders.
- The boundary conditions are: $U(x,0,t)=0; \quad U(x,d,t)=U(t)$

The Navier-Stokes equations for laminar, incompressible, tridimensional flow in Cartesian coordinates are given as:

$$\begin{aligned}
&\rho\left(\frac{\partial u}{\partial t}+u\frac{\partial u}{\partial x}+v\frac{\partial u}{\partial y}+w\frac{\partial u}{\partial z}\right)=-\frac{\partial p}{\partial x}+\rho g_x+\mu\left(\frac{\partial^2 u}{\partial x^2}+\frac{\partial^2 u}{\partial y^2}+\frac{\partial^2 u}{\partial z^2}\right)\\
&\rho\left(\frac{\partial v}{\partial t}+u\frac{\partial v}{\partial x}+v\frac{\partial v}{\partial y}+w\frac{\partial v}{\partial z}\right)=-\frac{\partial p}{\partial y}+\rho g_y+\mu\left(\frac{\partial^2 v}{\partial x^2}+\frac{\partial^2 v}{\partial y^2}+\frac{\partial^2 v}{\partial z^2}\right)\\
&\rho\left(\frac{\partial w}{\partial t}+u\frac{\partial w}{\partial x}+v\frac{\partial w}{\partial y}+w\frac{\partial w}{\partial z}\right)=-\frac{\partial p}{\partial z}+\rho g_z+\mu\left(\frac{\partial^2 w}{\partial x^2}+\frac{\partial^2 w}{\partial y^2}+\frac{\partial^2 w}{\partial z^2}\right)
\end{aligned} \tag{2.146}$$

The vector representation of equation (2.146) is:

$$\rho\frac{D\vec{V}}{Dt}=-\nabla p+\rho\vec{g}+\mu\nabla^2\vec{V} \tag{2.147}$$

The continuity equation is represented as:

$$\frac{\partial \rho}{\partial t}+\nabla(\rho\vec{V})=0 \tag{2.148}$$

Since for the present case the fluid density ρ is constant, the continuity equation will be reduced to:

$$\nabla\vec{V}=0=\frac{\partial u}{\partial x}+\frac{\partial v}{\partial y}+\frac{\partial w}{\partial z} \tag{2.149}$$

Since the flow is unidirectional, fluid velocity shall depend on $U=(y,t)$, equation (2.149) can further be reduced to:

$$\frac{\partial u}{\partial x}=0 \tag{2.150}$$

The Navier Stokes equations for two-dimensional flow shall be represented as:

$$\rho\left(\frac{\partial u}{\partial t}\right) = -\frac{\partial p}{\partial x} + \rho\vec{g}_x + \mu\left(\frac{\partial^2 u}{\partial y^2}\right)$$
$$0 = -\frac{\partial p}{\partial y} + \rho\vec{g}_y \tag{2.151}$$

The equations (2.151) can be reduced when combining pressure and gravitational forces. For a conservative field, it can be established:

$$\vec{g} = -\nabla\text{Ц} = -\left(\underset{\vec{g}_x}{\frac{\partial \text{Ц}}{\partial x}} i + \underset{\vec{g}_y}{\frac{\partial \text{Ц}}{\partial y}} j + \underset{\vec{g}_z}{\frac{\partial \text{Ц}}{\partial z}} k\right) \tag{2.152}$$

The parameter Ц is a scalar potential of gravitational forces. In a gravitational field the parameter Ц = g h The distance h is a vertical distance defined versus a coordinate system, h=h(y).

Substituting equation (2.152) into (2.151) and considering the value of Ц, it is obtained:

$$\rho\frac{\partial u}{\partial t} = -\frac{\partial p}{\partial x} - \rho g\frac{\partial h}{\partial x} + \mu\left(\frac{\partial^2 u}{\partial y^2}\right) = -\frac{\partial p^*}{\partial x} + \mu\frac{\partial^2 u}{\partial y^2} \tag{2.153}$$

$$0 = -\frac{\partial p}{\partial y} + \rho g\frac{\partial h}{\partial y} = -\frac{\partial p^*}{\partial y} \tag{2.154}$$

Notice that the term $p^* = p + \rho g h$ is defined as reduced pressure.

Equation (2.154) indicates that in y direction the fluid static low will prevail.

Subtracting the reduced pressure term from equation (2.153) it is obtained:

$$\frac{\partial p^*}{\partial x} = \mu\frac{\partial^2 u}{\partial y^2} - \rho\frac{\partial u}{\partial t} \tag{2.155}$$

It is important to notice that the left hand side of equation (2.155) depends on the position x; the right hand side depends on the position y and both parts may depend of time.

Assuming that the pressure variation along the x direction is constant, it can be said:

$$\frac{\partial p^*(t)}{\partial x} = -\frac{\Delta p^*}{\Delta L} = \frac{(p+\rho gh)_1 - (p+\rho gh)_2}{L} \tag{2.156}$$

Equation (2.156) defines the temporal pressure variation between two points separated a distance L. The negative sign indicates that pressure decreases as position x increases.

It is important to realize that the vector form of Navier Stokes equation for incompressible flow and as a function of the reduced pressure, takes the form:

$$\rho\frac{D\vec{u}}{Dt}=-\nabla p^{*}+\mu\nabla^{2}\vec{u} \tag{2.157}$$

2.7.1.1. Plane Couette - Poiseulle Flow

Consider the flow between two parallel plates defined in the previous case, at this point it will be considered that the upper plate moves at a time independent velocity U, the rest of hypothesis remain the same, therefore the new hypothesis shall be represented as:

The flow is time independent $\frac{\partial}{\partial t}=0$

Continuity and Navier-Stokes equations will be for the present conditions:
Continuity equation:

$$\nabla\vec{V}=0=\frac{\partial u}{\partial x}+\frac{\partial v}{\partial y}+\frac{\partial w}{\partial z}=\frac{\partial u}{\partial x} \tag{2.158}$$

Navier-Stokes equations:

$$0=-\frac{1}{\rho}\frac{\partial p^{*}}{\partial x}+\frac{\mu}{\rho}\frac{\partial^{2}u}{\partial y^{2}} \tag{2.159}$$

$$0=\frac{\partial p^{*}}{\partial y}\rightarrow p^{*}\Big|_{y}=cte \tag{2.160}$$

Integrating Navier-Stokes equation (2.159) along direction x and considering the pressure distribution along the x axis as constant $\frac{\partial p^{*}}{\partial x}=\frac{\Delta p^{*}}{\Delta x}=cte$, it is obtained:

$$\frac{1}{\rho}\frac{\Delta p^{*}}{\Delta x}=\frac{\mu}{\rho}\frac{\partial^{2}u}{\partial y^{2}} \tag{2.161}$$

$$\frac{\partial u}{\partial y}=\frac{\Delta p^{*}}{\Delta x}\frac{1}{\mu}y+C_{1} \tag{2.162}$$

$$u=\frac{\Delta p^{*}}{\Delta x}\frac{1}{\mu}\frac{y^{2}}{2}+C_{1}y+C_{2} \tag{2.163}$$

Integration constants C_1 and C_2 shall be obtained via using the following boundary conditions:

$y=0\ \ u=0$; Fluid velocity at the lower plate is cero.

$y=d\ \ u=U$; or $y=d\ \ u=0$; Fluid velocity at the upper plate is the velocity of the plate.

Since for the present cease the upper plate is moving with a constant velocity U, it is obtained:

$$C_2 = 0; \quad C_1 = \frac{U}{d} - \frac{\Delta p^*}{\Delta x}\frac{d}{2\mu}; \tag{2.164}$$

Substituting equation (2.164) into (2.163) it is obtained the equation which gives the flow velocity distribution as a function of the vertical axis y.

$$u = \frac{\Delta p^*}{\Delta x}\frac{y^2}{2\mu} - \frac{\Delta p^*}{\Delta x}\frac{dy}{2\mu} + \frac{U \cdot y}{d} = \frac{\Delta p^*}{\Delta x}\frac{y}{2\mu}[y-d] + \frac{U \cdot y}{d} \tag{2.165}$$

Notice that equation (2.165) provides the velocity distribution between two parallel plates when the lower plate is static, the upper plate moves with a constant velocity U and there is a reduced pressure differential between plate borders. From equation (2.164) two particular cases can be considered.

2.7.1.1.1. Couette Flow

Couette flow between two parallel plates is defined when considering that the reduced pressure between the two plate's borders is null. Flow movement is just due to the upper plate displacement.

$$\frac{\Delta p^*}{\Delta x} = 0 \tag{2.166}$$

As a result, the velocity distribution between the two plates will be:

$$u = \frac{U y}{d} \tag{2.167}$$

The force per unit surface which is opposing to the movement will take the form:

$$\tau = \mu \frac{\partial u}{\partial y} = \mu \frac{U}{d} \tag{2.168}$$

And the power per unit surface needed to maintain the upper plate moving at a constant velocity U will be:

$$\dot{W} = \tau u = \mu \frac{U^2}{d} \tag{2.169}$$

2.7.1.1.2. Hagen-Poiseulle or Plane Poiseulle Flow

For the present case, it shall be considered that both plates are static; the flow movement is due to the reduced pressure gradient between plate borders. From equation (2.165) it is obtained:

$$u = \frac{\Delta p^*}{\Delta x}\frac{y}{2\mu}(y-d) \tag{2.170}$$

Equation (2.170) is represented in figure 2.11 and defines a parabolic velocity distribution.

For a generic case, Couette-Poiseulle flow, the velocity distribution was already defined as equation (2.165). To obtain the shear stresses and the power needed to displace the upper plate, the same procedure as the one introduced in section 2.7.1.1.1 shall be followed. Figure 2.12 presents some of the possible different flow configurations appearing between the two plates and considering Couette-Poiseulle flow, notice that the flow configuration changes drastically when the reduced pressure is considered as positive or negative.

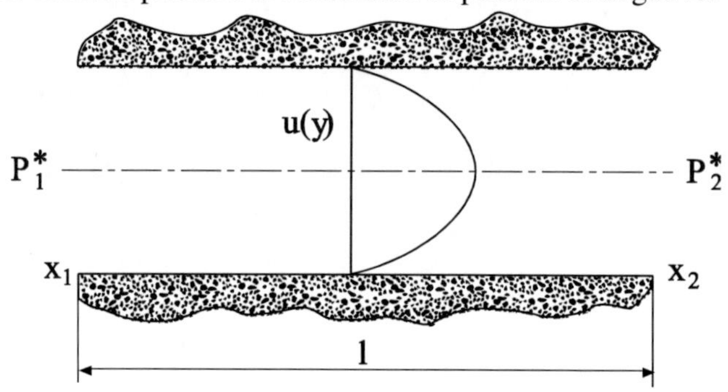

Figure 2.11. Parabolic velocity distribution between two parallel plates.

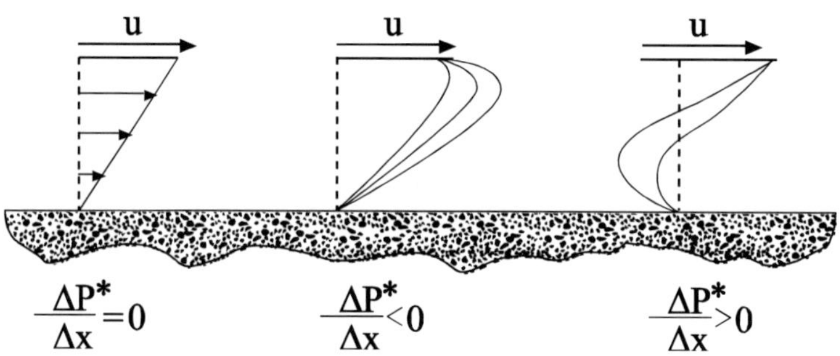

Figure 2.12. Several generic velocity distributions for Couette-Poiseulle flow.

2.7.2. Time Dependent Flow, Rayleich Flow

The present case is foccused in studding a horitzontal plate immmersed in a fluid, see figure 2.13, the plate will initially be static and will acelérate until reaching a constant velocity U. The idea is to study the temporal fluid movement as a function of the distance to the plate.

Hypothesis:

- Unidirectional flow $\frac{\partial u}{\partial x} = 0$.

- At t=0 fluid and plate are static, the plate accelerates until reaching a velocity U.
- Flow will be considered as incompressible.
- Flow will be seen as time dependent $\frac{\partial u}{\partial t} \neq 0$.
- There is no pressure gradient between plate borders. $\frac{\partial p^*}{\partial x} = 0$.

The Navier Sokes equation represented as a function of the reduced pressure takes the form:

$$\rho \frac{Du}{Dt} = -\nabla p^* + \mu \nabla^2 u \tag{2.171}$$

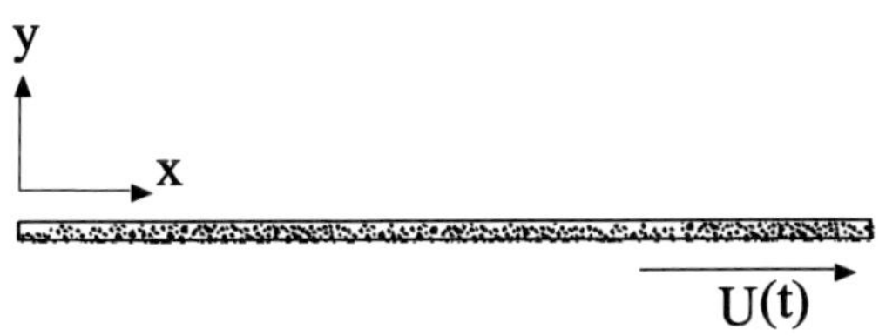

Figure 2.13. Rayleich flow.

According to the hypothesis established, equation (2.171) will be reduced to:

$$\rho \frac{\partial u}{\partial t} = 0 + \mu \frac{\partial^2 u}{\partial y^2} \tag{2.172}$$

The boundary conditions needed to solve the previous equation are:

Initially, for any value of the vertical position y it is established: $u(y,0) = 0$.

At any time t, the fluid velocity in a position y far away from the plate will be: $u(\infty, t) = 0$.

The fluid velocity in contact with the plate, will be the plate velocity: $u(0,t) = U$ (2.173)

From the differential equation (2.172) the following magnitudes relationship can be obtained:

$$\frac{U}{t_0} \approx \frac{\mu}{\rho} \frac{U}{\delta^2} = \nu \frac{U}{\delta^2} \tag{2.174}$$

Being:

δ = fluid height affected by the plate movement. δ is measured in y direction.

t_0 = generic time, measured versus the initial time.

Defining a similarity variable η, also called flow characteristic constant, as:

$$\delta^2 \approx \nu t_0 \rightarrow \delta \approx \sqrt{\nu t_0} \rightarrow \frac{\delta}{\sqrt{\nu t_0}} = \eta = \frac{y}{\sqrt{\nu t}} \tag{2.175}$$

The idea is transforming the fluid velocity u and giving it as a function of the variable η. This method is called similarity transformation method.

Calling $u^* = \frac{u}{U}$ and substituting this term in the Navier-Stokes equation, (2.172), it is obtained:

$$\frac{\partial u}{\partial t} = \frac{\partial (u^* U)}{\partial t} = U \frac{\partial u^*}{\partial t} \tag{2.176}$$

$$\frac{\partial^2 u}{\partial y^2} = \frac{\partial^2 (u^* U)}{\partial y^2} = \frac{\partial}{\partial y}\left(\frac{\partial}{\partial y}\left(u^* U\right)\right) = \frac{\partial}{\partial y}\left(U \frac{\partial u^*}{\partial y}\right) = U \frac{\partial^2 u^*}{\partial y^2} \tag{2.177}$$

Substituting (2.176) and (2.177) into (2.172) it can be established:

$$\frac{\partial u^*}{\partial t} = \frac{\partial u^*}{\partial \eta}\frac{\partial \eta}{\partial t} - \nu \frac{\partial^2 u^*}{\partial \eta^2}\frac{\partial \eta^2}{\partial y^2}; \tag{2.178}$$

Notice that: $\frac{\partial \eta^2}{\partial y^2} = \left(\frac{\partial \eta}{\partial y}\right)^2$ (2.179)

Recalling that $\eta = \frac{y}{\sqrt{\nu t}}$; its derivation versus time and the position y will be:

$$\frac{\partial \eta}{\partial t} = \frac{y}{\sqrt{\nu}} \cdot \left(-\frac{1}{2} t^{-\frac{3}{2}}\right) \tag{2.180}$$

$$\left(\frac{\partial \eta}{\partial y}\right)^2 = \frac{1}{\nu t} \tag{2.181}$$

Substituting equations (2.180) and (2.181) into equation (2.178) it is reached:

$$\frac{\partial u^*}{\partial \eta} \cdot \frac{y}{\sqrt{\nu}} \cdot \left(-\frac{1}{2} \cdot t^{-\frac{3}{2}}\right) = \nu \frac{\partial^2 u^*}{\partial \eta^2} \cdot \frac{1}{\nu t} \tag{2.182}$$

$$\frac{\partial u^*}{\partial \eta} \cdot \frac{y}{\sqrt{\nu t}} \frac{1}{t} \cdot \left(-\frac{1}{2}\right) = \frac{\partial^2 u^*}{\partial \eta^2} \cdot \frac{1}{t} \tag{2.183}$$

A further reduction of equation (2.183) gives birth to:

$$-\frac{\eta}{2}\frac{\partial u^*}{\partial \eta} = \frac{\partial^2 u^*}{\partial \eta^2} \tag{2.184}$$

and equation (2.184) can also be represented as:

$$-\frac{\eta}{2}=\frac{1}{\frac{\partial u^*}{\partial \eta}}\frac{\partial}{\partial \eta}\left(\frac{\partial u^*}{\partial \eta}\right);\equiv -\frac{\eta}{2}=\frac{1}{A}\frac{\partial A}{\partial \eta};\equiv\ -\frac{\eta}{2}d\eta=\frac{dA}{A} \tag{2.185}$$

The integration of the previous equation gives:

$$-\frac{\eta^2}{4}+C_0=\ln\left(\frac{\partial u^*}{\partial \eta}\right) \tag{2.186}$$

Considering that whenever $\eta = 0$, it can be said:

$$\eta=0\rightarrow\frac{\partial u^*}{\partial \eta}=\left(\frac{\partial u^*}{\partial \eta}\right)_0 \tag{2.187}$$

Therefore the integration constant C_0 will be:

$$C_0=\ln\left(\frac{\partial u^*}{\partial \eta}\right)_0 \tag{2.188}$$

Substituting equation (2.188) into equation (2.186) it is obtained:

$$-\frac{\eta^2}{4}=\ln\frac{\frac{\partial u^*}{\partial \eta}}{\left(\frac{\partial u^*}{\partial \eta}\right)_0} \tag{2.189}$$

removing the logarithms it is reached:

$$e^{-\frac{\eta^2}{4}}=\frac{\frac{\partial u^*}{\partial \eta}}{\left(\frac{\partial u^*}{\partial \eta}\right)_0};\qquad \frac{\partial u^*}{\partial \eta}=\left(\frac{\partial u^*}{\partial \eta}\right)_0\cdot e^{-\frac{\eta^2}{4}} \tag{2.190}$$

Integrating again it is obtained:

$$u^*=\left(\frac{\partial u^*}{\partial \eta}\right)_0\int_0^{\eta}e^{-\frac{\eta^2}{4}}d\eta+c_1 \tag{2.191}$$

To determine the constant c_1, the following boundary condition shall be employed:

for $y=0 \equiv \eta=0 \rightarrow u^* = u^*(0); \quad \frac{U}{U}=1$

It is to be noticed as well that if $\eta = 0; \equiv \int_0^{\eta} e^{-\frac{\eta^2}{4}} d\eta = \int_0^{0} e^{-0} d\eta = 0$

As a result, the constant C_1 will be having a value: $C_1 = u^*(0)$

Equation (2.191) will finally take the form:

$$u^* = u^*(0) + \left(\frac{\partial u^*}{\partial \eta}\right)_0 \int_0^{\eta} e^{-\frac{\eta^2}{4}} d\eta \tag{2.192}$$

The integral $\int_0^{\eta} e^{-\frac{\eta^2}{4}} d\eta$ is called the statistical error function, being its value:

$$\int_0^{\eta} e^{-\frac{\eta^2}{4}} d\eta = \sqrt{\pi}\, \text{erf}\left(\frac{\eta}{2}\right) \tag{2.193}$$

The velocity distribution will therefore take the form:

$$u^* = u^*(0) + \left(\frac{\partial u^*}{\partial \eta}\right)_0 \sqrt{\pi}\, \text{erf}\left(\frac{\eta}{2}\right) \tag{2.194}$$

In order to determine the value of $\left(\frac{\partial u^*}{\partial \eta}\right)_0$, it will be used the following boundary condition.

$$\text{for } y=\infty; \eta=\infty \rightarrow \text{erf}\left(\frac{\eta}{2}\right)=1; u^*=0=\frac{0}{U} \tag{2.195}$$

Obtaining:

$$0 = u^* = u^*(0) + \left(\frac{\partial u^*}{\partial \eta}\right)_0 \sqrt{\pi} \;; \Rightarrow \left(\frac{\partial u^*}{\partial \eta}\right)_0 = -\frac{u^*(0)}{\sqrt{\pi}}; \tag{2.196}$$

Substituting equation (2.196) into (2.194) the final resulting equation giving the velocity distribution will be:

$$u^* = \left(1 - \text{erf}\left(\frac{\eta}{2}\right)\right) u^*(0) \tag{2.197}$$

Notice that the term erf is the statistical error function.

2.7.3. Stationary Flow inside Circular Ducts

2.7.3.1. Poiseulle Flow

Figure 2.14 represents an inclined circular tube, fluid flow is moving from the left hand side, where the reduced pressure is maximum, towards the right hand side, where reduced pressure is minimum. It is important to remember that reduced pressure considers the static pressure plus the position versus a given plane, $p_i^* = p_i + \rho g h_i$. The equation giving the velocity distribution as a function of the radial position, is what a priory is needed.

The following hypothesis shall be considered.

- p= cte. Fluid is considered as incompressible.
- $\frac{\partial}{\partial t} = 0$; The flow is time independent, permanent.
- $v_r = v_\theta = 0$; Flow is considered as unidirectional and two-dimensional.
- $v_x = f(r)$; The velocity V_x is only radial dependant.

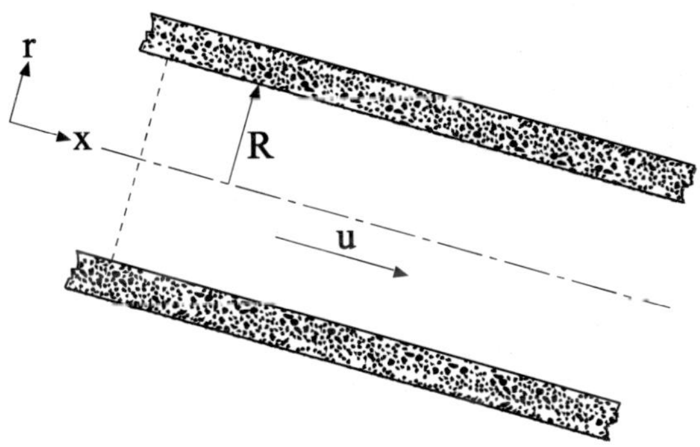

Figure 2.14. Poiseulle flow in cylindrical tubes.

The boundary conditions needed to solve the problem, are:

when $r = 0; \Rightarrow v_x = v_{max}$

when $r = R; \Rightarrow v_x = 0$

Remembering now the continuity and Navier-Stokes equations in cylindrical coordinates and for laminar flow, it can be said:

Continuity equation:

$$\frac{\partial \rho}{\partial t} + \frac{1}{r}\frac{\partial}{\partial r}(\rho\, r\, v_r) + \frac{1}{r}\frac{\partial}{\partial \theta}(\rho\, v_\theta) + \frac{\partial}{\partial x}(\rho\, v_x) = 0 \tag{2.198}$$

Navier stokes equations:

Radial direction.

$$\rho\left(\frac{\partial V_r}{\partial t}+V_r\frac{\partial V_r}{\partial r}+\frac{V_\theta}{r}\frac{\partial V_r}{\partial \theta}+V_x\frac{\partial V_r}{\partial x}-\frac{V_\theta^2}{r}\right)=-\frac{\partial p}{\partial r}+\rho\vec{g}_r+$$
$$\mu\left(\frac{\partial^2 V_r}{\partial r^2}+\frac{1}{r}\frac{\partial V_r}{\partial r}+\frac{1}{r^2}\frac{\partial^2 V_r}{\partial \theta^2}+\frac{\partial^2 V_r}{\partial x^2}-\frac{V_r}{r^2}-\frac{2}{r^2}\frac{\partial V_\theta}{\partial \theta}\right) \quad (2.199)$$

Angular direction.

$$\rho\left(\frac{\partial V_\theta}{\partial t}+V_r\frac{\partial V_\theta}{\partial r}+\frac{V_\theta}{r}\frac{\partial V_\theta}{\partial \theta}+V_x\frac{\partial V_\theta}{\partial x}+\frac{V_r V_\theta}{r}\right)=-\frac{1}{r}\frac{\partial p}{\partial \theta}+\rho\vec{g}_\theta+$$
$$\mu\left(\frac{\partial^2 V_\theta}{\partial r^2}+\frac{1}{r}\frac{\partial V_\theta}{\partial r}+\frac{1}{r^2}\frac{\partial^2 V_\theta}{\partial \theta^2}+\frac{\partial^2 V_\theta}{\partial x^2}+\frac{2}{r^2}\frac{\partial V_r}{\partial \theta}-\frac{V_\theta}{r^2}\right) \quad (2.200)$$

Axial direction.

$$\rho\left(\frac{\partial V_x}{\partial t}+V_r\frac{\partial V_x}{\partial r}+\frac{V_\theta}{r}\frac{\partial V_x}{\partial \theta}+V_x\frac{\partial V_x}{\partial x}\right)=-\frac{\partial p}{\partial x}+\rho\vec{g}_x+$$
$$\mu\left(\frac{\partial^2 V_x}{\partial r^2}+\frac{1}{r}\frac{\partial V_x}{\partial r}+\frac{1}{r^2}\frac{\partial^2 V_x}{\partial \theta^2}+\frac{\partial^2 V_x}{\partial x^2}\right) \quad (2.201)$$

It is also important to remember that in cylindrical coordinates, the gradient of a property is being defined as:

$$\nabla P=\frac{\partial P}{\partial r}\hat{r}+\frac{1}{r}\frac{\partial P}{\partial \theta}\hat{\theta}+\frac{\partial P}{\partial x}\hat{x} \quad (2.202)$$

$$\vec{g}=-\nabla \amalg=-\nabla(gh)=-\left[\frac{\partial(gh)}{\partial r}\hat{r}+\frac{1}{r}\frac{\partial(gh)}{\partial \theta}\hat{\theta}+\frac{\partial(gh)}{\partial x}\hat{x}\right]=\vec{g}_r+\vec{g}_\theta+\vec{g}_x \quad (2.203)$$

Considering the previous established hypothesis and the considerations defined in equations (2.202) and (2.203), the continuity and Navier-Stokes equations will be reduced to:

Continuity equation: $\frac{\partial V_x}{\partial x}=0$ (2.204)

Navier-Stokes equations:

Radial direction: $0=\frac{\partial P^*}{\partial r}$ (2.205)

Angular direction: $0=\frac{1}{r}\frac{\partial P^*}{\partial \theta}$ (2.206)

Axial direction: $0=-\frac{\partial P^*}{\partial x}+\mu\frac{1}{r}\frac{\partial V_x}{\partial r}+\mu\frac{\partial^2 V_x}{\partial r^2}=-\frac{\partial P^*}{\partial x}+\frac{\mu}{r}\frac{\partial}{\partial r}\left(r\frac{\partial V_x}{\partial r}\right)$ (2.207)

Equations in radial and angular direction, clarify that reduced pressure is independent of the radius and the angle chosen. Reduced pressure will only depend on the axial direction.

As in the previous cases, it has to be realized that the reduced pressure along the axial direction is constant.

$$\frac{\partial P^*}{\partial x} = \frac{\Delta P^*}{\Delta x} = \text{constant.}$$

In many cases, as well as in the previous case, the term $\frac{\partial P^*}{\partial x}$ has a negative term associated, since the pressure P^* decreases as the axial position x increases.

Integrating the Navier Stokes equation in axial direction, equation (2.207) it is obtained:

$$\int \frac{r}{\mu}\frac{\Delta P^*}{\Delta x}dr = \int d\left(r\frac{\partial V_x}{\partial r}\right) \tag{2.208}$$

$$\frac{1}{\mu}\frac{\Delta P^*}{\Delta x}\frac{r^2}{2} + C_0 = r\frac{dV_x}{dr} \tag{2.209}$$

Applying the first boundary condition to equation (2.209) it is obtained:

$$r = 0\,;\; v_x = v_{máx} \quad \Rightarrow C_0 = 0 \tag{2.210}$$

Integrating again equation (2.209) after considering (2.210) it is reached:

$$\int \frac{1}{\mu}\frac{\Delta P^*}{\Delta x}\frac{r^2}{2}\frac{dr}{r} = \int dV_x \tag{2.211}$$

$$\frac{\Delta P^*}{\Delta x}\frac{1}{\mu}\frac{r^2}{4} + C_1 = V_x \tag{2.212}$$

Using now the second boundary condition, it is obtained:

$$r = R;\;\; v_x = 0;\;\; C_1 = -\frac{\Delta P^*}{\Delta x}\frac{1}{\mu}\frac{R^2}{4} \tag{2.213}$$

Finally it is obtained:

$$V_x = \frac{\Delta P^*}{\Delta x}\frac{1}{\mu}\frac{1}{4}(r^2 - R^2) \tag{2.214}$$

Notice from equation (2.214) that the velocity V_x seems to be always negative, it must at this point be remembered that the term $\frac{\Delta P^*}{\Delta x}$ has a sign associated and in general this sign is negative.

The maximum velocity was defined to be at the central axis, $r=0;\quad V_x = V_{máx}$; therefore its value will be:

$$V_{max} = -\frac{\Delta P^*}{\Delta x}\frac{1}{4\mu}R^2 \tag{2.215}$$

Combining equations (2.215) with (2.214) it can be said:

$$V_x = V_{máx}\left(1-\left(\frac{r}{R}\right)^2\right) \tag{2.216}$$

Equation (2.216) characterizes the parabolic velocity distribution of fluid flow inside a cylindrical tube.

Once the velocity distribution is known, the volumetric flow will be found out via integrating the velocity distribution across the tube cross section.

$$Q = \int_0^R V_x \cdot 2\pi r\, dr = \int_0^R V_{max}\left(1-\left(\frac{r}{R}\right)^2\right)2\pi r\, dr = V_{max}\,2\pi\left(\frac{r^2}{2}-\frac{r^4}{R^2}\frac{1}{4}\right)_0^R =$$
$$V_{max}\,2\pi\frac{R^2}{4} = V_{max}\,\pi\frac{R^2}{2} = -\frac{\Delta P^*}{\Delta x}\frac{R^4\pi}{8\mu} \tag{2.217}$$

The fluid flow average velocity can be calculated by:

$$\bar{V} = \frac{Q}{S} = \frac{-\frac{\Delta P^*}{\Delta x}\frac{R^4\pi}{8\mu}}{\pi R^2} = -\frac{\Delta P^*}{\Delta x}\frac{R^2}{8\mu} = \frac{1}{2}V_{max} \tag{2.218}$$

2.7.4. Flow between Annular Tubes

Figure 2.15 presents two coaxial pipes, fluid is allowed to flow in the space between the two pipes, fluid moves due to the pipes relative movement, in the case present in figure 2.15, the internal tube has an axial displacement, while the external tube remains static. Fluid movement is also affected by the possible reduced pressure differential between the two tubes axial borders.

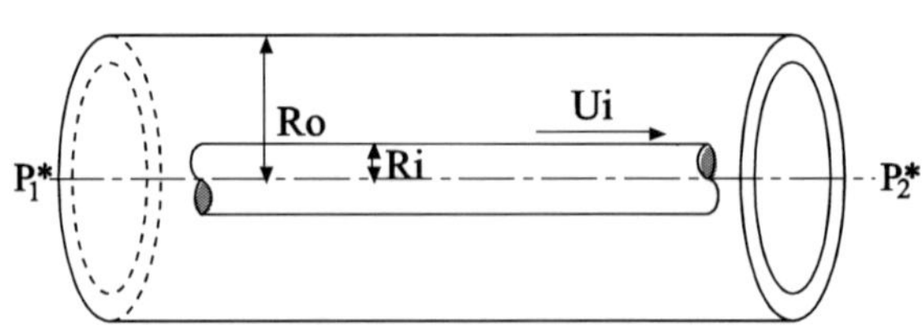

Figure 2.15. Flow between two concentric tubes.

For the present case, the working hypothesis are the same as the ones used for the previous case, therefore, the resultant differential equations will be the same as the ones already found, equations (2.204) to (2.207). In what follows the different possible boundary conditions applicable for the present case are presented.

Boundary conditions 1a and 1b.

1a. $r = R_0;\ V_x = 0$ 1b. $r = R_0;\ V_x = 0$

$r = R_i;\ V_x = U_i$ $r = R_i;\ V_x = U_i$

$\frac{\Delta P^*}{\Delta x} = 0$ $\frac{\Delta P^*}{\Delta x} \neq 0$

Boundary conditions 2a and 2b.

2a. $r = R_0;\ V_x = U_0$ 2b $r = R_0;\ V_x = U_0$

$r = R_i;\ V_x = U_i$ $r = R_i;\ V_x = U_i$

$\frac{\Delta P^*}{\Delta x} = 0$ $\frac{\Delta P^*}{\Delta x} \neq 0$

Boundary conditions 3a and 3b.

3a $r = R_0;\ V_x = U_0$ 3b. $r = R_0;\ V_x = U_0$

$r = R_i;\ V_x = 0$ $r = R_i;\ V_x = 0$

$\frac{\Delta P^*}{\Delta x} = 0$ $\frac{\Delta P^*}{\Delta x} \neq 0$

Boundary condition 4.

4 $r = R_0;\ V_x = 0$

$r = R_i;\ V_x = 0$

$\frac{\Delta P^*}{\Delta x} \neq 0$

Independently of the boundary conditions used, the differential equations characterizing the fluid performance are the same as the ones found in the previous case, equations (2.204) to (2.207). As in the previous case, it will now be considered: $\frac{\partial P^*}{\partial x} = \frac{\Delta P^*}{\Delta x} = \text{cte.}$

Integrating the Navier-Stokes equation in axial direction, equation (2.207) it is obtained:

$$\int \frac{\Delta P^*}{\Delta x} \frac{r}{\mu} dr = \int d\left(r \frac{\partial V_x}{\partial r} \right) \tag{2.219}$$

$$C_0 + \frac{\Delta P^*}{\Delta x} \frac{r^2}{2} \frac{1}{\mu} = r \frac{dV_x}{dr} \tag{2.220}$$

Integrating again it is reached:

$$V_x = \frac{\Delta P^*}{\Delta x}\frac{r^2}{4}\frac{1}{\mu} + C_0 \ln r + C_1 \tag{2.221}$$

The value of the constants C_0 and C_1 will be found via employing the previously defined boundary conditions. For the following example the boundary conditions employed are the 1a, there is no relative pressure gradient between pipe axial borders, the internal tube is moving axially and the external one remains static.

2.7.4.1. Example 1. Flow between Two Concentric Pipes. Boundary conditions 1a

Boundary conditions 2a.

$$r = R_0; \quad \Rightarrow V_x = 0$$
$$r = R_i; \quad \Rightarrow V_x = U_i$$
$$\frac{\Delta P^*}{\Delta x} = 0$$

Applying these conditions to the equation (2.221), it is obtained:

$$0 = C_0 \ln R_0 + C_1 \tag{2.222}$$

$$U_i = C_0 \ln R_i + C_1 \tag{2.223}$$

ubtracting equation (2.222) from (2.223) it is reached:

$$U_i = C_0\left(\ln R_i - \ln R_0\right) = C_0 \ln\frac{R_i}{R_0} \quad ; \quad C_0 = \frac{U_i}{\ln\frac{R_i}{R_0}} \tag{2.224}$$

Substituting equation (2.224) into (2.222) it is found:

$$0 = \frac{U_i}{\ln\frac{R_i}{R_0}} \ln R_0 + C_1 \tag{2.225}$$

$$C_1 = -\frac{U_i}{\ln\frac{R_i}{R_0}} \ln R_0 \tag{2.226}$$

Substituting equations (2.226) and (2.224) into equation (2.221) and remembering that for the present chose boundary conditions $\frac{\Delta P^*}{\Delta x} = 0$, it is reached:

$$V_x = \frac{U_1}{\ln\frac{R_i}{R_0}} \ln r - \frac{U_1}{\ln\frac{R_i}{R_0}} \ln R_0 \tag{2.227}$$

$$V_x = \frac{U_1}{\ln\frac{R_i}{R_0}} (\ln r - \ln R_0) = \frac{U_1}{\ln\frac{R_i}{R_0}} \ln\frac{r}{R_0} \tag{2.228}$$

Equation (2.228) gives the flow velocity distribution between the two concentric tubes. The volumetric flow existing between the two tubes can be determined as:

$$Q = \int_{R_i}^{R_0} V_x \cdot 2\pi r\,dr = \int_{R_i}^{R_0} \frac{U_1}{\ln\frac{R_i}{R_0}} \ln\frac{r}{R_0} 2\pi r\,dr \tag{2.229}$$

The integration of equation (2.229) produces the following result.

$$Q = \frac{1}{2}\pi R_0^2 U_1 \left[\frac{\left(\frac{R_i}{R_0}\right)^2 - 1}{\ln\left(\frac{R_i}{R_0}\right)} - 2\left(\frac{R_i}{R_0}\right)^2 \right] \tag{2.230}$$

The force needed to axially displace the internal cylinder, is calculated via determining the shear stresses acting over the cylinder surface.

$$\text{Shear stresses internal cylinder: } \tau_i = \mu \left.\frac{\partial V_x}{\partial r}\right|_{r=R_i} = \mu \frac{U_1}{\ln\frac{R_i}{R_0}} \frac{1}{R_i} \tag{2.231}$$

$$F_i = \tau_i \text{ Cylinder surface} = \tau_i\, 2\pi R_i\, L = \mu \frac{U_1}{\ln\frac{R_i}{R_0}} 2\pi L \tag{2.232}$$

where: L = Cylinders length.

The power needed to displace the central cylinder shall be obtained:

$$N_i = U_i F_i = \tau_i\, S_i\, U_i = \tau_i\, 2\pi R_i\, L\, U_i = \mu \frac{U_i^2}{\ln\frac{R_i}{R_0}} 2\pi L \tag{2.233}$$

2.7.4.2. Example 2. Flow between Two Concentric Pipes. Boundary conditions 4

Boundary conditions 4.

$$r = R_0; \quad \Rightarrow V_x = 0$$

$$r = R_i; \quad \Rightarrow V_x = 0$$

$$\frac{\Delta P^*}{\Delta x} \neq 0$$

Applying these boundary conditions to the generic velocity distribution equation, (2.221) it is obtained:

$$0 = \frac{\Delta P^*}{\Delta x}\frac{R_0^2}{4\mu} + C_0 \ln R_0 + C_1 \tag{2.234}$$

$$0 = \frac{\Delta P^*}{\Delta x}\frac{R_i^2}{4\mu} + C_0 \ln R_i + C_1 \tag{2.235}$$

Subtracting equation (2.235) from (2.234) it is reached:

$$0 = \frac{\Delta P^*}{\Delta x}\frac{1}{4\mu}\left(R_0^2 - R_i^2\right) + C_0\left(\ln\frac{R_0}{R_i}\right) \tag{2.236}$$

$$C_0 = -\frac{\Delta P^*}{\Delta x}\frac{\left(R_0^2 - R_i^2\right)}{4\mu}\frac{1}{\ln\left(\frac{R_0}{R_i}\right)} \tag{2.237}$$

Substituting equation (2.237) into (2.234) it gives:

$$C_1 = \frac{\Delta P^*}{\Delta x}\frac{R_0^2}{4\mu} + \frac{\Delta P^*}{\Delta x}\frac{R_0^2 - R_i^2}{4\mu}\frac{\ln R_0}{\ln\left(\frac{R_0}{R_i}\right)} \tag{2.238}$$

Substituting equations (2.238) and (2.237) into equation (2.221) and alter some rearrangement it is reached:

$$V_x = \frac{\Delta P^*}{\Delta x}\frac{1}{4\mu}\left[\left(r^2 - R_0^2\right) + \left(R_0^2 - R_i^2\right)\frac{\ln\left(\frac{R_0}{r}\right)}{\ln\left(\frac{R_0}{R_i}\right)}\right] \tag{2.239}$$

The radius, at which the flow velocity is maximum, will be obtained via deriving equation (2.239) versus the radius and equaling to zero $\frac{dV_x}{dr} = 0$, obtaining:

$$r=\left(\frac{\left(R_0^2-R_i^2\right)}{\ln\left(\frac{R_0}{R_i}\right)}\frac{1}{2}\right)^{1/2} \tag{2.240}$$

The fluid maximum velocity shall be obtained when substituting equation (2.240) into (2.239), being:

$$V_{max}=\frac{\Delta P^*}{\Delta x}\frac{1}{4\mu}\left[\frac{\left(R_0^2-R_i^2\right)}{\ln\left(\frac{R_0}{R_i}\right)}\frac{1}{2}-R_0^2+\left(R_0^2-R_i^2\right)\frac{\ln R_0-\ln\left(\left[\frac{\left(R_0^2-R_i^2\right)}{\ln\left(R_0/R_i\right)}\frac{1}{2}\right]^{1/2}\right)}{\ln\left(\frac{R_0}{R_i}\right)}\right] \tag{2.241}$$

The volumetric flow between the two cylinders shall be defined as:

$$Q=\int_{R_i}^{R_0}V_x\cdot 2\pi r\,dr=\int_{R_i}^{R_0}\frac{\Delta P^*}{\Delta x}\frac{2\pi}{4\mu}\left[\left(r^2-R_0^2\right)+\left(R_0^2-R_i^2\right)\frac{\ln\left(\frac{R_0}{r}\right)}{\ln\left(\frac{R_0}{R_i}\right)}\right]r\,dr= \tag{2.242}$$

$$Q=\frac{\Delta P^*}{\Delta x}\frac{2\pi}{4\mu}\left\{\left[\frac{r^3}{3}\right]_{R_i}^{R_0}-\left[\frac{r^2}{2}\right]_{R_i}^{R_0}R_0^2+\frac{\left(R_0^2-R_i^2\right)}{\ln\left(\frac{R_0}{R_i}\right)}\left\{\left[\frac{r^2}{2}\right]_{R_i}^{R_0}\ln R_0-\int_{R_i}^{R_0}\ln r\cdot r\,dr\right\}\right\} \tag{2.243}$$

$$Q=\frac{\Delta P^*}{\Delta x}\frac{2\pi}{4\mu}\left\{\left[\frac{r^3}{3}\right]_{R_i}^{R_0}-\left[\frac{r^2}{2}\right]_{R_i}^{R_0}R_0^2+\frac{\left(R_0^2-R_i^2\right)}{\ln\left(\frac{R_0}{R_i}\right)}\left\{\left[\frac{r^2}{2}\right]_{R_i}^{R_0}\ln R_0-\left[\frac{r^2}{2}\left(\ln r-\frac{1}{2}\right)\right]_{R_0}^{R_i}\right\}\right\} \tag{2.244}$$

Alter further arrangement it is reached:

$$Q=-\frac{\Delta P^*}{\Delta x}\frac{\pi R_0^4}{8\mu}\left(1-\left(\frac{R_i}{R_0}\right)^4-\frac{\left[1-\left(Ri/Ro\right)^2\right]^2}{\ln\left(\frac{Ro}{Ri}\right)}\right) \tag{2.245}$$

The flow average velocity can be determined according to the following expression:

$$\bar{V} = \frac{Q}{A} = \frac{Q}{\pi\left(R_0^2 - R_i^2\right)} = -\frac{\Delta P^* R_0^2}{\Delta x \; 8\mu}\left[1 + \left(\frac{Ri}{Ro}\right)^2 - \frac{1-(Ri/Ro)^2}{\ln\left(\frac{Ro}{Ri}\right)}\right] \quad (2.246)$$

In order to visualize several typical velocity distributions, figure 2.16 presents some particular cases, for each case, the boundary conditions are depicted.

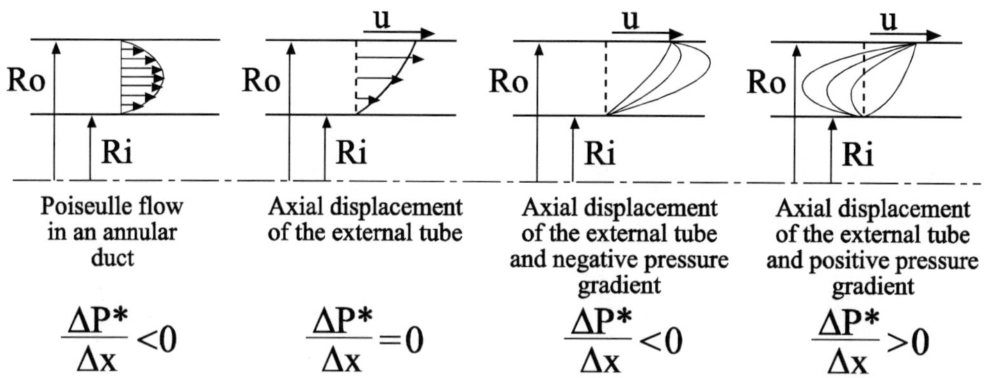

Figure 2.16. Velocity distribution between two concentric cylinders uder several boundary conditions.

2.7.4.3. Example 3. Flow between Two Concentric Pipes. Boundary conditions 2b

Boundary conditions 2b.

$$r = R_0; \quad \Rightarrow V_x = U_0$$

$$r = R_i; \quad \Rightarrow V_x = U_i$$

$$\frac{\Delta P^*}{\Delta x} \neq 0$$

It is important to notice, that for the present case both concentric cylinders will be moving towards the axial direction and at different velocities, also a reduced pressure differential between cylinder borders will exist.

Once again, the fluid flow velocity distribution is given by equation (2.221), via using the boundary conditions defined in the present case, the constants C_0 and C_1 shall be obtained:

Applying the boundary conditions to equation (2.221) it is reached:

$$U_0 = \frac{\Delta P^*}{\Delta x}\frac{R_0^2}{4}\frac{1}{\mu} + C_0 \ln R_0 + C_1 \quad (2.247)$$

$$U_i = \frac{\Delta P^*}{\Delta x}\frac{R_i^2}{4}\frac{1}{\mu} + C_0 \ln R_i + C_1 \quad (2.248)$$

Subtracting equation (2.248) from equation (2.247) it is found:

$$U_0 - U_i = \frac{\Delta P^*}{\Delta x}\frac{1}{4}\frac{1}{\mu}\left[R_0^2 - R_i^2\right] + C_0 \ln\frac{R_0}{R_i} \tag{2.249}$$

$$C_0 = \frac{(U_0 - U_i) - \frac{\Delta P^*}{\Delta x}\frac{1}{4}\frac{1}{\mu}\left[R_0^2 - R_i^2\right]}{\ln\frac{R_0}{R_i}} \tag{2.250}$$

Adding now equations (2.247) and (2.248) it is reached:

$$U_0 + U_i = \frac{\Delta P^*}{\Delta x}\frac{1}{4}\frac{1}{\mu}\left[R_0^2 + R_i^2\right] + C_0 \ln(R_0 R_1) + 2C_1 \tag{2.251}$$

$$C_1 = \frac{(U_0 + U_i) - \frac{\Delta P^*}{\Delta x}\frac{1}{4}\frac{1}{\mu}\left[R_0^2 + R_i^2\right] - C_0 \ln(R_0 R_i)}{2} \tag{2.252}$$

Substituting equation (2.250) into (2.252) the constant C_1 will take the value:

$$C_1 = \frac{(U_0 + U_i)}{2} - \frac{\Delta P^*}{\Delta x}\frac{1}{8}\frac{1}{\mu}\left[R_0^2 + R_i^2\right] - \frac{(U_0 - U_i) - \frac{\Delta P^*}{\Delta x}\frac{1}{4}\frac{1}{\mu}\left[R_0^2 - R_i^2\right]}{2\ln\frac{R_0}{R_i}}\ln(R_0 R_i) \tag{2.253}$$

And substituting now equations (2.253) and (2.250) into equation (2.221) it is obtained the final equation representing the velocity distribution as a function of the generic radius.

$$V_x = \frac{\Delta P^*}{\Delta x}\frac{r^2}{4}\frac{1}{\mu} + \frac{(U_0 - U_i) - \frac{\Delta P^*}{\Delta x}\frac{1}{4}\frac{1}{\mu}\left[R_0^2 - R_i^2\right]}{\ln\left(\frac{R_0}{R_i}\right)}\ln r + \frac{(U_0 + U_i)}{2}$$
$$- \frac{\Delta P^*}{\Delta x}\frac{\left(R_0^2 + R_i^2\right)}{8}\frac{1}{\mu} - \frac{(U_0 - U_i) - \frac{\Delta P^*}{\Delta x}\frac{1}{4}\frac{1}{\mu}\left[R_0^2 - R_i^2\right]}{2\ \ln\left(\frac{R_0}{R_i}\right)}\ln(R_0 R_i) \tag{2.254}$$

The force needed to axially displace the internal and external cylinders is calculated via determining the shear stresses onto the cylinders surface.

Internal cylinder: $$\tau_i = \mu\left.\frac{\partial V_x}{\partial r}\right|_{r=Ri} = \mu\left[\frac{\Delta P^*}{\Delta x}\frac{r}{2}\frac{1}{\mu} + \frac{C_0}{r}\right]_{r=R_i} = \mu\left[\frac{\Delta P^*}{\Delta x}\frac{R_i}{2}\frac{1}{\mu} + \frac{C_0}{R_i}\right] \tag{2.255}$$

$$F_i = \tau_i \,\text{Surface}_i = \tau_i \, 2\pi R_i \, L \tag{2.256}$$

L = Cylinder length.

$$\text{External cylinder: } \tau_0 = \mu \frac{\partial V_x}{\partial r}\bigg|_{r=R_0} = \mu \left[\frac{\Delta P^*}{\Delta x} \frac{r}{2} \frac{1}{\mu} + \frac{C_0}{r} \right]_{r=R_0} = \mu \left[\frac{\Delta P^*}{\Delta x} \frac{R_0}{2} \frac{1}{\mu} + \frac{C_0}{R_0} \right] \tag{2.257}$$

$$F_0 = \tau_0 \,\text{Surface}_0 = \tau_0 \, 2\pi R_0 \, L \tag{2.258}$$

The power needed to displace each tube shall be:

$$N_i = U_i \; F_i = \tau_i \; S_i \; U_i = \tau_i \; 2\pi R_i \; L \; U_i \tag{2.259}$$

$$N_0 = U_0 \; F_0 = \tau_0 \; S_0 \; U_0 = \tau_0 \; 2\pi R_0 \; L \; U_0 \tag{2.260}$$

The radius at which the fluid velocity is maximum will be determined when deriving the fluid velocity versus the radius and equaling the result to zero $\frac{dV_x}{dr} = 0$, obtaining:

$$r = \left[\frac{-C_0}{\frac{\Delta P^*}{\Delta x} \frac{1}{2\mu}} \right]^{\frac{1}{2}} \tag{2.261}$$

And finally, the volumetric flow between both tubes would be determined:

$$Q = \int_{R_i}^{R_0} V_x \, 2\pi r \, dr \tag{2.262}$$

2.7.5. Flow between Concentric Rotating Tubes

In the present section it will be studied the fluid flow between two concentric tubes, the fluid movement will just depend on the cylinders turning speed, which in figure 2.17 is represented as ω_i and ω_0 .

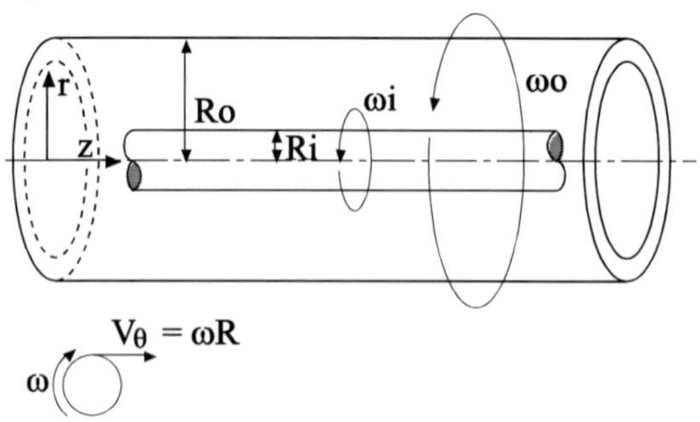

Figure 2.17. Flow between concentric rotating tubes.

The working hypothesis for the present case are very similar to the ones used in the previous studied cases, in what follows the hypothesis established are defined:

- ρ = cte. Fluid is considered as incompressible.
- $\frac{\partial}{\partial t} = 0$; The flow is time independent, permanent.
- $v_r = v_x = 0$; Flow is considered as unidirectional and two-dimensional.
- $v_\theta = f(r)$; The velocity V_θ depends only of the radius.
- $\frac{\Delta p^*}{\Delta x} = 0$; The reduced pressure between cylinders axial borders is zero.

Notice that if the fluid flow is radial dependent, it means the cylinders are supposed to have an infinite length. Notice as well that if the fluid flow is to be only radial dependent, the reduced pressure between cylinders axial borders must be always zero; otherwise the flow would not be unidirectional.

The different possible boundary conditions are:

Boundary conditions 1. Both cylinders turn.

$$\begin{array}{ll} r = R_i & \Rightarrow V_{\theta i} = \omega_i R_i \\ r = R_0 & \Rightarrow V_{\theta 0} = \omega_0 R_0 \end{array} \qquad \frac{\Delta p}{\Delta x} = 0 \; ; V_x = 0; \; V_r = 0$$

Boundary conditions 2. The external cylinder turn.

$$\begin{array}{ll} r = R_i & \Rightarrow V_{\theta i} = 0 \\ r = R_0 & \Rightarrow V_{\theta 0} = \omega_0 R_0 \end{array} \qquad \frac{\Delta p}{\Delta x} = 0 \; ; \; V_x = 0; \; V_r = 0$$

Boundary conditions 3. The internal cylinder turn.

$$\begin{array}{ll} r = R_i & \Rightarrow V_{\theta i} = \omega_i R_i \\ r = R_0 & \Rightarrow V_{\theta 0} = 0 \end{array} \qquad \frac{\Delta p}{\Delta x} = 0 \; ; \; V_x = 0; \; V_r = 0$$

The differential equations characterizing the fluid for the present case are the continuity and Navier-Stokes equations in cylindrical coordinates, equations previously presented as (2.198) to (2.201).

Considering the hypothesis established in the present case, equations (2.198) to (2.201) take the form:

$$\text{Continuity equation: } \frac{1}{r}\frac{\partial}{\partial \theta}(\rho V_\theta) = 0 \quad ; \quad \frac{\partial}{\partial \theta}(\rho V_\theta) = 0 \quad ; \quad \rho \frac{\partial V_\theta}{\partial \theta} = 0 \quad ; \quad \frac{\partial V_\theta}{\partial \theta} = 0 \qquad (2.263)$$

Navier Stokes equation:

Radial direction:

$$-\rho\frac{V_\theta^2}{r}=\rho g_r-\frac{\partial P}{\partial r}=-\rho g\frac{\partial h}{\partial r}-\frac{\partial P}{\partial r}=-\frac{\partial P^*}{\partial r} \tag{2.264}$$

$$\rho\frac{V_\theta^2}{r}=\frac{\partial P^*}{\partial r} \tag{2.265}$$

Angular direction: (It is assumed the reduced pressure in angular direction is zero $\frac{\Delta p^*}{\Delta\theta}=0$).

$$0=\frac{\partial}{\partial r}\left(\frac{1}{r}\frac{\partial}{\partial r}(rV_\theta)\right)\;;\;\frac{\partial}{\partial r}\left(\frac{V_\theta}{r}+\frac{\partial V_\theta}{\partial r}\right)=\frac{1}{r}\frac{\partial V_\theta}{\partial r}+V_\theta\frac{\partial}{\partial r}\left(\frac{1}{r}\right)+\frac{\partial^2 V_\theta}{\partial r^2}=\frac{1}{r}\frac{\partial V_\theta}{\partial r}-\frac{V_\theta}{r^2}+\frac{\partial^2 V_\theta}{\partial r^2} \tag{2.266}$$

Axial direction: $\frac{\partial P^*}{\partial x}=0$ (2.267)

Integrating the continuity equation it is obtained:

$\frac{\partial V_\theta}{\partial\theta}=0;\;\; V_\theta=\text{Constant};\;\; V_\theta$ is independent of the angular position θ

Integrating Navier-Stokes in radial direction it is reached: $\int dP^*=\int\rho\frac{V_\theta^2}{r}dr$ (2.268)

It must be noticed that in order to obtain the reduced pressure as a function of the radial direction, it is needed to know the angular velocity distribution as a function of the radial direction.

The velocity V_θ it is obtained when integrating the Navier-Stokes equation in angular direction.

$$0=\frac{\partial}{\partial r}\left(\frac{1}{r}\frac{\partial}{\partial r}(rV_\theta)\right) \tag{2.269}$$

$$\frac{1}{r}\frac{d}{dr}(rV_\theta)=C_1 \tag{2.270}$$

$$\frac{d}{dr}(rV_\theta)=C_1 r \tag{2.271}$$

$$rV_\theta=C_1\frac{r^2}{2}+C_2 \tag{2.272}$$

$$V_\theta=\frac{C_1}{2}r+\frac{C_2}{r} \tag{2.273}$$

To obtain the two integration constants, it will be used any of the boundary conditions previously defined.

2.7.5.1. Example 1. Case Boundary Conditions 1

As an example, the case defined by the boundary conditions 1, shall be first evaluated. Boundary conditions 1 were defined as:

$$\begin{array}{ll} r = R_i & \Rightarrow V_{\theta i} = \omega_i R_i \\ r = R_0 & \Rightarrow V_{\theta 0} = \omega_0 R_0 \end{array} \quad \frac{\Delta p}{\Delta x} = 0\,;\, V_x = 0;\; V_r = 0 \tag{2.274}$$

Applying the boundary conditions given by equation (2.274) to equation (2.273) it is obtained:

$$R_i \omega_i = \frac{C_1}{2} R_i + \frac{C_2}{R_i} \tag{2.275}$$

$$R_0 \omega_0 = \frac{C_1}{2} R_0 + \frac{C_2}{R_0} \tag{2.276}$$

Combining equations (2.275) and (2.276), the constants C_1 and C_2 for the present example, can be found:

$$C_1 = \left(R_0 \omega_0 - \frac{R_i^2 \omega_i}{R_0} \right) \frac{2}{R_0 - \dfrac{R_i^2}{R_0}} = \frac{\left(R_0^2 \omega_0 - R_i^2 \omega_i \right) 2}{R_0^2 - R_i^2} \tag{2.277}$$

$$C_2 = R_i^2 \omega_i - R_i^2 \frac{\left(R_0^2 \omega_0 - R_i^2 \omega_i \right)}{R_0^2 - R_i^2} \tag{2.278}$$

Substituting equations (2.277) and (2.278) into (2.273), it is found the fluid flow angular velocity distribution as a function of the radius.

$$V_\theta = \frac{\left(R_0^2 \omega_0 - R_i^2 \omega_i \right)}{R_0^2 - R_i^2} \cdot r + \frac{R_i^2 \omega_i}{r} - \frac{R_i^2}{r} \frac{R_0^2 \omega_0 - R_i^2 \omega_i}{R_0^2 - R_i^2} \tag{2.279}$$

$$V_\theta = \frac{R_0^2 \omega_0 - R_i^2 \omega_i}{R_0^2 - R_i^2} \left(r - \frac{R_i^2}{r} \right) + \frac{R_i^2 \omega_i}{r} \tag{2.280}$$

Once the tangential velocity distribution is being found, it can be used to obtain the radial reduced pressure distribution, recalling equation (2.268) it is established:

$$\int_{P_i^*}^{P_r^*} dP^* = \int_{R_i}^{r} \rho \frac{V_\theta^2}{r} dr = \int_{R_i}^{r} \frac{\rho}{r} \left[\frac{R_0^2 \omega_0 - R_i^2 \omega_i}{R_0^2 - R_i^2} \left(r - \frac{R_i^2}{r} \right) + \frac{R_i^2 \omega_i}{r} \right]^{\frac{1}{2}} dr \tag{2.281}$$

It is important to notice from equation (2.281), that in order to obtain the pressure distribution as a function of the radius, it is necessary to know the reduced pressure at a given radial position, this needs to be seen as an additional boundary condition.

2.7.5.2. Example 2. Case Boundary Conditions 2

Boundary conditions 2 were defined as:

$$\begin{aligned} r = R_i \quad &\Rightarrow V_{\theta i} = 0 \\ r = R_0 \quad &\Rightarrow V_{\theta 0} = \omega_0 R_0 \end{aligned} \qquad \frac{\Delta p}{\Delta x} = 0\,;\; V_x = 0\,;\; V_r = 0$$

Notice that boundary conditions 2 are a particular case of the one defined as boundary conditions 1, therefore the fluid tangential velocity distribution for the present case shall be obtained directly from equation (2.280) once applied the particular conditions, resulting:

$$V_\theta = \frac{R_0^2 \omega_0}{R_0^2 - R_i^2}\left(r - \frac{R_i^2}{r}\right) = \frac{\omega_0\left(r - \dfrac{R_i^2}{r}\right)}{1 - \left(\dfrac{R_i^2}{R_0}\right)} \tag{2.282}$$

The equation giving the radial reduced pressure distribution will be:

$$\rho\frac{V_\theta^2}{r} = \frac{dP^*}{dr} = \frac{\rho}{r}\left(\frac{\omega_0^2\left(r - \dfrac{R_i^2}{r}\right)^2}{\left[1 - \left(\dfrac{R_i}{R_0}\right)^2\right]^2}\right) \tag{2.283}$$

And its integration produces:

$$\int dP^* = \frac{\rho\omega_0^2}{\left[1 - \left(\dfrac{R_i}{R_0}\right)^2\right]^2}\int \frac{1}{r}\left(r - \frac{R_i^2}{r}\right)^2 dr = \frac{\rho\omega_0^2}{\left[1 - \left(\dfrac{R_i}{R_0}\right)^2\right]^2}\int\left(r - \frac{2R_i^2}{r} + \frac{R_i^4}{r^3}\right)dr = \tag{2.284}$$

$$P^* = \frac{\rho\omega_0^2}{\left[1 - \left(\dfrac{R_i}{R_0}\right)^2\right]^2}\left[\frac{r^2}{2} - 2R_i^2 \ln r + R_i^4 \frac{r^{-2}}{-2}\right] + C \tag{2.285}$$

A possible boundary condition needed to determine the constant C, shall be: for $r = R_i$ $\Rightarrow P^* = P_i^*$.

Applying this boundary condition to equation (2.285) it is obtained:

$$P_i^* = \frac{\rho\omega_0^2}{\left[1-\left(\frac{R_i}{R_0}\right)^2\right]^2}\left[\frac{R_i^2}{2} - 2R_i^2 \ln R_i - R_i^4 \frac{R_i^{-2}}{2}\right] + C \tag{2.286}$$

$$C = P_i^* - \frac{\rho\omega_0^2}{\left[1-\left(\frac{R_i}{R_0}\right)^2\right]^2}\left[\frac{R_i^2}{2} - 2R_i^2 \ln R_i - \frac{R_i^2}{2}\right] \tag{2.287}$$

$$P^* = P_i^* + \frac{\rho\omega_0^2}{\left[1-\left(\frac{R_i}{R_0}\right)^2\right]^2}\left[\frac{1}{2}(r^2 - R_i^2) - 2R_i^2 \ln\frac{r}{R_i} + \frac{R_i^4}{2}\left(\frac{1}{R_i^2} - \frac{1}{r^2}\right)\right] \tag{2.288}$$

Equation (2.288) is the final expression giving the radial pressure distribution.

When willing to determine the power needed to rotate the external cylinder, it will be first of all necessary to calculate the force due to the shear stresses in this particular cylinder.

Shear stresses for a radial dependent rotating flow, is given by the equation:

$$\tau_{r\theta} = 2\mu\phi_{r\theta} = 2\mu\,\frac{1}{2}\left(\frac{1}{r}\frac{\partial V_r}{\partial\theta} + r\frac{\partial}{\partial r}\left(\frac{V_\theta}{r}\right)\right) \tag{2.289}$$

Notice that the term $\phi_{r\theta}$ from equation (2.289), is in reality one of the terms of the deformation tensor defined in cylindrical coordinates.

For the present study, the velocity in radial direction does not exist; therefore equation (2.289) will be reduced to:

$$\tau_{r\theta} = r\mu\frac{d}{dr}\left(\frac{V_\theta}{r}\right)\Bigg|_{r=R_0} \tag{2.290}$$

Now, dividing the angular velocity distribution, equation (2.282) by the generic radius and deriving the resulting equation by the radius, equation (2.290) takes the form:

$$\tau_{r\theta} = r\,\mu\left[\frac{\omega_0}{1-\left(\frac{R_i}{R_0}\right)^2} R_i^2 \cdot 2r^{-3}\right]\Bigg|_{r=R_0} \tag{2.291}$$

Substituting the generic radius by the radius of the external cylinder, $r = R_0$, it is reached the final equation giving the shear stresses onto the external cylinder.

$$\tau_{r\theta}\Big|_{r=R_0} == \frac{\mu\,\omega_0\,2}{1-\left(\frac{R_i}{R_0}\right)^2}\left(\frac{R_i^2}{R_0^2}\right) \tag{2.292}$$

The next step is to calculate the torque generated by the shear stresses versus the cylinder central axis, the torque per unit length will be:

$$M_{R0} = \tau_{r\theta}\Big|_{r=R_o}\cdot 2\pi R_0\cdot 1\cdot R_0 = \frac{\mu\,\omega_0\,2}{1-\left(\frac{R_i}{R_0}\right)^2}\left(\frac{R_i^2}{R_0^2}\right)\cdot 2\pi R_0^2 = \frac{4\pi\mu\,\omega_0 R_i^2}{1-\left(\frac{R_i}{R_0}\right)^2} \tag{2.293}$$

And the power needed to maintain the external cylinder rotation is defined as:

$$N_{R0} = M_{R0}\,\omega_0 \tag{2.294}$$

2.7.5.3. Example 3. Case Boundary Conditions 3

Boundary conditions 3, are:

$$\begin{matrix} r = R_i & \Rightarrow V_{\theta i} = \omega_i R_i \\ r = R_0 & \Rightarrow V_{\theta 0} = 0 \end{matrix} \quad \frac{\Delta p}{\Delta x} = 0\,;\; V_x = 0\,;\; V_r = 0$$

Since Boundary conditions 3 are a particular case of boundary conditions 1, then, from equation (2.280) it is reached:

$$V_\theta = \frac{R_i^2\omega_i}{R_0^2 - R_i^2}\left(\frac{R_i^2}{r} - r\right) + \frac{R_i^2\omega_i}{r} \tag{2.295}$$

The angular velocity distribution given by equation (2.295) can also be presented as:

$$V_\theta = \frac{R_i^2\omega_i}{R_0^2 - R_i^2}\left(\frac{R_i^2}{r} - r + \frac{1}{r}\left(R_0^2 - R_i^2\right)\right) \tag{2.296}$$

$$V_\theta = \frac{R_i^2\omega_i}{R_0^2 - R_i^2}\left(\frac{1}{r}\left(R_i^2 + \left(R_0^2 - R_i^2\right)\right) - r\right) \tag{2.297}$$

$$V_\theta = \frac{R_i^2\omega_i}{R_0^2 - R_i^2}\left(\frac{R_0^2}{r} - r\right) \tag{2.298}$$

$$V_\theta = \frac{\omega_i \frac{R_i^2}{R_0^2}\left(\frac{R_0^2}{r} - r\right)}{1 - \frac{R_i^2}{R_0^2}} = \frac{-\omega_i\left(\frac{R_i}{R_0}\right)^2\left(r - \frac{R_0^2}{r}\right)}{1-\left(\frac{R_i}{R_0}\right)^2} \tag{2.299}$$

The radial reduced pressure distribution is to be determined using the same equation as in the previous cases.

$$\rho\frac{V_\theta^2}{r}=\frac{dp^*}{dr}(2.268) \tag{2.300}$$

$$\frac{dP^*}{dr}=\frac{\rho}{r}\frac{-\omega_i^2\left(\frac{R_i}{R_0}\right)^4}{\left[1-\left(\frac{R_i}{R_0}\right)^2\right]^2}\left(r-\frac{R_0^2}{r}\right)^2 \tag{2.301}$$

$$\frac{dP^*}{dr}=\frac{-\rho\,\omega_i^2\left(\frac{R_i}{R_0}\right)^4}{\left[1-\left(\frac{R_i}{R_0}\right)^2\right]^2}\frac{1}{r}\left(r^2-2R_0^2+\frac{R_0^4}{r^2}\right) \tag{2.302}$$

$$\int dP^*=\frac{-\rho\,\omega_i^2\left(\frac{R_i}{R_0}\right)^4}{\left[1-\left(\frac{R_i}{R_0}\right)^2\right]^2}\int\left(r-2\frac{R_0^2}{r}+\frac{R_0^4}{r^3}\right)dr \tag{2.303}$$

$$P^*=\frac{\rho\omega_i^2\left(\frac{R_i}{R_0}\right)^4}{\left[1-\left(\frac{R_i}{R_0}\right)^2\right]^2}\left(\frac{r^2}{2}-2R_0^2\ln r+\frac{R_0^4}{-2}r^{-2}\right)+C \tag{2.304}$$

To determine the integration constant, the following additional boundary condition will be used: $r=R_i\Rightarrow P=P_i$;

$$C=P_i+\frac{\rho\omega_i^2\left(\frac{R_i}{R_0}\right)^4}{\left[1-\left(\frac{R_i}{R_0}\right)^2\right]^2}\left(\frac{R_i^2}{2}-2R_0^2\ln R_i-\frac{R_0^4}{2}\frac{1}{R_i^2}\right) \tag{2.305}$$

As a result, the final form of the radial pressure distribution equation will be:

$$P^* = P_i + \frac{\rho\omega_i^2\left(\frac{R_i}{R_0}\right)^4}{\left[1-\left(\frac{R_i}{R_0}\right)^2\right]^2}\left(\frac{R_i^2}{2} - \frac{r^2}{2} + 2R_0^2 \ln\frac{r}{R_i} - \frac{R_0^4}{2}\left(\frac{1}{R_i^2} - \frac{1}{r^2}\right)\right) \tag{2.306}$$

The shear stresses onto the internal cylinder surface, $r = R_i$ are calculated:

$$\tau_{r\theta} = 2\mu\phi_{r\theta} = 2\mu\left(\frac{1}{2}\left(\frac{1}{r}\frac{\partial V_r}{\partial\theta} + r\frac{\partial}{\partial r}\left(\frac{V_\theta}{r}\right)\right)\right) \tag{2.307}$$

$$\tau_{r\theta} = r\mu\frac{d}{dr}\left(\frac{V_\theta}{r}\right) = r\mu\frac{d}{dr}\left[\frac{1}{r}\frac{-\omega_i\frac{R_i^2}{R_0^2}\left(r - \frac{R_0^2}{r}\right)}{1-\left(\frac{R_i}{R_0}\right)^2}\right]_{r=R_i} \tag{2.308}$$

$$\tau_{r\theta} = r\mu\frac{d}{dr}\left[\frac{-\omega_i\left(\frac{R_i}{R_0}\right)^2}{1-\left(\frac{R_i}{R_0}\right)^2}\left(1-\frac{R_0^2}{r^2}\right)\right]_{r=R_i} \tag{2.309}$$

$$\tau_{r\theta} = \frac{\omega_i\left(\frac{R_i}{R_0}\right)^2}{1-\left(\frac{R_i}{R_0}\right)^2}R_0^2 r\mu 2r^{-3}\Bigg|_{r=R_i} \tag{2.310}$$

$$\tau_{r\theta}\Big|_{r=R_i} = \frac{2\mu\omega_i\left(\frac{R_i}{R_0}\right)^2}{1-\left(\frac{R_i}{R_0}\right)^2}\left(\frac{R_0}{R_i}\right)^2 = \frac{2\mu\omega_i R_0^2}{R_0^2 - R_i^2} \tag{2.311}$$

And the torque per unit length would be:

$$M\Big|_{r=R_i} = 2\pi R_i \cdot R_i \cdot \tau_{r\theta}\Big|_{r=R_i} = \frac{4\pi\mu\omega_i R_i^2 R_0^2}{R_0^2 - R_i^2} \tag{2.312}$$

2.7.5.4. Example 4. Case Boundary Conditions 1 (Modified)

The present case is the one previously defined as boundary conditions 1, although now both cylinders shall turn with the same turning speed but at opposite directions.

The boundary conditions are defined as:

$$r = R_i \quad \Rightarrow \omega_i = -\frac{\omega}{2} \Rightarrow V_{\theta i} = -\frac{\omega R_i}{2} \qquad \frac{\Delta p}{\Delta x} = 0\,;\; V_x = 0;\; V_r = 0 \tag{2.313}$$
$$r = R_0 \quad \Rightarrow \omega_0 = \frac{\omega}{2} \Rightarrow V_{\theta 0} = \frac{\omega R_0}{2}$$

The angular direction generic velocity distribution was defined in Example 1 as:

$$V_\theta = \frac{R_0^2\omega_0 - R_i^2\omega_i}{R_0^2 - R_i^2}\left(r - \frac{R_i^2}{r}\right) + \frac{R_i^2\omega_i}{r} \; (2.280) \tag{2.314}$$

For the present particular case, the generic equation (2.314) will take the form:

$$V_\theta = \frac{R_0^2\frac{\omega}{2} + R_i^2\frac{\omega}{2}}{R_0^2 - R_i^2}\left(r - \frac{R_i^2}{r}\right) - \frac{R_i^2\omega}{r\,2} \tag{2.315}$$

After some arrangement equation (2.315) takes the form:

$$V_\theta = \frac{\omega}{2}\left[\frac{r\left[1+\left(\frac{R_i}{R_0}\right)^2\right] - 2\frac{R_i^2}{r}}{1-\left(\frac{R_i}{R_0}\right)^2}\right] \tag{2.316}$$

The equation giving the reduced pressure as a function of the radial direction will be:

$$\frac{dP^*}{dr} = \rho\frac{V_\theta^2}{r} = \frac{\rho}{r}\frac{\omega^2}{4}\frac{\left(r\left[1+\left(\frac{R_i}{R_0}\right)^2\right] - 2\frac{R_i}{r}\right)^2}{\left[1-\left(\frac{R_i}{R_0}\right)^2\right]^2} \tag{2.317}$$

$$\frac{dP^*}{dr} = \frac{\rho}{r}\frac{\omega^2}{4}\frac{r^2\left(1+\left(\frac{R_i}{R_0}\right)^2\right)^2 - 2r\left[1+\left(\frac{R_i}{R_0}\right)\right]2\frac{R_i}{r} + 4\left(\frac{R_i}{r}\right)^2}{\left[1-\left(\frac{R_i}{R_0}\right)^2\right]^2} \tag{2.318}$$

$$\int dP^* = \frac{\rho\frac{\omega^2}{4}}{\left[1-\left(\frac{R_i}{R_0}\right)^2\right]^2}\int\left[r\left(1+\left(\frac{R_i}{R_0}\right)^2\right)^2 - 4R_i\left[1+\left(\frac{R_i}{R_0}\right)\right]\frac{1}{r} + 4\frac{R_i^2}{r^3}\right]dr \tag{2.319}$$

$$P^* = \frac{\rho\frac{\omega^2}{4}}{\left[1-\left(\frac{R_i}{R_0}\right)^2\right]^2}\left[\left(1+\left(\frac{R_i}{R_0}\right)^2\right)^2\frac{r^2}{2}-4R_i\left[1+\left(\frac{R_i}{R_0}\right)\right]\ln r-2\frac{R_i^2}{r^2}\right]+C \tag{2.320}$$

Via applying the same extra boundary condition as in the previous examples, for $r = R_i \quad P = P_i$, it is obtained the value of the constant C.

$$C = P_i - \frac{\rho\frac{\omega^2}{4}}{\left[1-\left(\frac{R_i}{R_0}\right)^2\right]^2}\left[\left(1+\left(\frac{R_i}{R_0}\right)^2\right)^2\frac{R_i^2}{2}-4R_i\left[1+\left(\frac{R_i}{R_0}\right)\right]\ln R_i-2\right] \tag{2.321}$$

And substituting equation (2.321) into (2.320) it is reached:

$$P = P_i + \frac{\rho\frac{\omega^2}{4}}{\left[1-\left(\frac{R_i}{R_0}\right)^2\right]^2}\left[\begin{array}{l}\left(1+\left(\frac{R_i}{R_0}\right)^2\right)^2\left(\frac{r^2}{2}-\frac{R_i^2}{2}\right)- \\ 4R_i\left[1+\left(\frac{R_i}{R_0}\right)\right]\ln\left(\frac{r}{R_i}\right)-\left[2\left(\frac{R_i}{r}\right)^2-1\right]\end{array}\right] \tag{2.322}$$

The shear stresses for the internal and external rotating cylinders shall be calculated:

$$\tau_{r\theta}\Big|_{r=R_i} = 2\mu\,\phi_{r\theta}\Big|_{r=R_i} = \mu\, r\frac{d}{dr}\left(\frac{V_\theta}{r}\right)\Bigg|_{r=R_i} = \frac{2\mu\,\omega}{1-\left(\frac{R_i}{R_0}\right)^2} \tag{2.323}$$

$$\tau_{r\theta}\Big|_{r=R_0} = 2\mu\,\phi_{r\theta}\Big|_{r=R_0} = \mu\, r\frac{d}{dr}\left(\frac{V_\theta}{r}\right)\Bigg|_{r=R_0} = \frac{2\mu\,\omega\left(\frac{R_i^2}{R_0^2}\right)}{1-\left(\frac{R_i}{R_0}\right)^2} \tag{2.324}$$

The necessary torque per unit length needed to turn each of the two cylinders is:

$$M\Big|_{r=R_i} = 2\pi R_i \cdot R_i\,\tau_{r\theta}\Big|_{r=R_i} = \frac{4\pi\mu\,\omega R_i^2}{1-\left(\frac{R_i}{R_0}\right)^2} = \frac{4\pi\mu\,\omega R_i^2 R_0^2}{R_0^2 - R_i^2} \tag{2.325}$$

$$M|_{r=R_0} = 2\pi R_0 \cdot R_0 \tau_{r\theta}|_{r=R_0} = \frac{4\pi\mu\omega R_0^2 \left(\frac{R_i^2}{R_0^2}\right)}{1-\left(\frac{R_i}{R_0}\right)^2} = \frac{4\pi\mu\omega R_i^2 R_0^2}{R_0^2 - R_i^2} \quad (2.326)$$

It is interesting to notice that for the present case both torques produce the same results. The power per unit length needed to turn each cylinder it is obtained when multiplying the previous two equations by the respective angular velocity.

2.8. Introduction to Flow with Negligible Acceleration

2.8.1. Introduction

There exist two families of real flows in which the Navier-Stokes acceleration terms are negligible, these are the flow in narrow gaps and creeping flows. The second family embodies flow around spheres, cylinders and bluff bodies, being the Reynolds number small enough to consider the flow as laminar, Stokes flow is a characteristic creeping flow.

In the present sub section, the basis of flow in narrow gaps shall be established; working hypothesis as well as dimension considerations will be presented.

Flow in narrow gaps main working hypothesis are:

- ρ= cte; Fluid is regarded as incompressible.
- $\frac{\partial}{\partial t} = 0$; Fluid flow will be studied under permanent conditions.
- The local and convective acceleration terms shall be considered as irrelevant when compared with the viscous terms.
- A priory, fluid flow shall be considered as two directional and two dimensional.

Initially, the flow between two quasi parallel flat plates shall be evaluated.

Figure 2.18 presents two quasi parallel flat plates, the upper plate moves towards the right hand side with a constant velocity U, between the two plates there is a viscous flow. In what follows an evaluation of the most relevant terms will be undertaken.

Regarding the dimensions presented in figure 2.18 the following assumptions are established:

$$\frac{\delta}{L} << 1; \quad \alpha < \frac{\delta}{L}; \quad tg\alpha = \alpha = \frac{\delta'}{L}; \quad \frac{\delta'}{L} < \frac{\delta}{L}; \quad (2.327)$$

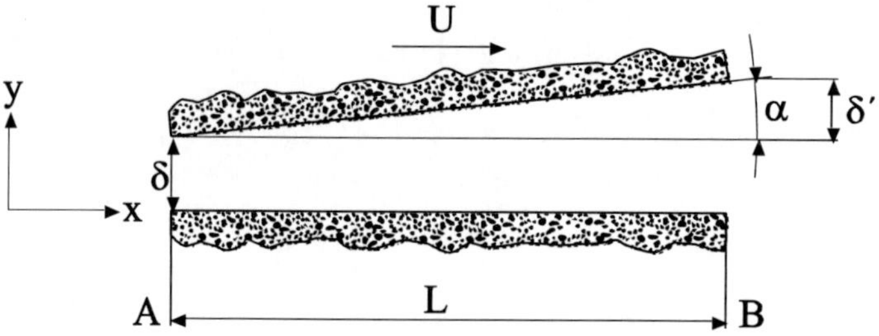

Figure 2.18. Flow between two quasi parallel flat plates.

Notice that the assumptions defined in (2.327) establish the flow as two dimensional, the plate length L is much larger than the distance between plates and the upper plate inclination angle is very small.

Defining as $\bar{u}$ the average mean x directional fluid velocity between the two plates, it can be mathematically represented as:

$$\bar{u} = \frac{1}{\delta}\int_0^\delta u\, dy \tag{2.328}$$

Applying now the mass conservation law between the plates axial borders it can be established:

$$\bar{u}_A \delta = \bar{u}_B (\delta + \delta') = \bar{u}_B (\delta + L\alpha) \tag{2.329}$$

And considering the upper plate inclination angle very small, $L \tan\alpha \cong L\alpha$, the previous equation takes the form:

$$0 = \bar{u}_B (\delta + L\alpha) - \bar{u}_A \delta \tag{2.330}$$

$$0 = \delta(\bar{u}_B - \bar{u}_A) + \bar{u}_B L\alpha \tag{2.331}$$

$$\bar{u}_B - \bar{u}_A = -\bar{u}_B \frac{L\alpha}{\delta} \tag{2.332}$$

On the other hand, considering absolute values, it can be said:

$$\left|\frac{\partial u}{\partial x}\right| \cong \left|\frac{\partial \bar{u}}{\partial x}\right| \cong \frac{\bar{u}_B - \bar{u}_A}{L} = -\bar{u}_B \frac{\alpha}{\delta} = \left|\bar{u}_B \frac{\alpha}{\delta}\right| \tag{2.333}$$

And considering the average velocity very similar at all axial points, $\bar{u}_B \cong \bar{u}$

$$\left|\frac{\partial u}{\partial x}\right| \cong \left|\bar{u} \frac{\alpha}{\delta}\right| \tag{2.334}$$

For two dimensional incompressible flow, the continuity equation in differential form, can be expressed as:

$$\frac{\partial u}{\partial x}+\frac{\partial v}{\partial y}=0; \quad \equiv \frac{\partial u}{\partial x}=-\frac{\partial v}{\partial y} \tag{2.335}$$

Considering equations (2.334) and (2.335) and when working in absolute values, it is reached:

$$\left|\frac{\partial v}{\partial y}\right|=\left|\frac{\bar{u}\alpha}{\delta}\right| \tag{2.336}$$

When looking at figure 2.18, it seems clear that the fluid velocity in y direction is zero, v = 0, when y = 0, and when y = δ. It can therefore be concluded that the maximum fluid velocity in y direction, $|V_{máx}|$, has to be found near the centre of the fluid film. As a result and working with absolute values, it can be established:

$$|V_{máx}| \approx \left|\frac{\partial v}{\partial y}\right|\frac{\delta}{2} \approx \left|\frac{\bar{u}\alpha}{2}\right| << \bar{u} \tag{2.337}$$

According to equation (2.337) it can be said that the average velocity in x direction is much bigger than the maximum velocity in y direction, which can be generalized saying v << u. This information will be used when finding the differential equations defining the present case.

According to Couette flow, the fluid velocity distribution between two parallel infinite plates when the upper plate moves axially at velocity U is defined as: (see equation 2.167).

$$u=\frac{U}{\delta}y \; ; \; \frac{u}{y}=\frac{U}{\delta}; \tag{2.338}$$

Based on equation (2.338) for the present case the following relation of magnitudes can be established:

$$\left|\frac{\partial u}{\partial y}\right| \cong \frac{\Delta u}{\Delta y} \cong \frac{U-0}{\delta} \cong \frac{\bar{u}}{\delta}; \tag{2.339}$$

Dividing equation (2.334) by (2.339) it is obtained:

$$\frac{\left|\frac{\partial u}{\partial x}\right|}{\left|\frac{\partial u}{\partial y}\right|} \cong \frac{\left|\frac{\bar{u}\alpha}{\delta}\right|}{\frac{\bar{u}}{\delta}} = \alpha < \frac{\delta}{L} << 1 \tag{2.340}$$

And considering that the upper plate inclination angle is very small it can be said:

$$\alpha \cong tg\alpha = \frac{\delta_B - \delta_A}{L} \tag{2.341}$$

It is important to realize that equation (2.340) establishes the relation between the fluid axial velocity versus the axial direction and the fluid axial velocity versus the y direction, and clarifies that the second term is much bigger than the first one. This information will be used when defining the characteristic differential equations applicable to the present case.

2.8.2. Reynolds Lubrication Theory. Hydrodynamic Plane Journal Bearings

In the present sub section, fluid forces generated between two quasi parallel flat plates will be determined. Figure 2.19 presents the two plates, notice that the upper one displaces axially at a constant velocity U.

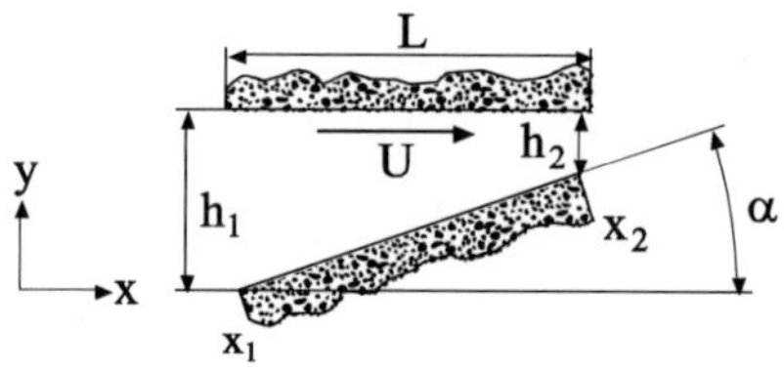

Figure 2.19. Hydrodynamic plane journal bearing.

The differential equations characterizing the present flow will be obtained from the Navier-Stokes equations, which in two-dimensional laminar flow and Cartesian coordinates take the form:

$$\rho\left(\frac{\partial u}{\partial t}+u\frac{\partial u}{\partial x}+v\frac{\partial u}{\partial y}+w\frac{\partial u}{\partial z}\right)=-\frac{\partial p}{\partial x}+\rho\vec{g}_x+\mu\left(\frac{\partial^2 u}{\partial x^2}+\frac{\partial^2 u}{\partial y^2}+\frac{\partial^2 u}{\partial z^2}\right) \tag{2.342}$$

$$\rho\left(\frac{\partial v}{\partial t}+u\frac{\partial v}{\partial x}+v\frac{\partial v}{\partial y}+w\frac{\partial v}{\partial z}\right)=-\frac{\partial p}{\partial y}+\rho\vec{g}_y+\mu\left(\frac{\partial^2 v}{\partial x^2}+\frac{\partial^2 v}{\partial y^2}+\frac{\partial^2 v}{\partial z^2}\right) \tag{2.343}$$

The continuity equation is taking the form:

$$\frac{\partial u}{\partial x}+\frac{\partial v}{\partial y}=0 \tag{2.344}$$

In the previous section, it was obtained several relations regarding which terms can be considered small when compared to some others, these relations were:

$$\frac{\left|\frac{\partial u}{\partial x}\right|}{\left|\frac{\partial u}{\partial y}\right|} \cong \alpha << 1\,; \tag{2.345}$$

And the second one was: $v << u$; (2.346)

The Navier-Stokes and continuity equations (2.342) to (2.344), when considering the relations established in (2.345) and (2.346), take the form:

Continuity: from equation (2.344) and (2.346) it is reached: $\frac{\partial u}{\partial x} = 0$ (2.347)

Navier-Stokes: from equations (2.342), (2.343) and (2.345) it is reached, after applying the concept of reduced pressure:

$$0 = -\frac{\partial p^*}{\partial x} + \mu \frac{\partial^2 u}{\partial y^2} \tag{2.348}$$

$$0 = -\frac{\partial p^*}{\partial y} \tag{2.349}$$

And considering the approximation $\frac{\partial p^*}{\partial x} \cong \frac{\Delta p^*}{\Delta x}$, equation (2.348) it can be given as:

$$\frac{\Delta p^*}{\Delta x} = \mu \frac{d^2 u}{dy^2} \tag{2.350}$$

The integration of equation (2.350) gives birth to:

$$u = \frac{\Delta P^*}{\Delta x} \frac{1}{2\mu} y^2 + c_1 y + c_2 \tag{2.351}$$

The constants c_1 and c_2 shall be found out when applying the following boundary conditions: at y = 0, the fluid velocity u = 0, and when y = d the fluid velocity will be u = U. obtaining:

$$C_2 = 0; \tag{2.352}$$

$$C_1 = \frac{U}{d} - \frac{1}{\mu} \frac{\Delta p^*}{\Delta x} \frac{d}{2} \tag{2.353}$$

Substituting equations (2.352) and (2.353) into equation (2.351) it is reached:

$$u = U \frac{y}{d} + \frac{\Delta p^*}{\Delta x} \frac{y}{2\mu} [y - d] \tag{2.354}$$

Equation (2.354) gives the fluid velocity distribution between the two quasi parallel plates, one of them is moving axially at a velocity U and it may exist a reduced pressure gradient between plate's borders.

It is important to realize that equation (2.354) is exactly the same as equation (2.165), which was characterizing the velocity distribution between two parallel plates and when considering Couette-Poiseulle flow. But for the present case, the distance d between plates, is not constant and depends on the axial position, to consider this fact, equation (2.354) needs to be rearranged and the variable d shall be substituted by $h_{(x)}$, obtaining:

$$u=\frac{\Delta P^{*}}{\Delta x}\frac{1}{2\mu}y(y-h_{(x)})+U\frac{y}{h_{(x)}} \tag{2.355}$$

From equation (2.355) can be calculated the mass flow per unit depth between the two quasi parallel plates:

$$\dot{m}=\int_{0}^{h(x)}\rho\,u\,dy=\int_{0}^{h(x)}\rho\left(\frac{\Delta P^{*}}{\Delta x}\frac{1}{2\mu}\left(y^{2}-yh_{(x)}\right)+U\frac{y}{h_{(x)}}\right)dy \tag{2.356}$$

$$\dot{m}=\rho\frac{\Delta P^{*}}{\Delta x}\frac{1}{2\mu}\left[\frac{1}{3}y^{3}-\frac{y^{2}}{2}h_{(x)}\right]_{0}^{h(x)}+\frac{\rho U}{h_{(x)}}\left[\frac{y^{2}}{2}\right]_{0}^{h(x)} \tag{2.357}$$

$$\dot{m}=(-)\rho\frac{\Delta P^{*}}{\Delta x}\frac{1}{2\mu}\frac{1}{6}h_{(x)}^{3}+\rho U\frac{h_{(x)}}{2} \tag{2.358}$$

$$\dot{m}=\rho U\frac{h_{(x)}}{2}-\rho\frac{\Delta P^{*}}{\Delta x}\frac{1}{12\mu}h_{(x)}^{3} \tag{2.359}$$

In order to calculate the load these particular journal bearings can hold, it is necessary to determine the fluid pressure as a function of the axial position, to do so, the following procedure shall be followed.

If the same mass flow given by equation (2.359) is to be obtained between two parallel flat plates separated an unknown distance h_0 and when considering Couette flow, one of the plates moves axially at a constant velocity U, the characteristic equation would take the form: Notice that the unknown distance h_0 is the necessary one to assure that both mass flows are the same.

$$\dot{m}=\int_{0}^{h_0}\rho U\frac{y}{h_{(x)}}dy=\rho\frac{U}{h_{(x)}}\left[\frac{y^{2}}{2}\right]_{0}^{h_0}=\rho U\frac{h_0}{2} \tag{2.360}$$

From equations (2.359) and (2.360) it is obtained:

$$\rho\frac{U}{2}h_0=\rho\frac{U}{2}h_{(x)}-\rho\frac{\Delta P^{*}}{\Delta x}\frac{1}{12\mu}h_{(x)}^{3} \tag{2.361}$$

$$\rho\frac{U}{2}\left(h_0-h_{(x)}\right)=-\rho\frac{\Delta P^{*}}{\Delta x}\frac{1}{12\mu}h_{(x)}^{3} \tag{2.362}$$

Calling $h_{(x)} = h$, the reduced pressure variation as a function of the axial direction will be:

$$\frac{\Delta P^*}{\Delta x} \cong \frac{dP^*}{dx} = 6\mu U \frac{(h - h_0)}{h^3} \quad (2.363)$$

Remembering that the inclination angle is given as $\alpha \cong \frac{dh}{dx}$, it can be established:

$$\frac{dP^*}{dx} = \frac{dP^*}{dh}\frac{dh}{dx} = \alpha \frac{dP^*}{dh} \quad (2.364)$$

$$\frac{dP^*}{dh} = \frac{1}{\alpha}\frac{dP^*}{dx} \quad (2.365)$$

Considering equation (2.365), equation (2.363) can be given as:

$$\frac{dP^*}{dh} = 6\mu \frac{U}{\alpha}\frac{(h - h_0)}{h^3} \quad (2.366)$$

$$P^* = \int 6\mu \frac{U}{\alpha}\left(\frac{1}{h^2} - \frac{h_0}{h^3}\right) dh = 6\mu \frac{U}{\alpha}\left(-\frac{1}{h} + \frac{h_0}{2}\frac{1}{h^2}\right) + C_1 \quad (2.367)$$

To determine the integration constant and the value of the distance h_0, the following boundary conditions shall be employed: figure 2.20 clarifies the boundary conditions used, notice that the pressure at both plates' borders is considered the same; this is what happens in reality.

When: $h = h_1, \Rightarrow P^* = P_0 = P_{atm.}$

When: $h = h_2, \rightarrow P^* = P_0 = P_{atm}$

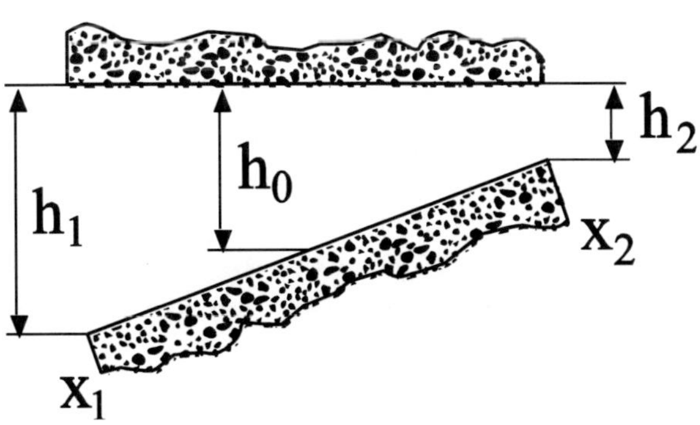

Figure 2.20. Definition of boundary conditions in a plane journal bearing.

When substituting the boundary conditions in equation (2.367) it is obtained:

$$P_0 = 6\mu\frac{U}{\alpha}\left(-\frac{1}{h_1}+\frac{h_0}{2}\frac{1}{h_1^2}\right)+C_1 \tag{2.368}$$

$$P_0 = 6\mu\frac{U}{\alpha}\left(-\frac{1}{h_2}+\frac{h_0}{2}\frac{1}{h_2^2}\right)+C_1 \tag{2.369}$$

Subtracting equation (2.369) from (2.368) it is reached:

$$0 = 6\mu\frac{U}{\alpha}\left(\frac{1}{h_2}-\frac{1}{h_1}+\frac{h_0}{2}\left(\frac{1}{h_1^2}-\frac{1}{h_2^2}\right)\right) \tag{2.370}$$

After some arrangement it is found:

$$\frac{h_0}{2} = \frac{(h_2-h_1)h_1h_2}{(h_2+h_1)(h_2-h_1)} = \frac{h_1h_2}{h_2+h_1} \tag{2.371}$$

On the other hand, the constant C_1 can be given as:

$$C_1 = P_0 - 6\mu\frac{U}{\alpha}\left(-\frac{1}{h_1}+\frac{h_0}{2}\frac{1}{h_1^2}\right) \tag{2.372}$$

Substituting equations (2.371) and (2.372) into equation (2.367) it is obtained:

$$P^* - P_0 = 6\mu\frac{U}{\alpha}\left(\frac{h-h_1}{hh_1}+\frac{h_1h_2}{h_1+h_2}\left(\frac{h_1^2-h^2}{h^2h_1^2}\right)\right) \tag{2.373}$$

Equation (2.373) gives pre pressure distribution in the gap between two quasi parallel plates and as a function of the axial position, which for the present case is represented by the variable distance h. After some further arrangement, equation (2.373) can be expressed as:

$$P^* - P_0 = 6\mu\frac{U}{\alpha}\frac{(h-h_1)(h-h_2)}{(h_1+h_2)h^2} \tag{2.374}$$

The mass flow between the two quasi parallel plates can finally be determined as: notice that according to the theory established, the mass flow can be obtained with equation (2.359) or (2.360).

$$\dot{m} = \frac{1}{2}\rho u h_0 = \rho u\frac{h_1h_2}{h_2+h_1}; \quad h_1 > h_0 > h_2 \tag{2.375}$$

The lift per unit depth shall be obtained: it is to be remembered the relation $\alpha = \frac{dh}{dx}$.

$$L = \int_{x_1}^{x_2} (P^* - P_0)\,dx = \frac{1}{\alpha}\int_{h_1}^{h_2} (P^* - P_0)\,dh \tag{2.376}$$

$$L = 6\mu\frac{U}{\alpha^2}\int_{h_1}^{h_2}\left[\frac{(h-h_1)(h-h_2)}{(h_1+h_2)h^2}\right]dh \tag{2.377}$$

After integration it is obtained:

$$L = 6\mu\frac{U}{\alpha^2}\left[\frac{2(h_2-h_1)}{(h_1+h_2)} - \ln\left(\frac{h_2}{h_1}\right)\right] \tag{2.378}$$

Equation (2.378) can be given as a function of a parameter k, (relating the film thicknesses at plates borders), and the plates length ℓ.

$$k = \frac{h_1}{h_2} > 1\,;\ \ell = x_2 - x_1 = \frac{1}{\alpha}(h_2 - h_1) \tag{2.379}$$

The final parametric equation takes the form:

$$L = \frac{6\mu U \ell^2}{(k-1)^2 h_2^2}\left[\ln k - \frac{2(k-1)}{k+1}\right] \tag{2.380}$$

The drag force per unit depth, necessary to maintain velocity from the upper plate, will be:

$$D = \int_{x_1}^{x_2} \tau dx = \int_{x_1}^{x_2} \mu\left(\frac{du}{dy}\right)_{y=h} dx = \int_{x_1}^{x_2} \mu\frac{d}{dy}\left(\frac{1}{2\mu}\left(\frac{dP^*}{dx}\right)y(y-h) + U\frac{y}{h}\right)_{y=h} dx \tag{2.381}$$

Considering the established relation $\alpha \cong \frac{dh}{dx}$, it is obtained:

$$D = \int_{h_1}^{h_2} \frac{\mu}{\alpha}\left[\frac{1}{2\mu}\frac{dP^*}{dx}(2y-h) + \frac{U}{h}\right]_{y=h} dh \tag{2.382}$$

After a further arrangement and considering the relation given in (2.365), it can be said:

$$D = \int_{h_1}^{h_2} \frac{\mu}{\alpha}\left(\frac{1}{2\mu}\frac{dP^*}{dh}\alpha h + \frac{U}{h}\right)dh = \int_{h_1}^{h_2}\left(\frac{dP^*}{dh}\frac{h}{2} + \frac{\mu}{\alpha}\frac{U}{h}\right)dh \tag{2.383}$$

And when considering equation (2.365) it is reached:

$$D=\int_{h_1}^{h_2}\left(h\frac{6\mu}{2}\frac{U}{\alpha}\frac{h-h_0}{h^3}+\frac{\mu}{\alpha}\frac{U}{h}\right)dh \tag{2.384}$$

$$D=\int_{h_1}^{h_2}\frac{\mu U}{\alpha}\left(3\frac{h-h_0}{h^2}+\frac{1}{h}\right)dh=\int_{h_1}^{h_2}\frac{\mu U}{\alpha}\left(3\left(\frac{1}{h}-\frac{h_0}{h^2}\right)+\frac{1}{h}\right)dh \tag{2.385}$$

$$D=\int_{h_1}^{h_2}\frac{\mu U}{\alpha}\left(\frac{4}{h}-\frac{3h_0}{h^2}\right)dh=\frac{\mu U}{\alpha}\left[4\ln\left(\frac{h_2}{h_1}\right)+3h_0\left(\frac{1}{h_2}-\frac{1}{h_1}\right)\right] \tag{2.386}$$

Modifying equation (2.386) via employing the parameters presented in equation (2.379) and considering the relation established in equation (2.371), it is obtained the parametric equation for the drag force.

$$D=\frac{2\mu U\ell}{(k-1)h_2}\left[2\ln k-\frac{3(k-1)}{k+1}\right] \tag{2.387}$$

The lift and drag parametric equations (2.380) and (2.387), allow to study the journal bearing dimensions for which the lift force is maximum, therefore when deriving equation (2.380) versus the parameter k, $\frac{dL}{dk}=0$, it is obtained: k = 2.2.

Substituting the value of k into equations (2.380) and (2.387), the optimum values for lift and drag are obtained:

$$L=0{,}4\mu UR^2; \qquad D=1{,}2\mu UR;$$

Being: $\frac{D}{L}=\frac{3}{R}$; $\quad R=\frac{2\ell}{h_1+h_2}$;

For these conditions it is considered that the plane journal bearing is optimized.

2.8.3. Reynolds Lubrication Equation in Cartesian Coordinates, Case Two Dimensional Flow

In many real applications, the fastest method to determine the pressure distribution along a particular direction is via using the Reynolds lubrication equation. In the next sections this equation applicable to different cases will be presented.

When studding the flow between two quasi parallel plates, equation (2.359) which was giving the mass flow per unit depth between the two plates was found.

$$\dot{m}_x=\rho U\frac{h_{(x)}}{2}-\rho\frac{\Delta P^*}{\Delta x}\frac{1}{12\mu}h_{(x)}^3 \ (2.359) \tag{2.388}$$

From the previous equation, the volumetric flow per unit depth shall take the form:

$$\left[\frac{m^3}{m\,s}\right] \quad \dot{\forall}_X = Q_x = U\frac{h_{(x)}}{2} - \frac{\Delta P^*}{\Delta x}\frac{1}{12\mu}h_{(x)}^3 \tag{2.389}$$

On the other hand, and for incompressible unidirectional flow, the continuity equation can be presented as:

$$\frac{\partial \dot{\forall}_X}{\partial x} = 0 \text{ ; The volumetric flow is constant along the x axis.} \tag{2.390}$$

Substituting equation (2.389) into (2.390) and after differentiating versus x, it is obtained:

$$6U\frac{\partial h}{\partial x} = \frac{\partial}{\partial x}\left(\frac{h^3}{\mu}\frac{\Delta P^*}{\Delta x}\right) \tag{2.391}$$

Equation (2.391) is the Reynolds lubrication equation for incompressible, time independent and unidirectional flow. Notice that $h_{(x)} = h$.

2.8.4. Reynolds Lubrication Equation in Cartesian Coordinates and for Two Directional Three Dimensional Time Independent Flow

In this section equation (2.391) shall be extended to be used in two dimensional cases.

The present case would be the one defined in figure 2.19 but the lower plate is now tilted in direction x and z, and the upper plate moves with a velocity U towards direction x and with a velocity W towards direction z.

Navier-Stokes equation versus x direction will take the form: (see equation (2.348))

$$0 = -\frac{\partial P^*}{\partial x} + \mu\frac{\partial^2 u}{\partial y^2} (2.348) \tag{2.392}$$

Navier-Stokes equation versus the z direction would be:

$$0 = -\frac{\partial P^*}{\partial z} + \mu\frac{\partial^2 w}{\partial y^2} \tag{2.393}$$

From equation (2.392), the velocity distribution, the mass and volumetric flows versus x direction have been previously obtained; see equations (2.354), (2.359) and (2.389).

In a homologous way, from equation (2.393) the volumetric flow equation versus the z direction shall be obtained, taking the form:

$$\dot{\forall}_Z = Q_z = W\frac{h}{2} - \frac{\Delta P^*}{\Delta z}\frac{1}{12\mu}h^3 \tag{2.394}$$

It is important to realize that for the present case, the film thickness between the two plates depends on x and z directions.

The volumetric flow between the two plates can be expressed in vector form as:

$$\vec{\dot{\forall}} = \vec{\dot{\forall}}_X \hat{i} + \vec{\dot{\forall}}_Z \hat{k} \tag{2.395}$$

And the continuity equation, for the case under consideration, is expressed as:

$$\frac{\partial \dot{\forall}_x}{\partial x} + \frac{\partial \dot{\forall}_z}{\partial z} = 0 \tag{2.396}$$

Now, differentiating equation (2.389) versus x direction and equation (2.394) versus z direction it is obtained:

$$\frac{\partial \dot{\forall}_x}{\partial x} = 0 = 6U\frac{\partial h}{\partial x} - \frac{\partial}{\partial x}\left(\frac{h^3}{\mu}\frac{\Delta P^*}{\Delta x}\right) \tag{2.397}$$

$$\frac{\partial \dot{\forall}_Z}{\partial z} = 0 = 6W\frac{\partial h}{\partial z} - \frac{\partial}{\partial z}\left(\frac{h^3}{\mu}\frac{\Delta P^*}{\Delta z}\right) \tag{2.398}$$

Substituting equations (2.397) and (2.398) into equation (2.396) it is obtained the Reynolds lubrication equation for two dimensional flow, notice that the height h is function of the directions x and z.

$$\left[\frac{\partial}{\partial z}\left(W\frac{h}{2}\right)\right] + \left[\frac{\partial}{\partial x}\left(U\frac{h}{2}\right)\right] = \frac{\partial}{\partial z}\left(\frac{\Delta P^*}{\Delta z}\frac{1}{12\mu}h^3\right) + \frac{\partial}{\partial x}\left(\frac{\Delta P^*}{\Delta x}\frac{1}{12\mu}h^3\right) \tag{2.399}$$

It is important to realize that equation (2.391) is a particular case of equation (2.399).

2.8.5. Reynolds Lubrication Equation in Cartesian Coordinates and for Two Directional Three Dimensional Time Dependent Flow

The present case is exactly the same as the previous one, but now the upper plate is not only moving towards the directions x and z with the respective U and W velocities, but it also fluctuates vertically, versus y direction. Figure 2.21 presents a fluid volume differential where the elemental mass flows versus directions x and z are depicted.

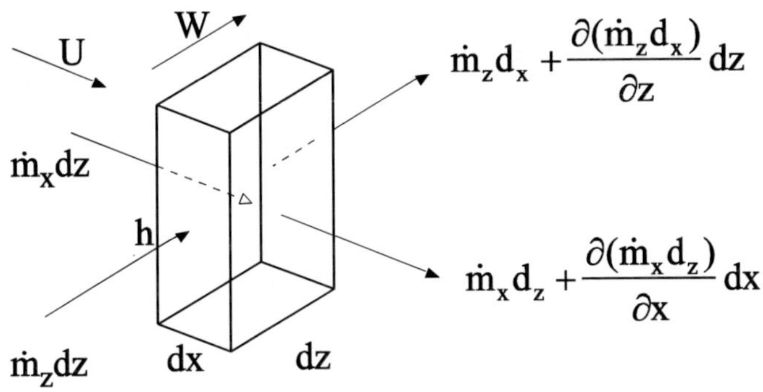

Figure 2.21. Fluid volume differential between the two quasi parallel plates, two directional flow.

Volume and mass differentials can be gives as: $d\forall = dx\,dz\,h$; $dm = \rho\,dx\,dz\,h$.

Calling $\dot{m}_i$ the mass flow per unit length towards a generic direction i and for a generic distance between plates h, the continuity equation can be given as:

$$\frac{\partial}{\partial t}\left(\text{mass}_{\text{control volume}}\right) + \dot{m}_{out} - \dot{m}_{in} = 0 \tag{2.400}$$

Considering the mass flows entering and leaving the control volume defined in figure 2.21, from equation (2.400) it can be established:

$$\frac{\partial}{\partial t}(\rho\,dx\,dz\,h) + \left[\dot{m}_z\,dx + \frac{\partial}{\partial z}(\dot{m}_z\,dx)dz\right] - \dot{m}_z\,dx + \left[\dot{m}_x\,dz + \frac{\partial}{\partial x}(\dot{m}_x\,dz)dx\right] - \dot{m}_z\,dx = 0 \tag{2.401}$$

Equation (2.401) can be reduced to:

$$\frac{\partial}{\partial t}(\rho\,dx\,dz\,h) + \left[\frac{\partial}{\partial z}(\dot{m}_z\,dx)dz\right] + \left[\frac{\partial}{\partial x}(\dot{m}_x\,dz)dx\right] = 0 \tag{2.402}$$

$$\frac{\partial}{\partial t}(\rho\,dx\,dz\,h) + \left[\frac{\partial}{\partial z}(\rho Q_z\,dx)dz\right] + \left[\frac{\partial}{\partial x}(\rho Q_x\,dz)dx\right] = 0 \tag{2.403}$$

It has to be understood that the differentials dx y dz do not depend of the directions x or z, or even the time t. Then, dividing equation (2.403) by dx dz, it is reached:

$$\frac{\partial}{\partial t}(\rho\,h) + \left[\frac{\partial}{\partial z}(\rho Q_z)\right] + \left[\frac{\partial}{\partial x}(\rho Q_x)\right] = 0 \tag{2.404}$$

It must be remembered that: $Q_x = \dot{\forall}_x$ and $Q_z = \dot{\forall}_z$.

The volumetric flows versus x and z directions were already defined by equations (2.389) and (2.394), substituting these two equations in (2.404) it is obtained:

$$\frac{\partial}{\partial t}(\rho\,h) + \left[\frac{\partial}{\partial z}\left(\rho W\frac{h}{2} - \frac{\Delta P^*}{\Delta z}\frac{\rho}{12\mu}h^3\right)\right] + \left[\frac{\partial}{\partial x}\left(\rho U\frac{h}{2} - \frac{\Delta P^*}{\Delta x}\frac{\rho}{12\mu}h^3\right)\right] = 0 \tag{2.405}$$

Equation which can be represented as:

$$\frac{\partial}{\partial t}(\rho\, h)+\left[\frac{\partial}{\partial z}\left(\rho\, W\frac{h}{2}\right)\right]+\left[\frac{\partial}{\partial x}\left(\rho\, U\frac{h}{2}\right)\right]=\frac{\partial}{\partial z}\left(\frac{\Delta P^{*}}{\Delta z}\frac{\rho}{12\mu}h^{3}\right)+\frac{\partial}{\partial x}\left(\frac{\Delta P^{*}}{\Delta x}\frac{\rho}{12\mu}h^{3}\right) \tag{2.406}$$

For incompressible fluid, the previous equation takes the form:

$$\frac{\partial}{\partial t}(h)+\left[\frac{\partial}{\partial z}\left(W\frac{h}{2}\right)\right]+\left[\frac{\partial}{\partial x}\left(U\frac{h}{2}\right)\right]=\frac{\partial}{\partial z}\left(\frac{\Delta P^{*}}{\Delta z}\frac{1}{12\mu}h^{3}\right)+\frac{\partial}{\partial x}\left(\frac{\Delta P^{*}}{\Delta x}\frac{1}{12\mu}h^{3}\right) \tag{2.407}$$

Equation (2.407) is the Reynolds lubrication equation for incompressible fluid, flow is regarded as two directional, three dimensional and time dependent. It is important to notice that equations (2.407) and (2.399) are the same equation, except for the temporal term incorporated in equation (2.407).

2.8.6. Flow with Negligible Acceleration, Case Cylindrical Journal Bearings Statically Loaded

In the present sub section, it will just be considered the case of infinitely long journal bearings, in reality this means that the journal bearing length is several orders of magnitude bigger than the film thickness existing between stator and rotor.

The working hypotheses used in the present case are nearly the same as the ones employed in plane journal bearings, section 2.8.1, and they are summarized as:

- $\rho = cte$; Fluid is regarded as incompressible.
- $\frac{\partial}{\partial t}=0$; Fluid flow will be studied under permanent conditions.
- The local and convective acceleration terms shall be considered as irrelevant when compared with the viscous terms.
- A priory, fluid flow shall be considered as two directional and two dimensional.

Figure 2.22 presents the cylindrical journal bearing under consideration, it must be noticed that the distance between stator and rotor is of the order of microns. The same figure presents several geometric relations between parameters.

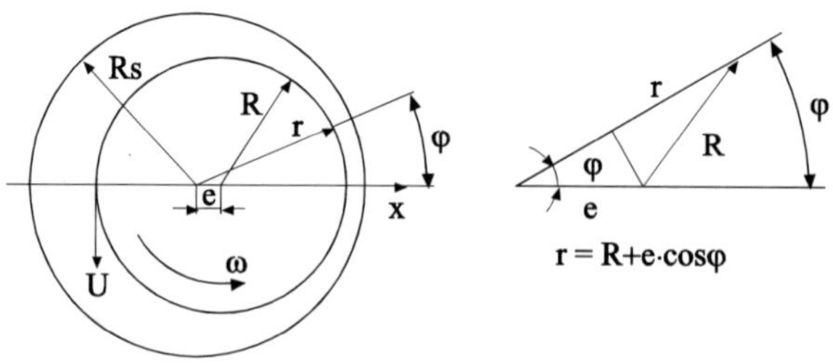

Figure 2.22. Cylindrical journal bearing of infinite length.

Defining $\bar{h}$ as the average distance between stator and rotor, it can be established:

$$R_s = R + \bar{h} = R\left(1 + \frac{\bar{h}}{R}\right) \tag{2.408}$$

From the previous equation, the non dimensional parameter ψ is defined as:

$$\frac{\bar{h}}{R} = \frac{R_s - R}{R} = \psi \tag{2.409}$$

From figure 2.22, the generic radius r can be expressed as:

$$r = R + e\ \cos\varphi = R\left(1 + \frac{e}{R}\cos\varphi\right) \tag{2.410}$$

Notice that e = is the journal bearing eccentricity.

The relation defined in equation (2.410) is particularly correct when the eccentricity tends to zero.

The generic film thickness between stator and rotor can be defined as:

$$h(\varphi) = R_s - r = R_s - R - e\cos\varphi = \bar{h} - e\cos\varphi = \bar{h}\left(1 - \frac{e}{\bar{h}}\cos\varphi\right) = \bar{h}\ (1 - \varepsilon\ \cos\varphi) \tag{2.411}$$

where: $\varepsilon = \frac{e}{\bar{h}}$

To be able to use some equations defined in previous sub sections, the cylindrical journal bearing will be opened and as a result it can be studied as two plates having a relative movement. Figure 2.23 represents the cylindrical journal bearing once being opened. Notice that the film thickness is still defined by equation (2.411).

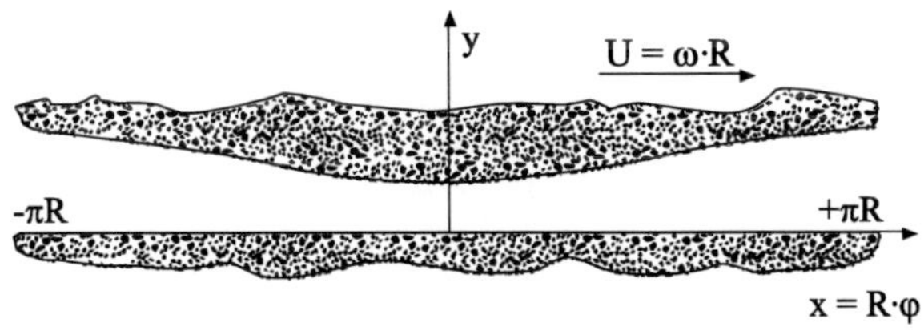

Figure 2.23. Cylindrical journal bearing once opened.

For the plane journal bearing defined in figure 2.23 it can be established: $dx = Rd\varphi$

At this point it is interesting to remember equation (2.389) which was defining the volumetric flow per unit depth between two quasi parallel plates. Notice that in reality, figure 2.23 is presenting two quasi parallel plates, being the hypothesis and boundary conditions the same as the ones defined in sub section 2.8.2. The conclusion is that equation (2.389) is fully applicable in the present sub section.

$$\dot{\forall}_X = Q = \frac{1}{2} U h_{(x)} - \frac{\partial P^*}{\partial x} \frac{1}{12\mu} h^3_{(x)} \text{ (2.389)} \tag{2.412}$$

When applying equation (2.412) to the present case, it must be considered: $h_x = h_{(\varphi)}$; $dx = Rd\varphi$;

From equation (2.412), the reduced pressure distribution along the angular position is defined as:

$$\frac{\dot{\forall}_X}{h^3_{(\varphi)}} = \frac{1}{2} \frac{R\omega}{h^2_{(\varphi)}} - \frac{\partial P^*}{\partial \varphi} \frac{1}{12\mu R} \tag{2.413}$$

$$\frac{\partial P^*}{\partial \varphi} = \left(\frac{1}{2} \frac{R\omega}{h^2_{(\varphi)}} - \frac{\dot{\forall}_X}{h^3_{(\varphi)}} \right) 12\mu R \tag{2.414}$$

Integrating equation (2.414) between 0 and 2π, it is obtained:

$$\int_{P^*=0}^{P^*=2\pi} dP^* = \int_0^{2\pi} \left(\frac{6R^2\omega\mu}{h^2_{(\varphi)}} - \frac{12\mu R \dot{\forall}_X}{h^3_{(\varphi)}} \right) d\varphi; \tag{2.415}$$

Notice that $P^*_{2\pi} = P^*_0$; therefore:

$$P^*_{2\pi} - P^*_0 = 0 = \int_0^{2\pi} \left(\frac{6R^2\omega\mu}{h^2_{(\varphi)}} - \frac{12\mu R \dot{\forall}_X}{h^3_{(\varphi)}} \right) d\varphi \tag{2.416}$$

$$\int_0^{2\pi} \left(\frac{R^2\omega\mu}{h^2_{(\varphi)}} \right) d\varphi = \int_0^{2\pi} \left(\frac{2\mu R \dot{\forall}_X}{h^3_{(\varphi)}} \right) d\varphi \tag{2.417}$$

And remembering that the volumetric flow is constant, it can be obtained:

$$\dot{\forall}_X = \int_0^{2\pi} \left(\frac{R^2\omega}{h^2_{(\varphi)}} \right) d\varphi \cdot \frac{1}{\int_0^{2\pi} \left(\frac{2R}{h^3_{(\varphi)}} \right) d\varphi} \tag{2.418}$$

Multiplying and dividing the right hand side of the previous equation by $\bar{h}^3$, it is obtained:

$$\dot{\forall}_X = \frac{\omega R}{2} \bar{h} \left[\int_0^{2\pi} \left(\frac{\bar{h}}{h_{(\varphi)}} \right)^2 d\varphi \cdot \frac{1}{\int_0^{2\pi} \left(\frac{\bar{h}}{h_{(\varphi)}} \right)^3 d\varphi} \right] \tag{2.419}$$

Remembering the relations established in (2.411), the two integrals defined in (2.419) can be given as:

$$I_2 = \int_0^{2\pi} (1-\varepsilon\cos\varphi)^{-2} d\varphi = \frac{2\pi}{(1-\varepsilon^2)^{\frac{3}{2}}} \tag{2.420}$$

$$I_3 = \int_0^{2\pi} (1-\varepsilon\cos\varphi)^{-3} d\varphi = \frac{\pi(2+\varepsilon^2)}{(1-\varepsilon^2)^{\frac{5}{2}}} \tag{2.421}$$

Substituting equations (2.420) and (2.421) into (2.419), and considering $\psi = \frac{\bar{h}}{R}$, it is reached:

$$\dot{\forall}_X = \frac{1}{2}\omega R^2 \psi \frac{I_2}{I_3} \tag{2.422}$$

Equation (2.422) defines the volumetric flow between the two quasi parallel plates presented in figure 2.23. Remembering now equation (2.414) the pressure distribution along the angular position can be defined as:

$$\frac{\partial P^*}{\partial x} = \frac{\partial P^*}{R\partial\varphi} = \left(\frac{\omega R\, h_{(\varphi)}}{2} - \dot{\forall}_X\right)\frac{12\mu}{h^3_{(\varphi)}} = \left(\frac{\omega R\, h_{(\varphi)}}{2} - \frac{1}{2}\omega R^2 \psi \frac{I_2}{I_3}\right)\frac{12\mu}{h^3_{(\varphi)}} \tag{2.423}$$

$$\frac{1}{R}\frac{\partial P^*}{\partial\varphi} = \frac{6\mu\omega R}{h^2_{(\varphi)}}\left(1 - R\psi\frac{I_2}{I_3}\frac{1}{h_{(\varphi)}}\right) \tag{2.424}$$

$$\frac{1}{R}\frac{\partial P^*}{\partial\varphi} = \frac{6\mu\omega R}{h^2_{(\varphi)}}\left(1 - \frac{\bar{h}}{h_{(\varphi)}}\frac{I_2}{I_3}\right) \tag{2.425}$$

The integration of equation (2.425) would give the reduced pressure distribution as a function of the angular position. This equation shall be integrated later, now, the vertical and horizontal forces acting on the cylindrical journal bearing will be defined, and their respective equations are: See figure 2.22.

$$F_y = -\int_0^{2\pi} P^* \sin\varphi\, R d\varphi \tag{2.426}$$

$$F_x = -\int_0^{2\pi} P^* \cos\varphi\, R d\varphi \tag{2.427}$$

Equations (2.426) and (2.427) represent the force per unit depth onto the cylindrical journal bearing. Notice that the reduced pressure P^* is angular dependent, therefore, the integration of equation (2.426) shall be done as follows.

The reduced pressure will be called a generic parameter u: $P^* = u$.

The rest of the angular dependant terms will be called: $\sin\varphi\, d\varphi = dV$

Differentiating the parameter u it is obtained: $du = \frac{\partial P^*}{\partial \varphi} d\varphi;$

And integrating the term dV it is reached: $V = -\cos\varphi;$

The solution of the integral will be:

$$F_y = -\left[u.V - \int V du\right]R \tag{2.428}$$

$$F_y = RP^* \cos\varphi\Big|_0^{2\pi} - R\int_0^{2\pi} \frac{\partial P^*}{\partial \varphi} \cos\varphi\, d\varphi \tag{2.429}$$

The first term of the right hand side, from equation (2.429) will become zero for the boundary conditions established, when substituting equation (2.425) into (2.429) it is reached:

$$F_y = -6\frac{\mu\omega R^3}{h_{(\varphi)}^2}\int_0^{2\pi}\left(1 - \frac{\bar{h}}{h_{(\varphi)}}\frac{I_2}{I_3}\right)\cos\varphi\, d\varphi \tag{2.430}$$

Multiplying and dividing the right hand side term of equation (2.430) by $\bar{h}^2$ it is obtained:

$$F_y = -\frac{6\mu\omega R}{\psi^2}\int_0^{2\pi}\left(\left(\frac{\bar{h}}{h_{(\varphi)}}\right)^2 - \left(\frac{\bar{h}}{h_{(\varphi)}}\right)^3 \frac{I_2}{I_3}\right)\cos\varphi\, d\varphi; \tag{2.431}$$

The remaining two integrals of equation (2.431) shall be solved as:

$$I_4 = \int_0^{2\pi} \cos\varphi (1 - \varepsilon\cos\varphi)^{-2} d\varphi = \frac{I_2 - I_1}{\varepsilon} \tag{2.432}$$

$$I_5 = \int_0^{2\pi} \cos\varphi (1 - \varepsilon\cos\varphi)^{-3} d\varphi = \frac{I_3 - I_2}{\varepsilon} \tag{2.433}$$

The integral I_1 is defined as:

$$I_1 = \int_0^{2\pi} (1 - \varepsilon\cos\varphi)^{-1} d\varphi = \frac{2\pi}{(1-\varepsilon^2)^{\frac{1}{2}}} \tag{2.434}$$

When substituting equations (2.432) and (2.433) into (2.431) it is reached:

$$F_y = \frac{\mu\omega R}{\psi^2} 6 \frac{I_2 I_5 - I_3 I_4}{I_3}; \tag{2.435}$$

This equation gives the vertical force per unit length of the cylindrical journal bearing. The non dimensional form of equation (2.435) is called Sommerfeld number, S_0, and takes the form:

$$S_0 = F_y \frac{\psi^2}{\mu \omega R} = 6 \frac{I_2 I_5 - I_3 I_4}{I_3}; \tag{2.436}$$

To determine the torque needed to turn the journal bearing rotor, it will be necessary to calculate the shear stresses onto the rotor surface.

Shear stresses shall be determined: (See figure 2.23).

$$\tau_{xy} = \mu \frac{\partial u}{\partial y}\bigg|_{y=h} = \mu \frac{\partial}{\partial y}\left(\frac{\partial P^*}{\partial x} \frac{1}{2\mu} y(y-h) + \frac{Uy}{h} \right)\bigg|_{y=h} \tag{2.437}$$

$$\tau_{xy} = \mu U \left(\frac{1}{h} + \frac{\partial P^*}{\partial x} \frac{h}{2\mu U} \right) \tag{2.438}$$

Remembering that $\frac{\partial P^*}{\partial x} = \frac{1}{R} \frac{\partial P^*}{\partial \varphi}$ is defined by equation (2.425), the shear stresses take the form:

$$\tau_{xy} = \mu U \left(\frac{1}{h} + \frac{h 6 \mu \omega R}{2 \mu h^2 U} \left(1 - \frac{\bar{h}}{h} \frac{I_2}{I_3} \right) \right); \tag{2.439}$$

It has to be noticed that: $h_{(\varphi)} = h$

Rearranging equation (2.439) it is obtained:

$$\tau_{xy} = \left(\frac{\mu U}{h} 4 - 3 \mu U \frac{\bar{h}}{h^2} \frac{I_2}{I_3} \right) \frac{\bar{h}}{\bar{h}} \tag{2.440}$$

$$\tau_{xy} = \frac{\mu \omega}{\psi} \left(\frac{\bar{h}}{h} 4 - 3 \frac{\bar{h}^2}{h^2} \frac{I_2}{I_3} \right) \tag{2.441}$$

Equation (2.441) represents the shear stresses acting onto the rotor surface.

The torque needed to turn the central axis at a velocity ω, will be defined as:

$$T = \int_0^{2\pi} \tau_{xy} \cdot R \, d\varphi \cdot 1 \cdot R \tag{2.442}$$

Substituting equation (2.441) into (2.442) it is reached:

$$T=\int_0^{2\pi}\frac{R^2\mu\omega}{\psi}\left(\frac{\bar{h}}{h}4-3\left(\frac{\bar{h}}{h}\right)^2\frac{I_2}{I_3}\right)d\varphi \tag{2.443}$$

$$T=\frac{R^2\mu\omega}{\psi}\left(4I_1-3\frac{I_2^2}{I_3}\right) \tag{2.444}$$

Notice that equations (2.422), (2.436) and (2.444) represent respectively the volumetric flow per unit length, the vertical force per unit length and the torque per unit length acting onto the cylindrical journal bearing. These three equations are given as a function of the integrals I_1; I_2; I_3; I_4; I_5. Substituting the value of the integrals in the three mentioned equations, it is reached:

$$\dot{\forall}_X\left(\omega R^2\psi\right)^{-1}=\frac{1-\varepsilon^2}{2+\varepsilon^2}; \tag{2.445}$$

$$S_0=F_y\psi^2(\mu\omega R)^{-1}=\frac{12\pi\varepsilon}{\sqrt{1-\varepsilon^2}\,(2+\varepsilon^2)} \tag{2.446}$$

$$T\psi(\mu\omega R^2)^{-1}=\frac{4\pi(1+2\varepsilon^2)}{\sqrt{1-\varepsilon^2}\,(2+\varepsilon^2)} \tag{2.447}$$

It is important to realize that equations (2.445); (2.446) and (2.447) are non dimensional and depend on the non dimensional eccentricity ε and the turning speed ω.

From equation (2.446) it is seen that the vertical force a journal bearing can handle, is directly proportional to the axis turning speed. An increase of eccentricity also produces an increase of the vertical force.

From equation (2.447), when eccentricity is null, it is reached the Petroff expression.

$$T=2\pi\mu\omega\frac{R^2}{\psi} \tag{2.448}$$

Notice as well that the higher the fluid viscosity, the higher the necessary torque and the higher the vertical force.

To better understand the flow performance inside the journal bearing, it is interesting to evaluate the fluid pressure as a function of the angular position. The differential equation giving the reduced pressure variation as a function of the angle was presented as equation (2.425), in what follows this equation will be integrated.

$$\frac{1}{R}\frac{\partial P^*}{\partial\varphi}=\frac{6\mu\omega R}{h_{(\varphi)}^2}\left(1-\frac{\bar{h}}{h_{(\varphi)}}\frac{I_2}{I_3}\right)\ (2.425) \tag{2.449}$$

$$\frac{\partial P^*}{\partial\varphi}=6\mu\omega R^2\left(\frac{1}{h_\varphi^2}-\frac{\bar{h}}{h^3_{(\varphi)}}\frac{I_2}{I_3}\right)\frac{\bar{h}^2}{\bar{h}^2} \tag{2.450}$$

$$\int dP^* = \frac{6\mu\omega}{\psi^2}\int_0^{\varphi}\left[\left(\frac{\bar{h}}{h_{(\varphi)}}\right)^2 - \left(\frac{\bar{h}}{h_{(\varphi)}}\right)^3 \frac{I_2}{I_3}\right] d\varphi \tag{2.451}$$

And remembering the relation $\frac{h_{\varphi}}{\bar{h}} = 1 - \varepsilon \cos\varphi$; it is obtained:

$$\int dP^* = \frac{6\mu\omega}{\psi^2}\int_0^{\varphi}\left[(1-\varepsilon\cos\varphi)^2 - (1-\varepsilon\cos\varphi)^3 \frac{I_2}{I_3}\right] d\varphi \tag{2.452}$$

After integration it is obtained:

$$P^* = C - 6\frac{\mu\omega}{\psi^2}\varepsilon\frac{\sin\varphi(2-\varepsilon\cos\varphi)}{(2+\varepsilon^2)(1-\varepsilon\cos\varphi)^2} \tag{2.453}$$

To obtain the remaining integration constant, it is necessary to use a boundary condition, a possible one would be; at a particular angular position, the reduced pressure is known: when $\varphi = \varphi_{(A)}$; $P^* = P^*_{(A)}$

In figure 2.24, equation (2.453) is plotted as a function of the angular position. Notice that the reduced pressure is higher than the boundary condition pressure at positions located below the cylinder horizontal axis, but the film thickness located above the journal bearing horizontal axis, might be submitted to cavitation conditions. The reduced pressure used as boundary condition has o be selected to prevent the appearance of cavitation.

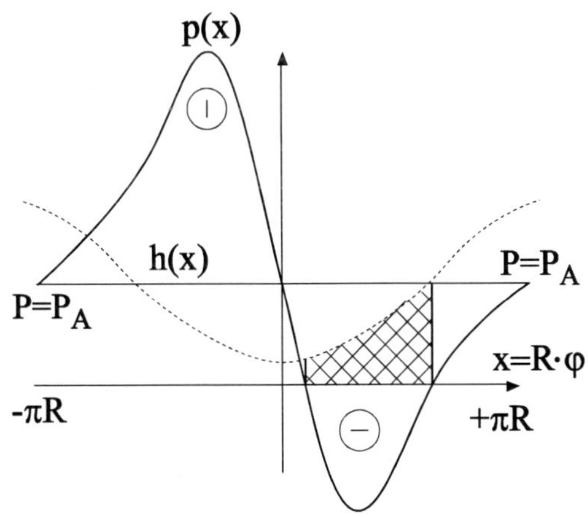

Figure 2.24. Generic pressure distribution as a function of the angular position.

2.8.7. Reynolds Equation of Lubrication in Cylindrical Coordinates

In sections 2.8.3 to 2.8.5, the Reynolds equation in Cartesian coordinates has been presented. In the present section equation (2.407) which was defined for incompressible fluid, being the flow regarded as two directional, three dimensional, and time dependent, will be modified to be used in cylindrical coordinates. The modification will consist of an axis transformation; the procedure is described as follows, see figure 2.25.

The generic positions x and z, defined as a function of the radial and angular coordinates, take the form: $x = r\cos\theta; \quad z = r\ \sin\theta$. It must also be fulfilled: $x^2 + z^2 = r^2$.

Differentiating the equations of x and z versus x, it is obtained:

$$\frac{\partial(x)}{\partial x} = 1 = \cos\theta\frac{\partial r}{\partial x} - r\sin\theta\frac{\partial\theta}{\partial x} \tag{2.454}$$

$$\frac{\partial(z)}{\partial x} = 0 = \sin\theta\frac{\partial r}{\partial x} + r\cos\theta\frac{\partial\theta}{\partial x} \tag{2.455}$$

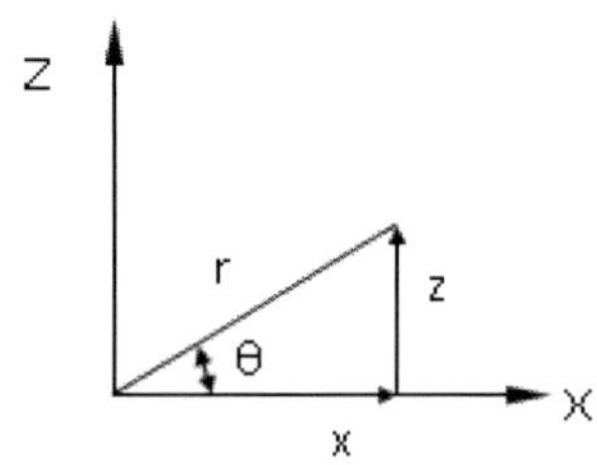

Figure 2.25. Cartesian and Polar coordinate systems.

And differentiating the equations of x and z versus z, it is reached:

$$\frac{\partial(x)}{\partial z} = 0 = \cos\theta\frac{\partial r}{\partial z} - r\sin\theta\frac{\partial\theta}{\partial z} \tag{2.456}$$

$$\frac{\partial(z)}{\partial z} = 1 = \sin\theta\frac{\partial r}{\partial z} + r\cos\theta\frac{\partial\theta}{\partial z} \tag{2.457}$$

From the combination of equations (2.454) and (2.455) it is obtained:

$$\frac{\partial r}{\partial x} = \cos\theta; \qquad \frac{\partial\theta}{\partial x} = -\frac{\sin\theta}{r}; \tag{2.458}$$

From equations (2.456) and (2.457) it is reached:

$$\frac{\partial r}{\partial z} = \sin\theta; \qquad \frac{\partial\theta}{\partial z} = \frac{\cos\theta}{r}; \tag{2.459}$$

At this point, a generic function f which may be dependent on the parameters r, θ or x, z, shall be defined. The partial derivative of this function f versus x and z shall give:

$$\frac{\partial f}{\partial x} = \frac{\partial f}{\partial r}\frac{\partial r}{\partial x} + \frac{\partial f}{\partial\theta}\frac{\partial\theta}{\partial x} \tag{2.460}$$

$$\frac{\partial f}{\partial x} = \cos\theta \frac{\partial f}{\partial r} - \frac{\sin\theta}{r}\frac{\partial f}{\partial \theta} \tag{2.461}$$

$$\frac{\partial f}{\partial z} = \frac{\partial f}{\partial r}\frac{\partial r}{\partial z} + \frac{\partial f}{\partial \theta}\frac{\partial \theta}{\partial z} \tag{2.462}$$

$$\frac{\partial f}{\partial z} = \sin\theta \frac{\partial f}{\partial r} + \frac{\cos\theta}{r}\frac{\partial f}{\partial \theta} \tag{2.463}$$

The second derivatives of the function f versus x and z will take the form:

$$\frac{\partial^2 f}{\partial x^2} = \frac{\partial}{\partial x}\left(\frac{\partial f}{\partial x}\right) \tag{2.464}$$

$$\frac{\partial^2 f}{\partial z^2} = \frac{\partial}{\partial z}\left(\frac{\partial f}{\partial z}\right) \tag{2.465}$$

When substituting equation (2.460) into (2.464) and (2.462) into (2.465) and after some arrangement it is obtained:

$$\frac{\partial^2 f}{\partial x^2} = \frac{\partial}{\partial r}\left(\frac{\partial f}{\partial x}\right)\cos\theta + \frac{\partial}{\partial \theta}\left(\frac{\partial f}{\partial x}\right)\left(-\frac{\sin\sigma}{r}\right) \tag{2.466}$$

$$\frac{\partial^2 f}{\partial z^2} = \frac{\partial}{\partial r}\left(\frac{\partial f}{\partial z}\right)\sin\theta + \frac{\partial}{\partial \theta}\left(\frac{\partial f}{\partial z}\right)\left(\frac{\cos\theta}{r}\right) \tag{2.467}$$

Substituting now equations (2.461) and (2.463) respectively into equations (2.466) and (2.466) and after some long manipulation it is obtained:

$$\frac{\partial^2 f}{\partial x^2} + \frac{\partial^2 f}{\partial z^2} = \frac{\partial^2 f}{\partial r^2} + \frac{1}{r}\frac{\partial f}{\partial r} + \frac{1}{r^2}\frac{\partial^2 f}{\partial \theta^2} \tag{2.468}$$

The Reynolds equation of lubrication in Cartesian coordinates was having the form.

$$\frac{\partial}{\partial t}(h) + \left[\frac{\partial}{\partial z}\left(W\frac{h}{2}\right)\right] + \left[\frac{\partial}{\partial x}\left(U\frac{h}{2}\right)\right] = \frac{\partial}{\partial z}\left(\frac{\Delta P^*}{\Delta z}\frac{1}{12\mu}h^3\right) + \frac{\partial}{\partial x}\left(\frac{\Delta P^*}{\Delta x}\frac{1}{12\mu}h^3\right) \quad (2.407) \tag{2.469}$$

Considering the fluid viscosity as constant, this equation can also be given as:

$$\frac{\partial}{\partial z}\left(\frac{\partial P^*}{\partial z}h^3\right) + \frac{\partial}{\partial x}\left(\frac{\partial P^*}{\partial x}h^3\right) = 6\mu\left\{\left[\frac{\partial}{\partial z}(Wh)\right] + \left[\frac{\partial}{\partial x}(Uh)\right] + 2\frac{\partial h}{\partial t}\right\} \tag{2.470}$$

The left hand side of equation (2.470) can take the form:

$$h^3\frac{\partial^2 P^*}{\partial x^2}+h^3\frac{\partial^2 P^*}{\partial z^2}+3h^2\frac{\partial h}{\partial x}\frac{\partial P^*}{\partial x}+3h^2\frac{\partial h}{\partial z}\frac{\partial P^*}{\partial z} \tag{2.471}$$

Equation (2.470) can be presented as:

$$h^3\left[\frac{\partial^2 P^*}{\partial x^2}+\frac{\partial^2 P^*}{\partial z^2}+\frac{3}{h}\left(\frac{\partial h}{\partial x}\frac{\partial P^*}{\partial x}+\frac{\partial h}{\partial z}\frac{\partial P^*}{\partial z}\right)\right] \tag{2.472}$$

Substituting equations (2.468), (2.461) and (2.463) into (2.472), and considering the generic function f as h or P^* it is obtained:

$$h^3\left[\frac{\partial^2 P^*}{\partial r^2}+\frac{1}{r}\frac{\partial P^*}{\partial r}+\frac{1}{r^2}\frac{\partial^2 P^*}{\partial \theta^2}+\frac{3}{h}\left\{\begin{matrix}\left(\cos\theta\frac{\partial h}{\partial r}-\frac{\sin\theta}{r}\frac{\partial h}{\partial \theta}\right)\left(\cos\theta\frac{\partial P^*}{\partial r}-\frac{\sin\theta}{r}\frac{\partial P^*}{\partial \theta}\right)+\\ \left(\sin\theta\frac{\partial h}{\partial r}+\frac{\cos\theta}{r}\frac{\partial h}{\partial \theta}\right)\left(\sin\theta\frac{\partial P^*}{\partial r}+\frac{\cos\theta}{r}\frac{\partial P^*}{\partial \theta}\right)\end{matrix}\right\}\right] \tag{2.473}$$

After some development equation (2.473) takes the form:

$$\frac{1}{r^2}\left[r^2h^3\frac{\partial^2 P^*}{\partial r^2}+r\,h^3\frac{\partial P^*}{\partial r}+h^3\frac{\partial^2 P^*}{\partial \theta^2}+3h^2r^2\frac{\partial h}{\partial r}\frac{\partial P^*}{\partial r}+3h^2\frac{\partial h}{\partial \theta}\frac{\partial P^*}{\partial \theta}\right] \tag{2.474}$$

The second and fourth terms of equation (2.474) can be given as:

$$r\,h^3\frac{\partial P^*}{\partial r}+3h^2r^2\frac{\partial h}{\partial r}\frac{\partial P^*}{\partial r}=r\left(h^3\frac{\partial P^*}{\partial r}+3h^2r\frac{\partial h}{\partial r}\frac{\partial P^*}{\partial r}\right)=r\frac{\partial P^*}{\partial r}\left(\frac{\partial}{\partial r}\left(h^3 r\right)\right) \tag{2.475}$$

And the first term of equation (2.474) plus equation (2.475) can be given as:

$$r^2h^3\frac{\partial^2 P^*}{\partial r^2}+r\frac{\partial P^*}{\partial r}\left(\frac{\partial}{\partial r}\left(h^3 r\right)\right)=r\frac{\partial}{\partial r}\left(r\,h^3\frac{\partial P^*}{\partial r}\right) \tag{2.476}$$

The third and fifth terms of equation (2.474) can be presented as:

$$h^3\frac{\partial^2 P^*}{\partial \theta^2}+3h^2\frac{\partial h}{\partial \theta}\frac{\partial P^*}{\partial \theta}=\frac{\partial}{\partial \theta}\left(h^3\frac{\partial P^*}{\partial \theta}\right) \tag{2.477}$$

Considering equations (2.476) and (2.477), equation (2.474) takes the form:

$$\frac{1}{r^2}\left(\frac{\partial}{\partial \theta}\left(h^3\frac{\partial P^*}{\partial \theta}\right)+r\frac{\partial}{\partial r}\left(r\,h^3\frac{\partial P^*}{\partial r}\right)\right)=\frac{1}{r^2}\frac{\partial}{\partial \theta}\left(h^3\frac{\partial P^*}{\partial \theta}\right)+\frac{1}{r}\frac{\partial}{\partial r}\left(\left(h^3 r\right)\frac{\partial P^*}{\partial r}\right) \tag{2.478}$$

Equation (2.478) represents the transformation to cylindrical coordinates of the left hand side of equation (2.470). The right hand side of Reynolds equation of lubrication in Cartesian coordinates, equation (2.470) was taking the form.

$$6\mu\left\{\frac{\partial}{\partial z}(Wh)+\frac{\partial}{\partial x}(Uh)+2\frac{\partial h}{\partial t}\right\} \tag{2.479}$$

To transform equation (2.479) to cylindrical coordinates, it is necessary to give the U and W velocities, which respective directions are x and z, as a function of the radial L and tangential T velocities, see figure 2.26, obtaining:

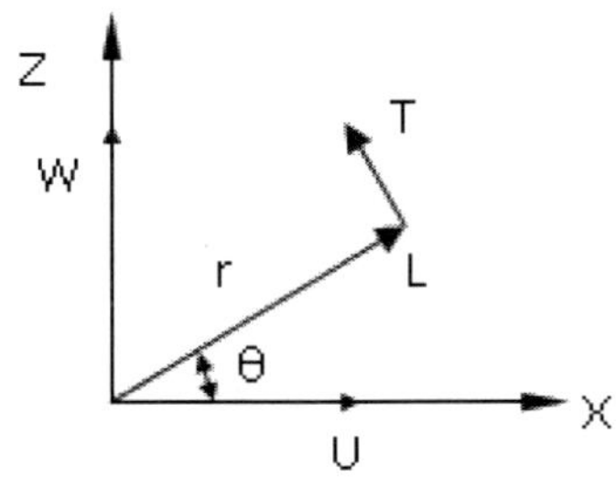

Figure 2.26. Relation between velocities in Cartesian and cylindrical coordinates.

$$U = L\cos\theta - T\sin\theta \tag{2.480}$$

$$W = L\sin\theta + T\cos\theta \tag{2.481}$$

According to figure 2.26 can also be established:

$$T = W\cos\theta - U\sin\theta \tag{2.482}$$

$$L = U\cos\theta + W\sin\theta \tag{2.483}$$

Calling the generic function f the terms (Wh) and (Uh), and considering the equations (2.461) and (2.463), equation (2.479) takes the form:

$$6\mu\left\{\cos\theta\frac{\partial(Uh)}{\partial r}-\frac{\sin\theta}{r}\frac{\partial(Uh)}{\partial\theta}+\sin\theta\frac{\partial(Wh)}{\partial r}+\frac{\cos\theta}{r}\frac{\partial(Wh)}{\partial\theta}+2\frac{\partial h}{\partial t}\right\} \tag{2.484}$$

Substituting equations (2.480) and (2.481) into equation (2.484) it is found:

$$6\mu\left\{\begin{array}{l}\cos\theta\dfrac{\partial(h(L\cos\theta-T\sin\theta))}{\partial r}-\dfrac{\sin\theta}{r}\dfrac{\partial(h(L\cos\theta-T\sin\theta))}{\partial\theta}+\\ \sin\theta\dfrac{\partial(h(L\sin\theta+T\cos\theta))}{\partial r}+\dfrac{\cos\theta}{r}\dfrac{\partial(h(L\sin\theta+T\cos\theta))}{\partial\theta}+2\dfrac{\partial h}{\partial t}\end{array}\right\} \tag{2.485}$$

After some further rearrangement, equation (2.485) takes the form:

$$6\mu\left\{\frac{\partial(hL)}{\partial r}+\frac{hL}{r}+\frac{1}{r}\frac{\partial(hT)}{\partial\theta}+2\frac{\partial h}{\partial t}\right\} \tag{2.486}$$

This term can also be presented as:

$$\frac{6\mu}{r}\left\{\frac{\partial(r\,h\,L)}{\partial r}+\frac{\partial(hT)}{\partial\theta}+2r\frac{\partial h}{\partial t}\right\} \tag{2.487}$$

Equations (2.482) and (2.483) represent the tangential and radial velocities of a system displacing with velocities U and W towards the corresponding directions x and z. If the system under study is both spinning with a rotational velocity ω and undergoing a translation, as represented in figure 2.27, the tangential velocity T would take the form:

$$T = W\cos\theta - U\sin\theta + r\,\omega \tag{2.488}$$

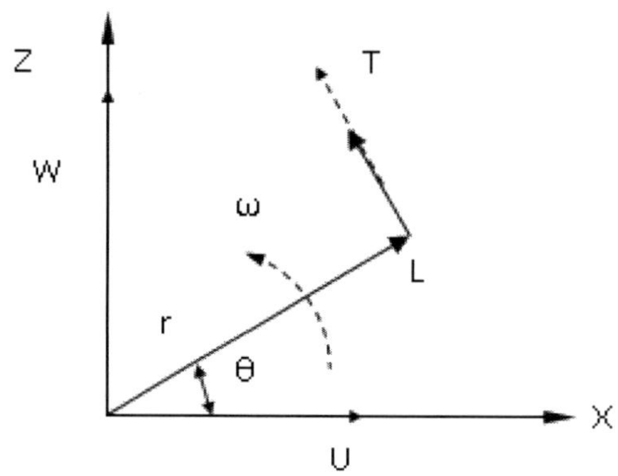

Figure 2.27. Tangential and radial velocities when considering spin and translation.

It is to be noticed that equation (2.487) represents the right hand side of Reynolds lubrication equation in cylindrical coordinates, the radial and tangential velocities, for a system undergoing translation and rotation, were defined by equations (2.483) and (2.488), when substituting these two equations in (2.487) it is obtained:

$$\frac{6\mu}{r}\left\{\frac{\partial(r\,h\left(U\cos\theta+W\sin\theta\right))}{\partial r}+\frac{\partial(h\left(W\cos\theta-U\sin\theta+r\,\omega\right))}{\partial\theta}+2r\frac{\partial h}{\partial t}\right\} \tag{2.489}$$

After some rearrangement equation (2.488) takes the form:

$$6\mu\left\{\frac{\partial h}{\partial r}\left(U\cos\theta+W\sin\theta\right)+\frac{1}{r}\frac{\partial h}{\partial\theta}\left(W\cos\theta-U\sin\theta\right)+\frac{\partial(h\,\omega)}{\partial\theta}+2\frac{\partial h}{\partial t}\right\} \tag{2.490}$$

Equation (2.490) is the final form of the right hand side of the Reynolds lubrication equation in cylindrical coordinates. Now, uniting the left hand side and the right hand side of Reynolds lubrication equation, which are respectively given by equations (2.478) and (2.490), it is obtained the final Reynolds equation of lubrication in cylindrical coordinates. Please notice that this equation considers system translation and spin.

$$\frac{1}{r^2}\frac{\partial}{\partial\theta}\left(h^3\frac{\partial P^*}{\partial\theta}\right)+\frac{1}{r}\frac{\partial}{\partial r}\left(\left(h^3 r\right)\frac{\partial P^*}{\partial r}\right)=6\mu\left\{\begin{array}{l}\frac{\partial h}{\partial r}\left(U\cos\theta+W\sin\theta\right)+\\ \frac{1}{r}\frac{\partial h}{\partial\theta}\left(W\cos\theta-U\sin\theta\right)+\frac{\partial(h\,\omega)}{\partial\theta}+2\frac{\partial h}{\partial t}\end{array}\right\} \quad (2.491)$$

2.9. NOMENCLATURE

$\vec{a}$	Acceleration. (m/s^2).
B	Generic fluid property.
b	Generic fluid property per unit mass.
D	Drag force per unit length. (N/m).
d	Generic distance. (m).
$\frac{D\phi}{Dt}$	Material derivative of a property ϕ.
e	Eccentricity. (m).
E	Energy. (N m).
E_c	Kinetic energy. (N m).
E_p	Potential energy. (N m).
f	Generic function.
F	Force. (N).
g	Gravitational acceleration. (m/s^2).
h	Enthalpy. Generic distance. (J/Kg).
$L = \ell$	Generic Length. (m).
L	Lift force per unit depth. (N/m).
M	Momentum associated to the fluid. (N m).
M	Angular momentum. Torque. (N m).
m	Mass. (m).
$\dot{m}$	Mass flow. (Kg/s).
N	Power. (W).
$\hat{n}; \hat{x}; \hat{r}; \hat{\theta}$	Normal vector.
P	Pressure. (N/m^2).
P^*	Reduced pressure. (N/m^2).
Q	Heat flux. Volumetric flow. (J). (m^3/s).
$\dot{Q}$	Heat flow per unit time. (J/s).
$R = r$	Generic radius. (m).
S	Surface. (m^2).
S_o	Sommerfeld number.

T	Torque. (N m).
t	Time. (s).
u	Internal energy. (J/Kg).
$V = v$	Generic velocity. (m/s).
$U = u = V_x$	Velocity component X direction. (m/s).
$V = v = V_y$	Velocity component Y direction. (m/s).
$W = w = V_z$	Velocity component Z direction. (m/s).
W	Work. (N m).
$\dot{W}$	Power. (W).
V_r	Velocity component radial direction. (m/s).
V_θ	Velocity component angular direction. (m/s).
V_ϕ	Velocity component angular direction. (m/s).
$X = x$	Generic direction. Generic distance. (m).
$Y = y$	Generic direction. Generic distance. Energy per unit mass. (m). (J/Kg).
$Z = z$	Generic direction. Generic distance. (m).
α	Inclination angle. (rad).
δ	Generic distance. (m).
δ'	Generic distance. (m).
ε	Non dimensional eccentricity.
η	Efficiency. Rayleich flow non dimensional variable.
λ	Second viscosity coefficient.
π	Number Pi.
$\forall$	Volume. (m^3).
$\dot{\forall} = Q$	Volumetric flow. (m^3/s).
θ	Generic angle. (rad).
ϕ	Generic angle. (rad).
ρ	Density. (Kg/m^3).
σ	Generic angle. (rad).
σ_i	Normal stresses. (N/m^2).
$\overline{\overline{\tau}}$	Stress tensor. (N/m^2).
τ_{ij}	Shear stresses (ij direction). (N/m^2).
μ	Dynamic viscosity. (Kg/(m s)).
ν	Kinematic viscosity. (m^2/s).
ψ	Non dimensional fluid film thickness.
$\upsilon = \frac{1}{\rho}$	Specific volume. (m^3/Kg).
φ	Generic angle. (rad).
$\amalg$	Scalar potential of gravitational forces. (m^2/s^2).
$\omega = \Omega$	Angular velocity. (rad/s).

2.10. REFERENCES

[1] Anderson John D. Jr. (2001). *Fundamentals of Aerodynamics*. New York. McGraw-Hill.

[2] Barrero A, Perez-Saborid M. (2005). *Mecànica de Fluidos*. Madrid. McGraw Hill.

[3] Bergadà Josep M. (2012). *Mecánica de Fluidos*. Breve introducción teórica con problemas resueltos. Barcelona. Iniciativa digital politécnica UPC.

[4] Cameron A. (1966). *The principles of lubrication*. London. Longmans.

[5] Douglas JF, Gasiorek JM, Swaffield JA. (1998). *Fluid Mechanics* 3rd edition. Singapore. Longman.

[6] Pnueli D, Gutfinger C. (1997). Cambridge. Cambridge University Press.

[7] Spurk Joseph H. (1997). *Fluid Mechanics*. Berlin-Heidelberg. Springer.

[8] White Frank M. (2004). *Mecánica de Fluidos*. 5th edition. Madrid. McGraw Hill.

Chapter 3

INTRODUCTION TO COMPUTER FLUID DYNAMICS (CFD)

While doing physical modeling in fluid mechanics, we often encounter complicated partial deferential equations which can't be solved analytically without making certain approximations, which results in decrease in accuracy. For many industrial applications, researchers try to make these intelligent approximations in order to solve these complicated mathematical equations without losing much in accuracy. But with the increasing demand in efficiency of industrial machines and improving quality of personal computers, solving such equations numerically with greater accuracy is becoming more and more feasible.

This chapter is focused on computational aspects of solving a partial differential equation numerically with an example of axial piston pump. First, we will present basic equations used in Fluid dynamics from numerical stand point, then time and space discretization using finite volume, finite difference and finite element method. Afterwards we will present an example of a finite volume method being used instead of finite element method by using coordinate transformation. We will finish the chapter with convergence criteria and with some comments and remarks regarding mesh topology.

3.1. STEP BY STEP NUMERICAL FORMULATION

In order to successfully solve a fluid dynamic problem numerically, it is important to define a few check points to structure the problem correctly.

3.1.1. Selecting an Appropriate Grid and Integration Formulation

When deciding a coordinate system for a given problem there are two things that need to be kept in mind, complexity of the numerical equations and complexity of physical geometries of the problem under consideration.

Let's consider the fluid flow problem shown in figure 3.1. It is possible to solve this problem using Finite element method through a triangular mesh as shown in figure 3.1.a or by using finite volume method on a transformed grid, as shown in figure 3.1.b. Finite Volume method is based on conservation of physical properties, therefore, different terms of equation provide a straightforward understanding of the results. On the other hand, finite element method approximates the Navier-Stokes equation in its variational form. It is highly accurate but less intuitive. The choice of method, one over the other, is based on personal taste. The details of the method are, however, the subject of much controversial discussion concerning the pros and cons of the various methods and their variants. However, this conflict is partially resolved in many cases, as the differences between the methods often disappear on general meshes. In fact, some of the FVMs can be interpreted as variants of certain "mixed" FEMs.

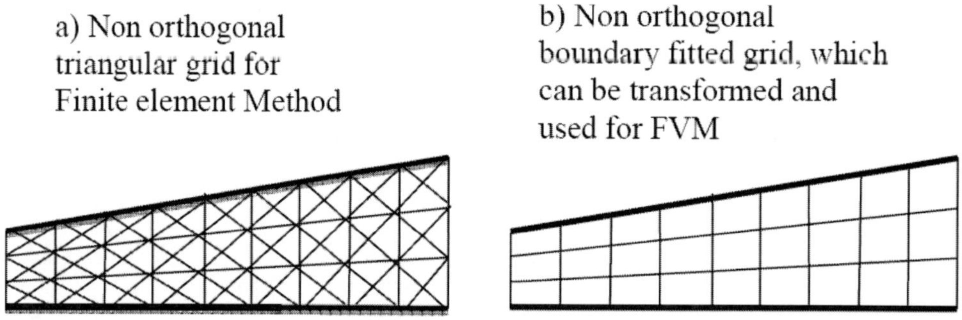

Figure 3.1. Non orthogonal domain and grid choices.

3.1.2. Selection of an Appropriate Reference of Frame for the Problem

Consider the problem shown is figure 3.2, a slipper running over a swash plate. If it is required to simulate the fluid flow in slipper swash plate pocket, a problem can be considered when coordinate's axis are attached to swash plate center or in the other case, coordinate's axis are attached to the slipper center.

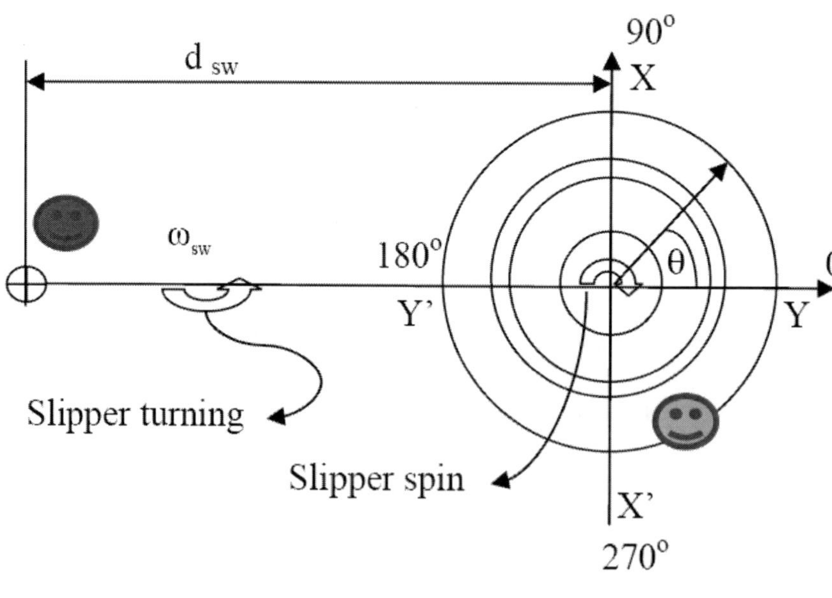

Figure 3.2. Diagram of slipper running over swash plat axis.

If coordinate's axis are attached to swash plate axis, we don't have to consider the pseudo forces such as centripetal forces. But at the same time, it becomes much more difficult to understand the vorticity inside the slipper pocket. On the other hand, when coordinate's axis are assumed to be attached to slipper's axis, the momentum equation will have some extra source terms of pseudo forces but the understanding of fluid flow stream lines will become much more easier. Therefore, as shown in figure 3.2, green face (coordinates axis attached to slipper center) is a correct reference frame for the problem formulation.

When axis of the computational domain is considered to be attached to the swash plate centre, movement of swash plate can be introduced by velocity boundary conditions. Therefore, the problem can be treated in an inertial frame of reference and corresponding NVS equation can be written as (3.1)

$$\rho\left[\frac{\partial u}{\partial t}+u\cdot\nabla u\right]=\mu\,\Delta u+S \tag{3.1}$$

On the other hand, if the coordinate axis are attached to the slipper centre, which makes the reference frame as non-inertial, in this case, corresponding NVS equation can be written as (3.2), which considers Coriolis and centripetal forces.

$$\rho\left[\frac{\partial u}{\partial t}+u\cdot\nabla u+2\omega\times u+\omega\times\left(\omega\times R\right)\right]=\mu\,\Delta u+S \tag{3.2}$$

3.1.3. Selecting Appropriate Boundary Conditions for the Problem

Uniqueness and convergence of the solution of any partial differential equation fully depends on the appropriate boundary conditions. Boundary conditions are specified into two main categories, Dirichlet type and Neumann type. In Dirichlet type condition, the exact value of the unknown field is specified at the boundary. In Neumann type condition, specific values of the derivative of the unknown field are specified at the boundary.

Following are the boundary conditions that are usually encountered when approximating boundaries of a fluid flow.

- A no slipping boundary condition at the walls. (Dirichlet type)

$$V_z\big|_{\Gamma_{in}}=0 \tag{3.3}$$

- The pressure at inlet and outlet boundaries is specified or the pressure difference across the boundary is specified. (Dirichlet type)

$$P\big|_{\Gamma_{in}}=P_{in};\ P\big|_{\Gamma_{out}}=P_{out}; \tag{3.4}$$

- The flow variables at inlet and outlet boundaries are specified by zero normal derivatives (Neumann type).

$$\left.\frac{\partial(r.V_r)}{\partial r}\right|_{\Gamma_{in}\,or\,\Gamma_{out}} = 0; \qquad \left.\frac{\partial V_\theta}{\partial r}\right|_{\Gamma_{in}\,or\,\Gamma_{out}} = 0; \qquad \left.\frac{\partial V_z}{\partial r}\right|_{\Gamma_{out}} = 0; \tag{3.5}$$

- Inlet velocity is specified (Dirichlet type).

$$\left.V_r\right|_{\Gamma_{in}\,or\,\Gamma_{out}} = V \tag{3.6}$$

The important point to be noticed here is the exact number of boundary conditions required to solve a problem correctly and to provide a unique solution. For example, the inlet and outlet flow velocity can't be specified simultaneously because one will depend on the other and specifying both of them may result in divergence in solution. Sometimes boundary conditions are given in the form of Reynolds number or some other flow property, such boundary condition can be transformed easily into velocity or pressure form.

3.2. Basic Fluid Dynamic Equations and Their Physical Interpretation

Fluid dynamics is basically a study of fluids in motion. All fundamental fluid dynamics equations, in one form or other, can be derived from three physical principles, conservation of mass, momentum and energy. Conservation of mass results in Continuity equation, conservation of momentum results in Navier-Stokes equation, which is known to be the fundamental governing equation for fluid motion and conservation of energy results in energy equation.

Any fluid dynamics equation can be developed/understood in two forms, differential form and integral form. When an equation is developed using an infinitesimally small element, it results into differential form of the equation and such a form is used in finite difference method. On the other hand, when a finite size control volume is used in developing fluid flow equation, the resulting equation is called integral form of the equation and such equations are used in finite volume formulations. One form can easily be transformed into another with simple mathematical manipulations.

One of the major differences between integral and differential form of the governing equation is that the integral form of the equation allows for the presence of discontinuities inside the control volume. However, in the differential form of the governing equation, the flow properties are differentiable, hence continuous. This consideration becomes of particular importance when calculating a flow with real discontinuities such as shock wave.

In order to solve these equations numerically for a given practical problem, it is a wiser practice to first understand the physical meaning of each individual term in the equation separately. This will allow us to understand the effect of each operation while estimating them numerically.

3.2.1. Understanding Momentum Equation as Flux Equation

In this section, we will develop a conservation of momentum equation as a conservation of flux, which is easier to understand and interpret. Change in momentum flux in control volume over a period of time can be represented as a sum of three physical phenomena, convection, diffusion, and source generation due to body forces etc. Equation (3.7) represent these quantities mathematically, as shown in figure 3.3.

$$\underbrace{\rho\frac{\partial\phi}{\partial t}}_{\text{Change in flux}} = \overbrace{\{-\rho\nabla(u.\phi)\}}^{\text{Convection of flux}} + \overbrace{\{\mu\nabla^2\phi\}}^{\text{Diffusion of flux}} + \overbrace{\{S_\phi\}}^{\text{Source generation of flux}} \tag{3.7}$$

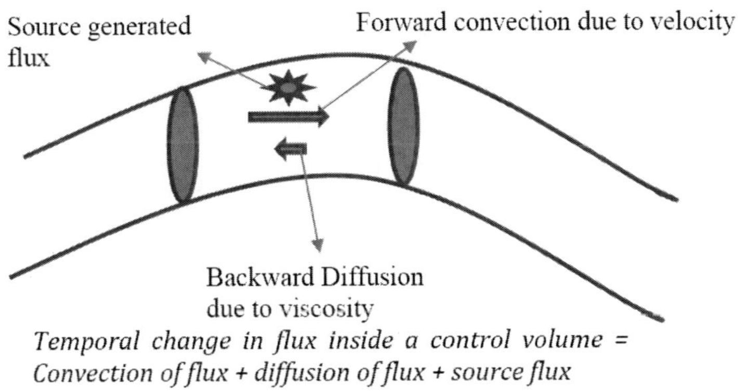

Figure 3.3. Control volume showing change in flux as a combination of convection, diffusion and source.

When extending equation (3.7) in three dimensional cylindrical coordinates system, it will take the form of equation (3.8), known as Momentum equation or Navier Stokes Equation.

$$\underbrace{\rho\frac{\partial\phi}{\partial t}}_{\text{Change in flux}} + \overbrace{\rho\left[\frac{1}{r}\frac{\partial}{\partial r}(r.V_r.\phi)+\frac{1}{r}\frac{\partial}{\partial\theta}(V_\theta.\phi)+\frac{\partial}{\partial z}(V_z.\phi)\right]}^{\text{Convection of flux in same direction} \approx \nabla(u.\phi)} = \overbrace{\mu\left[\frac{\partial}{\partial r}\left(\frac{1}{r}\frac{\partial(r.\phi)}{\partial r}\right)+\frac{1}{r^2}\frac{\partial^2\phi}{\partial\theta^2}+\frac{\partial^2\phi}{\partial z^2}\right]}^{\text{Diffusion of flux} \approx \nabla^2\phi} + \overbrace{S_\phi}^{\text{Source}} \tag{3.8}$$

Therefore, a general methodology for the numerical approximation of each term can be developed, which later can be implemented for all three dimensions. The source term of each direction in cylindrical coordinate system is given in the table 3.1.

The development of continuity equation is far simpler and has already been explained in previous chapter, therefore, here we will directly present the equation,

$$\frac{\partial \rho}{\partial t} + \nabla.(\rho u) = 0 \tag{3.9}$$

Table 3.1. The values of S_ϕ for different ϕ

ϕ	S_ϕ
V_r	$-\frac{\partial P}{\partial r} - \frac{2\mu}{r^2}\frac{\partial V_\theta}{\partial \theta} - \frac{\mu V_r}{r^2} + \frac{V_\theta^2 \cdot \rho}{r}$
V_θ	$-\frac{1}{r}\frac{\partial P}{\partial \theta} + \frac{2\mu}{r^2}\frac{\partial V_r}{\partial \theta} - \frac{\mu V_\theta}{r^2} - \frac{\rho V_r V_\theta}{r}$
V_z	$-\frac{\partial P}{\partial z}$

3.3. Discretization of Momentum Equation

Before attempting the numerical solution of a problem, first let's re-write the generalized form of the flow equation with appropriate boundary conditions for an incompressible flow in domain Ω between time interval 0 to T. Γ represent the boundary of the domain Ω.

$$\begin{array}{lll} \rho\, \partial_t u + \rho\, u . \nabla u - \mu\, \Delta u = S & \text{In } \Omega \times (0,T) & \text{NVS} \\ \nabla . u = 0 & \text{In } \Omega \times (0,T) & \text{Continuity} \\ u = 0 & \text{In } \Gamma \times (0,T) & \text{(Boundary Condition)} \\ u = u_o & \text{In } \Omega, t = 0 & \text{(Initial Condition)} \end{array} \tag{3.10}$$

Discretization of these partial differential equations involve estimating the continuous 1st and 2nd order derivative terms as a function of discrete nodal values. As it can be seen from equation (3.10), Navier-stokes equation has one time derivative term and rest space derivative terms, therefore a separate formulation needs to be done for both of them. First we will present the time discretization followed by spatial discretization. For spatial discretization, there are mainly three methods which are frequently used in the fluid flow simulation, namely Finite difference method, Finite volume method and Finite element method. We will present the discretization of momentum equation by using each of these methods.

3.3.1. Temporal Discretization of Generalized Momentum Equation

Let's rewrite the NVS equation in the form of equation (3.11). Left hand side of equation (3.11) contains the time derivative term and right hand side contains all space derivative terms which include convection, diffusion and source, represented by F. F is a function of velocity and pressure but it does not contain any time derivative term.

$$\rho \frac{du}{dt} = F(u,p) \qquad \text{In } \Omega \times (0,T) \tag{3.11}$$

There are three types of schemes which can be used for time discretization,

$$\rho \frac{u^{t+1} - u^{t}}{\Delta t} = F(u^{t}, p^{t}) \qquad \text{fully explicit} \tag{3.12}$$

$$\rho \frac{u^{t+1} - u^{t}}{\Delta t} = F(u^{t+1}, p^{t}) \qquad \text{fully implicit} \tag{3.13}$$

$$\rho \frac{u^{t+1} - u^{t}}{\Delta t} = \alpha\, F(u^{t+1}, p^{t}) + (1-\alpha)\, F(u^{t}, p^{t}) \qquad \alpha = 0.5, \text{crank nikolson scheme} \tag{3.14}$$

When solving the flow problem with fully explicit scheme, convergence of the solution depends highly on the time step used. For most of the problems a very small time step is required for the convergence, therefore, resulting in a very slow convergence rate. On the other hand, a comparatively bigger time step can be used when solving a problem as a fully implicit scheme, therefore, resulting in a higher convergence rate.

At the same time it is also important to point out that using explicit scheme results in simple additive iterative equations which are computationally very fast. On the other hand, implicit scheme results in a system of linear equations and a solution of the form $X = A^{-1}B$ which needs to be computed at each time step. But even with this higher computational overhead, fully implicit scheme takes less computational time in general.

Another interesting thing that needs to be pointed out is when looking inside right hand side (function F), there exists a nonlinear convective term. When using a fully implicit or semi-implicit scheme this nonlinear term needs to be linearized, on the other hand, in fully explicit scheme it is not required as it become constant.

3.3.2. Spatial Discretization of Generalized Momentum Equation Using Finite Volume Method

To understand the physical flow phenomena and implement the computational technique more precisely, NVS equation (3.8) can be treated as a combination of four different terms, Transient term, Convection term (T_1), Diffusion term (T_2) and Source term (S). As a result of this treatment, now different numerical approximations can be implemented for each different term to attain more accuracy when integrating it numerically.

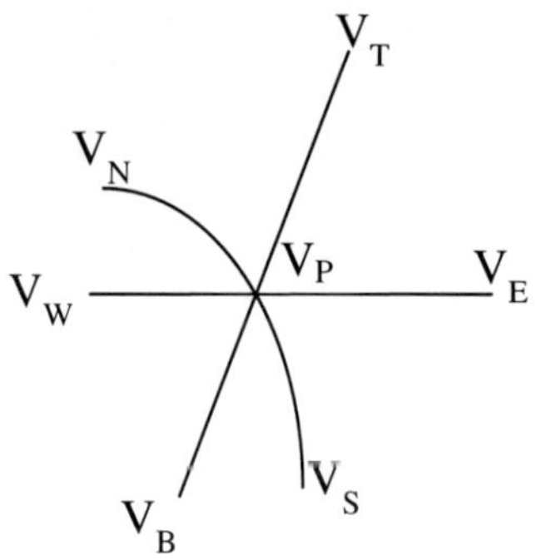

Figure 3.4. r directional grid nomenclature.

$$\int_\Omega T_1 = \int_\Omega \rho\, u \,.\, \nabla\phi = \int_\Omega \rho\left[\frac{1}{r}\frac{\partial}{\partial r}(r.V_r.\phi) + \frac{1}{r}\frac{\partial}{\partial \theta}(V_\theta.\phi) + \frac{\partial}{\partial z}(V_z.\phi)\right]$$

$$= \rho\left[\begin{array}{l} \dfrac{1}{R_p}\dfrac{(R_E\left\|-V_r^E,0\right\|.\phi_E - R_W\left\|V_r^W,0\right\|.\phi_W)}{2\Delta r} + \\ \dfrac{1}{R_p}\dfrac{(\left\|-V_\theta^N,0\right\|.\phi_N - \left\|V_\theta^S,0\right\|.\phi_S)}{2\Delta\theta} + \dfrac{1}{R_p}\dfrac{(\left\|-V_z^T,0\right\|.\phi_T - \left\|V_z^B,0\right\|.\phi_B)}{2\Delta z} \end{array}\right] R\Delta r\Delta\theta\Delta z \tag{3.15}$$

$$\int_\Omega T_2 = \int_\Omega \mu\,\Delta\phi = \int_\Omega \mu\left[\frac{\partial}{\partial r}\left(\frac{1}{r}\frac{\partial(r.\phi)}{\partial r}\right) + \frac{1}{r^2}\frac{\partial^2\phi}{\partial\theta^2} + \frac{\partial^2\phi}{\partial z^2}\right]$$

$$= \mu\left[\frac{\left.\dfrac{1}{r}\dfrac{\partial(r.\phi)}{\partial r}\right|_e - \left.\dfrac{1}{r}\dfrac{\partial(r.\phi)}{\partial r}\right|_w}{\Delta r} + \frac{\left.\dfrac{\partial\phi}{\partial\theta}\right|_n - \left.\dfrac{\partial\phi}{\partial\theta}\right|_s}{R_p^2\Delta\theta} + \frac{\left.\dfrac{\partial\phi}{\partial z}\right|_t - \left.\dfrac{\partial\phi}{\partial z}\right|_b}{\Delta z}\right] R\Delta r\Delta\theta\Delta z$$

$$= \mu\left[\frac{\dfrac{R_E\phi_E - R_P\phi_P}{R_E} - \dfrac{R_P\phi_P - R_W\phi_W}{R_W}}{\Delta r^2} + \frac{(\phi_N - \phi_P) - (\phi_P - \phi_S)}{R_p^2\Delta\theta^2} + \frac{(\phi_T - \phi_P) - (\phi_P - \phi_B)}{\Delta z^2}\right] R\Delta r\Delta\theta\Delta z \tag{3.16}$$

Figure 3.4 shows the nomenclature of the grid which has been used in FVM formulation. When integrating different NVS term using the control volumes formulation, transient term is treated by implicit scheme for faster convergence, as described in equation (3.13).

An upwind scheme is implemented for the convection term to maintain the positivity of the coefficients. Let's say we need to calculate the r direction flow velocity at control volume east and west face. If west (w) to east (e) is considered to be a positive flow direction in the coordinate axis, then according to the upwind scheme, r direction velocity at west face can be written as max(V_W, 0) and in the similar manner at east face velocity can be written as max(-V_E,0).

According to the current formulation, velocity at east face will become function of east point if the flow starts flowing from east to west. A max function is implemented to maintain the positivity of the coefficient otherwise negative coefficient at the diagonal of the matrix will result in a non-invertible system of linear equations. Power law scheme is used as a shape function to integrate the diffusion term and source term is treated by implicit scheme. Equations (3.15) and (3.16) represent the convection and diffusion term integration formulation.

Once all the terms are integrated, final discrete momentum equation can be written in the form of equation (3.17) and its coefficients are given by equations (3.18) to (3.26), where p represents the center of the control volume under consideration and e, w, n, s, t, b corresponds to east, west, north, south, top and bottom neighbor grid points on the grid as shown in figure 3.4.

$$a_p \Phi_p^* = \sum_{nb} a_{nb} \Phi_{nb}^* + b_\Phi \tag{3.17}$$

where,

$$a_E = D_e + \| -F_e, 0 \| \quad \text{where } D_e = \frac{\mu R_e \Delta\theta \Delta z}{\Delta r} \quad \& \; F_e = \rho R_e V_e^r \Delta\theta \Delta z \tag{3.18}$$

$$a_W = D_w + \| F_w, 0 \| \quad \text{where } D_w = \frac{\mu R_w \Delta\theta \Delta z}{\Delta r} \quad \& \; F_w = \rho R_w V_w^r \Delta\theta \Delta z \tag{3.19}$$

$$a_N = D_n + \| -F_n, 0 \| \quad \text{where } D_n = \frac{\mu \Delta r \Delta z}{R_p \Delta\theta} \quad \& \; F_n = \rho V_n^\theta \Delta r \Delta z \tag{3.20}$$

$$a_S = D_s + \| F_s, 0 \| \quad \text{where } D_s = \frac{\mu \Delta r \Delta z}{R_p \Delta\theta} \quad \& \; F_s = \rho V_s^\theta \Delta r \Delta z \tag{3.21}$$

$$a_T = D_t + \| -F_t, 0 \| \quad \text{where } D_t = \frac{\mu R_p \Delta\theta \Delta r}{\Delta z} \quad \& \; F_t = \rho R_p V_t^z \Delta\theta \Delta r \tag{3.22}$$

$$a_B = D_b + \| F_b, 0 \| \quad \text{where } D_b = \frac{\mu R_p \Delta\theta \Delta r}{\Delta z} \quad \& \; F_b = \rho R_p V_b^z \Delta\theta \Delta r \tag{3.23}$$

$$a_P^o = \frac{\rho R_p \Delta\theta \Delta r \Delta z}{\Delta t} \tag{3.24}$$

$$b_\Phi = a_P^o \Phi_{p.old}^* + S_{\Phi.C} R_p \Delta\theta \Delta r \Delta z \tag{3.25}$$

$$a_P = a_E + a_W + a_N + a_S + a_T + a_B + a_P^o - S_{\Phi.P} R_p \Delta\theta \Delta r \Delta z \tag{3.26}$$

3.3.2.1. Source Term Linearization

Source term of r and θ momentum equations are also a function of corresponding velocities. There exists two types of treatments when approximating them numerically. Source term can be treated as constant by using the computed velocity from previous iteration. Although this method will slow down the convergence due to its explicit nature. Another better approach is, linearizing it with a negative gradient to maintain the positivity of the coefficient according to the equation (3.27) and table 3.2.

$$S_\phi = S_{\phi.C} + S_{\phi.P}\,\phi \tag{3.27}$$

Among the negative-slope lines, the tangent to the curve represented by S_ϕ is usually the best choice. Steeper lines are acceptable but would normally lead to slower convergence. Less steep lines are undesirable, as they fail to incorporate the given rate of fall of S_ϕ with ϕ. By keeping all these points in mind, the value of $S_{\phi.C}$ and $S_{\phi.P}$ chosen are given in table 3.2. During each iteration cycle these values are recalculated from the new available data.

Table 3.2. The Values of $S_{\phi.C}$ and $S_{\phi.P}$ for different ϕ

ϕ	$S_{\phi.C}$	$S_{\phi.P}$
V_r	$-\frac{\partial P}{\partial r} - \frac{2\mu}{r^2}\frac{\partial V_\theta}{\partial \theta} + \frac{V_\theta^2 \cdot \rho}{r}$	$-\frac{\mu}{r^2}$
V_θ	$-\frac{1}{r}\frac{\partial P}{\partial \theta} + \frac{2\mu}{r^2}\frac{\partial V_r}{\partial \theta} - \frac{\rho V_\theta}{r}\max[-V_r, 0]$	$-\frac{\mu}{r^2} - \frac{\rho}{r}\max[V_r, 0]$
V_z	$-\frac{\partial P}{\partial z}$	0

By taking all the above mentioned points into consideration, the final discrete momentum equations in all three dimension can be obtained from general momentum equation (3.17) for different values of ϕ, as given in equations (3.28, 3.29 and 3.30). These equations are written by implementing an under relaxation factor α_v as suggested by Patankar [1]. Pressure term is separated from $S_{\Phi.C}$, as it plays an important role in developing momentum continuity coupling.

$$\frac{a^r_{i,j,k}}{\alpha_v} V^{r*}_{i,j,k} = \sum_{nb} a^r_{nb} V^{r*}_{nb} + b_r + r\,\Delta\theta\,\Delta z\left(P^*_{i,j,k} - P^*_{i+1,j,k}\right) + \frac{(1-\alpha_v)}{\alpha_v} a^r_{i,j,k} V^{r*(n-1)}_{i,j,k} \tag{3.28}$$

$$\frac{a^\theta_{i,j,k}}{\alpha_v} V^{\theta*}_{i,j,k} = \sum_{nb} a^\theta_{nb} V^{\theta*}_{nb} + b_\theta + \Delta r\,\Delta z\left(P^*_{i,j+1,k} - P^*_{i,j-1,k}\right) + \frac{(1-\alpha_v)}{\alpha_v} a^\theta_{i,j,k} V^{\theta*(n-1)}_{i,j,k} \tag{3.29}$$

$$\frac{a^{z}_{i,j,k}}{\alpha_{v}} V^{z*}_{i,j,k} = \sum_{nb} a^{z}_{nb} V^{z*}_{nb} + b_{z} + r\,\Delta\theta\,\Delta r\left(P^{*}_{i,j,k-1} - P^{*}_{i,j,k+1}\right) + \frac{(1-\alpha_{v})}{\alpha_{v}} a^{z}_{i,j,k} V^{z*(n-1)}_{i,j,k} \tag{3.30}$$

3.3.3. Spatial Discretization of Generalized Momentum Equation Using Finite Difference Method

Finite difference method involves replacing the partial derivatives with a suitable algebraic difference expression in a small finite size element. Most common finite difference representation of derivatives are based on Taylor's series expansion, details of which can be found in all standard CFD text books Anderson [2], therefore, we will not repeat the same exercise here. On the other hand, we will present a complete finite difference discretize scheme for Navier Stokes equation in a cylindrical coordinate system. The most commonly used forms of the derivative expression are given below in equations (3.31-3.33),

$$\frac{\partial\phi}{\partial x} = \frac{\phi_{x+\Delta x} - \phi_{x-\Delta x}}{2\Delta x} \tag{3.31}$$

$$\frac{\partial^2\phi}{\partial x^2} = \frac{\phi_{x+\Delta x} + \phi_{x-\Delta x} - 2\phi_{x}}{(\Delta x)^2} \tag{3.32}$$

$$\frac{\partial^2\phi}{\partial x\partial y} = \frac{\phi_{x+\Delta x,y+\Delta y} + \phi_{x-\Delta x,y-\Delta y} - \phi_{x-\Delta x,y+\Delta y} - \phi_{x+\Delta x,y-\Delta y}}{4\Delta x\Delta y} \tag{3.33}$$

If we replace all the derivative terms of NVS equation according to equation (3.31-3.32), we will have a discrete NVS equation (3.34) through finite difference formulation. Important thing which needs to be pointed out is that if we are using a fully implicit scheme in time, then velocity multiplication in convective term needs to be linearized. A very straight forward practice which is commonly used in literature is that, convective velocity is replaced by the current best known velocity, i.e. velocity from previous iteration. Another interesting thing which reader will have in mind is, in finite difference we have not focused much on positivity of the coefficients i.e. upwind scheme. First it is possible to formulate such scheme and it will result in faster convergence. Secondly, in finite difference method, positivity of the coefficient can also be maintained using finer grid. Equation (3.34) represents corresponding discrete NVS equation and its coefficients are given by equations (3.35-3.43).

$$a_{p}\Phi^{*}_{p} = \sum_{nb} a_{nb}\Phi^{*}_{nb} + b_{\Phi} \tag{3.34}$$

where,

$$a_E = \frac{\mu}{\Delta r^2} + \rho\, V_e^r \tag{3.35}$$

$$a_W = \frac{\mu}{\Delta r^2} + \rho\, V_w^r \tag{3.36}$$

$$a_N = \frac{\mu}{\Delta r^2} + \rho\, V_n^\theta \tag{3.37}$$

$$a_S = \frac{\mu}{\Delta r^2} + \rho\, V_s^\theta \tag{3.38}$$

$$a_T = \frac{\mu}{\Delta r^2} + \rho\, V_t^z \tag{3.39}$$

$$a_B = \frac{\mu}{\Delta r^2} + \rho\, V_b^z \tag{3.40}$$

$$a_P^o = \frac{\rho}{\Delta t} \tag{3.41}$$

$$b_\Phi = a_P^o\, \Phi^*_{p.old} + S_{\Phi.C} \tag{3.42}$$

$$a_P = a_E + a_W + a_N + a_S + a_T + a_B + a_P^o - S_{\Phi.P} \tag{3.43}$$

As explained in finite volume method, source term can be linearized with positive gradient and different momentum equations can be obtained from general momentum equation (3.34) for different values of ϕ as represented by equations (3.44, 3.45 and 3.46).

$$\frac{a^r_{i,j,k}}{\alpha_v} V^{r*}_{i,j,k} = \sum_{nb} a^r_{nb} V^{r*}_{nb} + b_r + \frac{\left(P^*_{i,j,k} - P^*_{i+1,j,k}\right)}{\Delta r} + \frac{(1-\alpha_v)}{\alpha_v} a^r_{i,j,k} V^{r*(n-1)}_{i,j,k} \tag{3.44}$$

$$\frac{a^\theta_{i,j,k}}{\alpha_v} V^{\theta*}_{i,j,k} = \sum_{nb} a^\theta_{nb} V^{\theta*}_{nb} + b_\theta + \frac{\left(P^*_{i,j+1,k} - P^*_{i,j-1,k}\right)}{R_p \Delta\theta} + \frac{(1-\alpha_v)}{\alpha_v} a^\theta_{i,j,k} V^{\theta*(n-1)}_{i,j,k} \tag{3.45}$$

$$\frac{a^z_{i,j,k}}{\alpha_v} V^{z*}_{i,j,k} = \sum_{nb} a^z_{nb} V^{z*}_{nb} + b_z + \frac{\left(P^*_{i,j,k-1} - P^*_{i,j,k+1}\right)}{\Delta z} + \frac{(1-\alpha_v)}{\alpha_v} a^z_{i,j,k} V^{z*(n-1)}_{i,j,k} \tag{3.46}$$

3.3.4. Pressure and Velocity Coupling for Finite Volume and Finite Difference Method

We have established discrete momentum equations by using different techniques as described in the previous two sections. Now these momentum equations can be solved only when the pressure field is given or is somehow estimated. Unless the correct pressure is employed, the resulting velocity will not satisfy the continuity equation. So an improvement in guessed pressure (P^*) is required which is denoted by P^c and is implemented in equation (3.47).

$$P = P^* + P^c \tag{3.47}$$

To estimate the change in behavior of velocity due to this pressure correction, corresponding velocity correction can be introduced in the similar manner, as represented in equation (3.48).

$$V^{\phi} = V^{\phi.*} + V^{\phi.c} \tag{3.48}$$

By implementing these corrections into discrete momentum equation (here finite volume discrete momentum equations are considered, similar formulation can developed for finite difference method), equations for corrected velocities are developed. These corrected velocities needs to satisfy continuity equation too. While introducing these corrected velocities in continuity, the term $\sum_{nb} a_{nb} V^c_{nb}$ is dropped from momentum equations to avoid coupling of all grid points. Equations (3.49-3.51) represent the velocity correction formulas.

$$V^r_{i,j,k} = V^{r*}_{i,j,k} + \frac{r\,\Delta\theta\,\Delta z}{a^r_{i,j,k}}\left(P^c_{i,j,k} - P^c_{i+1,j,k}\right) \tag{3.49}$$

$$V^{\theta}_{i,j,k} = V^{\theta*}_{i,j,k} + \frac{\Delta r\,\Delta z}{a^{\theta}_{i,j,k}}\left(P^c_{i,j+1,k} - P^c_{i,j-1,k}\right) \tag{3.50}$$

$$V^z_{i,j,k} = V^{z*}_{i,j,k} + \frac{r\,\Delta\theta\,\Delta r}{a^z_{i,j,k}}\left(P^c_{i,j,k-1} - P^c_{i,j,k+1}\right) \tag{3.51}$$

This corrected velocity field is implemented in continuity and a pressure correction formula is developed as described by Patankar [1], this method is named as SIMPLE (Semi-Implicit Method for Pressure-Linked Equation) based algorithm. Equation (3.52) represents the final pressure correction formula, which coefficients are given in equations (3.53-3.59).

$$A_p P^c_p = \sum_{nb} A_{nb} P^c_{nb} + B_p \tag{3.52}$$

where,

$$A_E = \frac{\rho\left(R_p\,\Delta\theta\,\Delta z\right)^2}{\left(a_{r.P}\right)_e} \tag{3.53}$$

$$A_W = \frac{\left(R_p\,\Delta\theta\,\Delta z\right)^2}{\left(a_{r.P}\right)_w} \tag{3.54}$$

$$A_N = \frac{(\Delta r\ \Delta z)^2}{(a_{\theta.P})_n} \tag{3.55}$$

$$A_S = \frac{(\Delta r\ \Delta z)^2}{(a_{\theta.P})_s} \tag{3.56}$$

$$A_T = \frac{(R_p\ \Delta\theta\ \Delta r)^2}{(a_{z.P})_t} \tag{3.57}$$

$$A_B = \frac{(R_p\ \Delta\theta\ \Delta r)^2}{(a_{z.P})_b} \tag{3.58}$$

$$B_p = \left(R_e V_e^r - R_w V_w^r\right)\Delta\theta\ \Delta z + \left(V_n^\theta - V_s^\theta\right)\Delta r\ \Delta z + \left(V_t^z - V_b^z\right) R_P\ \Delta\theta\ \Delta r \tag{3.59}$$

When improving the guessed pressure using the calculated correction from equation (3.52), an under-relaxation factor α_p is implemented, as shown in equation (3.60).

$$P = P^* + \alpha_p P^c \tag{3.60}$$

Although the value of the under-relaxation factor suggested by Patankar [1] is 0.8, often, this value can be as low as a 0.1 as discussed by Anderson [2], especially in finite difference formulation.

3.3.5. Spatial Discretization of Generalized Momentum Equation Using Finite Element Method

Finite element method is a technique which tries to approximate the solution of a partial differential equation using variational calculus. The basic idea is to minimize an error function to formulate a stable solution to the equation. The important thing to be stated about the finite element method is that it does not impose any constrain on the grid, therefore, any complex geometry can be treated by using structured or unstructured grid. There are several finite element algorithms used in literature, one of the very often used algorithm is Galerkin approximation which we will describe in this section.

Let's recall equation (3.10) for the FEM formulation with pressure source term separated from source S , remaining source term is represented by f.

$$
\begin{array}{llll}
\rho\,\partial_t u + \rho\, u.\nabla u - \mu\,\Delta u + \rho\,\Delta p = f & \text{In } \Omega\times(0,T) & \text{NVS} & \\
\nabla.u = 0 & \text{In } \Omega\times(0,T) & \text{Continuity} & (3.61) \\
u = 0 & \text{In } \Gamma\times(0,T) & \text{(Boundary Condition)} & \\
u = u_o & \text{In } \Omega, t=0 & \text{(Initial Condition)} &
\end{array}
$$

Functional space of the solution where these equations are meaningful can be given by equation (3.62),

$$
\begin{aligned}
& u \in L^2(0,T;H_0^1(\Omega)) \cap L^\infty(0,T;L^2(\Omega)) \\
& p \in L^1(0,T;L_0^2(\Omega)) \\
& \text{where} \\
& L^p(0,T;X) := \left\{ f:(0,T) \to X \middle| \int_0^T \|f\|_X^p \, dt \lhd \infty \right\}
\end{aligned}
\qquad (3.62)
$$

For setting up the spatial FE approximation, the first step is to rewrite the generalized vector form of NVS and continuity equations in its variational or weak form.

3.3.5.1. Weak form of NVS

Now let's assume a pair of function (v, q) is a solution to the Navier Stokes equations, then it will satisfy these equations in point wise manner. Certainly, this pair will still be a solution if we multiply the Navier Stokes equations by any functions we like. In particular, we could multiply the momentum equation by any velocity basis function v and the continuity equation by any pressure basis function q. Therefore,

$$
\begin{aligned}
& \rho\int_\Omega v.\partial_t u + \rho\int_\Omega v.(u.\nabla)u - \mu\int_\Omega v.\Delta u + \rho\int_\Omega v.\Delta p = \int_\Omega v.f \\
& \int_\Omega q.\Delta.u = 0
\end{aligned}
\qquad (3.63)
$$

Using integral by parts,

$$
\begin{aligned}
& \int_\Omega v.\Delta u = -\int_\Omega \nabla v : \nabla u + \int_\Gamma v.n.\nabla u \\
& \int_\Omega v.\Delta p = -\int_\Omega p.\nabla v + \int_\Gamma p\,n.v
\end{aligned}
\qquad (3.64)
$$

Now in order to approximate the boundary integral term of equation (3.64), let's divide our boundary such that $\Gamma = \Gamma_D \cup \Gamma_N$. If v is such that v=0 on Γ_D and we make use of the Neumann boundary condition:

$$\int_{\Omega} v.\Delta u + \int_{\Omega} v.\Delta p = -\int_{\Omega} \nabla v:\nabla u - \int_{\Omega} p.\nabla v + \int_{\Gamma_N} v.h \tag{3.65}$$

where h is the orthogonal projection of Chauchy stress on Neumann boundary which can be introduced through boundary conditions.

Introducing the notation,

$$\begin{aligned} &\int_{\Omega}(v.u) = (u,v) && L^2 \text{ inner product} \\ &\rho\int_{\Omega} v.(u.\nabla)u = c(u,u,v) && \text{convective term} \\ &\mu\int_{\Omega}(\nabla v:\nabla u) = a(u,v) && \text{Viscus term} \\ &\int_{\Omega}(p\nabla.v) = b(p,v) \\ &\int_{\Omega}\rho f.v + \int_{\Gamma} v.h = l(v) \end{aligned} \tag{3.66}$$

We can rewrite the weak form of Navier Stokes equation as,

$$\rho\frac{d}{dt}(u,v) + c(u,u,v) + a(u,v) - b(p,v) = l(v) \tag{3.67}$$

$$b(q,u) = 0 \tag{3.68}$$

The interesting thing to be noticed between Normal NVS equation and its weak form is that, weak form contains maximum to the first order derivatives of the velocity. Now in order to discretize weak form of NVS in space, different types of velocity and pressure interpolation or basis function can be chosen according the choice of element or grid. There are basically two sorts of interpolation which are used in literature, discontinuous pressure interpolation and continuous pressure interpolation. Different types of element with pressure-velocity interpolation are classified in the figure bellow.

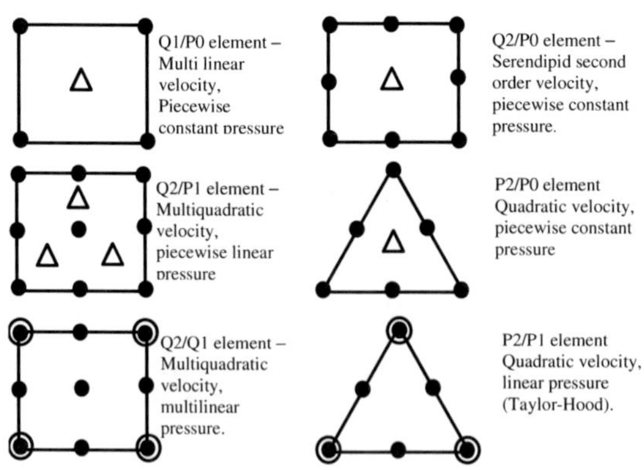

Figure 3.5. Different types of elements and velocity pressure interpolation used in finite element method.

3.3.5.2. Galerkin Finite Element Approximation

Galerkin finite element approximation involves the construction of the finite dimensional subspace $V^N \subset V$. We will call it V_h. Let Ω^e be a finite element partition of the domain Ω, with index e ranging from 1 to the number of elements n_{el}. We denote with a subscript *h* finite element quantities associated to Ω^e, with

$$h = \max_{e=1.......n_{el}} \dim(\Omega^e) \tag{3.69}$$

The discrete finite element problem is: find $u_h \in v_h$ such that

$$\rho\frac{d}{dt}(u_h, v_h) + c(u_h, u_h, v_h) + a(u_h, v_h) - b(p, v_h) = 1(v_h) \quad \forall v_h \in V_h \tag{3.70}$$

Let N^a be the standard shape (basis) function of node a, then functions in V_h can be interpolated by using these shape functions:

$$u_h = \sum_{a=1}^{n_{pts}} N^a(X)U^a \qquad v_h = \sum_{a=1}^{n_{pts}} N^a(X)V^a \tag{3.71}$$

Therefore,

$$a(u_h, v_h) = \sum_{a=1}^{n_{pts}} V^b a(N^a, N^b)U^a \tag{3.72}$$

$$1(v_h) = \sum_{a=1}^{n_{pts}} V^b 1(N^b) \tag{3.73}$$

$$c(u_h, u_h, v_h) = \sum_{a=1}^{n_{pts}} V^b c(N^a, a\, N^a, N^b)U^a \tag{3.74}$$

$$\rho\frac{d}{dt}(u_h, v_h) = \frac{d}{dt}\sum_{a=1}^{n_{pts}} V^b (N^a, N^b)U^a \tag{3.75}$$

$$b(p, v_h) = \sum_{a=1}^{n_{pts}} P^a (N^a, N^b)V^b \tag{3.76}$$

$$\begin{aligned} &\frac{d}{dt}\sum_{a=1}^{n_{pts}} V^b (N^a, N^b)U^a + \sum_{a=1}^{n_{pts}} V^b c(N^a, a\, N^a, N^b)U^a + \\ &\sum_{a=1}^{n_{pts}} V^b a(N^a, N^b)U^a - \sum_{a=1}^{n_{pts}} P^a (N^a, N^b)V^b = \sum_{a=1}^{n_{pts}} V^b 1(N^b) \end{aligned} \tag{3.77}$$

N^a, N^b are the basis function therefore, they only depend on space. Due to this fact, when putting the value of N^a, N^b into the equation (3.77), it will transform into a linear system, represented by (3.78) which can be solved by using iterative process.

$$\mathrm{AU} = \mathrm{F} \tag{3.78}$$

This is a very simple Galerkin formulation of FEM. When going into deep, there are further interesting topics that need to be understood in order to solve a problem through finite element approximation such as stability issue, convergence issue of FEM and coupling of continuity through fractional step method. Details of these topics can be found in O.C. Zienkiewicz [3].

3.4. Solving a Finite Element Problem via Finite Volume Through Coordinate Transformation

One of the major drawback of using finite volume technique is the requirement of the orthogonality of the grid. If the control volume faces are not orthogonal to each other then the component of the flux derivative parallel to control volume face also needs to be considered. Their computation will be much more complicated. It will make more grid points coupled together, making the pressure correction formulation next to impossible. Therefore, an alternative to such a situation is grid transformation. The idea is to transform the non-orthogonal physical domain into orthogonal computational domain. Here we present an example for complete formulation of Navier Stokes using finite volume method in such a non-orthogonal physical domain.

Figure 3.6 represents a tilted slipper swash plat assembly. As it can be seen from figure the domain under consideration is a non-orthogonal domain (Ω_{out}), all the equations (Navier Stokes and continuity) need to be transformed according to the transformation given in table 3.3 before implementing control volume formulation. The non-orthogonal domain will be represented by $\Omega_\mathrm{p}\left\{\Omega:\Re^4\left(r,\theta,z,t\right)\right\}$ namely the physical domain and the orthogonal domain will be represented by $\Omega_\mathrm{c}\left\{\Omega:\Re^4\left(R,\xi,\eta,\tau\right)\right\}$ namely the computational domain.

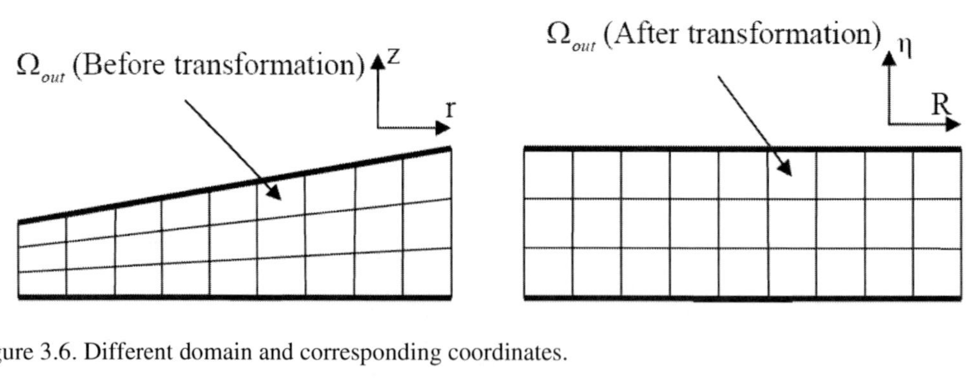

Figure 3.6. Different domain and corresponding coordinates.

Navier Stokes equation and continuity equations, which are represented by equations (3.8) and (3.9) on physical domain Ω_p are transformed into equations (3.79) & (3.80) on domain Ω_c. The value of source term S_ϕ corresponding to the value of ϕ is given in table 3.4. The transformation is performed by using covariant property of derivatives.

$$\rho\left[\frac{\partial\phi}{\partial\tau}+\frac{1}{R}\frac{\partial(RV_R\phi)}{\partial R}+\frac{1}{R}\frac{\partial(V_\xi\phi)}{\partial\xi}+\frac{1}{H}\frac{\partial(V_\eta\phi)}{\partial\eta}+\frac{\eta\alpha}{H}\left\{\sin\xi\frac{\partial(V_\xi\phi)}{\partial\eta}-\cos\xi\frac{\partial(V_R\phi)}{\partial\eta}\right\}\right]=$$
$$\mu\left[\frac{1}{R}\frac{\partial}{\partial R}\left(R\frac{\partial\phi}{\partial R}\right)+\frac{1}{R^2}\frac{\partial^2\phi}{\partial\xi^2}+\frac{1+(\eta\alpha)^2}{H^2}\frac{\partial^2\phi}{\partial\eta^2}+\frac{2\eta\alpha^2}{H^2}\frac{\partial\phi}{\partial\eta}+\frac{2\eta\alpha}{H}\left\{\frac{\sin\xi\,\partial^2\phi}{R\,\partial\xi\partial\eta}-\frac{\cos\xi\,\partial^2\phi}{\partial R\partial\eta}\right\}\right]+S_\phi \tag{3.79}$$

$$\left[\frac{1}{R}\frac{\partial(RV_R)}{\partial R}-\frac{\eta\alpha\cos\xi}{H}\frac{\partial V_R}{\partial\eta}\right]+\left[\frac{1}{R}\frac{\partial V_\xi}{\partial\xi}+\frac{\eta\alpha\sin\xi}{H}\frac{\partial V_\xi}{\partial\eta}\right]+\frac{1}{H}\frac{\partial V_\eta}{\partial\eta}=0 \tag{3.80}$$

Table 3.3. $\Omega_p\left\{\Omega:\Re^4(r,\theta,z,t)\right\}\rightarrow\Omega_c\left\{\Omega:\Re^4(R,\xi,\eta,\tau)\right\}$

Physical domain.	Computational domain.
R	R
θ	ξ
Z	$\frac{z}{(ho+\alpha r\cos\theta)}$
T	τ

Table 3.4. Value of S_ϕ for different ϕ in computational domain Ω_c

ϕ	S_ϕ
V_R	$\left[-\frac{\partial P}{\partial R}+\frac{\eta\alpha\cos\xi}{H}\frac{\partial P}{\partial\eta}\right]-\frac{\mu V_R}{R^2}+\frac{\rho V_\xi^2}{R}-\frac{2\mu}{R^2}\frac{\partial V_\xi}{\partial\xi}-\frac{2\mu\eta\alpha\sin\xi}{RH}\frac{\partial V_\xi}{\partial\eta}$
V_ξ	$\left[-\frac{1}{R}\frac{\partial P}{\partial\xi}-\frac{\eta\alpha\sin\xi}{H}\frac{\partial P}{\partial\eta}\right]-\frac{\mu V_\xi}{R^2}-\frac{\rho V_R V_\xi}{R}+\frac{2\mu}{R^2}\frac{\partial V_R}{\partial\xi}+\frac{2\mu\eta\alpha\sin\xi}{RH}\frac{\partial V_R}{\partial\eta}$
V_η	$-\frac{1}{H}\frac{\partial P}{\partial\eta}$

3.4.1. Source Term Linearization for Transformed NVS equations

We have already explained the source term linearization of the NVS in section 3.3.2. A similar concept of negative gradient is applied while linearizing the source terms presented in table 3.4 and corresponding values of $S_{\phi.C}$ and $S_{\phi.P}$ are given in table 3.5.

Table 3.5. $S_{\phi.C}$ and $S_{\phi.P}$ **for different** ϕ

ϕ	$S_{\phi.C}$	$S_{\phi.P}$
V_r	$\left[-\frac{\partial P}{\partial R}+\frac{\eta\alpha\cos\xi}{H}\frac{\partial P}{\partial\eta}\right]+\frac{\rho V_\xi^2}{R}-\frac{2\mu}{R^2}\frac{\partial V_\xi}{\partial\xi}-\frac{2\mu\eta\alpha\sin\xi}{RH}\frac{\partial V_\xi}{\partial\eta}$	$-\frac{\mu}{R^2}$
V_θ	$\left[-\frac{1}{R}\frac{\partial P}{\partial\xi}-\frac{\eta\alpha\sin\xi}{H}\frac{\partial P}{\partial\eta}\right]+\frac{2\mu}{R^2}\frac{\partial V_R}{\partial\xi}+\frac{2\mu\eta\alpha\sin\xi}{RH}\frac{\partial V_R}{\partial\eta}-\frac{\rho V_\xi}{R}\max[-V_R,0]$	$-\frac{\mu}{R^2}-\frac{\rho}{R}\max[V_R,0]$
V_z	$-\frac{1}{H}\frac{\partial P}{\partial\eta}$	0

3.4.2. Spatial Discretization of Generalized Transformed Momentum Equation

As described earlier, for better understanding of the flow phenomena and to implement the computational technique more precisely, NVS equation is divided into a combination of different terms, Transient term, Convection term (T_1), Diffusion term (T_2) and Source term (S).

In the equations presented in this section, the transient term is treated by implicit scheme for faster convergence; an upwind scheme is implemented for the convection term to maintain the positivity of the coefficients, central difference scheme is implemented for the diffusion term and the source term is treated by implicit scheme. Convection and diffusion terms are represented by equation (3.81) and (3.82).

$$T_1=\rho\left[\frac{1}{R}\frac{\partial(RV_R\phi)}{\partial R}+\frac{1}{R}\frac{\partial(V_\xi\phi)}{\partial\xi}+\frac{1}{H}\frac{\partial(V_\eta\phi)}{\partial\eta}+\frac{\eta\alpha}{H}\left\{\sin\xi\frac{\partial(V_\xi\phi)}{\partial\eta}-\cos\xi\frac{\partial(V_R\phi)}{\partial\eta}\right\}\right] \tag{3.81}$$

$$T_2=\mu\left[\frac{1}{R}\frac{\partial}{\partial R}\left(R\frac{\partial\phi}{\partial R}\right)+\frac{1}{R^2}\frac{\partial^2\phi}{\partial\xi^2}+\frac{1+(\eta\alpha)^2}{H^2}\frac{\partial^2\phi}{\partial\eta^2}+\frac{2\eta\alpha^2}{H^2}\frac{\partial\phi}{\partial\eta}+\frac{2\eta\alpha}{H}\left\{\frac{\sin\xi\,\partial^2\phi}{R\,\partial\xi\partial\eta}-\frac{\cos\xi\,\partial^2\phi}{\partial R\partial\eta}\right\}\right] \tag{3.82}$$

Different terms of equation (3.79) are discretized by control volume formulation over a staggered grid. The final discrete equation can be written in the form of equation (3.83) and its coefficients are given by equations (3.84) to (3.107).

$$\frac{a_p}{\alpha_v}\Phi_p^*=\sum_{nb}a_{nb}\Phi_{nb}^*+b_\Phi+\frac{(1-\alpha_v)}{\alpha_v}a_{\phi.p}\Phi_{p.old}^* \tag{3.83}$$

where:

$$C_1=\frac{\rho\,\eta_P\,\alpha\,R_P\,\Delta\xi\,\Delta R}{2} \tag{3.84}$$

$$F_e = \rho R_e V_e^R H_P \Delta\xi \Delta\eta \qquad D_e = \frac{\mu R_e H_P \Delta\xi \Delta\eta}{\Delta R} \tag{3.85}$$

$$a_E = D_e + \max(-F_e, 0) + C_1 \left[\max(\cos\xi, 0) \max(-V_b^R, 0) + \max(-\cos\xi, 0) \max(-V_t^R, 0) \right] \tag{3.86}$$

$$F_w = \rho R_w V_w^R H_P \Delta\xi \Delta\eta \qquad D_w = \frac{\mu R_w H_P \Delta\xi \Delta\eta}{\Delta R} \tag{3.87}$$

$$a_W = D_w + \max(F_w, 0) + C_1 \left[\max(-\cos\xi, 0) \max(V_b^R, 0) + \max(\cos\xi, 0) \max(V_t^R, 0) \right] \tag{3.88}$$

$$F_n = \rho V_n^\xi H_P \Delta R \Delta\eta \qquad D_n = \frac{\mu H_P \Delta R \Delta\eta}{R_P \Delta\xi} \tag{3.89}$$

$$a_N = D_n + \max(-F_n, 0) + C_1 \left[\max(\sin\xi, 0) \max(-V_t^\xi, 0) + \max(-\sin\xi, 0) \max(-V_b^\xi, 0) \right] \tag{3.90}$$

$$F_s = \rho V_s^\xi H_P \Delta R \Delta\eta \qquad D_s = \frac{\mu H_P \Delta R \Delta\eta}{R_P \Delta\xi} \tag{3.91}$$

$$a_S = D_s + \max(F_s, 0) + C_1 \left[\max(-\sin\xi, 0) \max(V_t^\xi, 0) + \max(\sin\xi, 0) \max(V_b^\xi, 0) \right] \tag{3.92}$$

$$D_t = D_b = \frac{\mu \Delta R R_P \Delta\xi}{H_P \Delta\eta} \qquad F_t = \rho V_t^\eta \Delta R R_P \Delta\xi \qquad F_b = \rho V_b^\eta \Delta R R_P \Delta\xi \tag{3.93}$$

$$a_T = D_t \left[1 + (\eta_t \alpha)^2 \right] + \max(-F_t, 0) \tag{3.94}$$

$$a_B = D_b \left[1 + (\eta_b \alpha)^2 \right] + \max(F_b, 0) \tag{3.95}$$

$$C_2 = \frac{\mu \eta_P \alpha R_P}{2} \tag{3.96}$$

$$a_{ET} = C_1 \max(-\cos\xi, 0) \max(-V_t^R, 0) - C_2 \cos\xi R \Delta\xi \tag{3.97}$$

$$a_{WT} = C_1 \max(\cos\xi, 0) \max(V_t^R, 0) + C_2 \cos\xi R \Delta\xi \tag{3.98}$$

$$a_{NT} = C_1 \max\left(\sin\xi, 0\right) \max\left(-V_t^{\xi}, 0\right) + C_2 \sin\xi \, \Delta R \tag{3.99}$$

$$a_{ST} = C_1 \max\left(-\sin\xi, 0\right) \max\left(V_t^{\xi}, 0\right) - C_2 \sin\xi \, \Delta R \tag{3.100}$$

$$a_{EB} = C_1 \max\left(\cos\xi, 0\right) \max\left(-V_b^{R}, 0\right) + C_2 \cos\xi \, R \, \Delta\xi \tag{3.101}$$

$$a_{WB} = C_1 \max\left(-\cos\xi, 0\right) \max\left(V_b^{R}, 0\right) - C_2 \cos\xi \, R \, \Delta\xi \tag{3.102}$$

$$a_{NB} = C_1 \max\left(-\sin\xi, 0\right) \max\left(-V_b^{\xi}, 0\right) - C_2 \sin\xi \, \Delta R \tag{3.103}$$

$$a_{SB} = C_1 \max\left(\sin\xi, 0\right) \max\left(V_b^{\xi}, 0\right) + C_2 \sin\xi \, \Delta R \tag{3.104}$$

$$a_P^o = \frac{\rho \, R_p \, \Delta\xi \, \Delta R \, \Delta\eta}{\Delta\tau} \tag{3.105}$$

$$b_{\Phi} = a_P^o \, \Phi^*_{p.old} + S_{\Phi.C} \, R_p \, \Delta\xi \, \Delta R \, \Delta\eta \tag{3.106}$$

$$a_P = \sum_{nb} a_{nb} + a_P^o - S_{\Phi.P} \, R_p \, \Delta\theta \, \Delta r \, \Delta z \tag{3.107}$$

3.4.3. Different Transformed Momentum Equations in Discrete Form

Different momentum equations can be obtained from general momentum equation (3.82) for different values of ϕ as represented in equations (3.107, 3.108 and 3.109).

$$a^{r}_{i,j,k} V^{R*}_{i,j,k} = \sum_{nb} a_{nb} V^{R*}_{nb} + R_{m(i)} \Delta\xi \left[\left(P^*_{i,j,k} - P^*_{i+1,j,k} \right) H^R_{i,j} \Delta\eta + \frac{\eta_k \alpha \cos\xi_j \Delta R}{4} \left(P^*_{i,j,k+1} + P^*_{i+1,j,k+1} - P^*_{i,j,k-1} - P^*_{i+1,j,k-1} \right) \right] \tag{3.108}$$

$$a^{\xi}_{i,j,k} V^{\xi*}_{i,j,k} = \sum_{nb} a_{nb} V^{\xi*}_{nb} + \Delta R \left[\left(P^*_{i,j,k} - P^*_{i,j+1,k} \right) H^{\xi}_{i,j} \Delta\eta - \frac{\eta_k \alpha \sin\xi^m_j \Delta\xi}{4} \left(P^*_{i,j,k+1} + P^*_{i,j+1,k+1} - P^*_{i,j,k-1} - P^*_{i,j+1,k-1} \right) \right] \tag{3.109}$$

$$a^{\eta}_{i,j,k} V^{\eta*}_{i,j,k} = \sum_{nb} a_{nb} V^{\eta*}_{nb} + \left(P^*_{i,j,k+1} - P^*_{i,j,k} \right) R_i \Delta\xi \Delta R \tag{3.110}$$

3.4.4. Pressure and Velocity Coupling for Transformed Equation

When implementing similar formulation which was explain in section 3.3.4, different velocity correction equations can be given as in equations (3.111-3.113). The major difference as it can be seen when comparing the equations (3.111-3.113) to the equations (3.49-3.51) is that now pressure at any grid point p is not only a function of 8 neighbour points but 24 neighbouring points. Figure 3.7 shows these points and their nomenclature used in this chapter.

$$V^{R}_{i,j,k}=V^{R*}_{i,j,k}+\frac{R_{m(i)}\Delta\xi}{a^{r}_{i,j,k}}\left[\left(P^{c}_{i,j,k}-P^{c}_{i+1,j,k}\right)H^{R}_{i,j}\Delta\eta+\frac{\eta_{k}\alpha\cos\xi_{j}\Delta R}{4}\left(P^{c}_{i,j,k+1}+P^{c}_{i+1,j,k+1}-P^{c}_{i,j,k-1}-P^{c}_{i+1,j,k-1}\right)\right] \quad (3.111)$$

$$V^{\xi}_{i,j,k}=V^{\xi*}_{i,j,k}+\frac{\Delta R}{a^{\xi}_{i,j,k}}\left[\left(P^{c}_{i,j,k}-P^{c}_{i,j+1,k}\right)H^{\xi}_{i,j}\Delta\eta-\frac{\eta_{k}\alpha\sin\xi^{m}_{j}\Delta\xi}{4}\left(P^{c}_{i,j,k+1}+P^{c}_{i,j+1,k+1}-P^{c}_{i,j,k-1}-P^{c}_{i,j+1,k-1}\right)\right] \quad (3.112)$$

$$V^{\eta}_{i,j,k}=V^{\eta*}_{i,j,k}+\frac{R_{i}\Delta\xi\Delta R}{a^{\eta}_{i,j,k}}\left(P^{c}_{i,j,k+1}-P^{c}_{i,j,k}\right) \quad (3.113)$$

These velocities are implemented into the continuity equation (3.80) to obtain the pressure correction formula. The final pressure correction equation can be written in the form of equation (3.114) and its coefficients are given by equations (3.115-3.139).

$$A_{p}P^{c}_{p}=\sum_{nb}A_{nb}P^{c}_{nb}+B_{p} \quad (3.114)$$

where,

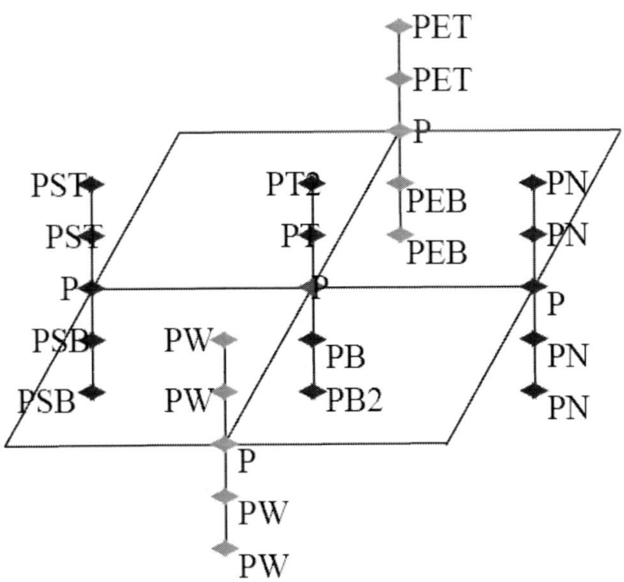

Figure 3.7. Nomenclature of Pressure crrection neighbor points on Transformed grid.

$$A_{PET2} = \frac{\eta_{k+1}\eta_k R_{m(i)} R_i \left(\alpha \cos\xi_j \Delta R \Delta\xi\right)^2}{16\, arp_{(i,j,k+1)}} \tag{3.115}$$

$$A_{PEB2} = \frac{\eta_{k-1}\eta_k R_{m(i)} R_i \left(\alpha \cos\xi_j \Delta R \Delta\xi\right)^2}{16\, arp_{(i,j,k-1)}} \tag{3.116}$$

$$A_{PE} = \frac{H_{i,j} R^2_{m(i)} \Delta\eta \left(\Delta\xi\right)^2}{arp_{i,j,k}} - \left(A_{PET2} + A_{PEB2}\right) \tag{3.117}$$

$$A_{PWT2} = \frac{\eta_{k+1}\eta_k R_{m(i-1)} R_i \left(\alpha \cos\xi_j \Delta R \Delta\xi\right)^2}{16\, arp_{(i-1,j,k+1)}} \tag{3.118}$$

$$A_{PWB2} = \frac{\eta_{k-1}\eta_k R_{m(i-1)} R_i \left(\alpha \cos\xi_j \Delta R \Delta\xi\right)^2}{16\, arp_{(i-1,j,k-1)}} \tag{3.119}$$

$$A_{PW} = \frac{H_{i,j} R^2_{m(i-1)} \Delta\eta \left(\Delta\xi\right)^2}{arp_{i-1,j,k}} - \left(A_{PWT2} + A_{PWB2}\right) \tag{3.120}$$

$$A_{PNT2} = \frac{\eta_{k+1}\eta_k \sin\xi^m_j \sin\xi_j \left(\alpha\, R_i \Delta R \Delta\xi\right)^2}{16\, ayp_{(i,j,k+1)}} \tag{3.121}$$

$$A_{PNB2} = \frac{\eta_{k-1}\eta_k \sin\xi^m_j \sin\xi_j \left(\alpha\, R_i \Delta R \Delta\xi\right)^2}{16\, ayp_{(i,j,k-1)}} \tag{3.122}$$

$$A_{PN} = \frac{H^{\xi}_{i,j} H_{i,j} \left(\Delta R \Delta\eta\right)^2}{ayp_{i,j,k}} - \left(A_{PNT2} + A_{PNB2}\right) \tag{3.123}$$

$$A_{PST2} = \frac{\eta_{k+1}\eta_k \sin\xi^m_{j-1} \sin\xi_j \left(\alpha\, R_i \Delta R \Delta\xi\right)^2}{16\, ayp_{(i,j-1,k+1)}} \tag{3.124}$$

$$A_{PSB2} = \frac{\eta_{k-1}\eta_k \sin\xi^m_{j-1} \sin\xi_j \left(\alpha\, R_i \Delta R \Delta\xi\right)^2}{16\, H_{i,j}\ ayp_{(i,j-1,k-1)}} \tag{3.125}$$

$$A_{PS} = \frac{H^{\xi}_{i,j-1} H_{i,j} \left(\Delta R \Delta \eta\right)^2}{ayp_{i,j-1,k}} - \left(A_{PST2} + A_{PSB2}\right) \tag{3.126}$$

$$A_{PT2} = A_{PET2} + A_{PWT2} + A_{PNT2} + A_{PST2} \tag{3.127}$$

$$A_{PB2} = A_{PEB2} + A_{PWB2} + A_{PNB2} + A_{PSB2} \tag{3.128}$$

$$A_{PEB} = \frac{\eta_k \alpha\, \Delta R\, \Delta\eta \cos\xi_j \left(\Delta\xi\right)^2}{4} \left(\frac{H_{i,j} R^2_{m(i)}}{arp_{i,j,k}} + \frac{H^R_{i,j} R_{m(i)} R_i}{arp_{i,j,k-1}} \right) \tag{3.129}$$

$$A_{PET} = -A_{PEB} \tag{3.130}$$

$$A_{PWB} = -\frac{\eta_k \alpha\, \Delta R\, \Delta\eta \cos\xi_j \left(\Delta\xi\right)^2}{4} \left(\frac{H_{i,j} R^2_{m(i-1)}}{arp_{i-1,j,k}} + \frac{H^R_{i-1,j} R_{m(i-1)} R_i}{arp_{i-1,j,k-1}} \right) \tag{3.131}$$

$$A_{PWT} = -A_{PWB} \tag{3.132}$$

$$A_{PNT} = \frac{\eta_k \alpha R_i\, \Delta\xi\, \Delta\eta \left(\Delta R\right)^2}{4} \left(\frac{\sin\xi^m_j H_{i,j}}{ayp_{i,j,k}} + \frac{\sin\xi_j H^{\xi}_{i,j}}{ayp_{(i,j,k+1)}} \right) \tag{3.133}$$

$$A_{PNB} = -A_{PNT} \tag{3.134}$$

$$A_{PST} = -\frac{\eta_k \alpha R_i\, \Delta\xi\, \Delta\eta \left(\Delta R\right)^2}{4} \left(\frac{\sin\xi^m_{j-1} H_{i,j}}{ayp_{i,j-1,k}} + \frac{\sin\xi_j H^{\xi}_{i,j-1}}{ayp_{(i,j-1,k+1)}} \right) \tag{3.135}$$

$$A_{PSB} = -A_{PST} \tag{3.136}$$

$$\begin{aligned} A_{PT} = {} & \frac{\eta_k \alpha \cos\xi_j \Delta R \Delta\eta \left(\Delta\xi\right)^2}{4} \left[H_{i,j} \left(\frac{R^2_{m(i-1)}}{arp_{i-1,j,k}} - \frac{R^2_{m(i)}}{arp_{i,j,k}} \right) + R_i \left(\frac{H^R_{i,j} R_{m(i)}}{arp_{i,j,k+1}} - \frac{H^R_{i-1,j} R_{m(i-1)}}{arp_{i-1,j,k+1}} \right) \right] \\ & + \frac{\eta_k \alpha R_i \left(\Delta R\right)^2 \Delta\eta \Delta\xi}{4} \left[H_{i,j} \left(\frac{\sin\xi^m_j}{ayp_{i,j,k}} - \frac{\sin\xi^m_{j-1}}{ayp_{i,j-1,k}} \right) + \sin\xi_j \left(\frac{H^{\xi}_{i,j-1}}{ayp_{(i,j-1,k+1)}} - \frac{H^{\xi}_{i,j}}{ayp_{(i,j,k+1)}} \right) \right] + \frac{\left(R_i \Delta R \Delta\xi\right)^2}{azp_{(i,j,k+1)}} \end{aligned} \tag{3.137}$$

$$A_{PB} = -A_{PT} + \frac{\left(R_i \Delta R \Delta\xi\right)^2}{azp_{(i,j,k+1)}} + \frac{\left(R_i \Delta R \Delta\xi\right)^2}{azp_{(i,j,k-1)}} \tag{3.138}$$

$$A_P = \sum_{nb} A_{nb} \tag{3.139}$$

3.5. Convergence Criteria

When doing simulation for a flow problem, an important thing to notice is the convergence of the solution. Convergence means that the solution is improving along with the iterations meaning residual of different equations is decreasing with time.

There are several measures which can be used to decide the convergence in a simulation i.e. residue of momentum equations, residue of continuity and the value of pressure correction. Equation (3.140) represents the residual of momentum equation.

$$R_{p.\phi} = a_{\phi.p} V_{\phi.p} - \sum_{nb} a_{\phi.nb} V_{\phi.nb} - b_{\phi} - (\Delta P) A_n \tag{3.140}$$

Once we know that solution is improving in iteration, another interesting thing need to be established is, how to identify if the convergence is achieved, meaning solution stops improving or improves at a very slow rate General practice is that once the improvement in any normalized quantity in successive iterations reaches below 10^{-6}, solution is assumed to have reached the convergence. In certain cases stream line plot can also be used as a convergence criteria.

After convergence is achieved, obtained simulation results need to be tested for their accuracy. One of the best way is to directly compare them against experimental results but that is not a feasible option in all the cases. An alternative option is to run some bench mark test for simplified problem and compare them against the published results in literature. In any case, one of the test which needs to be done in each and every simulation is, ensuring that the simulation results does not depends on the grid size.

3.5.1. Grid Independency Test

Grid independency test involves, running the simulation over different grid sizes and comparing the results. For a successful simulation, such results needs to be independent of grid sizes used.

For example in figure 3.8, 2D stream line plots are shown for different grid sizes inside the groove of a slipper. Figure 3.8.a and 3.8.b correspond to 30-36-60 and 30-36-80 grid size in r-θ -z direction inside the grove and 312-36-20 outside the groove, Figure 8.c presents same results when grid refinement is in r direction, with 45-36-60 grid points inside the groove and 468-36-20 outside the groove to maintain uniform grid in r direction. Figure 8.d shows the results with 30-36-60 grid points inside the groove and 312-36-30 grid points outside the groove. From all four figures, it can be concluded that the solution is independent of the grid size chosen.

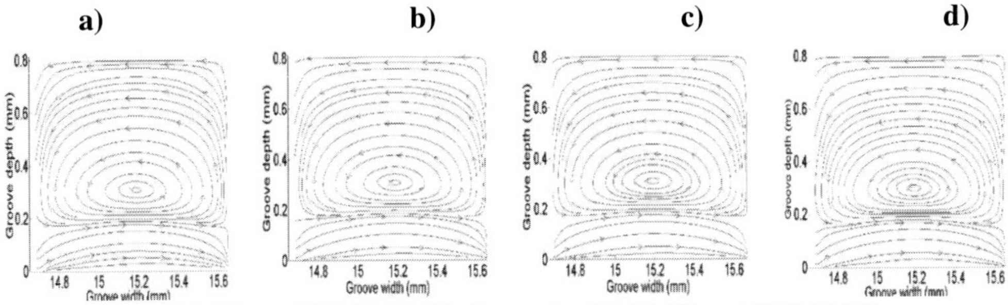

(a). Groove in- [30-36-60], out-[312-36-20] (b). Groove in- [30-36-80], out-[312-36-20]
(c). Groove in- [45-36-60], out-[468-36-20] (d). Groove in- [30-36-60], out-[312-36-30]

Figure 3.8. Stream lines inside the groove for different grid size in static conditions at 13 MPa inlet pressure and 20 micron clearance.

3.6. Closing Remarks

We have covered three major computational techniques which are used in CFD to solve different practical problems. There are few more topics which require further attention to solve a CFD problem successfully.

3.6.1. Solving a Steady and Transient Flow Problem

When developing an analytical equation for a steady state flow, we assume the time varying component of the equation to be zero, therefore, simplify the equation to a great extent. On the other hand when solving a steady state problem numerically, time step works as an iteration gap between two iterations, therefore, even when simulating the steady flow we discretize the transient NVS equation.

3.6.2. Mesh Topology

Numerical methods such as the finite difference method, finite-volume method, and finite element method were defined on meshes of data points. In such a mesh, each point has a fixed number of predefined neighbors, and this connectivity between neighbors can be used to define mathematical operators like the derivative. These operators are then used to construct the equations to be simulated, such as the Navier–Stokes equations.

The origin of the term mesh (or grid) goes back to early days of CFD when most analyses were done in 2D. For 2D analysis, a domain split into elements resembles a wire mesh, hence the name. As CFD has developed, better algorithms and more computational power has become available to researchers. One of the direct results of this development has been the expansion of available mesh elements and mesh connectivity.

The elements in a mesh can be classified either by their shape or by their connectivity. Common elements in 2D are triangles or rectangles, and common elements in 3D are tetrahedra or bricks. From connectivity point of view, mesh can be classified as structured or unstructured. A structured mesh is characterized by regular connectivity that can be expressed as a two or three dimensional array. An unstructured mesh is characterized by irregular connectivity of the nodes and is not readily expressed as a two or three dimensional array in computer memory. The storage requirements for an unstructured mesh can be substantially larger since the neighborhood connectivity must be explicitly stored.

3.6.3. Mesh less Method

In the field of numerical simulation, meshfree methods are those which do not require a mesh connecting the data points of the simulation domain. These methods are in particular interest in structure mechanics but nowadays they are also catching up in fluid dynamics. Mesh less method are very helpful when boundary of the domain can deform and the connectivity of the mesh can be difficult to maintain without introducing error into the simulation. If the mesh becomes tangled or degenerate during simulation, the operators defined on it may no longer give correct values. The mesh may be recreated during simulation (a process called re-meshing), but this can also introduce error, since all the existing data points must be mapped onto a new and different set of data points which results in a new CFD fiend known as data transfer between meshes. There are several methods which are used in literature for this purpose such as, super convergent patch recovery (SPR), equilibrium patch recovery (EPR) etc. But mapping of data on new mesh can introduce a significant error if the deformation in material is huge.

3.7. NOMENCLATURE

Γ	Computation domain boundary (m).
$\Gamma_{in}, \Gamma_{out}, \Gamma_w$	Inlet, outlet and wall computation domain boundary (m).
Γ_N, Γ_D	Dirichlet and Neumann computation domain boundary (m).
Ω	Computation domain (m^3).
ρ	Density of fluid (Kg/m^3).
μ	Dynamics viscosity (Kg/m/s).
ϕ	Flux vector (m/s).
P, p	Pressure (Pa).
S_ϕ, f	Source term in momentum Eq for corresponding ϕ ($Kg/m^2/s^2$).
V_r, V_z, V_θ	Velocity in different direction (m/s).
u	Velocity (vector form m/s).
u_o	Initial velocity (m/s).
Ho	Slipper/plate central clearance (m).
H	Generic slipper/plate clearance (m).
r, θ, z	Cylindrical coordinates vector (m, rad, m).
x, y, z	Cartesian coordinates vector (m, rad, m).

R, ξ, η	Transformed cylindrical coordinates vector (m, rad, non dimensional).
α	Slipper tilt or a non-dimensional factor between 0 and 1.
T, t	Time (s).
d_s	Swash plate radius (m).
ω_{sw}	Swash plate turning speed (rad/s)
ω	Angular velocity (rad/s)
F	A generalize function of velocity and pressure ($Kg/m^2/s^2$)

Non dimensional filed and Functional Space

H	Functional space (Hilbert space).
L	Functional space.
g	Generalized function.
v	Velocity basis function.
q	Pressure baiss function.
h	Orthogonal projection of chauchy stress.
h	Finite element size representation.
a,b,c	Different bilinear products as defined in their equation.
N	Basis function.
U^a, V^a	Nodal value.

Subscript

r, θ, z	Component of vector in r, θ and z direction.
R, ξ, η	Component of vector in R, ξ and η direction.
e, w, n, s, t, b, p	East, west, north, south, top, bottom, center point of a control volume.
*	Intermediate filed.
nb	Neighbors points.
nel	No of elements in domain.
h	finite element representation.

3.8. REFERENCES

[1] Patankar Subhas V. (1980). *Numerical Heat Transfer and Fluid Flow*. Taylor & Francis.

[2] Anderson John D. (1995). *Computational Fluid Dynamics*, The Basic with Applications. Mc Graw Hill.

[3] O. C. Zienkiewic, Robert L taylor, J. Z. Zhu (2005). The Finite Element Method: Its basis and Fundamentals, Sixth Edition. Mc Graw Hill.

Chapter 4

VALVES

4.1. INTRODUCTION

In previous chapters, an introduction to fluid properties, fluid mechanics and CFD has been undertaken. The present chapter is the first one completely focused in presenting several original research linked with the fluid power technology. Initially, a brief description of the different types of valves used in fluid power will be presented, next theoretical and experimental original research regarding the stability on pressure relieve valves shall be introduced. Static experimental performance curves measured on a proportional valve will also be introduced. The final part will be, explaining mathematically and experimentally some performance characteristics based on original research and related to four nozzle two flapper servovalves.

The different valves used in fluid power systems, fall in three main groups, directional control valves, pressure control valves and flow control valves. As its name indicates, directional control valves are used to direct fluid to any part of the circuit. Pressure control valves allow establishing sequences in hydraulic operations via scaling the pressure in different parts of the system. Flow control valves are designed to regulate the flow to any part of the circuit, and they can compensate the effect of fluid temperature and or pressure. What it is also interesting to highlight is that all these different sort of valves can have a proportional actuation. Figure 4.1 presents the main different existent sort of valves. The idea is just presenting the different valves without introducing any performance curves or technical characteristics of them, since the present book chapter is mostly focused in introducing original research work and valve characteristics can be found in many existing books.

Pressure control valves, are used to select pressure levels at which particular parts of the circuit must work, they control actuator force, avoid hydraulic system damage, power wastage and circuit overheating. More in detail, it can be said that they perform the following functions.

A. Limiting the maximum pressure in a given circuit, Relief valves or security valves.
B. Reduce the pressure in a particular part of the circuit. Pressure reducing valves.
C. Redirecting fluid to tank while maintaining pressure in the circuit. Unloading valves.
D. Redirecting fluid to tank and not maintaining pressure in the circuit, Offloading valves.

E. Offering resistance to the fluid at selectable pressure levels. Brake and counterbalance valves.
F. Allowing the flow entering parts of the circuit at selected pressure levels. Pressure-sequence valves.

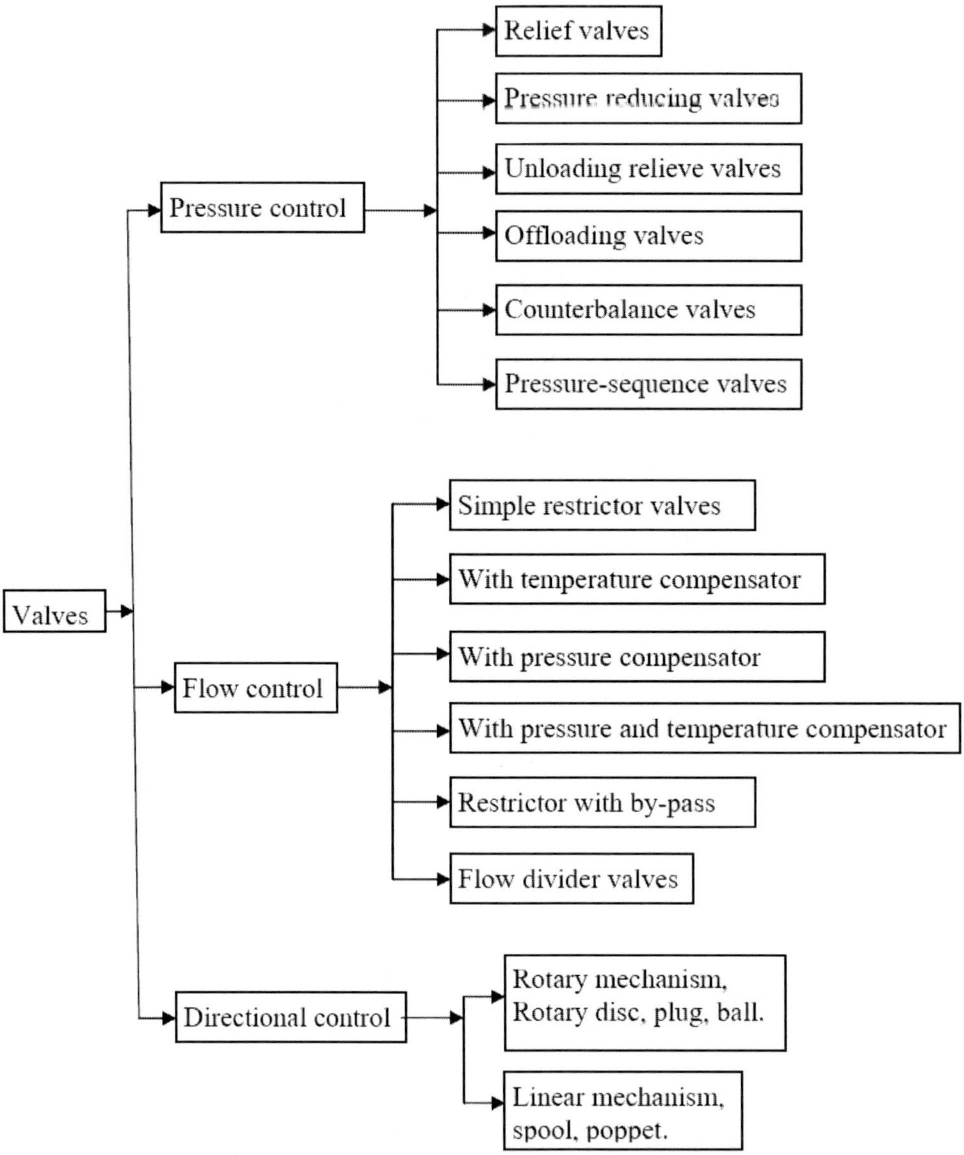

Figure 4.1. Main description of hydraulic valves.

For pressure control valves to be able to perform all these functions, valves need to be whether pilot operated, remote operated or having a several stage configuration.

Flow control valves are used to regulate the pump flow to other branches of the circuit, and this means regulating the maximum speed of actuators and even the power available. It is very important to realize that simple restrictor flow control valves regulate the flow via increasing the pressure upstream in the circuit and forcing the pressure relief valve to open and direct part of the fluid to tank. The function of flow control valves with temperature compensator is to maintain the volumetric flow constant to a branch of the circuit independently of the fluid temperature. Pressure compensator allows maintaining constant the volumetric flow in a circuit branch regardless of the upstream pressure. Upstream pressure must always be higher than the required pressure downstream. Flow control valves having a by-pass in parallel, acts as a restrictor when flow goes in one direction and allows free flow in the opposite direction. A flow divider splits the incoming flow in two equal or non equal parts.

Directional control valves, provide control functions in a circuit, like controlling actuators motion direction, select alternative circuits and in general perform logic control functions. Regarding its internal control elements there exist two different mechanisms to direct the flow, the one based on a rectilinear movement, generally a spool is used, and the one based on a turning movement, a rotary ball, disc or similar. These sort of valves, have several switching positions, commonly two or three, the number of connecting ports is usually four, although other configurations are also common. To switch form one position to another there are several actuation methods, hydraulic, electrical, manual, mechanical and even pneumatic.

All these valves have one or several springs which establish a given position when no actuation is taken, the valve is at its rest position. Depending on the volumetric flow crossing the directional valves, these ones might be having two or even three stages.

It must be noticed that all sort of valves have the possibility of having a proportional actuation, although the most common ones with proportional actuation are the directional control ones. Proportional actuation means that the electric solenoid responsible for switching from one position to another is capable of accepting variable intensities and therefore the valve position will not switch from position A to position B, then in reality for a proportional actuation there will be infinite positions between these two. Proportional actuation opens the door to evaluate valve dynamics, since proportional valves accept intensity inputs at different frequencies and amplitudes, but the central part of the proportional valve, usually a spool, has problems in following the input dynamics, specially at high frequencies, when comparing the output spool dynamic displacement with the input dynamic current applied to the valve, the Bode diagram of the valve dynamics can be obtained, where phase shift and amplitude ratio between inlet and outlet signals is evaluated.

There is no need to say that as input frequency increases, outlet amplitude will decrease and phase shift will increase, as a result, each proportional valve will have a frequency limit at which valve performance can be regarded as acceptable.

Whenever a valve needs to perform at high frequencies, nowadays few hundred Hertz, the best choice is to use servovalves. Servovalves have usually more than one stage and its first stage is often based on a flapper nozzle configuration. There exist two different flapper nozzle configurations, a single flapper with two nozzles and a double flapper with four nozzles, the second one has better dynamic performance.

The authors recommend to the reader to go over the reference books to deeply understand the main valves performance and characteristics. In what follows original research undertaken in conical seat relief valves, proportional seat and servovalves will be presented.

4.2. Conical Seat Relief Valves

Relief valves are mostly used in hydraulic circuits as safety valves; they are normally closed and open whenever circuit pressure overcomes a certain value. In order to insure fluid tightness and therefore minimize leakage, poppet is usually conical or spherical. In all existing cases the seats where poppet rest have a small land, on the other hand it is also known that relief valves tend to vibrate. The following research will link the flow momentum with the poppet conical seat land and the valve stability.

4.2.1. Previous Research on Conical Seat Relief Valves

Pressure relief valve poppets used in oil-hydraulic industry have traditionally been of the cone-seated type since they are reliable from a leakage point of view and are dynamically more stable. To prevent leakage, the seats have a flat land, usually referred to as a chamfered land, and with lengths generally smaller than 2 mm. Such a configuration also helps to minimise permanent marks on the spool. For the last 40 years poppet valves have undergone a very small evolution, and the following study enhances this evolution and the way the theory can be extended to obtain suitable design equations.

Stone [1] in 1960 presented a study on discharge coefficients and steady state flow forces on poppet valves. His work was mainly experimental and focused on valves with a sharp-edged seat, the working fluid being mineral oil. In the first part of the paper, Stone applied the momentum equation to determine the theoretical force acting on the valve spool. He assumed a uniform velocity distribution at the inlet and outlet and a linear pressure drop approximation along the conical seat, in the absence of the exact solution which is derived in this study. From the experiments he found that when the flow exhausted into a chamber filled with fluid, it tended to be unstable and oscillated at a frequency of 1300 Hz. The diameter of the exhaust chamber did not affect the frequency although the valve performance seemed to be more unstable with the larger diameter chamber. He found a low pressure region underneath the poppet and realised that in this area there was a release of dissolved air from the oil which seemed to be related to valve stability. In fact just before the valve started to oscillate, the number of bubbles increased. He concluded that there were several regimes of flow which needed to be further studied and three dimensional effects of the fluid should also be taken into account. He deduced that discharge coefficients needed to be much better understood, effects of viscosity, cavitation and fluid rotation needed to be studied, otherwise no real work could be done on flow forces.

Urata [2], developed a high-level mathematical approach, where he studied such valves with and without laps, and under laminar and turbulent conditions. He assumed that the velocity distribution across the gap followed boundary layer theory. Therefore when integrating the Navier-Stokes and continuity equations along the gap he used the laminar velocity distribution in the boundary layer. For turbulent flow he applied the turbulent velocity distribution across the boundary layer and as a result obtained the pressure distribution along the valve seats and therefore the theoretical force over the valve. In order to find the force he used the momentum theory, finding that when the valves have laps, integration of the pressure along the seat is needed in order to have an accurate estimation of the force. This idea is consistent with that postulated by Stone[1]. In his experiments, Urata used mineral oil and water and cone angles of 40 and 90 degrees were studied. One of the interesting results obtained when studying the pressure distribution along the poppet valves with lap, was that the pressure at the inlet of the conical chamfered seat was in some cases very slightly negative and therefore cavitation was likely to occur. This decrease of pressure at the inlet was postulated to be caused by the contraction of the flow streamlines due to the boundary layer at the entrance to the flow path.

The dynamic characteristics of a hydraulic pressure relief valve with a conical seat, was studied by Scheffel [3]. He established the dynamic equations which took into account the damping coefficient of the system and also the pressure forces on the back of the valve, relating them to the bulk modulus, although he considered this of minor importance when considering flow forces. He found out that as the angle of the cone on the spool increased, so did the pressure stability, noticing a similar effect with the increase of the cone length. He also pointed out the flow dependence on the fluid temperature and suggested the necessity to maintain a constant temperature system for experimental purposes.

In a further study, Scheffel [4] showed some plots which demonstrated what was suggested in the previous paper. This indicated that as the angle of the cone seat, and/or the length of the chamfered seat increased, so did the stability of the spool by decreasing the overpressure which appeared to be independent of the volumetric flow across the valve. The non-dimensional pressure on the conical seat was determined as a function of the conical angle, and its minimum appeared to be for a cone semi-angle of 90 degrees. Concerning pressure relief valve instability, Scheffel postulated that performance improvement could be achieved by increasing the oil volume between the pump and the relief valve, increasing the spring constant, increasing the cone angle, increasing the oil bulk modulus, increasing the length of the cone seat. On the other hand, the instability was not improved when the flow across the valve increased, and/or the damping constant was decreased. He also noticed that changing the spool mass did not create any appreciable effect on the valve behaviour.

In a further paper Scheffel [5] studied the dynamic behaviour of a pressure relief valve with a conical seat and an additional damping spool. He considered the dynamic equations when the damping effects acted on one side or both sides of the extra damping spool. He then discovered that the dynamic behaviour of this particular pressure relief valve improved with the increase of the cone angle and the inlet pressure, the optimum behaviour being obtained for cone angles between 40 and 90 degrees. He also noticed an improvement in the dynamic behaviour when decreasing the oil volume on the back side of the additional damping spool. The dynamics of the valve was also considered for damping coefficients depending on the valve direction, that is, the damping coefficient when the valve opened was arranged to be much smaller than the coefficient for a closing valve. It was demonstrated that as the damping coefficient increased in the closing direction, the overlap pressure decreased, and consequently the stability of the valve increased. This effect due to directional damping has also been studied analytically and experimentally by Watton [6] for a spool-type poppet.

One of the most interesting studies done on conical seated valves, was developed by Kollek [7] who experimentally investigated the flow behaviour, finding the Reynolds number limit between laminar and turbulent flow and its variation with cone angle. This study took into account the length of the cone seat, deducing that as the length increased for a constant cone angle, the Reynolds number also had to increase to achieve turbulent flow. Also as the cone angle increased, the Reynolds number had to increase to achieve turbulent flow. The fluid used was a chemical mixing between $HOCH_2CHOHCH_2$ and Titanoxid, cone semi angles of 30, 40 and 60 degrees were used, the cone lengths studied were, 0, 0.5, 1, 2 mm. The method he used to differentiate the two kinds of flow was via visualization, particularly in relation to the vortex that was created at the outlet of the cone. It was noticed that the flow path angle at the cone exit did not follow the cone direction, and therefore the flow forces due to change in momentum did vary from what was being assumed in theory. He demonstrated that at low Reynolds number the flow angle at the cone exit, was higher than the angle of the cone, and this angle increased sharply with the cone length. When studying the discharge coefficients, Kollek used the traditional equation as many authors before [1]. The variation of discharge coefficients versus Reynolds number could clearly be seen and as the cone angle increased the discharge coefficient also increased. For a given cone angle the discharge coefficients decreased as the cone length increased. This effect can be explained in terms of the flow path length affecting the real flow rate while the theoretical turbulent flow rate equation does not take cone length into account.

Johnston et al. [8] carried out a substantial experimental programme on poppet and disk valves, focussing on discharge coefficients, flow visualisation, pressure drop, and cavitating conditions for high Reynolds number. The fluid used in all cases was water. This paper gave a an introduction to previous work and pointed out that until 1992 very few people attempted to describe mathematically the characteristics of disk and popped valves with chamfered seats [2,9,10]. In all these studies a set of assumptions had to be considered, and the research was mainly focussed on the thrust acting over the valve. As a result, it was concluded that the thrust was highly dependent on the reattachment or separation of the flow, as suggested by Stone [1] and Kollek [7]. At small openings, in relation of the lap width, it was considered that the valve behaved as a long orifice since the jet tended to reattach itself to the seat. Flow visualisation of flow separation on a conical seat for different valve openings, lengths of the chamfered seat and cone angles, showed that at small gaps the flow tended to reattach to the valve body while as the gap increased the flow reattached to the conical spool. When studying the valves with chamfered seats he found that at small gaps the flow forces increased with the increase of the cone seat length, which was in accordance with that established by Urata [2]and Stone[1]. It was found that the discharge coefficients decreased with the cone length, the same result also being reported by Kollek [7]. The discharge coefficients and the flow forces decrease at higher openings. The chamfered popped valve was found to be less prone to cavitation than the sharp-edged one which differed from that found by Urata [2] who showed experimentally and analytically that cavitation was more likely to occur when using chamfered seats. It has to be said that Johnston refers to the exit of the valve seat, where vorticity is created, while Urata was trying to explain the pressure distribution along the valve seat and focussing at the valve inlet. On the other hand, the value of the negative pressure in Urata's experimental findings is very small.

A state of the art in pressure relief valve design was given by Petherick [11], which gave accurate information of the different fields in which such valves may be used, and also, the main point of interest of the different researchers. They pointed out that Sallet et al [12] performed a large number of tests which covered compressible and incompressible flow and several configuration of valve seating. They described the fluid dynamics phenomena that may lead to valve disc vibrations, although tests were not performed for conical seated valves. Petherick et al found that the pressure relief valve designs have barely changed for more than thirty years despite the instabilities and other problems related to its shape. Johnston et al [8], for example, stated that for valves in reciprocating pumps, where low pressure drop and hence a high discharge coefficient was desirable, a popped valve with a chamfered seat would give a suitably high discharge coefficient over a wide range of openings. All researchers seem to agree that flow instabilities are due to the change in flow pattern at the valve outlet.

Vaughan et al [13] performed a two dimensional CFD simulation for a range of conical seat valve geometries, the flow being turbulent and incompressible. The aim of the work was to study the reattachment and separation of the flow, the vortex created at the outlet, and its relation with the flow forces acting on the spool. They found that a poppet valve with a lip on its outer edge had the effect of minimising the flow force at small openings, but a change in the flow pattern occurred at larger openings, accompanied by an increase in the magnitude of the flow force. Recirculation was found to be an important feature of the flow through conical valves, but it appeared that the overall pressure/flow characteristics were less sensitive to the large-scale recirculation which occurred downstream of the valve rather than to the small scale but intense recirculation which may occur in the orifice region. In this region the high velocities and streamline curvatures caused high body forces on the fluid.

Mokhtarzadeh-Dehghan et al [14] studied the laminar flow of oil through a hydraulic pressure relief valve used in a variable compression ratio internal combustion engine. The work was mainly focussed on a two dimensional computational model of the valve seat passage. The clearance was in the range 0,2 to 1 mm, the inlet pressure between 100 and 200 bar, the flow was reported to be laminar in all cases which is in accordance with what was established by Stone [1]. It was found that the pressure drop under the conical seat increased as the gap decreased, and for small gaps around 0.2 mm a small pressure drop appeared at the entrance of the conical seat followed by a pressure recovery. This effect was already reported by Urata [2] although he was working with much smaller pressures and the decrease of inlet pressure was due to a small decrease of the vena contracta at the entrance of the seat. The results showed that the forces acting on the valve spool increased with the flow passage gap, the flow being reattached for small gaps but becoming separated as the distance between plates increased.

More recently, some new developments on poppet and relief valves have been presented by Chenvisuwat [15] which improved the dynamic behaviour of a small size of a poppet valve used in railway brake system, via compensating the force on the displacement device, and improving the performance against wear. Yanping [16] proposed a relief valve with a bridge of hydraulic resistances in order to produce a valve insensitive to flow rate.

From all these investigations, it can be concluded that a large number of studies have been focused on the dynamics stability of the valve [3,4,5], some others have visualised the flow along the conical gap and at the seat exit [1,7,8,12,13,14], which happened to be linked with the discharge coefficients and the flow forces. In other studies, the influence of the length of the conical seat has been taken into consideration [1,2,3,4,5,7,8] relating in some cases the length with the valve stability and or flow separation. In almost all studies the flow is assumed to be turbulent, or no consideration is made upon this point. Therefore little attention has been paid to the fact that at small gaps the flow could be laminar, [1,2,10] especially when the fluid is mineral oil. It seems that only Urata [2] has considered a theoretical basis for the pressure distribution along conical seats. He developed his theory for laminar and turbulent flow, assuming a velocity distribution across the conical gap which followed the boundary layer theory in both cases. Due to this assumption, it is difficult to establish an explicit mathematical relationship for the flow across the gap and the flow through the valve is not reported in his paper.

As chamfered conical seats increase in length, and the clearance decreases, laminar flow is expected to appear and some of the main characteristics of such valves would be:

- improved stability
- sharp reduction of the vortex creation at the outlet
- very small flow separation
- less prone to cavitation resulting in improved reliability
- a linear pressure/flow characteristic, preferable from a control point of view
- the concept of discharge coefficient variation with flow rate does not exist since the flow is laminar; the effect of temperature variation on the pressure/flow characteristic for laminar flow is well known and predictable.

In the work that follows, a theoretical and experimental study has been performed on a conical valve, especially regarding the behaviour of such valves for long chamfered seats, pressure distribution along the seat, flow through the valve and theoretical flow forces. The theoretical results from the new set of equations will be compared with CFD simulation and experimental test, validating the theory proposed. The equations developed are generic and the fluid considered is mineral oil ISO 32.

4.2.2. Mathematical Development Based on Laminar Flow across a Conical Valve Seat

For small displacements of the conical poppet, the gap between the conical seat and the conical poppet will be small and considering as well that the chamfered seats studied here have a much longer length than the ones found in conventional valves, it can be stated that flow will be laminar. As a result the following theory can be applied.

4.2.2.1. Theoretical Background

Consider the conical poppet valve shown in figure 4.2.

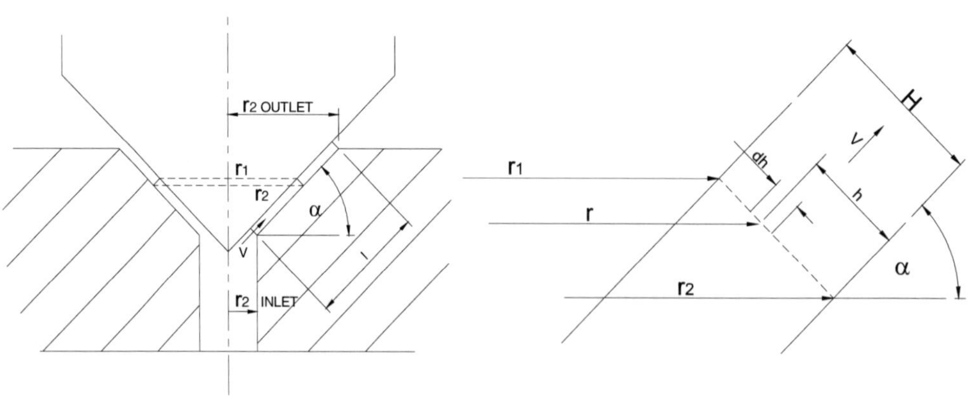

Figure 4.2. General view of the conical device.

The volumetric flow through the gap can be defined as:

$$Q=\int_{0}^{H} V\,2\pi(r_2 - h\cos(90-\alpha))\,dh \tag{4.1}$$

when $r = r_2$, $h = 0$ and $r = r_1$, $h = H$

The velocity distribution between the two surfaces according to Poiseulle equations is given as:

$$V=-\frac{1}{\mu}\frac{dp}{dx}\frac{h}{2}(H-h) \tag{4.2}$$

The flow along the cone will be:

$$Q=\int_{0}^{H} -\frac{1}{\mu}\frac{dp}{dx}\frac{h}{2}(H-h)\,2\pi(r_2 - h\cos(90-\alpha))\,dh \tag{4.3}$$

and its integration yields:

$$Q=-\frac{\pi}{\mu}\frac{dp}{dx}\left[r_2 H^3\frac{1}{6}-H^4\cos(90-\alpha)\frac{1}{12}\right] \tag{4.4}$$

the pressure distribution along the seat will be:

$$\int_{0}^{l}\frac{dx}{r_2 H^3\frac{1}{6}-H^4\cos(90-\alpha)\frac{1}{12}}=\int_{P_{inlet}}^{P_{outlet}}-\frac{\pi}{\mu}\frac{dp}{Q} \tag{4.5}$$

Knowing that: $x=0$, $r_2 = r_2(\text{inlet})$; and $x=l$, $r_2 = r_2\ (\text{outlet})$.

$$\cos\alpha=\frac{r_2(\text{outlet})-r_2(\text{inlet})}{l}\qquad \cos\alpha=\frac{r_2-r_2(\text{inlet})}{x}\qquad r_2=\cos\alpha\, x+r_2(\text{inlet}) \tag{4.6}$$

And therefore the pressure distribution is:

$$\int_{0}^{l}\frac{dx}{(\cos\alpha\, x+r_2(\text{inlet}))H^3\frac{1}{6}-H^4\cos(90-\alpha)\frac{1}{12}}=\int_{P_{inlet}}^{P_{outlet}}-\frac{\pi}{\mu}\frac{dp}{Q} \tag{4.7}$$

After integration and some arrangement is obtained:

$$Q=(P_{(inlet)}-P_{(outlet)})\frac{\pi}{\mu}\cos\alpha\frac{H^3}{6}\frac{1}{\ln\left[\dfrac{l\cos\alpha\dfrac{H^3}{6}+r_{2(inlet)}\dfrac{H^3}{6}-H^4\cos(90-\alpha)\dfrac{1}{12}}{r_{2(inlet)}\dfrac{H^3}{6}-H^4\cos(90-\alpha)\dfrac{1}{12}}\right]} \tag{4.8}$$

If the pressure at a generic point is P, located a distance x from the co-ordinate axis origin, then:

$$Q=(P_{(inlet)}-P)\frac{\pi}{\mu}\cos\alpha\frac{H^3}{6}\frac{1}{\ln\left[\dfrac{x\cos\alpha\dfrac{H^3}{6}+r_{2(inlet)}\dfrac{H^3}{6}-H^4\cos(90-\alpha)\dfrac{1}{12}}{r_{2(inlet)}\dfrac{H^3}{6}-H^4\cos(90-\alpha)\dfrac{1}{12}}\right]} \tag{4.9}$$

From equations (4.8) and (4.9) it can be concluded:

$$P=P_{(inlet)}-(P_{(inlet)}-P_{(outlet)})\frac{\ln\left[\dfrac{x\cos\alpha\dfrac{H^3}{6}+r_{2(inlet)}\dfrac{H^3}{6}-H^4\cos(90-\alpha)\dfrac{1}{12}}{r_{2(inlet)}\dfrac{H^3}{6}-H^4\cos(90-\alpha)\dfrac{1}{12}}\right]}{\ln\left[\dfrac{l\cos\alpha\dfrac{H^3}{6}+r_{2(inlet)}\dfrac{H^3}{6}-H^4\cos(90-\alpha)\dfrac{1}{12}}{r_{2(inlet)}\dfrac{H^3}{6}-H^4\cos(90-\alpha)\dfrac{1}{12}}\right]} \tag{4.10}$$

Equation (4.10) gives the pressure distribution along the conical seat. The representation of this equation for a mineral oil ISO 32, a gap of 3 microns, a pressure differential of 100 bar and for a range of cone angles and cone seat lengths is plotted in figure 4.3. It can be noticed that this distribution is not linear and as the angle of the cone increases, the pressure drop at any point along the seat increases. This behaviour is clearly understandable since for small cone angles the area change with the radius is smaller than for large cone angles, therefore the resistance increases. It will also be seen, as expected, from figure 4.3 that the pressure rate of decrease with distance is highest at the entrance to the flow gap.

Equation (4.8), which gives the flow through the poppet valve, is represented in figure 4.4 for a set of clearances, pressure differentials and cone angles. Notice that as the cone angle increases, while maintaining the same cone length (30mm) and the same distance between plates, the flow through the poppet valve increases as well. This effect can be easily explained since as the cone angle increases the flow passage increases much more rapidly and as a result, the flow increases. Nevertheless, it must also be remembered that as the cone angle increases the momentum change increases as well, and also the streamlines will compress at the cone inlet. These effects will tend to slightly decrease the flow through the valve and create a small pressure drop followed by a pressure recovery at the valve inlet as reported by Urata [2] and also by Mokhtarzardeh-Dehghan [14].

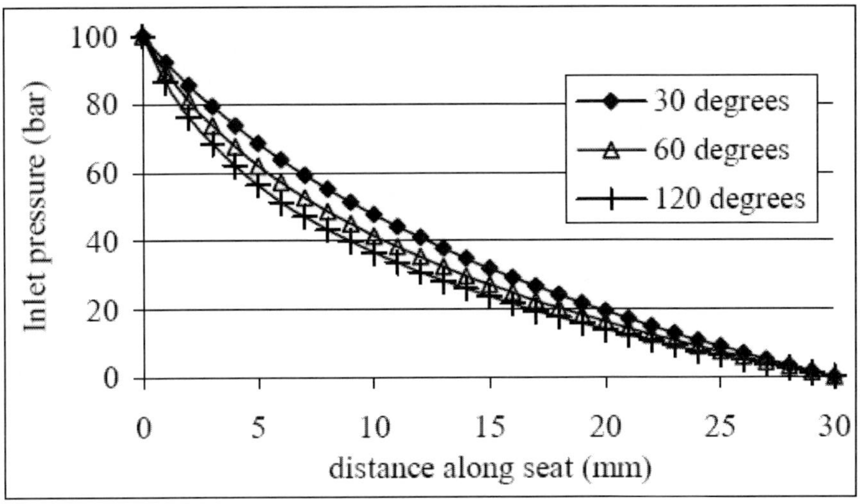

(a) Cone angle 30°, 60°, 120°

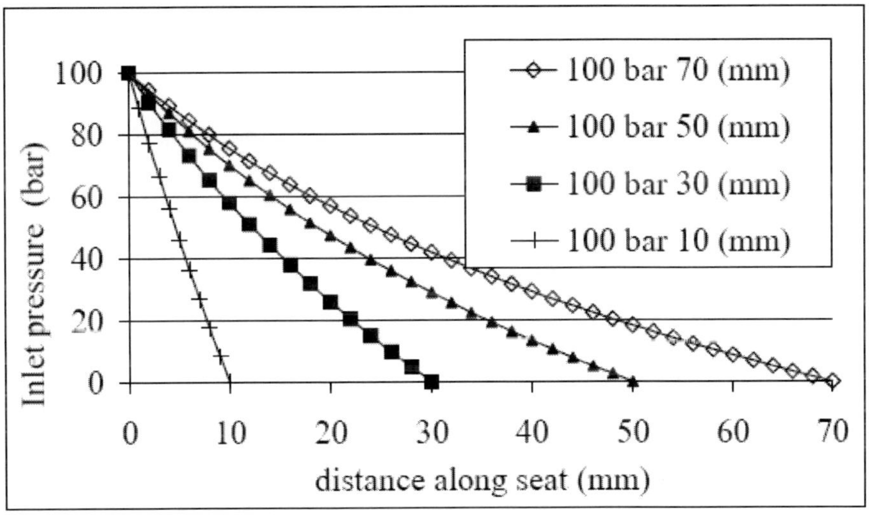

(b) Cone seat lengths of 10, 30, 50, 70mm, $\alpha=45^{\circ}$

Figure 4.3. Variation of the pressure distribution along the cone. Oil ISO 32 Pressure =100bar, clearance = 3 microns.

The theory established has considered the viscosity as constant. To check this assumption the maximum oil temperature increase that could exist experimentally can be estimated. Results presented consider the valve performance near to the closed position such that the maximum flow rate is about 1.1l/min with a maximum pressure of 150bar. The maximum temperature increase will be 4.1°C and the change in viscosity over this temperature, and pressure range considered, is negligible. With water cooling and heat transfer from the valve and pipework, the actual temperature increase experienced by the oil is reduced from the theoretical value.

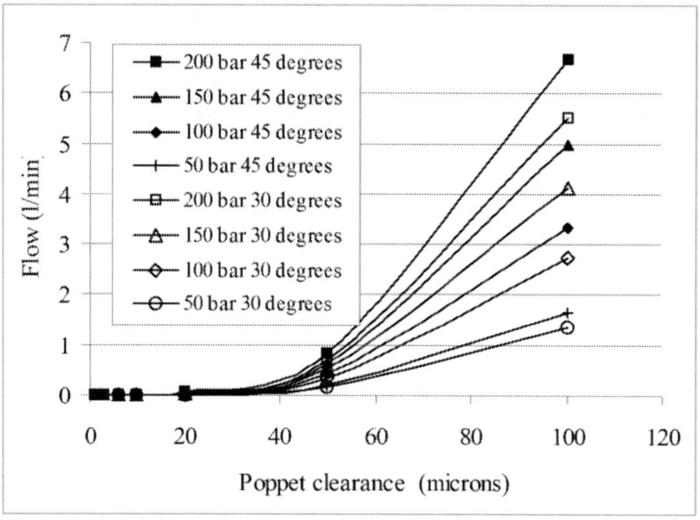

Figure 4.4. Flow characteristic for a set of poppet clearances α= 45° and 30°.

4.2.2.2. Force on a Conical Poppet Assuming Laminar Flow

The momentum equation in vectorial form is applied to the control volume in the axial (j) direction:

$$\int_{A\ inlet} P_{inlet}\, d\vec{A} + \int_{A\ outlet} P_{outlet}\, d\vec{A}\cos(90-\alpha) + \int_{A\ cone} P_{conesurface}\, d\vec{A}\,\cos\alpha + \int_{A\ cone\,seat} P_{coneseat}\, d\vec{A}\cos\alpha + \int_{A\ cone\&seat} \vec{\tau}\, d\vec{A}\sin\alpha + m\vec{g} = \int_{A\ outlet} \rho \vec{V}_j\, \vec{V}\, d\vec{A} + \int_{A\ inlet} \rho \vec{V}_j\, \vec{V}\, d\vec{A} \tag{4.11}$$

Neglecting the gravitational forces, the axial force over the conical spool will be:

$$F_{spool\hat{j}} = -\int_{A\ cone} P_{conesurface}\, dA\cos\alpha - \int_{A\ cone\&seat} \tau\, dA\sin\alpha = \int_{A\ outlet} \rho V_j\, V\, dA - \int_{A\ inlet} \rho V_j\, V\, dA - \int_{A\ inlet} P_{inlet}\, dA + \int_{A\ outlet} P_{outlet}\, dA\cos(90-\alpha) - \int_{A\ cone\,seat} P_{coneseat}\, dA\cos\alpha \tag{4.12}$$

Stone [1] and Urata [2] indicated that to evaluate the flow forces on the valve poppet, the pressure distribution along the valve seat needs to be taken into account. Notice that the last term of equation (4.12) gives this force. Since the pressure distribution along the cone seat is given by equation (4.10) its integration along the conical wall will give the pressure forces over the seat. When solving equation (4.12), the velocity distribution is assumed parabolic at the inlet and outlet, since laminar flow is proposed. An approximate assessment of boundary layer development from an initially assumed, hypothetical, uniform velocity profile at the inlet can be done using laminar flow theory for flow between flat plates. This suggests that a parabolic distribution would be established at distances less than 0.1mm for the maximum flows and clearances experienced in the practical tests. It should be recalled that the cone seating is 30mm long. The assumption of a parabolic profile at the inlet is considered sufficient for theoretical purposes and has been validated by observation of the CFD results. At the inlet the velocity distribution is therefore assumed to be:

$$V = V_{max}\left[1-\left(\frac{r}{r_{2\,inlet}}\right)^2\right] \tag{4.13}$$

Therefore the momentum force at the inlet is:

$$-\int_{A\,inlet}\rho V_j\, V\,dA = -\int_0^{r_{2\,inlet}} \rho V^2\, 2\pi r\,dr = -\rho\pi V_0^2\, r_{2inlet}^2\,\frac{4}{3} \tag{4.14}$$

where V_o is the inlet velocity. The momentum force at the outlet will be, from equations (4.2) and (4.10):

$$\frac{dp}{dx} = \frac{p_{outlet} - P_{inlet}}{K_3}\,\frac{K_1}{x\,K_1 + K_2} \tag{4.15}$$

where K_1, K_2 and K_3 are constants given by:

$$K_1 = \cos\alpha\,\frac{H^3}{6} \tag{4.16}$$

$$K_2 = r_{2inlet}\,\frac{H^3}{6} - H^4\cos(90-\alpha)\frac{1}{12} \tag{4.17}$$

$$K_3 = \ln\left[\frac{1\cos\alpha\,\frac{H^3}{6} + r_{2inlet}\,\frac{H^3}{6} - H^4\cos(90-\alpha)\frac{1}{12}}{r_{2inlet}\,\frac{H^3}{6} - H^4\cos(90-\alpha)\frac{1}{12}}\right] \tag{4.18}$$

It then follows that:

$$\int_{A\,outlet}\rho V_j\, V\,dA = \int_0^H \rho\, V^2 \sin\alpha\,(r_{2outlet} - h\cos(90-\alpha))\,2\pi\,dh =$$

$$\rho\, 2\pi \mathrm{Sin}(\alpha)\frac{1}{\mu^2}\left[\frac{K_1^2\,(P_{outlet}^2 + P_{inlet}^2 - 2P_{outlet}\,P_{inlet})}{K_3^2\,(1K_1 + K_2)^2}\right]\int_0^H \frac{h^2}{4}(H-h)^2(r_{2\,outlet} - h\cos(90-\alpha)dh \tag{4.19}$$

and after integration:

$$\int_{A\,outlet}\rho V_j\, V\,dA = \rho\, 2\pi \mathrm{Sin}(\alpha)\frac{1}{\mu^2}\left[\frac{K_1^2\,(P_{outlet}^2 + P_{inlet}^2 - 2P_{outlet}P_{inlet})}{K_3^2\,(1K_1 + K_2)^2}\right](\frac{H^5}{120}\,r_{2outlet} - \cos(90-\alpha)\frac{H^6}{240}) \tag{4.20}$$

The force due to the inlet pressure will be:

$$-\int_{A\,inlet} P_{inlet}\, dA = -\int_0^{r_{2\,inlet}} P_{inlet}\, 2\pi r\, dr = -P_{inlet}\, \pi r_{2\,inlet}^2 \tag{4.21}$$

and the force due to the outlet pressure:

$$\begin{aligned}\int_{A\,outlet} P_{outlet}\, dA \cos(90-\alpha) &= \int_0^H P_{outlet}\, 2\pi (r_{2\,outlet} - h\cos(90-\alpha)) \cos(90-\alpha)\, dh = \\ &= P_{outlet}\, 2\pi \cos(90-\alpha)\left[r_{2\,outlet} H - \frac{H^2}{2}\cos(90-\alpha)\right]\end{aligned} \tag{4.22}$$

Taking into account that $P_{cone\ seat}$ is being given by equation (4.10) the forces on the cone seat will be given as:

$$-\int_{A\ cone\,seat} P_{coneseat}\, dA \cos\alpha = -\int_0^l P_{coneseat}\, 2\pi \cos\alpha\, (\cos\alpha\ x + r_{2\ inlet})\, dx =$$

$$= \int_0^l -\left[2\pi\cos\alpha\,(\cos\alpha\ x + r_{2\ (inlet)})\right] \left\{ P_{(inlet)} - (P_{(inlet)} - P_{(outlet)}) \frac{\ln\left[\dfrac{x\cos\alpha\dfrac{H^3}{6} + r_{2(inlet)}\dfrac{H^3}{6} - H^4\cos(90-\alpha)\dfrac{1}{12}}{r_{2(inlet)}\dfrac{H^3}{6} - H^4\cos(90-\alpha)\dfrac{1}{12}}\right]}{\ln\left[\dfrac{l\cos\alpha\dfrac{H^3}{6} + r_{2(inlet)}\dfrac{H^3}{6} - H^4\cos(90-\alpha)\dfrac{1}{12}}{r_{2(inlet)}\dfrac{H^3}{6} - H^4\cos(90-\alpha)\dfrac{1}{12}}\right]} \right\} dx \tag{4.23}$$

Using the same constants K_1, K_2 and K_3 as represented in equations (4.16), (4.17), (4.18), the result of the integration then gives the following equation for the pressure force on the cone face:

$$\begin{aligned}-\int_{A\ coneseat} P_{coneseat}\, dA \cos\alpha = &-2\pi\cos\alpha \{ P_{inlet}\, r_{2\,inlet}\, l + P_{inlet}\cos\alpha \frac{l^2}{2} + \\ &+ \frac{r_{2\,inlet}}{k_3}[(l+\frac{k_2}{k_1})\ln(l\frac{k_1}{k_2}+1) - l]\,[P_{outlet} - P_{inlet}] \\ &+ \frac{\cos\alpha}{k_3}[\frac{1}{2}(l^2 - \frac{k_2^2}{k_1^2})\ln(l\frac{k_1}{k_2}+1) - \frac{1}{2}(\frac{l^2}{2} - \frac{k_2}{k_1}l)]\,[P_{outlet} - P_{inlet}]\}\end{aligned} \tag{4.24}$$

The axial force acting on the cone then requires the addition of the momentum terms and the static pressure terms, and becomes:

$$F_{spool\,j} = \rho 2\pi \sin(\alpha) \frac{1}{\mu^2}\left[\frac{K_1^2 (P_{outlet}^2 + P_{inlet}^2 - 2P_{outlet}P_{inlet})}{K_3^2 (l K_1 + K_2)^2}\right] (\frac{H^5}{120} r_{2outlet} - \cos(90-\alpha)\frac{H^6}{240})$$

$$-\rho\pi V_0^2 r_{2inlet}^2 \frac{4}{3} - P_{inlet}\pi r_{2inlet}^2 + P_{outlet} 2\pi\cos(90-\alpha)\left[r_{2outlet}H - \frac{H^2}{2}\cos(90-\alpha)\right]$$

$$-2\pi\cos\alpha P_{inlet} r_{2inlet} l - 2\pi\cos\alpha P_{inlet}\cos\alpha\frac{l^2}{2} - 2\pi\cos\alpha\frac{r_{2inlet}}{k_3}\left[\left(1+\frac{k_2}{k_1}\right)\ln\left(l\frac{k_1}{k_2}+1\right)-l\right][P_{outlet}-P_{inlet}]$$

$$-2\pi\cos\alpha\frac{\cos\alpha}{k_3}\left[\frac{1}{2}\left(l^2-\frac{k_2^2}{k_1^2}\right)\ln\left(l\frac{k_1}{k_2}+1\right)-\frac{1}{2}\left(\frac{l^2}{2}-\frac{k_2}{k_1}l\right)\right][P_{outlet}-P_{inlet}]$$

(4.25)

In equation (4.25) it is noted that:

- the first term represents the momentum forces at the outlet of the control volume
- the second term gives the momentum forces at the inlet
- the third and fourth terms represents the pressure forces on the inlet and outlet respectively
- the rest of the terms are linked with the forces on the conical seat

Unless explicitly stated, all further figures will have the same common parameters $r_{2\ inlet} = 2$mm, $\alpha = 45$ degrees. All forces represented maintain the same sign as has been given in equation (4.25) and therefore in reality represent the reaction forces.

Figure 4.5 gives the outlet and inlet momentums versus the cone seat length for several pressures and poppet clearances. Notice that for small seat lengths, typically a few millimetres, the outlet momentum increases sharply as the length decreases. In fact, the outlet momentum increases with the inlet pressure and with the clearance. The same effect is being seen for the inlet momentum, although the sign is the opposite of the outlet momentum. Comparing the momentum graphs it can be clearly seen that magnitude of the inlet momentum is much smaller than at the outlet.

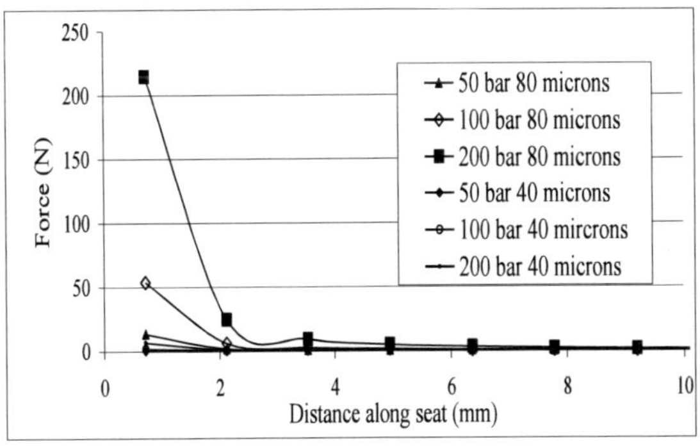

(a) Outlet

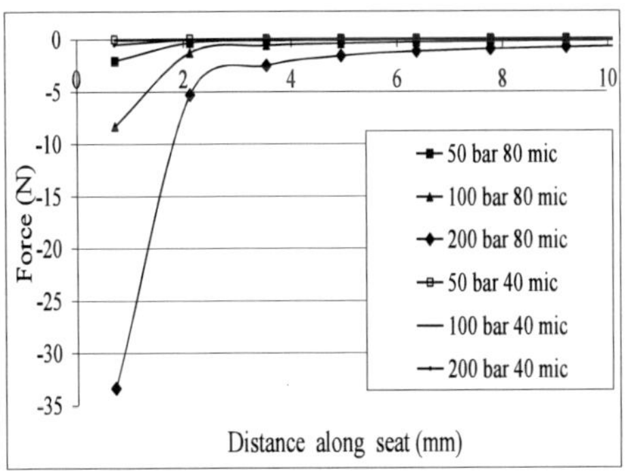

(b) Inlet

Figure 4.5. Momentum/cone seat length, $r_{2\ inlet} = 2$ mm, clearance = 80 microns $\alpha = 45$ degrees, inlet pressures of 50, 100, 200bar.

Both moments change with pressure and clearance and therefore the fluctuation for small conical lengths is very high. The momentum forces alone would tend to close the valve, and this would especially happen at small cone seat lengths, high pressures and large clearances.

In these cases flow force instability is more likely to happen. To appreciate the significance of these plots, it must be recalled that poppet valves are usually built with seat lengths of less than 3mm.

A very important term to take into account, according to Urata [2], is the one which gives the pressure forces on the cone. This term, which in fact is the addition of the last four terms in equation (4.25), or the entire equation (4.24), has been represented in figure 4.6. In the same plot are avaluated as well, the different forces acting on the spool, all terms of equation (4.25). Notice that the total force on the conical spool is mainly due to the force acting on the conical seat and the inlet area, as the cone seat length increases, the force on the cone seat becomes the most significant. On the other hand, when the cone seat lengths are smaller than 2 mm aproximately, the momentum forces play a decisive role forcing the total force curve to follow the net momentum trajectory, in some cases, see figure 4.6(b), this leads to a change of sign of the total force. Although not specifically stated in figure 4.6, it must be pointed out that forces on the cone seat remains reasonably constant for different gaps.

On the other hand, cone seat forces change decisively with the inlet pressure. This effect could be predicted if the pressure distribution along the conical seat is checked, since the pressure distribution remains quite constant for a set of clearances, and changes drastically with the inlet pressure.

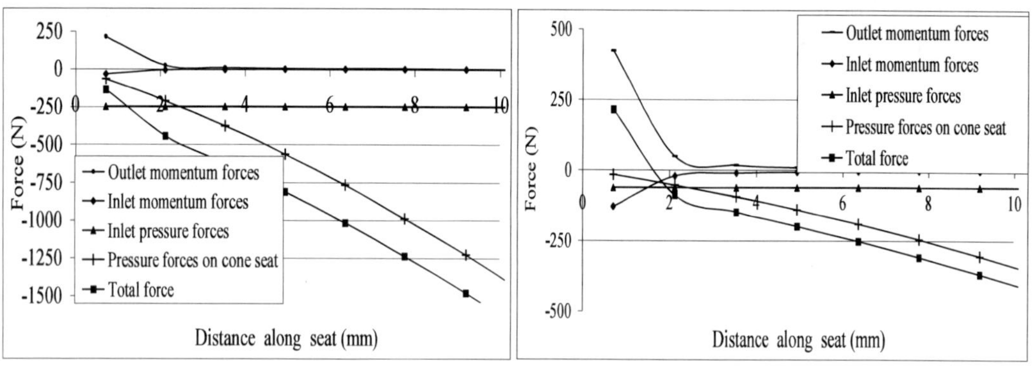

(a) Inlet pressure 200 bar, cone/cone seat distance 80 microns. (b) 50 bar, 160 microns.

Figure 4.6. Axial forces term by term on the conical spool.

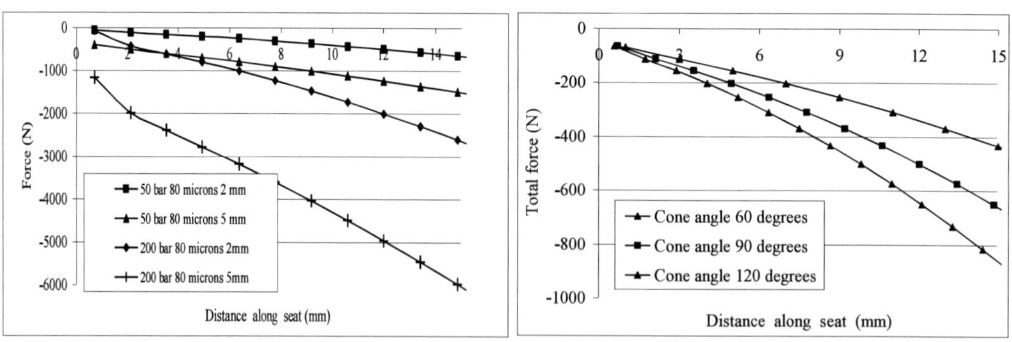

(a) Inlet radius, 2, 5 mm. (b) Cone angles 60,90,120, inlet pressure 50 bar.

Figure 4.7. Total force on the cone for a set on inlet diameters and cone angles, distance cone/cone seat 80 microns.

The effect of inlet diameter and cone seat angle is studied in figure 4.7, where it can be seen that the total force pattern is affected when the inlet diameter is increased. When the inlet diameter is increased, the force on the poppet valve increases and the rate of change of force with position also increases, as can be seen in figure 4.7 (a).

Figure 4.7 (b) shows the total force on the spool with the cone angle. Notice that as the cone angle increases, the force on the conical spool increases as well. To better understand this graph it has to be remembered that for a given cone seat length, the larger the cone angle the larger the outlet radius ($r_{2\ outlet}$), see figure 4.2. Therefore the force component in axial direction will also be larger and as a result a bigger force on the spool must be expected.

4.2.3. CFD modelling

Although Stone [1], amongst others, suggested that a proper study of the flow had to be three-dimensional, most of the computational models found in the published literature use two-dimensional models. The computed velocity at the conical seat exit was higher than in reality in all these cases.

In what follows, a three-dimensional CFD model of the conical seat is presented. Since the aim of this model is to check the equations developed, the conical flow path was modelled and for three different distances between plates, 0.025, 0.037, 0.050mm. The cone seat length was maintained constant at 30mm with an inlet diameter of 4 mm. For every distance three inlet pressures were studied, 44, 94, and 144 bar and in all cases the outlet pressure was set to 100 Pa. In all cases the pressure distribution and the flow across the conical valve were studied. Although the flow was expected to be laminar, in order to utilise the Navier Stokes equations for turbulent flow the standard K-ε model of turbulence was used within the FLUENT CFD package. The idea behind this method was to use the most generic equations and check which flow regime evolved from the simulation. In all cases the Reynolds number used to check laminar flow was computed from:

$$Re = \frac{V\, D_h\, \rho}{\mu} \tag{4.26}$$

D_h = hydraulic diameter at the conical seat inlet, defined as:

$$D_h = \frac{4\left[2\pi r_{2(inlet)}\, H - 2\pi \cos(90-\alpha)\frac{H^2}{2}\right]}{2\pi r_{2\,(inlet)} + 2\pi\,(r_{2(inlet)} - H\cos(90-\alpha))} \tag{4.27}$$

μ = Fluid dynamic viscosity, taken for oil ISO 32 as 0.028 Kg / (m s)

ρ = Fluid density, taken for oil ISO 32 as 875 kg / m^3

V = Fluid velocity at the conical seat inlet defined as:

$$V = \frac{Q}{2\pi r_{2(inlet)} H - 2\pi \cos(90-\alpha)\frac{H^2}{2}} \tag{4.28}$$

The theoretical Reynolds number for any clearance and any pressure differential can be found when equation (4.8) is substituted in equation (4.28). All cases studied had a Reynolds number well below 2000.

In order to gather a maximum amount of information from the CFD model the grid created was very dense, six cells were located between cone and cone seat and double precision was employed. One quarter of the conical seat was modelled, bringing the number of cells down to 720,000. Figure 4.8(a) compares the pressure distribution along the valve seat found via theoretical equation (4.10) and via the CFD model. Notice the excellent agreement. The same sort of agreement was reached for the other two gaps studied. The CFD modelisation also showed that the pressure distribution along the seat remains constant for different cone seat distances, which is exactly what is obtained by the equations proposed, therefore the pressure distribution depends almost exclusively on the pressure differential, this being a key factor when the valve stability is considered. At this point, it must be noted that Mokhtarzadeh-Dehghan [14] observed when working with poppet valves and laminar flow, an increase of force on the valve with the increase of the distance cone/cone seat. It has to be understood that the valve used by Mokhtarzadeh had a lip at the end of the cone, and the force on his spool was very much driven by the reattachment and separation of the flow below that lip, while on the other hand, the study presented in this paper is focused on the cone seat path. When comparing the flow through the valve found via CFD and equation (4.8), figure 4.8(b), it can be noticed that although there is a very good agreement between them the flow found via CFD is slightly smaller than the theoretical one, especially at higher flows. This is due to the effect of including a wall roughness of 0.5 microns in the model computed. In reality, the flow given by equation (4.8), will always be larger than the real one, since the equations do not take into account the wall roughness and also the effect of small yet real reattachment of the streamlines at the cone seat entrance.

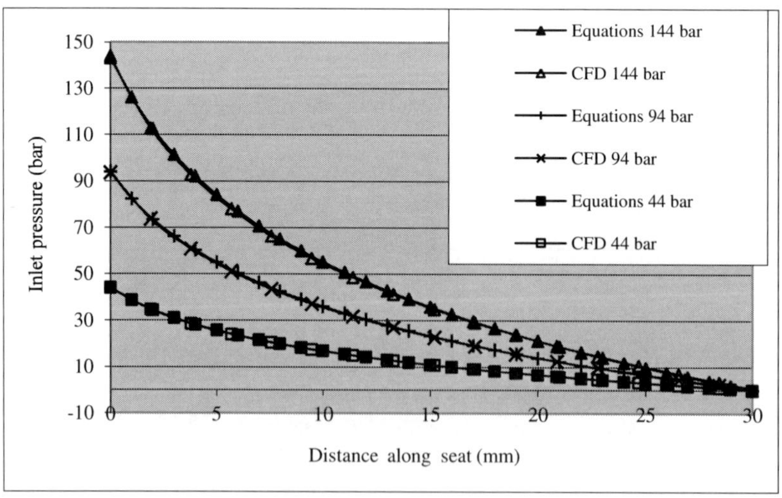

a) Pressure distribution. Cone seat length =30mm, distance between cone/cone seat 37 microns.

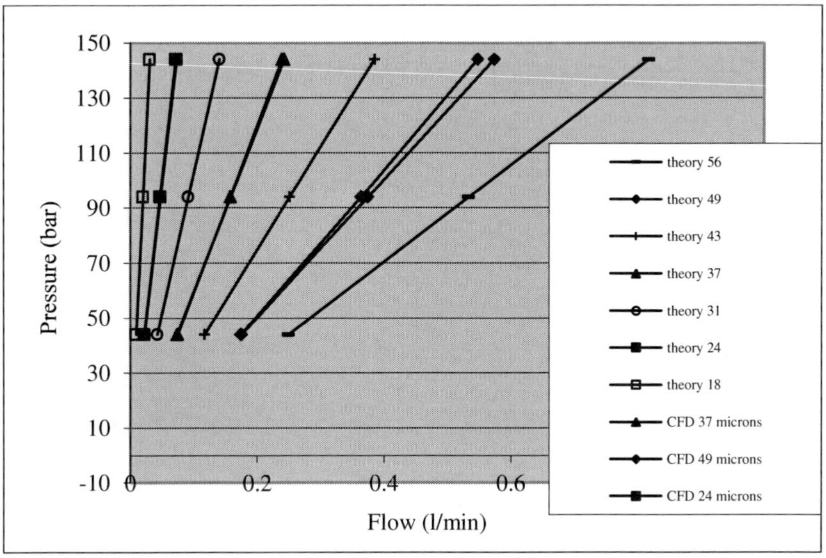

b) Relation pressure flow across the valve, for a set of distances between cone and cone seat.

Figure 4.8. Comparison of pressure distribution along the conical seat and flow though the valve, obtained by CFD and via theoretical solution.

4.2.4. Experimental Results

The test rig used to measure the pressure distribution and flow across the conical device is represented schematically in figure 4.9 along with a photograph. The inlet connecting block (1) and seat unit (2) are connected to the poppet stage (3) via a simple micrometer-thread (40 threads per 25.4mm) adjuster. A dial place on top of the adjustment was used to determine the rotation and hence the poppet clearance. Pressure tappings were machined around the poppet cone, via brass inserts on the face, and each tapping was connected to a calibrated 250mm Bourdon test gauge.

Tests were repeated by re-positioning the clearance each time and average pressures and flow rates were calculated. Due to the large forces acting on the test poppet, movements of up to 16microns were measured, via a micrometer gauge coupled across the test unit, due to the fine thread deformation. An effective poppet increased clearance was therefore created during testing, but this was easily taken into account when collating the data. This created longer testing times as a consequence of building a large poppet with a fine adjuster to validate the theory.

The flow rate was determined using a volume counter manufactured by Kracht. Tests were performed for clearances of 6, 12, 19, 25, 31, 37, 44, 50, microns and three different inlet pressures, 44, 94, 144 bar. The oil temperature was also checked at every test, the temperature at the beginning of every test was always 38°C and the end of each test temperature was between 42°C and 46°C. The effect on viscosity change was also taken into account when collating the test data. The cone seat length and the inlet diameter were 30 mm and 4 mm respectively.

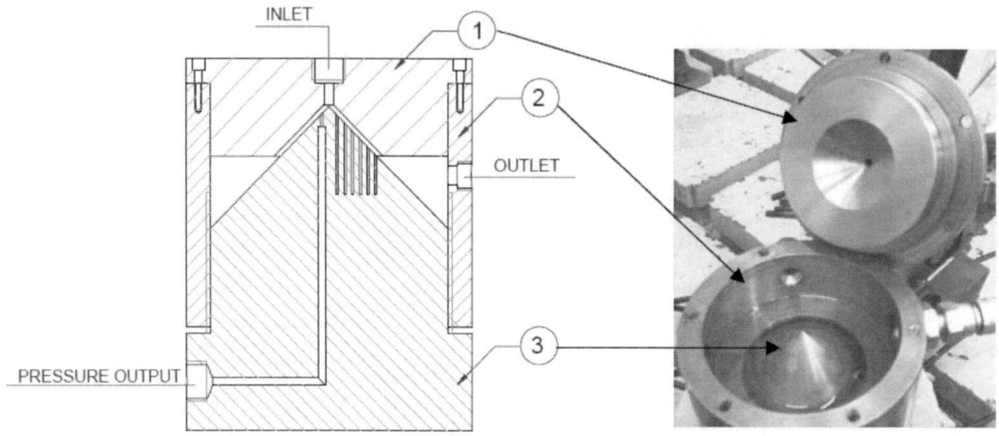

Figure 4.9. Schematic of the test unit and general view of the inside part of the test rig.

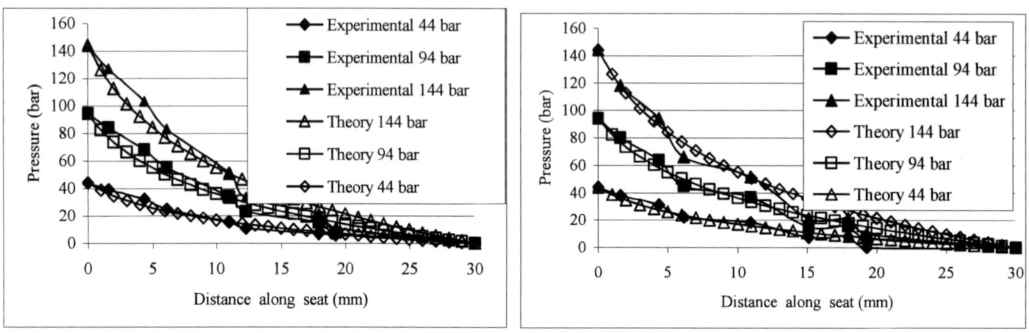

(a) Clearance 25 microns. (b) Clearance 50 microns.

Figure 4.10. Experimental pressure distribution along the cone length and comparisons with theoretical results.

Figure 4.10 shows the experimental pressure along the conical seat for some representative inlet pressures and distances cone/cone seat. Notice that the experimental pressure distribution is as suggested by the theory, quite independent of the distance between plates and completely dependent on the inlet pressure. A good agreement between theory and experimentation is established.

The gap corrections for thread displacement and fluid corrections for temperature are crucial for comparing flow rate results, although of course not so important for the previously discussed pressure distribution. This gap correction results from the simple test technique used, utilising mechanical micrometer screw adjustment. A large poppet design was selected to aid the experimental validation of theory, but this does introduce large forces that are transmitted to the adjuster fine thread. Using a micrometer dial gauge the displacement of the upper part of the test rig, parts (1) and (2) of figure 4.9, versus the base, part (3) of figure 4.9, was carefully measured for a set of different inlet pressures and different distances between cone and cone seat. These measurements, represented by thread rotation, are plotted in figure 4.11 where for every curve a third order polynomic regression curve has been computed. The dial gauge scale smallest division visible, 0.5microns, resulted in readings estimated to be within an accuracy of ±0.125microns. The repeatability of the initial displacement of the poppet is then estimated to be within an accuracy of ±0.25microns.

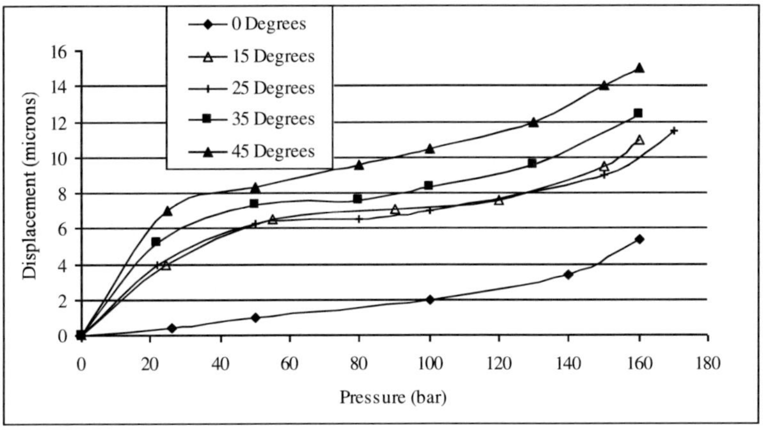

Figure 4.11. Poppet adjuster movement/inlet pressure for a set of adjuster rotations from the poppet closed position.

From all five polynomic equations, a single equation has been produced to take into account the thread displacement as a function of the pressure differential between the valve ports and the angular position of the device, which is directly related to the distance between plates. This equation takes the form:

$$\begin{aligned} d = &[2.439\ 10^{-10}\ \theta^3 - 2.168\ 10^{-8}\ \theta^2 + 6.826\ 10^{-7}\ \theta + 2.465\ 10^{-6}]\ p^3 \\ &+ [-8.113\ 10^{-8}\ \theta^3 + 7.131\ 10^{-6}\ \theta^2 - 2.166\ 10^{-4}\ \theta - 4.007\ 10^{-4}]\ p^2 \\ &+ [9.155\ 10^{-6}\ \theta^3 - 7.557\ 10^{-4}\ \theta^2 + 2.127\ 10^{-2}\ \theta + 3.490\ 10^{-2}\ ]\ p \\ &+ [-1.084\ 10^{-5}\ \theta^3 + 9.423\ 10^{-4}\ \theta^2 - 9.963\ 10^{-3}\ \theta - 7.603\ 10^{-2}\ ] \end{aligned} \qquad (4.29)$$

where d = thread compression (microns), p = pressure differential (bar), θ = angle turned (degrees).

The temperature variation during testing varied between 39°C and 45°C and the change in dynamic viscosity between 30°C and 50°C, and for mineral oil ISO 32, is given by the equation:

$$\mu = 4.38\ 10^{-5}\ T^2 - 4.64\ 10^{-3}\ T + 1.44\ 10^{-1}\ \text{kg/m s} \qquad (4.30)$$

Figure 4.12 represents the theoretical and experimental flow once the theoretical flow given originally in figure 4.8(b) has been properly corrected. This means that the poppet is originally set up for the required clearance, but as pressure is increased the change in clearance is used in the theoretical equation (counter-rotating the adjuster is almost impossible under load). Thus the actual pressure/flow relationship will not be linear due to the test technique adopted. Small flow rates have been specifically considered since the characteristic near to the closed position is important from a control point of view, the reason for this study.

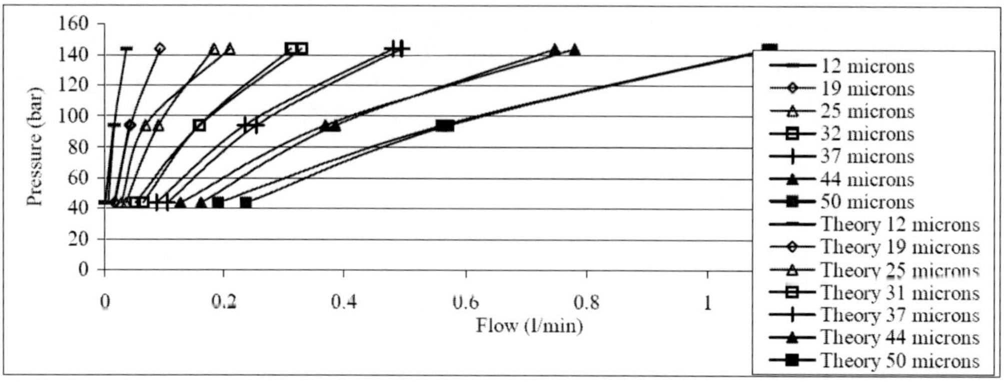

Figure 4.12. Correlation between experimental and theoretical results.

4.2.5. Conclusion

A set of new equations have been successfully developed to establish the pressure drop and flow through a conical valve with a long seat. The equations are valid for laminar flow, which is the kind of flow to be expected on such valves when operating at small openings.

An improoved force equation has been developed based on the new flow equations and therefore taking into account the true pressure distibution. Further experimental research on this force characteristic is needed to validate its applicability for dynamic performance analysis.

To validate the pressure and flow equations, a CFD computer simulation and an experimental test rig were created. Results have been compared showing an excellent agreement. As the flow increases, this agreement between theory and experimentation shows a very small difference and may be caused by secondary effects due to the contraction of the streamlines at the cone inlet.

As pointed by Urata [2], the pressure forces on the seat play a decisive role especially as the seat length increases, and this has been validated mathematically in more detail. For seat lenghts smaller than 3 mm, it is mathemathically demonstrated that the momentum forces increase drastically changing mainly with pressure drop and clearance. This change tends to create an undesirable force characteristic on the valve poppet, but on the other hand as the cone length increases the momentum forces remain much smaller and relatively insignificant. This is a particularly useful outcome of the analysis.

A set of graphs have been presented showing the effect of the force on the spool when changing physical parameters such as cone length, inlet diameter and cone angle. These graphs and the accompanying theory may be used to establish cone designs to minimise variations in total force with poppet lift.

It has been demonstrated that the particular test rig used required careful consideration of adjuster thread compression, particularly for ostensibly small gaps and higher pressures. The accuracy of poppet position estimation is actually more significant when comparing results at low pressures. This may well explain the differences between experimental and computed flow rates observed at the lower pressure and small displacement end of the study. In addition oil temperature variation effects must also be again emphasised for laminar flow conditions expected for the poppet design being considered.

As a final conclusion, it has to be said that the poppet valve momentum change effect would not need to be considered if the length of the chamfered seat is increased to typically greater than 5 mm. In addition, for the poppet under study, this effect is also negligible for clearances beyond 40microns. It is proposed that the developed theory may be used for general cone-seated poppet valve design.

4.2.6. References

[1] Stone JA. (1960). Discharge coefficients and steady-state flow forces for hydraulic poppet valves. *Transactions of the ASME*. 144-154.

[2] Urata E. (1969). Thrust of poppet valve. Bull. Japan. Soc. Mech. Engrs. vol 12 num. 53; 1099-1109.

[3] Scheffel G. (1978). Dynamisches Verhalten eines directgesteurten Kegelsitzventils unter dem Einfluss der Geometrie des Schliesselementes. *Ölhydraulic und pneumatik N* 5; 280-282.

[4] Scheffel G. (1978). Dynamisches Verhalten eines directgesteurten Kegelsitzventils unter dem Einfluss der Geometrie des Schliesselementes. *Ölhydraulic und pneumatik N* 8; 445-448.

[5] Scheffel G. (1978). Einfluss des hydraulischen Schwingungsdämpfers auf das dynamische Verhalten eines Druckbegrenzungsventils. *Ölhydraulic und pneumatik* 10; 583-586.

[6] Watton J. (1988). The design of a single-stage relief valve with directional damping. *Journal of Fluid Control,* vol 18, No 2; 22-35.

[7] Kollek W. Kudzma Z. (1988). Untersuchung des Einflusses von Konstruktionsparametern auf Strömungserscheinungen in Sitzventilen mit kegelförmigen sperrsystem. *Konstruktion* 40; 267-271.

[8] Johnston DN, Edge KA, Vaughan ND. (1991). Experimental investigation of flow and force characteristics of hydraulic poppet and disk valves. *Proc. Instn. Mech. Engrs.* vol 205; 161-171.

[9] Takenaka T, Yamane R, Iwamizu T. (1964). Thrust of the Disk Valves. *Bull. Japan. Soc. Mech. Engrs.* 1964, vol 7 No. 27; 558-566.

[10] Hagiwara T. (1962). Studies on the Characteristics of Radial Flow Nozzles (1st report). Bull. Japan. *Soc. Mech. Engrs. vol* 5 No 20; 656-663.

[11] Petherick PM, Birk AM. (1991). State of the Art Review of Pressure Relief Valve Design, Testing and Modelling. *Journal of pressure Vessel Technology. ASME.* Vol 113; 46-54.

[12] Sallet DW, Nastoll W, Knight RW, Palmer M E, Singh A. (1981). An experimental investigation of the internal pressure and flow fields in a safety valve. Winter Annual Meeting, November 15-20. Washington DC. ASME 81-WA/NE-19; 1-8.

[13] Vaughan ND, Johnston DN.; Edge KA. (1992). Numerical simulation of fluid flow in poppet valves. *Proceedings of the Institution of Mechanical Engineers.* vol 206 1992, 119-127.

[14] Mokhtarzadeh-Dehghan MR, Ladommatos N, Brennan TJ. (1997). Finite element analysis of flow in a hydraulic pressure valve. *Appl. Math. Modelling* vol 21; 437-445.

[15] Chenvisuwat T, Park S, Kitagawa A. (2002). A development of a poppet type brake pressure control valve for a friction brake of rolling stock. *Fifth JFPS International Symposium on Fluid Power*, Nara, Japan vol 3; 733-738.

[16] Yanping Hu, Deshum Liu. (2002). Static Characteristics of relief valve with pilot G-π bridge hydraulic resistances network. *Fifth JFPS International Symposium on Fluid* Power, Nara, Japan vol 3; 739-744.

[17] Bergada JM, Watton J. (2004). A direct solution for flowrate and force along a cone-seated poppet valve for laminar flow conditions. *Proc. Instn Mech Engrs. Part I. J. Systems and Control Engineering*. vol 218; 197-210.

4.3. Some Measured Steady State Characteristics on Proportional Directional Control Valves

Proportional control valves have the advantage of accepting dynamic current inputs, and therefore they are capable of more properly regulating the required variable, pressure, flow, etc, than conventional valves. Proportional directional control valves, should in theory produce a linear relationship between the current applied to the solenoids and the volumetric flow crossing the valve, providing the pressure differential between the pressure and tank ports remains constant. Figure 4.13 presents the measured pressure flow characteristics of a directional proportional control spool valve. It must be clarified that the valve, when performing the measurements, was still under development and further improvements had to be undertaken. The valve was having four connecting ports (ways), and three switching positions, and was having a single stage with a spool. The voltage applied to the valve electronic card was 24V DC, pressure differential between the pressure and tank ports was maintained constant at 1MPa. To perform the test, current was applied first to one of the two valve solenoids, starting from null and increasing it to the maximum value 2A, and then returned back to null. The same process was followed when applying current to the second valve solenoid. At particular points, the volumetric flow through the valve was measured. From figure 4.13 several valve characteristics can be clearly seen, first, it is noticed that the valve requires a current of about 0.75A to allow the flow to go thought, which clearly indicates that the spool overlap might be too large, on one hand this is beneficial because minimizes leakage when the valve is at its centered position, but on the other hand this fact will drastically reduce the valve dynamic response. Another point to be noticed is that he valve has some hysteresis, the flow flowing through the valve when current is increasing will be different from the one obtained when current is decreasing. Notice as well that the valve is not symmetric.

In order to better clarify valve symmetry, a second set of experiments was undertaken. As in the previous case, current was applied first to one valve solenoid and then to the other. Now a constant pressure differential of 0.5 MPa, was maintained between ports P-A and also between ports B-T, the same pressure differential was used when fluid was going from ports P-B and A-T, the flow through the valve and between ports P-A and B-T and when applicable between P-B and A-T was measured independently.

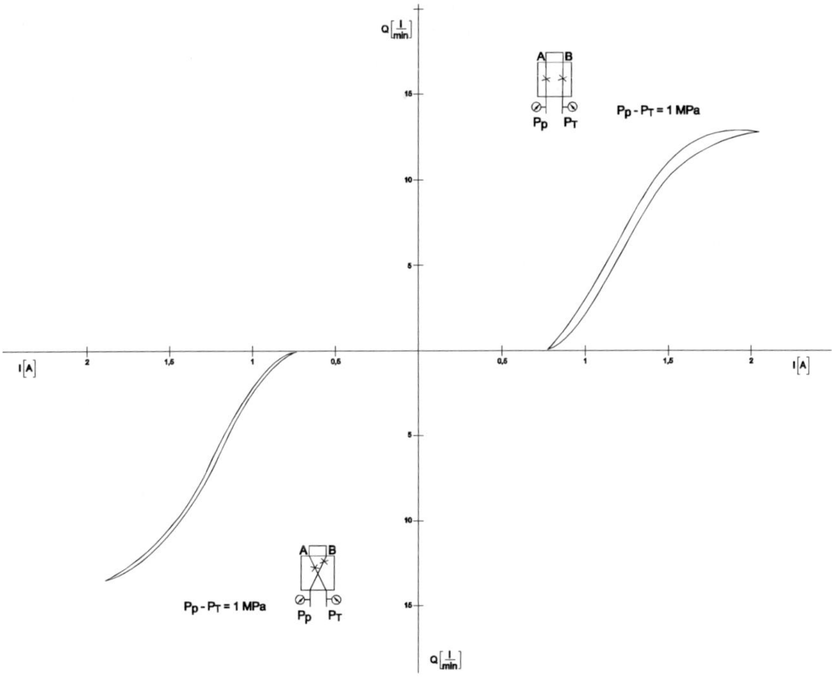

Figure 4.13. Directional Control proportional valve, static characteristics.

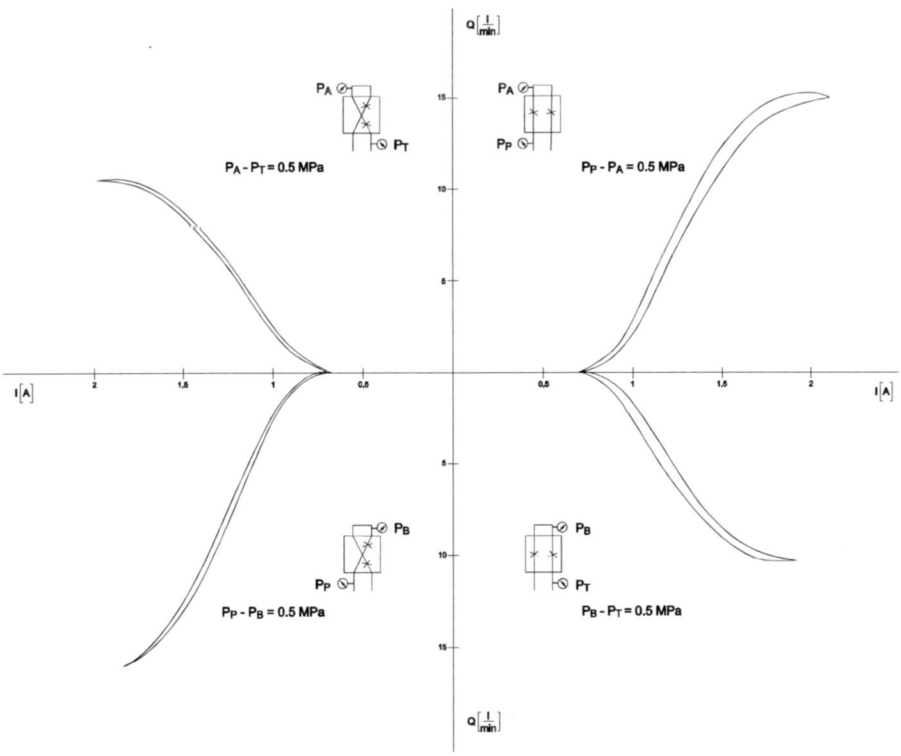

Figure 4.14. Directional Control proportional valve, static characteristics measured between two consecutive ports.

Figure 4.14 shows the results obtained. Notice that for this particular proportional valve, the restriction the fluid is facing when the flow goes from B to T and from A to T appears to be much higher than the one flow is facing when going from P to A and from P to B, then for the same pressure differential the volumetric flow is much smaller in the first two cases than in the second two.

In order to clearly see, which is the pressure drop between two consecutive ports and when current applied to the valve solenoids was increasing or decreasing, the first experiment described in figure 4.13 was repeated, pressure between ports P and T was maintained at 1MPa, and now at each intensity applied, pressure at port A, which was connected to port B, was measured. Figure 4.15 presents the results when current was increasing and figure 4.16 presents the results when current was decreasing. As already was seen in figure 4.14, figures 4.15 and 4.16 show that pressure drop when flow goes from ports A to T and from B to T is nearly for all volumetric flows, higher than the pressure drop necessary for the flow to go from P to A and P to B.

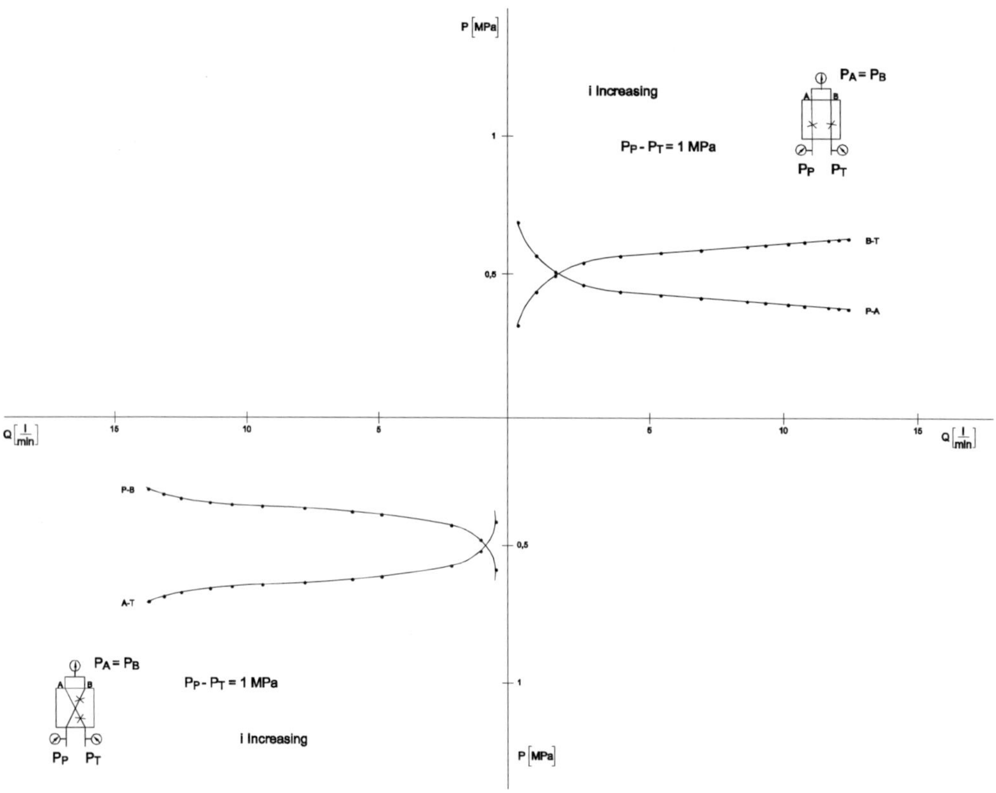

Figure 4.15. Relation pressure drop versus flow for a directional proportional valve. Solenoid intensity increasing.

As previously defined the dynamic characteristics of a proportional valve are given as a Bode diagram plot. For the present valve, and due to the problems already defined, it was expected a poor dynamic performance, therefore dynamic characteristics were not measured.

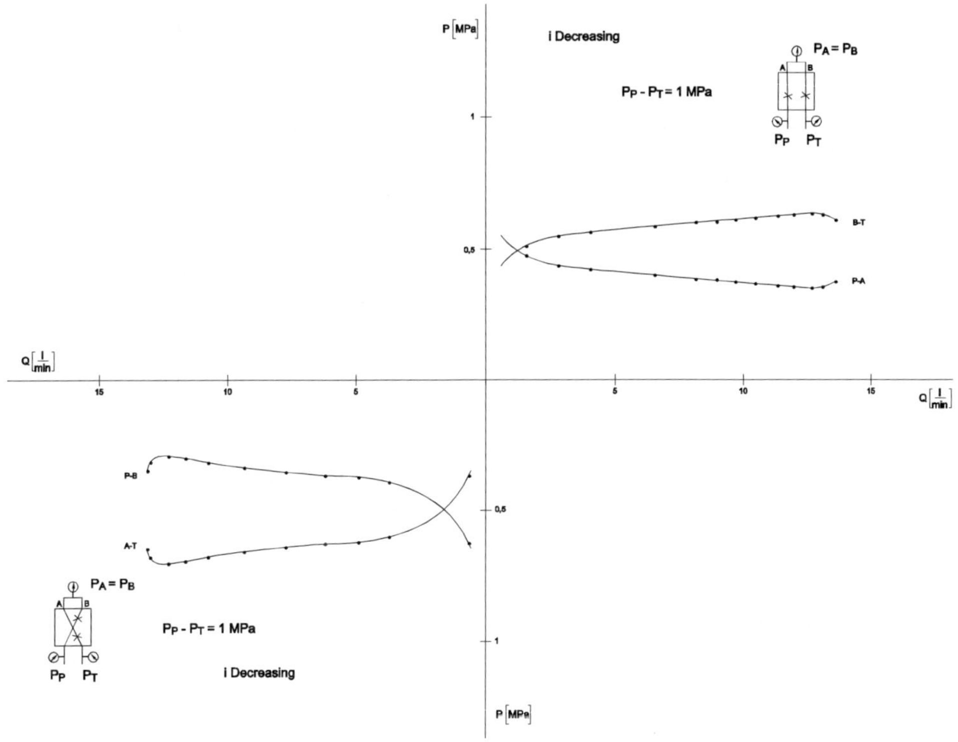

Figure 4.16. Relation pressure drop versus flow for a directional proportional valve. Solenoid intensity decreasing.

4.4. Servovalve Performance

In the present section the first stage of a flapper nozzle servovalve will be deeply studied. First an analytical study will be presented, later servovalve erratic behavior will be analyzed, and possible solutions to the problems described will be given, the section will end via presenting the measured static performance curves of the servovalve.

4.4.1. Introduction to the Four Nozzle Two Flapper Single Stage Servovalve

The four nozzle two flapper flow divider was initially described by Williams [1] in 1965. In his paper were presented the static and dynamic characteristics of this first stage servovalve, some design characteristics were evaluated in order to improve the flow divider performance and decrease the torque motor power requirements. Tshouprakov [2] on late seventies, extensively studied the nozzles used in servovalves, he analyzed the theoretical static characteristics of the different flow dividers. The flow configuration in the flapper nozzle gap was also carefully studied, he highlighted the necessity of considering the discharge coefficients dependent on Reynolds number and distance flapper nozzle. Kassem and Arafa [3] presented in 1982, a theoretical model which allowed to determine the static and dynamic characteristics of the flow divider introduced by Williams [1], and compare them with the ones of a two nozzle first stage servovalve. They indicated that the modification of nozzle internal diameter and length resulted in the variation of the servovalve dynamic performance. Arafa el al [4] implemented the work done in [3] including a comparison between experimental and simulated results, again highlighted the importance of oil volumes inside the servovalve regarding servovalve dynamic behavior. During experiments they appreciated cavitation effects.

Kassem and Arafa [5] study the flow forces acting onto the flappers, they indicate that when flappers are centered the resulting force is null, but when flappers are in a generic position reaction forces tend to displace the flappers from this position, similar results were also found by Williams [1]. Bahr [6] further analyze the work done in [5] and found that the valve working limits were due to flow cavitation. Via studding the dynamic performance indicated which should be the most appropriate internal volumes.

Elgamil [7] evaluated the servovalve performance whenever a perfect symmetry was not fulfilled; he concluded that the overall torque acting onto the armature tended to destabilize it, stability could be improved via using bigger centering springs and or reducing working pressure.

Some other relevant works related to servovalve improvements although not focused on the four nozzles two flapper flow divider are:

Duggins et al. [8] analyzed a scale model of the nozzle flapper gap, they found a negative pressure zone just when the fluid starts entering the channel formed by the flapper and nozzle. Hayashi et al [9] studied the nozzle flapper flow via using finite elements analysis. They observed different flow types for different distances nozzle flapper, areas of negative pressure were also spotted. They indicated that Flow Reynolds number was usually smaller than 500. Good agreement with experimental results was also found. Kirshner and Schmidlin [10] indicated that in the jet stability limit, the Strouhal number is proportional to the square rood of Reynolds number. Capdevila [11] developed a double nozzle pneumatic servovalve, its performance characteristics improved with the valve dimensions reduction. Flow/torque motor instabilities were also noticed. Lebrun et al [12] simulated a double stage servovalve with mechanical feedback, static and dynamic characteristics were compared with experimental results obtaining a good agreement. In the simulation, discharge coefficients were considered Reynolds dependent. Hayashi et al [13] studied the oscillations stability in a system nozzle flapper, they found several oscillation modes which could lead to instability; oscillation frequency is to be independent of the nozzle length. Oscillation amplitude is limited by the nozzle flapper mechanical contact. Watton [14], studied the dynamic performance of a double nozzle servovalve and found a high frequency vibration. Instability is highly dependent on inlet pressure and nozzle flapper distance. He said that oil damping coefficient plays an important role on the stability. Nakada [15] based on maximum flow and 90º phase lag, defines the control margins of servovalves. Lin and Akers [16] studied the static and dynamic performance of two nozzle servovalves, as some authors before they realized that servovalve dynamic characteristics depend on the inlet pressure and on the servovalve internal channels. They propose the use of a modified flapper to reduce instability. In a further paper, Lin and Akers [17] further studied the effect of a modified flapper called "squeeze film damper" saying that stability is reached when using the proposed device. They recommend a working pressure to minimize instabilities. In Lin and Akers [18], study the effect of the damping system tolerances on the dynamic servovalve performance. Akers et al [19] studding again a single flapper double nozzle servovalve, proposed the use of a new configuration servovalve which includes the use of the "squeeze film damper" in the first stage and two spools on the second stage. Akers and Lin [20, 21], further studied the proposed stability mechanism and the second stage double spool configuration servovalve, finding out the appropriate dimensions of the damping mechanism, they compared the new configuration servovalve with the conventional one reporting clear benefits on the new one. Tsai et al [22;24] and Lin and Akers [23], further develop the new servovalve concept defined in [16-21], carefully studding the static and dynamic characteristics of the new configuration.

4.4.2. Directional Control Four Nozzle Two Flapper First Stage Servovalve

The first stage servovalve under study is presented in figure 4.17, it has two flappers and four nozzles, nozzles are connected to the four ports, P= pressure, T=tank, A and B, notice that two different flow configurations are introduced, configuration a) and b), both configurations are possible, yet it appears as if one of them might be more stable, regarding the forces acting onto the flappers, than the other. The particular sort of servovalve presented here has the advantage of being able to work at very high frequencies. In order to find out which flow configuration is producing smaller resultant forces onto the valve armature, the two flow configurations are clearly defined in figure 4.18, in what follows, the momentum equation will be applied to each case.

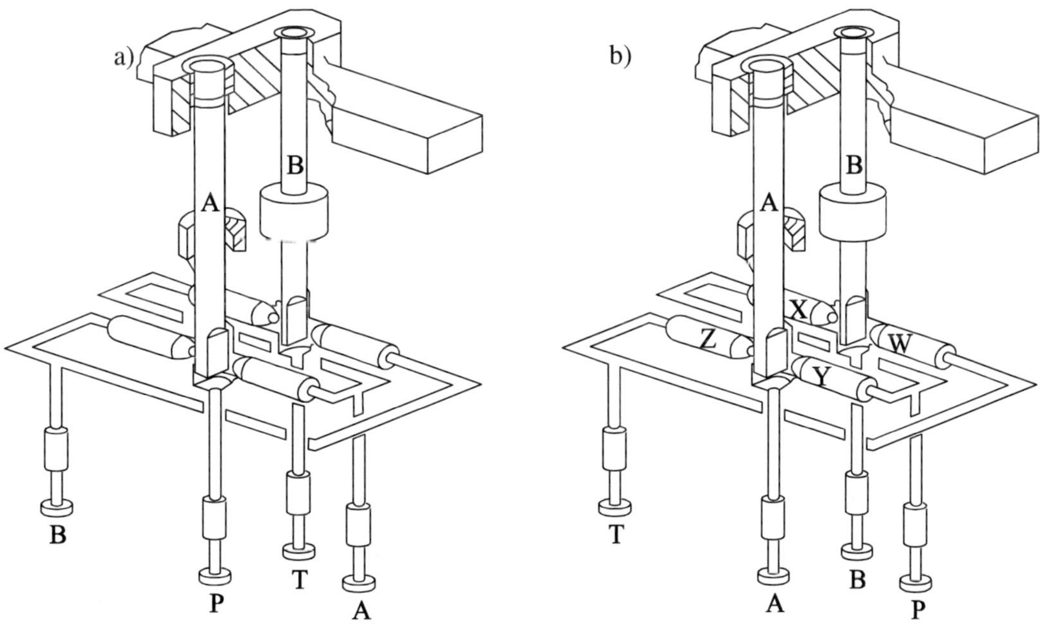

Figure 4.17. Four nozzle two flapper first stage servovalve. Two different flow configurations.

4.4.2.1. Forces Acting onto the Flappers

To determine the flow and pressure forces acting onto the two flappers and for a generic distance nozzle flapper, momentum equation will be applied to each nozzle flapper configuration. The integral form of momentum equation it is expressed as:

$$-\int_{SE} P\, d\vec{S} - \int_{SS} P\, d\vec{S} - \int_{SL} P\, d\vec{S} + \int_{SL} \bar{\bar{\tau}}\, dS - \int_{\forall c} \rho\, \vec{g}\, d\forall = \frac{\partial}{\partial t}\int_{\forall c} \rho\, \vec{V_1}\, d\forall + \oint_{SC} \rho\, \vec{V_1}\, \vec{V_r}\, d\vec{S} \tag{4.31}$$

This expression will now be applied to a control volume defined in figure 4.19, which corresponds to the right hand upper flapper/nozzle of configuration a). Gravitational forces will not be considered and flow is regarded as incompressible and steady.

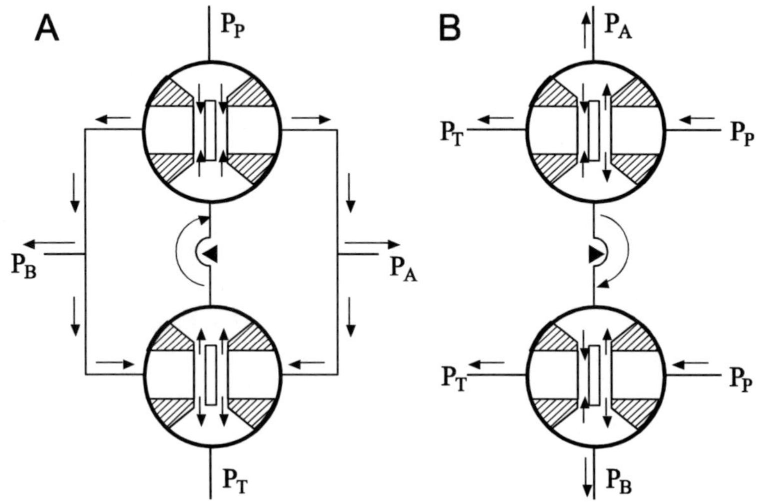

Figure 4.18. Scheme of the two different flow configurations.

Configuration A). Right hand side upper flapper/nozzle. See figure 4.18.

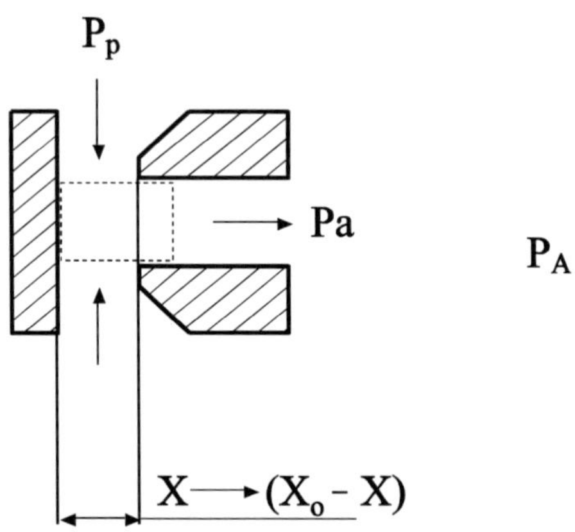

Figure 4.19. Configuration A). Right hand side upper flapper/nozzle.

When applying equation (4.31) to the control volume defined in figure 4.19, dot lines, it must be considered that P_p is the inlet pressure, Pa is the pressure at the nozzle entrance and P_A is the pressure at port A, when the flapper is centered, its distance versus any nozzle is X_0, the generic displacement of the flapper versus its central position is defined as X, then, for the present case, the distance flapper nozzle will be (X_0-X). The nozzle diameter is defined as D. As a conclusion, the reaction force in x direction acting onto the flapper is defined as:

$$F_x = \rho \cdot Q \cdot v_a + P_a \cdot \frac{\pi \cdot D^2}{4} \tag{4.32}$$

The relation between pressure at port P_P and pressure at the nozzle entrance P_a could be given as:

$$P_P = P_a + \rho \frac{v_a^2}{2} \tag{4.33}$$

The volumetric flow entering the nozzle will be a function of the pressure differential, the hydraulic diameter and the discharge coefficient, resulting:

$$Q = C_{d_{FN}} \cdot \pi \cdot D \cdot (x_o - x) \cdot \sqrt{\frac{2 \cdot (P_P - P_A)}{\rho}} \tag{4.34}$$

The fluid average velocity inside the nozzle is:

$$v_a = \frac{Q}{\frac{\pi \cdot D^2}{4}} = \frac{4 \cdot C_{d_{FN}} \cdot (x_o - x)}{D} \cdot \sqrt{\frac{2 \cdot (P_P - P_A)}{\rho}} \tag{4.35}$$

Substituting equations (4.33), (4.34) and (4.35) in (4.32) it is obtained:

$$F_x = 4 \cdot C_{d_{FN}}{}^2 \cdot \pi \cdot (x_o - x)^2 \cdot (P_P - P_A) + P_P \cdot \frac{\pi \cdot D^2}{4} \tag{4.36}$$

Equation (4.36) represents the reaction force in direction x, acting onto the flapper and due to the right hand upper flapper/nozzle part.

Configuration A), right hand lower nozzle/flapper part. See figure 4.18.

Applying equation (4.31) to the control volume defined in figure 4.20 and following the same process as in the previous case, it is found:

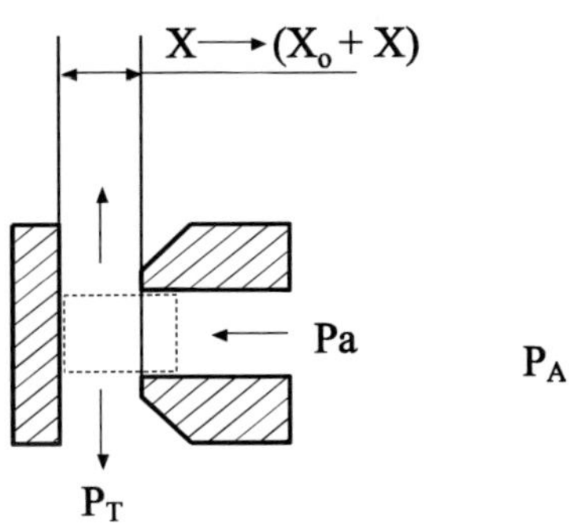

Figure 4.20. Configuration A). Right hand side lower nozzle/flapper part.

$$F_x = \rho \cdot Q \cdot v_a + P_a \cdot \frac{\pi \cdot D^2}{4} \tag{4.37}$$

$$P_A = P_a + \rho \frac{v_a^2}{2} \tag{4.38}$$

$$Q = C_{d_{NF}} \cdot \pi \cdot D \cdot (x_o + x) \cdot \sqrt{\frac{2 \cdot (P_A - P_T)}{\rho}} \tag{4.39}$$

$$v_a = \frac{4 \cdot C_{d_{NF}} \cdot (x_o + x)}{D} \cdot \sqrt{\frac{2 \cdot (P_A - P_T)}{\rho}} \tag{4.40}$$

$$F_x = 4 \cdot C_{d_{NF}}^{\ 2} \cdot \pi \cdot (x_o + x)^2 \cdot (P_A - P_T) + P_A \cdot \frac{\pi \cdot D^2}{4} \tag{4.41}$$

Equation (4.41) represents the reaction force in direction x, acting onto the flapper and due to the right hand side lower flapper/nozzle part. It is important to realize that the forces represented by equations (4.36) and (4.41) point to opposite directions, see figures 4.17 and 4.18.

Configuration a). Left hand side lower nozzle/flapper part. See figure 4.18.

In a similar way as it has been done in the previous two cases, equation (4.31) will now be applied to the control volume defined in figure 4:21, obtaining.

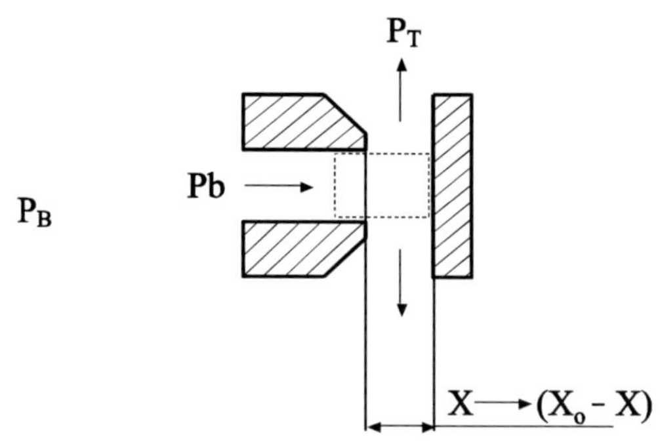

Figure 4.21. Configuration A) Left hand side lower nozzle/flapper part.

$$F_x = -\rho \cdot Q \cdot v_b - P_b \cdot \frac{\pi \cdot D^2}{4} \tag{4.42}$$

The relation between P_B and P_b will be:

$$P_B = P_b + \rho \frac{v_b^2}{2} \tag{4.43}$$

The volumetric flow going from the nozzle towards the flapper, flow going from port B to tank, as a function of the pressure differential, hydraulic diameter and the discharge coefficient will be expressed as.

$$Q = C_{d_{NF}} \cdot \pi \cdot D \cdot (x_o - x) \cdot \sqrt{\frac{2 \cdot (P_B - P_T)}{\rho}} \tag{4.44}$$

The average flow velocity will be:

$$v_b = \frac{4 \cdot C_{d_{NF}} \cdot (x_o - x)}{D} \cdot \sqrt{\frac{2 \cdot (P_B - P_T)}{\rho}} \tag{4.45}$$

And substituting equations (4.43), (4.44) and (4.45) in (4.42) it is reached:

$$F_x = -4 \cdot C_{d_{TP}}{}^2 \cdot \pi \cdot (x_o - x)^2 \cdot (P_B - P_T) - P_B \cdot \frac{\pi \cdot D^2}{4} \tag{4.46}$$

Equation (4.46) represents the reaction force acting onto the flapper and due to the present case.

The final flapper/nozzle assembly from configuration a) is the one presented in figure 4.22. Notice that now flow goes from the flapper to the nozzle, from pressure port to port B, and the distance flapper nozzle is (X_0+X).

Configuration A). Left hand side upper flapper/nozzle part. See figure 4.18.

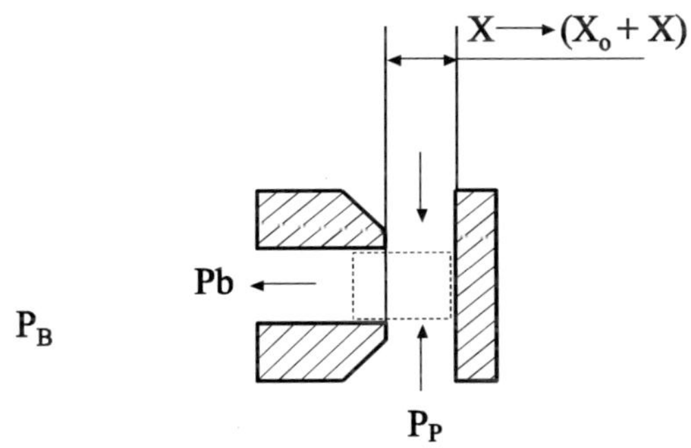

Figure 4.22. Configuration A) Left hand side upper flapper/nozzle part.

Following the process already established in the previous three cases it is obtained:

$$F_x = -\rho \cdot Q \cdot v_b - P_b \cdot \frac{\pi \cdot D^2}{4} \tag{4.47}$$

$$P_P = P_b + \rho \frac{v_b^2}{2} \tag{4.48}$$

$$Q = C_{d_{FN}} \cdot \pi \cdot D \cdot (x_o + x) \cdot \sqrt{\frac{2 \cdot (P_P - P_B)}{\rho}} \tag{4.49}$$

$$v_b = \frac{4 \cdot C_{d_{FN}} \cdot (x_o + x)}{D} \cdot \sqrt{\frac{2 \cdot (P_P - P_B)}{\rho}} \tag{4.50}$$

$$F_x = -4 \cdot C_{dFN}{}^2 \cdot \pi \cdot (x_o + x)^2 \cdot (P_P - P_B) - P_P \cdot \frac{\pi \cdot D^2}{4} \tag{4.51}$$

Considering now the reaction forces defined by equations (4.36), (4.41), (4.46) and (4.51), and considering as well the real position of the nozzles, defined in figure 4.17a), the overall reaction force acting onto the two flappers will be: notice that equations (4.41) and (4.46) have the opposite sign than the one presented, then the flow direction shown in figure 4.18a) is for these two nozzles the opposite than the real one, presented in figure 4.17a.

$$\begin{aligned} F_x &= 4 \cdot C_{dFN}{}^2 \cdot \pi \cdot (x_o - x)^2 \cdot (P_P - P_A) - 4 \cdot C_{dNF}{}^2 \cdot \pi \cdot (x_o + x)^2 (P_A - P_T) - P_A \cdot \frac{\pi \cdot D^2}{4} \\ &+ 4 \cdot C_{dNF}{}^2 \cdot \pi \cdot (x_o - x)^2 \cdot (P_B - P_T) + P_B \cdot \frac{\pi \cdot D^2}{4} - 4 \cdot C_{dFN}{}^2 \cdot \pi \cdot (x_o + x)^2 \cdot (P_P - P_B) \end{aligned} \tag{4.52}$$

Regarding configuration A) it is important to notice that as flow enters the valve, goes from the pressure port and enters to the nozzles, the flow direction is from flapper to nozzle, as will later be demonstrated, whenever flow enters a nozzle, flow instability is likely to appear and since the incoming pressure is high, the entire valve might vibrate.

Using now the same procedure to the second configuration, configuration B) it is found:

Configuration B) figure 4.18, right hand side upper nozzle/flapper.

Applying the momentum equation to the control volume defined in figure 4.23, the following equations are obtained.

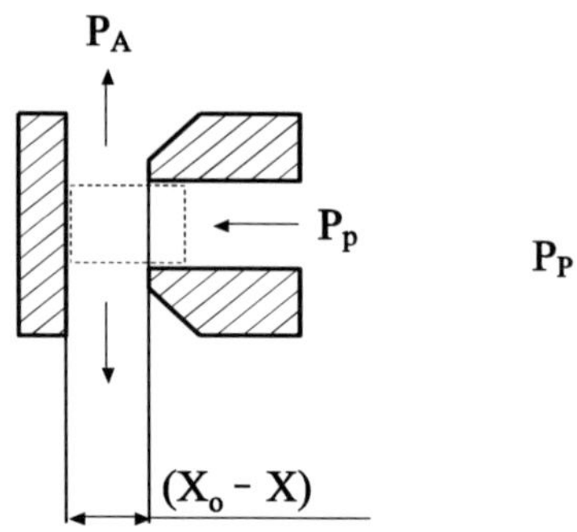

Figure 4.23. Configuration B). Right hand side upper nozzle/flapper.

$$F_x = \rho \cdot Q \cdot v_p + P_p \cdot \frac{\pi \cdot D^2}{4} \tag{4.53}$$

$$P_P = P_p + \rho \frac{v_p^2}{2} \tag{4.54}$$

$$Q = C_{dNF} \cdot \pi \cdot D \cdot (x_o - x) \cdot \sqrt{\frac{2 \cdot (P_P - P_A)}{\rho}} \tag{4.55}$$

$$v_p = \frac{4 \cdot C_{dNF} \cdot (x_o - x)}{D} \cdot \sqrt{\frac{2 \cdot (P_P - P_A)}{\rho}} \tag{4.56}$$

$$F_x = 4 \cdot C_{d_{NF}}^{\ 2} \cdot \pi \cdot (x_o - x)^2 \cdot (P_P - P_A) + P_p \cdot \frac{\pi \cdot D^2}{4} \tag{4.57}$$

Equation (4.57) gives the reaction force acting onto the flapper defined in figure 4.23.

Configuration B). Right hand side lower nozzle/flapper.

Applying now the momentum equation to the control volume defined in figure 4.24, it is obtained:

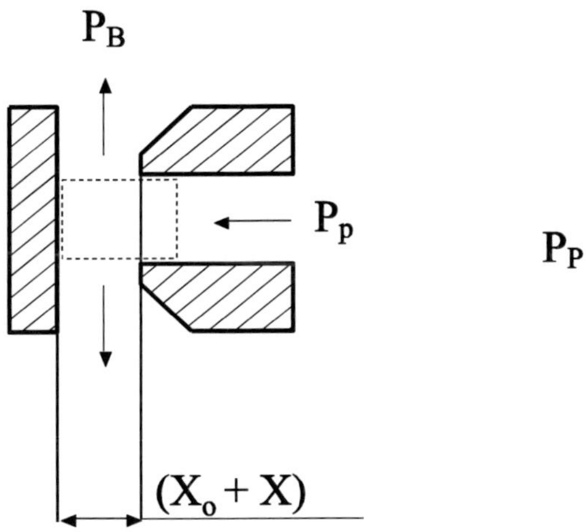

Figure 4.24. Configuration b). Right hand side lower nozzle/flapper.

$$F_x = \rho \cdot Q \cdot v_p + P_p \cdot \frac{\pi \cdot D^2}{4} \tag{4.58}$$

$$P_P = P_p + \rho \frac{v_p^2}{2} \tag{4.59}$$

$$Q = C_{d_{NF}} \cdot \pi \cdot D \cdot (x_o + x) \cdot \sqrt{\frac{2 \cdot (P_P - P_B)}{\rho}} \tag{4.60}$$

$$v_p = \frac{4 \cdot C_{d_{NF}} \cdot (x_o + x)}{D} \cdot \sqrt{\frac{2 \cdot (P_P - P_B)}{\rho}} \tag{4.61}$$

$$F_x = 4 \cdot C_{d_{NF}}^{\ 2} \cdot \pi \cdot (x_o + x)^2 \cdot (P_P - P_B) + P_p \cdot \frac{\pi \cdot D^2}{4} \tag{4.62}$$

Equation (4.62) evaluates the reaction force acting onto the flapper defined in figure 4.24.

Configuration B). Left hand side lower flapper/nozzle.

From the application of the momentum equation to the control volume defined in figure 4.25, it is reached:

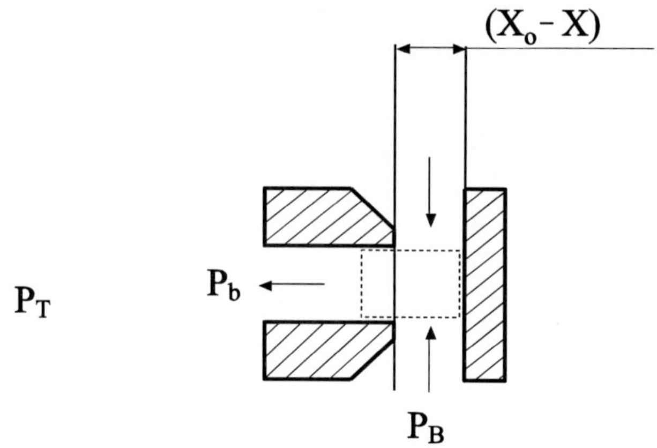

Figure 4.25. Configuration B). Left hand side lower nozzle/flapper.

$$F_x = -\rho \cdot Q \cdot v_b - P_b \cdot \frac{\pi \cdot D^2}{4} \tag{4.63}$$

$$P_B = P_b + \rho \frac{v_b^2}{2} \tag{4.64}$$

$$Q = C_{d_{FN}} \cdot \pi \cdot D \cdot (x_o - x) \cdot \sqrt{\frac{2 \cdot (P_B - P_T)}{\rho}} \tag{4.65}$$

$$v_P = \frac{4 \cdot C_{d_{FN}} \cdot (x_o - x)}{D} \cdot \sqrt{\frac{2 \cdot (P_B - P_T)}{\rho}} \tag{4.66}$$

$$F_x = -4 \cdot C_{d_{FN}}^{\ 2} \cdot \pi \cdot (x_o - x)^2 \cdot (P_B - P_T) - P_B \cdot \frac{\pi \cdot D^2}{4} \tag{4.67}$$

The last nozzle to be evaluated from configuration B) is:

Configuration B). Left hand side upper nozzle/flapper.

$$F_x = -\rho \cdot Q \cdot v_a - P_a \cdot \frac{\pi \cdot D^2}{4} \tag{4.68}$$

$$P_A = P_a + \rho \frac{v_a^2}{2} \tag{4.69}$$

$$Q = C_{d_{FN}} \cdot \pi \cdot D \cdot (x_o + x) \cdot \sqrt{\frac{2 \cdot (P_A - P_T)}{\rho}} \tag{4.70}$$

$$v_T = \frac{4 \cdot C_{d_{FN}} \cdot (x_o + x)}{D} \cdot \sqrt{\frac{2 \cdot (P_A - P_T)}{\rho}} \tag{4.71}$$

$$F_x = -4 \cdot C_{d_{FN}}^{\ 2} \cdot \pi \cdot (x_o + x)^2 \cdot (P_A - P_T) - P_A \cdot \frac{\pi \cdot D^2}{4} \tag{4.72}$$

It is important to remember that all forces determined are the reaction forces.

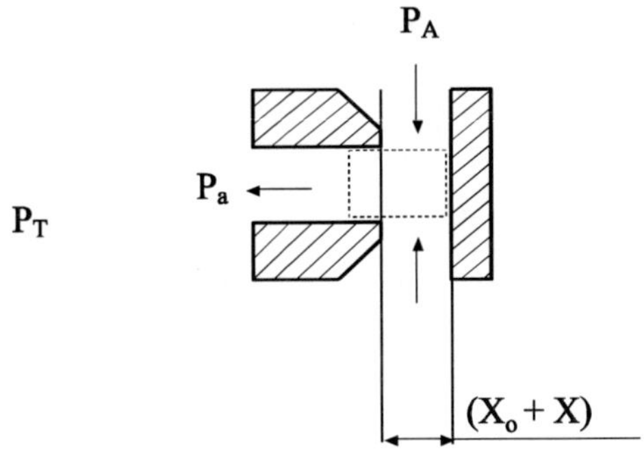

Figure 4.26. Configuration B). Left hand side upper nozzle/flapper.

Equations (4.57), (4.62), (4.67) and (4.72), need to be added together considering its sign to find out the overall reaction force of configuration b).

At this point it must be remembered that figure 4.18b) is not exactly matching the real valve configuration presented in figure 4.17b), then in reality the flow going from pressure port to port B, figure 4.24, and the flow going from port B to tank, figure 4.25, point to the opposite direction than what happens in reality, see figure 4.17b), as a result the sign of equations (4.62) and (4.67) needs to be changed.

The overall reaction force for configuration B) will be:

$$F_x=4C_{d_{NF}}{}^2\pi(x_o-x)^2(P_P-P_A)-4C_{d_{NF}}{}^2\pi(x_o+x)^2(P_P-P_B)$$
$$+4C_{d_{FN}}{}^2\pi(x_o-x)^2(P_B-P_T)+P_B\frac{\pi D^2}{4}-4C_{d_{FN}}{}^2\pi(x_o+x)^2(P_A-P_T)-P_A\frac{\pi D^2}{4} \quad (4.73)$$

When comparing equations (4.52) with (4.73), it is observed that both equations look very much alike, just the discharge coefficients, are generating some differences between these two equations. Notice that two different discharge coefficients are detected, $C_{d_{FN}}$ which it is seen as the discharge coefficient when the flow enters the nozzle from the flapper side, and $C_{d_{NF}}$ which is the discharge coefficient when the flow leaves the nozzle in direction to the flapper. If it is assumed that $C_{d_{NF}}=C_{d_{FN}}$, then both configurations generate the same force onto the flappers. Nevertheless, and in order to be able to clearly see the differences between both configurations the force acting on each single flapper is now evaluated. Using the previous force equations acting on each nozzle can be obtained:

Configuration A). Overall reaction force acting on flapper A. See figure 4.17.

$$F_{x,a)A}=4\cdot C_{d_{FN}}{}^2\cdot\pi\cdot(x_o-x)^2\cdot(P_P-P_A)-4\cdot C_{d_{FN}}{}^2\cdot\pi\cdot(x_o+x)^2\cdot(P_P-P_B) \quad (4.74)$$

Configuration A). Overall reaction force acting on flapper B. See figure 4.17.

$$F_{x,a)B}=-4\cdot C_{d_{NF}}{}^2\cdot\pi\cdot(x_o+x)^2(P_A-P_T)-P_A\cdot\frac{\pi\cdot D^2}{4}+4\cdot C_{d_{NF}}{}^2\cdot\pi\cdot(x_o-x)^2\cdot(P_B-P_T)+P_B\cdot\frac{\pi\cdot D^2}{4} \quad (4.75)$$

Configuration B). Overall reaction force acting on flapper A. See figure 4.17.

$$F_{x,b)A}=4C_{d_{NF}}^{2}\pi(x_o-x)^2(P_P-P_A)+P_P\frac{\pi D^2}{4}-4C_{d_{FN}}^{2}\pi(x_o+x)^2(P_A-P_T)-P_A\frac{\pi D^2}{4} \quad (4.76)$$

Configuration B). Overall reaction force acting on flapper B. See figure 4.17.

$$F_{x,b)B}=-4C_{d_{NF}}^{2}\pi(x_o+x)^2(P_P-P_B)-P_P\frac{\pi D^2}{4}+4C_{d_{FN}}^{2}\pi(x_o-x)^2(P_B-P_T)+P_B\frac{\pi D^2}{4} \quad (4.77)$$

Comparing now equations (4.74) with (4.76) and (4.75) with (4.77) it can clearly be seen that the forces acting on each flapper in configuration B) are much higher than the ones acting on configuration a). As a result, the torque motor used in configuration B) should be more powerful than the one used in configuration A), and therefore configuration B) is capable of better holding possible instabilities. The stiffness in configuration B) is higher than the one in configuration a) as a result it can be said that configuration B) is more stable.

The next interesting step is to evaluate the role discharge coefficients play in the stability of each configuration.

Configuration A), from equation (4.52) and considering $P_A = P_B$; $P_T = 0$; the overall reaction force acting onto both flappers can be given as:

$$F_x = -16 \cdot x_o \cdot x \cdot \pi\left[P_P \cdot C_{d_{FN}}^{2} + P_A\left(C_{d_{NF}}^{2} - C_{d_{FN}}^{2}\right)\right] \quad (4.78)$$

Configuration B), from equation (4.73) and having the same considerations as in the previous case $P_A = P_B$; $P_T = 0$; the overall reaction force will now take the form:

$$F_x = -16 \cdot x_o \cdot x \cdot \pi\left[P_P \cdot C_{d_{NF}}^{2} + P_A\left(C_{d_{FN}}^{2} - C_{d_{NF}}^{2}\right)\right] \quad (4.79)$$

Equations (4.78) and (4.79) clarify the important effect of discharge coefficients on the overall reaction force, indicating that if they were different for different flow directions, the overall forces would also be different for the two different servovalve configurations. Discharge coefficients; depend on the nozzle-flapper distance, and also on the flow direction. In order to demonstrate what it has just been said, the following experiment was undertaken.

4.4.2.2. Servovalve Discharge Coefficients

Using the test rig presented in figure 4.27, flow was allowed to go through a single nozzle and whether in direction from flapper to nozzle or nozzle to flapper. For a set of pressure differentials ranging from 1 to 10MPa every 1MPa, flow was volumetrically measured. Before starting the measurements, the servovalve armature was hold in position, preventing any flapper displacement, then distance flapper nozzle was measured with an accuracy of 1 Micron using the position transducer presented in figure 4.27, finally the system was pressurized and flow measurements undertaken. Fluid used was Hydraulic oil ISO 32.

Discharge coefficient is defined as $C_D = \dfrac{\text{Volumetric flow, measured}}{\text{Volumetric flow, theoretical}} = \dfrac{Q_{measured}}{Q_{theoretical}}$;

For a given distance X* between flapper and nozzle, a given pressure differential and a given fluid, the theoretical volumetric flow was calculated as:

$$Q_{theoretical} = \pi \cdot D \cdot X_* \cdot \sqrt{\frac{2 \cdot \Delta P}{\rho}} \tag{4.80}$$

Reynolds number was defined as:

$$Re = 2 \cdot X_* \cdot \frac{\sqrt{\dfrac{2 \cdot \Delta P}{\rho}}}{\nu} \tag{4.81}$$

ν is the fluid kinematic viscosity [m^2/s^2]
ρ is the fluid density [Kg/m^3]
ΔP is the pressure differential [Pa]
X_* is the nozzle flapper distance [m]
D is the nozzle internal diameter [m]
Q volumetric flow [m^3/s]

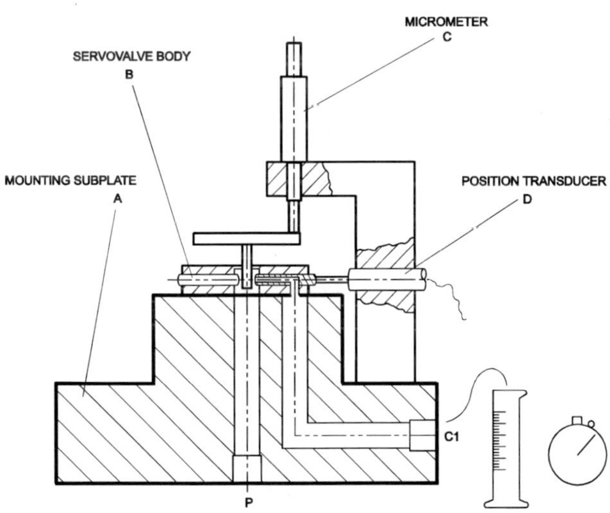

Figure 4.27. Test rig used to find out the flapper-nozzle and nozzle-flapper discharge coefficients.

In figures (4.28) and (4.29) are presented the discharge coefficients found experimentally and for the two flow directions, flapper-nozzle and nozzle flapper.

It is clearly seen that discharge coefficients depend on Reynolds number, flow direction and distance flapper-nozzle. It is noticed that regardless of flow direction the general tendency is similar, yet discharge coefficients when flow goes from nozzle to flapper are slightly smaller than when flow goes from flapper to nozzle. The conclusion is that the overall forces acting onto the two flappers will in reality be different for the two servovalve configurations.

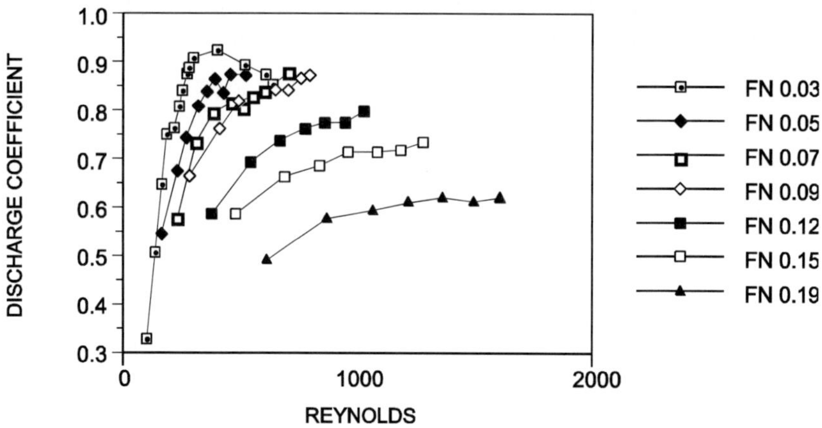

Figure 4.28. Discharge coefficients for flow direction flapper to nozzle.

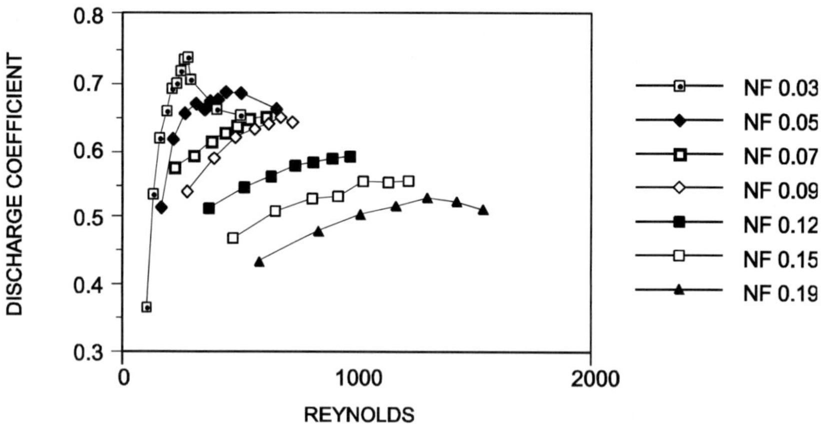

Figure 4.29. Discharge coefficients for flow direction nozzle to flapper.

4.4.2.3. Flow Instability

One of the particular characteristics of the first stage servovalve presented in this section is that flow direction, in two of the nozzles enters from the flapper side. The flow is defined as flow from flapper to nozzle and so are the discharge coefficients. Experimentally it was found that whenever the flow has the flapper to nozzle direction, flow tends to be unstable; an effect similar to Coanda effect appears. In this sub section, a 2-D simulation using the package Fluent and when the flow has both the flapper to nozzle and nozzle to flapper direction is presented in figure 4.30a) b). Fluid used was water, pressure differential 2 Mpa, distance flapper nozzle 0.1 mm.

The simulation shows that whenever the flow enters the nozzle coming from the flapper side, it theoretically generates a jet flowing along the nozzle centerline, figure 4.30a), in reality nevertheless, the jet flips over and tends to be unstable, this instability is not shown in the figure. Figure 4.30b) shows the flow performance when fluid goes from the nozzle towards the flapper, no instabilities were found in this configuration.

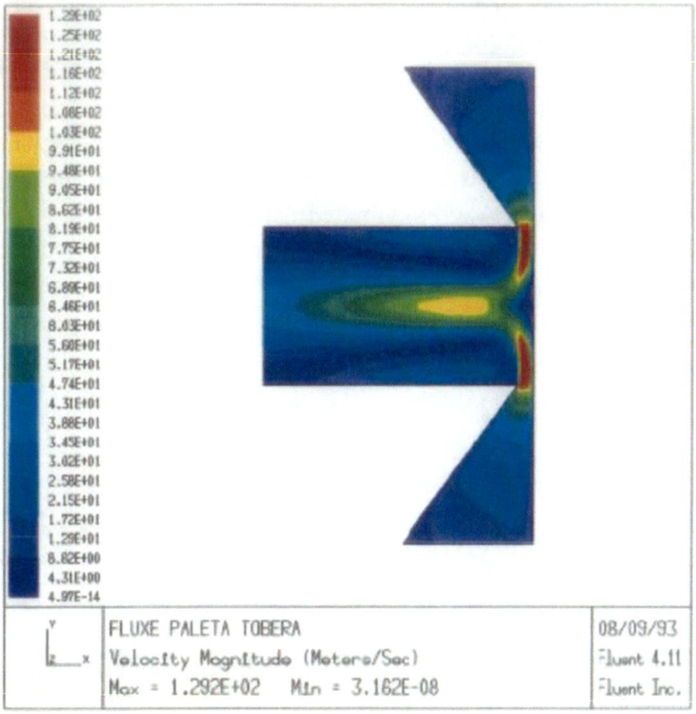

a

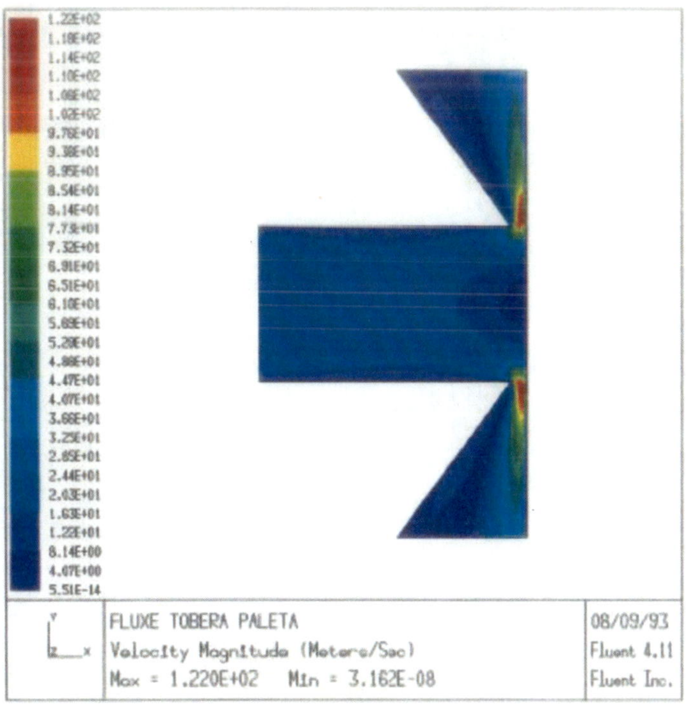

b

Figure 4.30. Theoretical flow performance when fluid goes from: a) Flapper to nozzle. b) Nozzle to flapper.

4.4.2.4. Servovalve Erratic Performance

In order to clarify flow instabilities and its effect onto the servovalve dynamics a set of experiments were undertaken. The first one to be introduced is schematically presented in figure 4.31, notice that the test rig consisted of a mounting sub-plate, where the servovalve armature was being held. The initial distance between flapper and nozzle was measured using a position transducer, flow was allowed to go from flapper to nozzle and through a single nozzle, servovalve armature was allowed to vibrate freely.

Vibrations were recorded via using a sonometer and also via the pressure transducers located at the inlet and outlet. Tests were performed at flapper nozzle distances ranging from 0.01 to 0.1 mm and every 0.01 mm. At each distance, pressure differentials ranging from 1 to 10 MPa were employed. Results are presented in table 4.1.

Table 4.1.

X[mm]	a) Pressure differential between inlet and outlet (MPa). b) Frequency of maximum amplitude (Hz). In some cases there are two main frequencies. c) Amplitude of the main frequency or frequencies (dB).									
0.01	0.99 587 12	2.2 587 14.7	3.21 2262 5.5	4.22 2312 10.2	5.27 2350 11.1	6.29 2412 8.6	7.31	8.27	9.31	10.34
0.02	1.01 587 47.5	2.17 587 57.3	3.2 625 2275 42 43.4	4.26 2312 47.4	5.24 2375 48.5	6.25	7.31	8.27	9.31	10.34
0.03	1.1 587 60.1	2.19 587 2212 42 43.2	3.2 2250 51.9	4.24 2262 55	5.22 2350 63.4	6.31 2362 67.2	7.31 2362 68.4	8.27	9.31	10.34
0.04	1.2 587 51.6	2.2 587 42.1	3.22 2125 39.8	4.24 2050 63.3	5.24 2025 63.6	6.27 2012 61.5	7.31 2037 61.2	8.27	9.31 2025 41.3	10.34
0.05	1.2 587 47.8	2.17 2150 33.2	3.2 2137 37.3	4.27 2112 40.7	5.27 2087 43.9	6.32 2075 46	7.29 2075 44.9	8.31 1987 62.8	9.31 2012 62.7	10.41 2112 51
0.06	1.16 587 43.9	2.2 2150 37	3.2 2137 36	4.24 2100 42.4	5.24 2075 47.7	6.28 2062 46.9	7.31 2050 45.8	8.33 2050 45.8	9.34 2037 43.6	10.48 1975 63.7
0.07	1.16 587 1750 38.4 39.7	2.17	3.2 2137 31.5	4.24 2125 32.3	5.23 2087 41.8	6.28 2050 46.8	7.3 2037 44.8	8.33 2025 43	9.32 2025 42	10.34 2025 42
0.08	1.16	2.19	3.2	4.24 2125 28.1	4.64 2100 31.7	6.26 2062 42.4	7.27 2037 46.4	8.33 2025 44.6	9.31 2012 43.3	10.34 2000 42.3
0.09	1.16	2.19	3.2	4.23	5.25 2112 30.8	6.28 2075 34.3	7.28 2050 37	8.33 2025 40.8	9.32 2012 42.5	10.34 2000 41.2
0.1	1.17	2.18	3.2	4.22	5.26 2132 17.6	6.29 2112 23.1	7.29 2062 31.1	8.31 2037 35.3	9.32 2025 36.1	10.34 2000 34.2

One of the first things to be noticed from table 4.1 is that two main frequencies appear, the first one has about 587 Hz and the second one is of 2200 Hz. At some points, the amplitudes of both frequencies are relevant, which clarifies that for these particular conditions both instabilities have a similar order of magnitude. As an example of the spectra obtained during the experiments, figure 4.32 presents some plots for two distances flapper nozzle and several pressure differentials. For the physical conditions defined in figure 4.32a, the most relevant frequency is the 587 Hz one, figure 4.32b, is characterized by a main frequency of about 2200 Hz, being the 587 Hz frequency completely missing. In figure 4.32c both frequencies have similar amplitude.

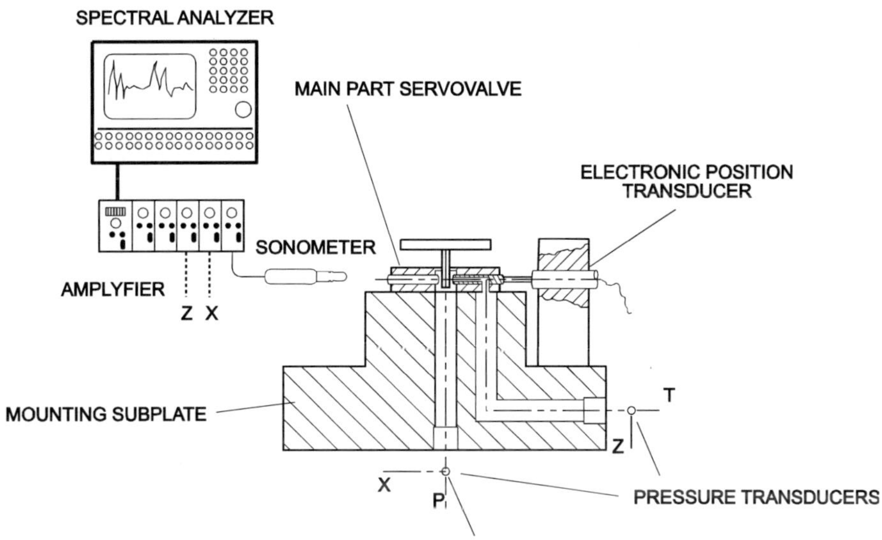

Figure 4.31. Test rig used to evaluate flow instabilities and its effect onto the servovalve armature.

The theoretical calculation of the servovalve armature natural frequency, gave a frequency of 602Hz, and during experimentation was observed that whenever the frequency of 587Hz was appearing, the armature was vibrating. As a conclusion, the frequency around 587Hz is the natural frequency of the servovalve armature.

Regarding the frequency of about 2200 Hz, from table 4.1 it can be seen that it is not constant, it varies depending on the distance flapper nozzle and pressure differential between 1975 and 2375 Hz.

In order to find out the origin of the tonal oscillations occurring around 2200Hz, initially, a diagram indicating the area where instabilities appeared was generated. In figure 4.33, which is based on the information presented in table 4.1, it can be seen that once the distance flapper nozzle increases, the pressure range in which instability appears also increases. Notice that the upper part of the instability area defined in figure 4.33 is in reality limited by the maximum inlet pressure used in the experiments 10.31 MPa.

A very interesting graph is the one presented in figure 4.34, which represents the oscillation frequency of about 2200 Hz versus pressure differential and as a function of the distance flapper nozzle. It must be noticed that for distances flapper nozzle lower than 0.03

mm frequency increases with the increase of pressure, while at bigger flapper nozzle distances, frequency decreases as pressure differential increase.

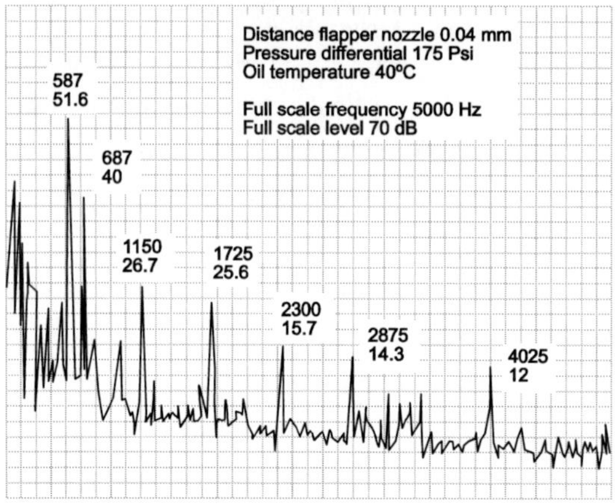

a

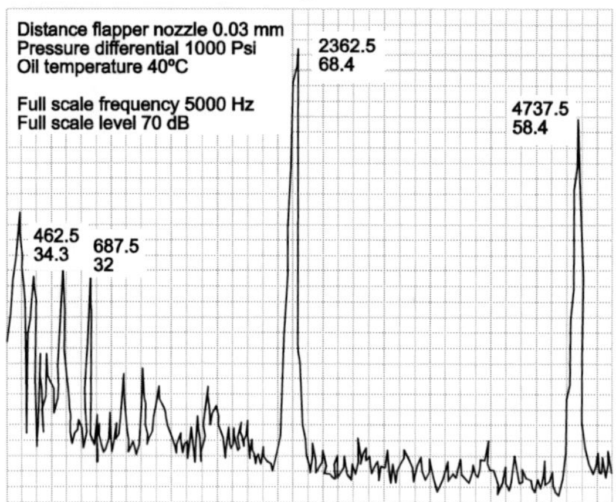

b

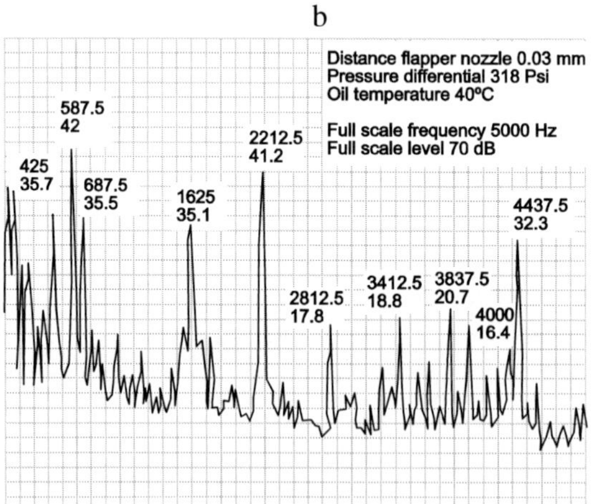

c

Figure 4.32. a, b, c. some characteristic vibration modes obtained during experimentation.

Nevertheless, it is interesting to realize that both sets of curves tend to converge in a particular frequency point, indicating that all those instabilities of frequencies around 2200 Hz have the same origin. To check this statement, figure 4.35 presents in a double logarithmic diagram, the Strouhal number versus Reynolds number and as a function of flapper nozzle distance. Reynolds number was defined in equation (4.81) and Strouhal number is defined as:

$$St = \frac{f\,X}{\sqrt{\frac{2 \cdot \Delta P}{\rho}}} \tag{4.82}$$

Being f, the oscillation frequency measured.

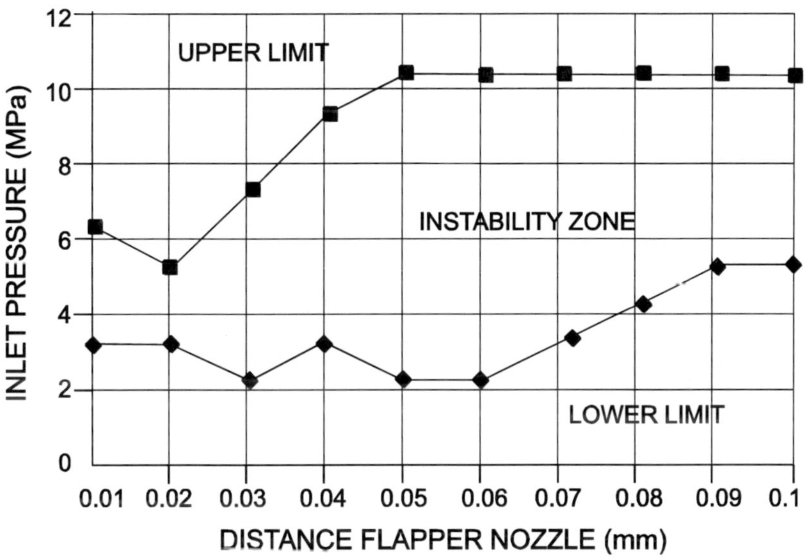

Figure 4.33. Instability zone for the frequency of 2200 Hz.

From figure 4.35 it is clearly seen that all frequency lines are parallel, frequency decreases as Reynolds number increases, and again this graph suggests the possible existence of a single oscillating phenomena. It must at this point be highlighted that if a given frequency is studied, for small distances flapper nozzle and low pressures, oscillation amplitude reaches its maximum, for any given oscillation frequency, oscillation amplitude decreased as pressure differential and or distance flapper nozzle increased. It is also interesting to realize when looking at table 4.1, that the frequency of 587 Hz associated to the vibration of servovalve armature appears at small distances flapper nozzle and small pressures.

According to this information the following hypothesis is established, the origin of all oscillation frequencies is the flow instability when the fluid goes from flapper to nozzle, as the amplitude of a given instability frequency is higher at low distances flapper nozzle and low pressures, it forces the servovalve armature to vibrate at its natural frequency.

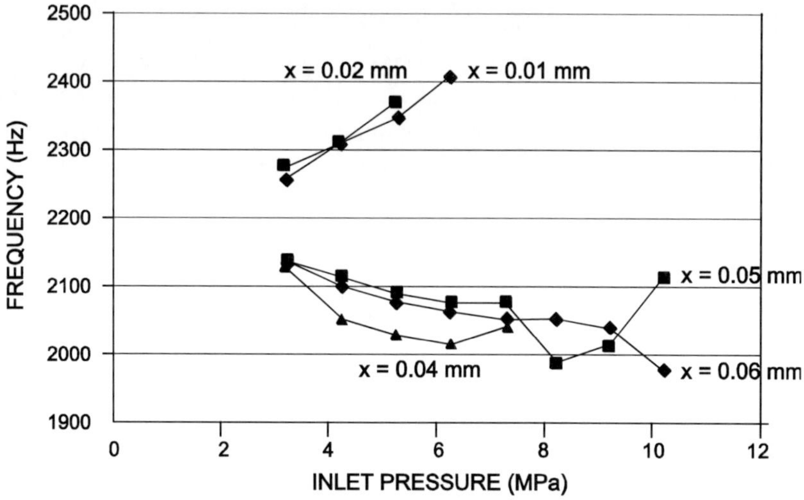

Figure 4.34. Frequency versus inlet pressure for different distances flapper nozzle.

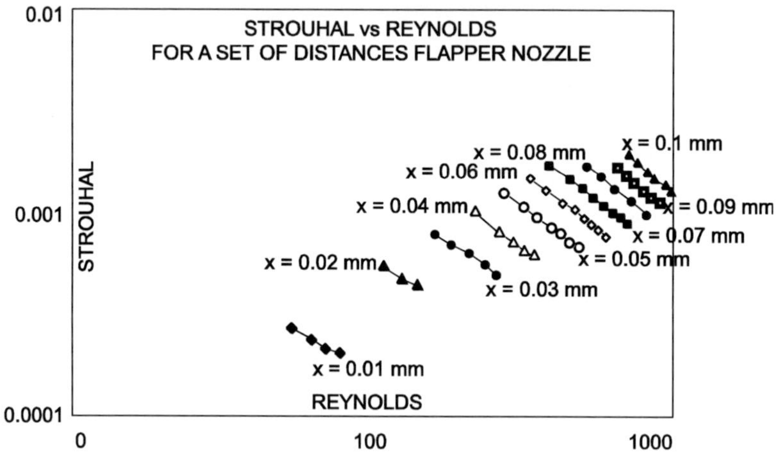

Figure 4.35. Strouhal versus Reynolds in a double logarithmic diagram and as a function of the distance flapper nozzle.

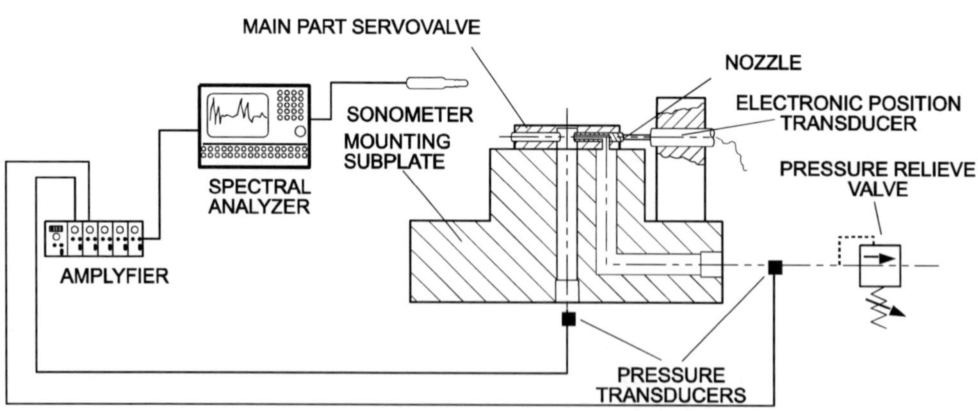

Figure 4.36. Test rig without flapper.

In order to validate the previous hypothesis, the following experiment was undertaken. Using the same test rig as the one presented in figure 4.31, a pressure reduced valve was introduced at the outlet and the armature was removed, this modified test rig is presented in figure 4.36. Flow was allowed to go from flapper to nozzle, pressure differential studied ranged from 1 to 10 MPa. Downstream pressure was modified via using the pressure reduced valve. Table 4.2 presents the results obtained, and it is very interesting to notice that even without the flapper, flow in direction flapper to nozzle is unstable, frequencies increase versus the ones obtained with flapper, and amplitudes decrease. Figure 4.37 presents a typical spectrum obtained with this test rig. It is interesting to point out that in any of the experiments undertaken using the test rig presented in figure 4.36 audible noise was obtained.

From the information obtained using test rigs, figure 4.31 and figure 4.36 the following can be stated:

1. The flow when going from flapper to nozzle is unstable even if there is no flapper.
2. The flapper has an effect of increasing flow instabilities.
3. The chamber formed by the nozzle and flapper tends to amplify the instabilities.

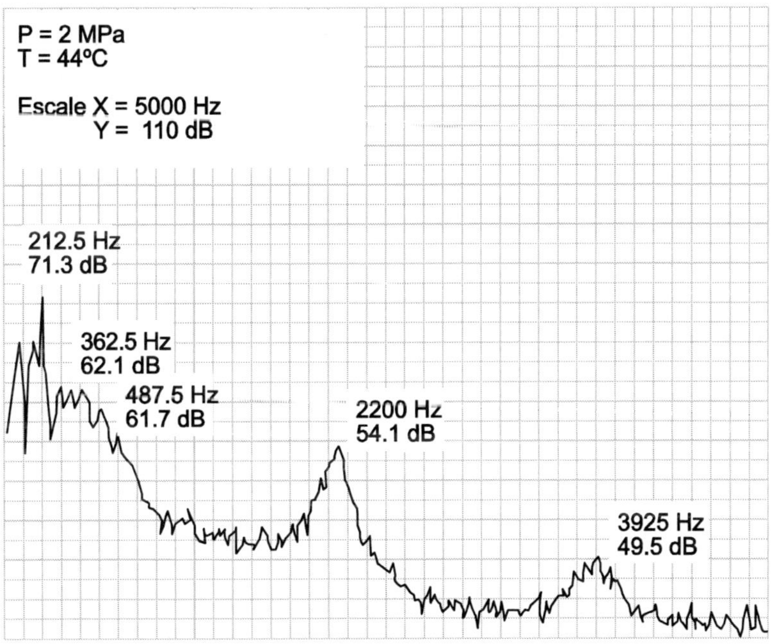

Figure 4.37 .Typical spectrum obtained via using the test rig described in figure 4.36.

The amplifying effect of the nozzle chamber was evaluated using test rig presented in figure 4.38. In reality this test rig is the same as the one presented in figure 4.31, the difference resides in that now the mounting sub-plate outlet was blocked and the nozzle was drilled in order to allow the fluid to leave axially. Tests were done for distances flapper nozzle ranging from 0.01 to 0.1 mm, and pressure differentials ranging from 1 to 10 Mpa. The outcome of this experiment showed no vibration or fluid instability of any kind, and therefore could be concluded that:

Table 4.2. Results obtained via using test rig from figure 4.36

Inlet pressure MPa	Outlet pressure MPa	Pressure differential MPa	a.- Frequency (Hz) b.- Amplitude (dB)
2	1.5	0.5	2362 40.9
	1	1	2262 47.1
	0.5	1.5	2300 53.7
	0	2	2212.5 54.1
4	3.5	0.5	2700 49.8
	3	1	2625 62.5
	2	2	2675 63.5
	1	3	2575 65.6
	0.5	3.5	2575 65.6
	0	4	2487 69.8
5	0	5	2675 74.8
6	0	6	2675 75.2
7	0	7	2750 76.9

a) The onset of flow instability is located at the entrance of the nozzle, the flapper plays an important role regarding the instability onset.
b) The amplifying effect of nozzle chamber is essential for having self sustained oscillations and audible noise. In reality the nozzle chamber seems to be acting as a resonator chamber.

In order to identify the phenomena occurring at the nozzle chamber, a brief review of the self sustained oscillations is required.

According to Powell [25] and Black and Powell [26], to have flow tone generation, three basic steps must be accomplished:

1. It is necessary to have the presence of an unstable flow.
2. The flow has to generate a secondary perturbation which propagates backwards.
3. A feedback mechanism must exist, which amplifies and arranges the oscillations.

A priory, the sound in the present case, can be generated whether by some sort of flow tone generator or by a flow tone resonator, see figures 4.39 a,b.

On the other hand, regarding the onset of flow instability, the unstable flow past cavities is groped in three categories Rockwell and Naudascher [27]. See figure 4.40. It must be pointed out that the work done in [27] was mainly focused in studding rectangular cavities.

A.-Fluid dynamic oscillations.
B.-Fluid resonant oscillations.
C.-Fluid elastic oscillations.

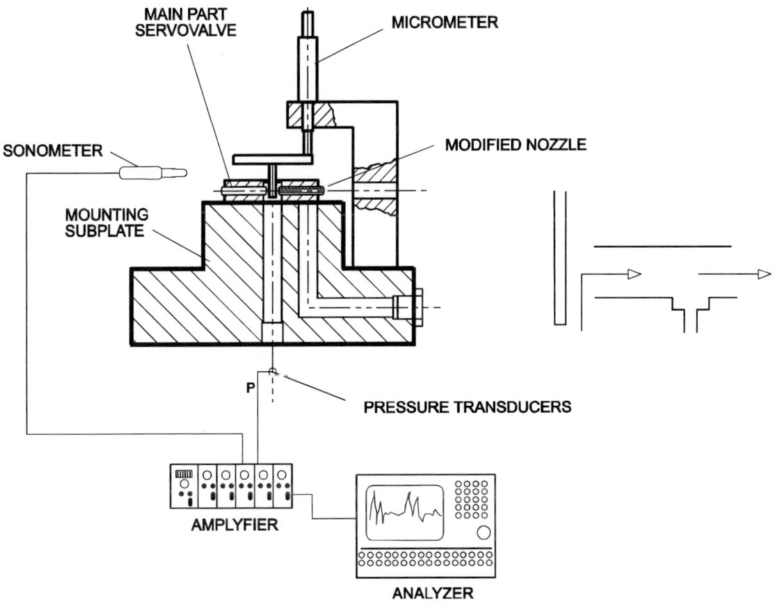

Figure 4.38. Test rig used to evaluate the effect of nozzle chamber regarding the amplification of fluid oscillations.

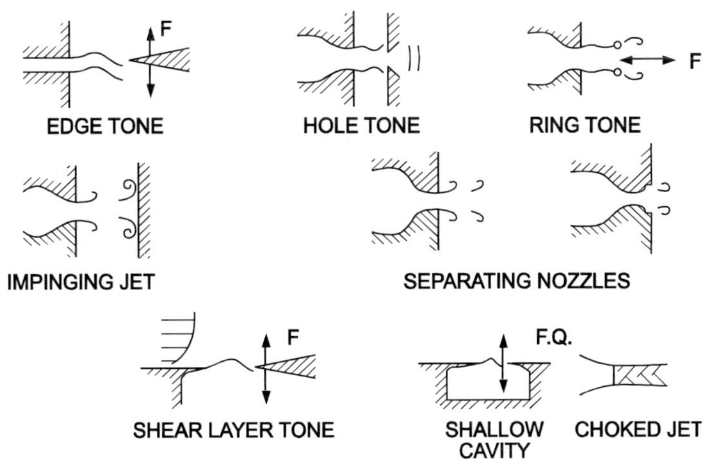

a

Figure 4.39. Continued on next page.

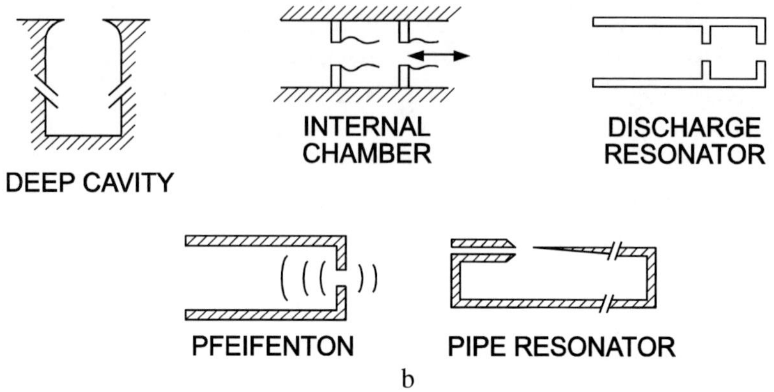

Figure 4.39. a) Basic flow tone generators. Blake & Powell. b) Some flow tone resonators. Blake & Powell.

A. Fluid Dynamic Oscillations

The fluid dynamic category is attributable to instability of the cavity shear layer and enhanced through a feedback mechanism. The oscillations onset, arise from inherent oscillations of the flow.

Purely fluid dynamic oscillations can occur if the ratio of the cavity length to the acoustic wavelength is very small; in the case of liquids no free surface wave effects are present.

The primary mechanism for excitation is the amplification of unstable disturbances in the cavity shear layer; the oscillation is strongly enhanced by the presence of the downstream edge of the cavity. This type of oscillation has some features in common with the jet edge type, involving impingement of a free jet upon an edge.

Two aspects of this fluid-dynamic excitation mechanism should be emphasized.

The amplification conditions of the shear layer instability.
The feedback condition.

The selective amplification characteristic of shear flow causes certain disturbances to be amplified more than others, is a necessary but not sufficient condition for coherent oscillating flow to be produced. An important additional condition for the generation of large amplitude oscillations is an effective feedback. This feedback, which is essentially the upstream propagation of disturbances, is enhanced by the presence of downstream cavity edge. The pressure perturbations emanating from (or in the vicinity of) the downstream cavity edge produce vorticity fluctuations near the sensitive shear layer origin, which in turn provide enhanced disturbances to be further amplified in the shear layer, and so on.

B. Fluid Resonant Oscillations

Fluid resonant oscillations are governed by resonance conditions associated with compressibility or free surface phenomena. For this class of oscillations, the frequencies are sufficiently high that the corresponding acoustic wavelength is of the same order of magnitude or smaller than the cavity characteristic length, L or W, see figure 4.40.

For the ideal case of an organ pipe resonator, resonance is possible if the acoustic wavelength (λ) is 2L for a closed pipe end and 4L for an open pipe end. When the ratio W/L > 1 according to Heller et al [28], longitudinal standing waves may exist, the cavity is termed as a shallow cavity. When W/L < 1, transverse waves may be present, the cavity is denoted as a deep cavity.

Strictly speaking, fluid resonant cavity oscillation occurs only for certain values of λ/W, corresponding to a resonant standing wave in the cavity.

Cavity oscillations are linked with sufficiently high speeds and consequently to sufficient high frequencies. A central feature of shallow cavity oscillations, and in a less grade, of a deep cavity and fluid dynamic oscillations, is the coexistence of several periodic or quasi periodic frequencies at a given value of Mach number. It is plausible therefore, that this multiple resonance effect can be traced to the simultaneous existence of several frequency components in the unstable shear layer. The typical frequency response of a Helmholtz and organ pipe resonators, is presented in figure 4.41.

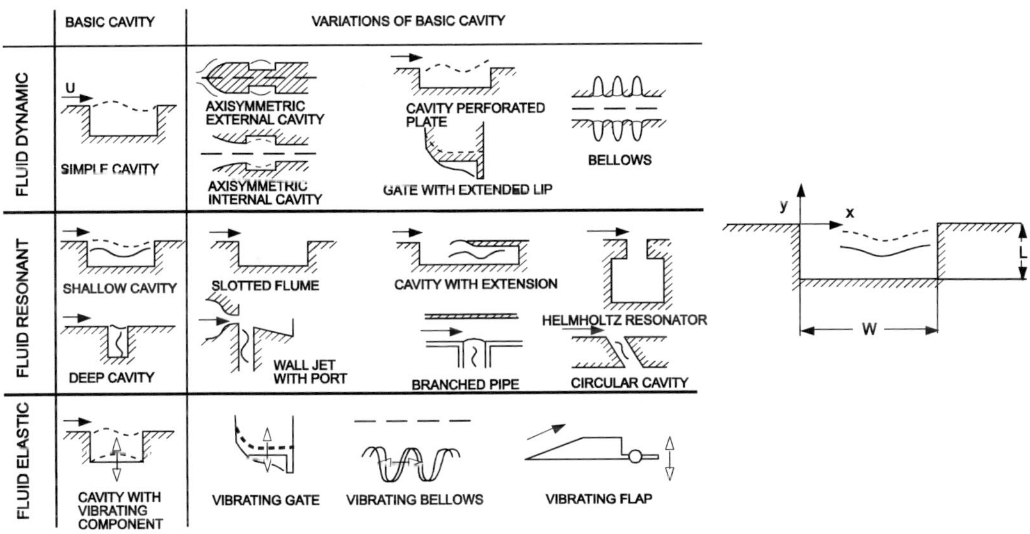

Figure 4.40. Categorization of fluid dynamic, fluid resonant and fluid elastic types of cavity oscillations. D. Rockwell, E. Naudascher.

C. Fluid Elastic Oscillations

Fluid elastic oscillations are primarily controlled by the elastic displacement of a solid boundary and are dependent upon the elastic, inertial and damping properties of the structural system.

As a rather crude but conceptually helpful analogy, it can be assumed that the vibrating structural part has much the function that the resonating wave has in the case of fluid-resonant oscillation. The frequency response of the system therefore, can be represented by a diagram very similar to the one in figure 4.41.

Another possible cause of vibration would be the existence of a flow tone generator, due to a jet, which flows axially along the pipe or by a jet, which impinges on a wall, impinging jet. This wall could be in the present study the end of the nozzle, see figure 4.42.

According to the work done by Powell [25] in which studied jets at high Reynolds numbers, and to the work done by Nossier and Ho (1982), figure 4.43, where tones created by impinging jets were analyzed, the relation Strouhal versus Reynolds has to increase as the Reynolds number increases, see figure 4.43.

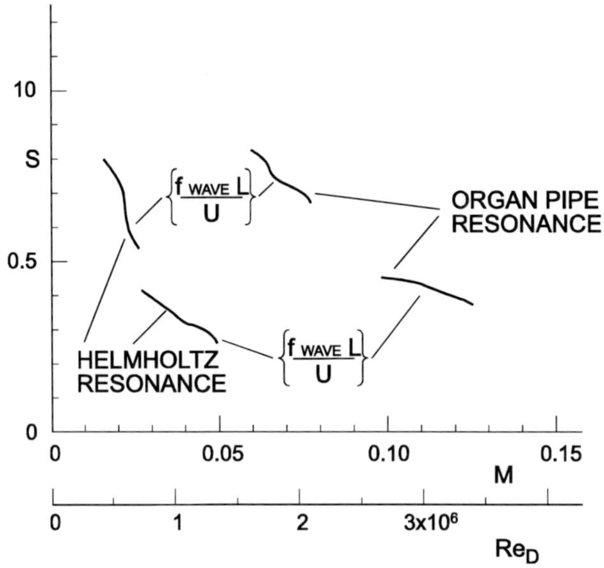

Figure 4.41. Tendencies Strouhal Reynolds for different resonators. Rockwell and Naudasher.

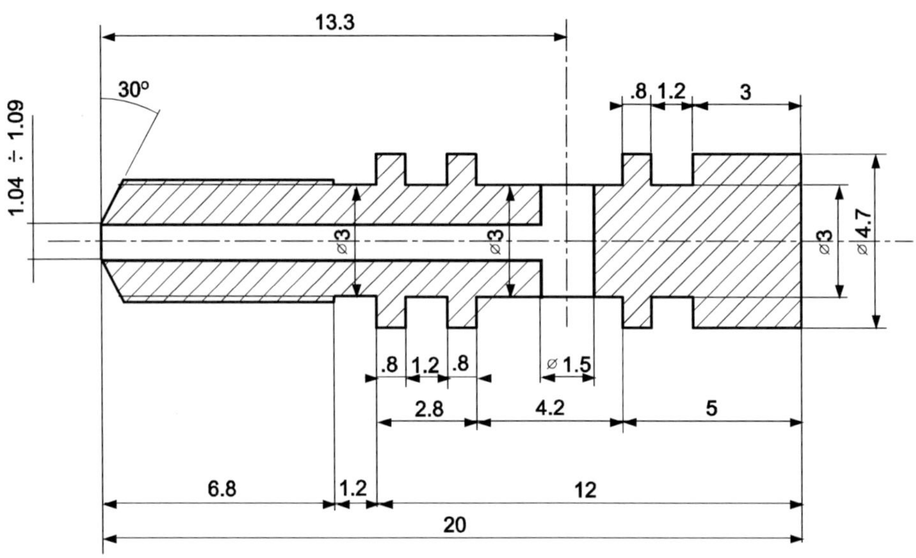

Figure 4.42. One of the nozzles of the servovalve. (Units in mm).

Comparing the results presented in figure 4.35 where Strouhal versus Reynolds number for the present study were introduced, with figure 4.43, it can be concluded that no jet tone or impinging jet tone is appearing in the present study, since for these cases the Strouhal number increases with the increase of Reynolds number and in the present study the opposite happens. On the other hand, taking into account the results found with the test rig introduced in figure 4.36, where there was no flapper, it must be said that the onset of the perturbation is not due to a fluid elastic displacement.

As a resultant of this exposition, is possible to affirm that the tone generated must be created by fluid dynamic or fluid resonant oscillations.

Fluid dynamic oscillations can occur according to [27] if:

- -Exist inherent oscillations of the flow. (Test presented in figure 4.36 shows a clear instability of the flow).
- -The ratio cavity length (L) versus acoustic wavelength (λ) is small. For the present case: $\frac{L}{\lambda}=\frac{L}{CT}=\frac{L}{C\frac{1}{f}}=\frac{L}{\sqrt{\frac{\beta}{\rho}}\frac{1}{f}}=\frac{0.014}{\sqrt{\frac{1.6\cdot 10^{9}}{875}}\frac{1}{2200}}=0.022$; which is reasonably small.
- -No free surface effects are present.

It seems clear at this point that fluid dynamic oscillations are in the present study possible. The question now is to find the feedback mechanism which amplifies the instability generated at the nozzle entrance. This feedback mechanism must be related to the nozzle chamber, since the results obtained using the test rig introduced in figure 4.38, showed no fluid oscillations of any kind.

At this point, and as a hypothesis, can be said that the nozzle chamber acts as an amplifying mechanism of the perturbations created at the nozzle entrance. If this is true, it must be some relation between the chamber natural frequency and the perturbation frequency. As a first approximation, considering the nozzle as a closed or open chamber, its natural frequency will be given as:

$$f(Hz)=\frac{C}{2\pi\lambda};\ C=\sqrt{\frac{\beta}{\rho}}; \tag{4.83}$$

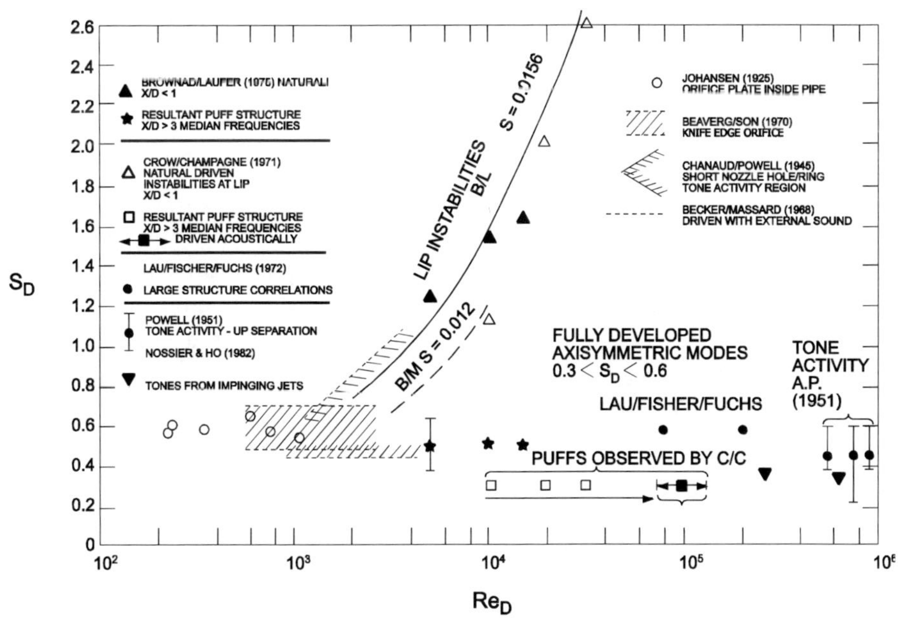

Figure 4.43. Typical plot Strouhal number versus Reynolds number for disturbance-sensitive circular jets from orifice plates and short nozzles. Blake and Powell.

Using equation 4.83, the natural frequency of the ideal nozzle was obtained to be 6928 Hz for a closed end and 3464 for an open end. Notice that the third sub-harmonic of the closed end nozzle frequency is 2309 Hz which is approximately the flow instability frequency found during experimentation. See table 4.1 flapper nozzle distance 0.01 mm. It is to be highlighted as well that, the twelfth sub-harmonic of the ideal closed nozzle or the sixth sub-harmonic of the ideal open nozzle, gives a frequency of 577Hz, which is almost the natural frequency of the servovalve armature.

Looking again at the work done by Rockwell & Naudascher [27] and Powell [25] on Helmholtz and organ pipe resonators, and comparing their conclusions detailed in figure 4.41 with figure 4.35 obtained from the authors experiments, can be concluded that the nozzle acts as a resonator chamber.

Vogel [29] investigating a reed organ, which consisted of a resonator pipe and a blade, found that flow oscillations can presumably be controlled by both fluid resonant and fluid elastic mechanisms. He also found that coupling between the oscillations of the blade and the fluid in the resonator, produces control conditions which are completely different from those of any one of the elementary systems, (the purely fluid resonator or the purely fluid elastic). He also explained that although the effect of coupling can substantially alter the overall nature of the oscillation, it is evident that considerable insight into a system undergoing mixed excitation can be gained by synthesizing it into categories corresponding to each of the basic types of excitation.

4.4.2.5. Conclusion

The self-excited oscillations are due to the flow instability. The onset of flow instability is located at the entrance of the nozzle, where the flow is alternatively separated and reattached. One of the functions of the flapper is amplifying the inlet instability, the other function is to help in creating a resonator chamber.

The resonator is the main amplifying mechanism of the phenomena.

From the tests undertook using the third test rig, can be concluded that although there is flow instability at the entrance of the nozzle, the feedback mechanism is essential to maintain fluid dynamic oscillations.

The frequencies presented in table 4.2, are higher than the ones introduced in table 1, they also have a wider range of variation. This might be explained when considering that the resonator chamber can with some difficulties amplify and direct the flow instabilities, and so the instabilities created at the nozzle inlet have a higher freedom degree to oscillate.

In the first test, the resonator effect is maximum, the oscillations are directed by the resonator chamber, notice that when the chamber is closed (small distances flapper nozzle), the vibration frequencies are higher, and for wider distances flapper nozzle the vibration frequencies decrease. This effect follows the tendency of an ideal organ pipe for closed and open end.

Table 4.1 presented two main oscillation peaks, the 587 Hz one associated to small distances flapper nozzle and low pressure differentials, which is liked to fluid elastic oscillations, meaning that the flow oscillations created in the flapper nozzle gap are governed by a fluid elastic movement. The peak of around 2200 Hz is due to the resonant amplification of the inlet instabilities.

As defined by Vogel [29], both phenomena might appear simultaneously, this happens in the first test, table 1 for a distance flapper nozzle of 0.03 mm and an inlet pressure of 2.19 MPa.

The oscillation frequency changes with:

- The distance flapper nozzle. Which affect the inlet perturbations.
- The pressure. Which affect the inlet perturbations and also the natural frequency of the nozzle via changing the bulk modulus.

To solve the problem, a priory, two approaches are possible, whether acting on the onset of the instability, flapper nozzle entrance, or on the feedback mechanism. Since very few can be done at the inlet point, the solution to the problem has to be obtained via destroying the feedback mechanism as demonstrated in the latest introduced test rig.

4.4.2.6. Servovalve Static Performance Curves

In section 4.3 static characteristics of a, still under development, spool proportional valve were presented. In the present section the static performance curves of the four nozzle two flapper first stage servovalve shall also be introduced.

Three static performance curves are going to be presented, the pressure versus current, flow versus current and leakage flow versus current.

To determine the curve pressure versus current, see figure 4.44, the ports A and B will be blocked, simulating the case the load is blocked, and pressure differential between both ports is measured for all intensities range applied to the servovalve. Inlet pressure is maintained constant at 7 MPa.

Via using the same configuration and measuring the flow leaving the servovalve, the flow leakage curve versus intensity current applied to the servovalve is to be found, see figure 4.45.

To determine the curve flow-current, the flowmeter needs to be installed between ports A and B, as in the previous two cases, inlet pressure is to be maintained constant at 7 Mpa, figure 4.46 presents the curve obtained.

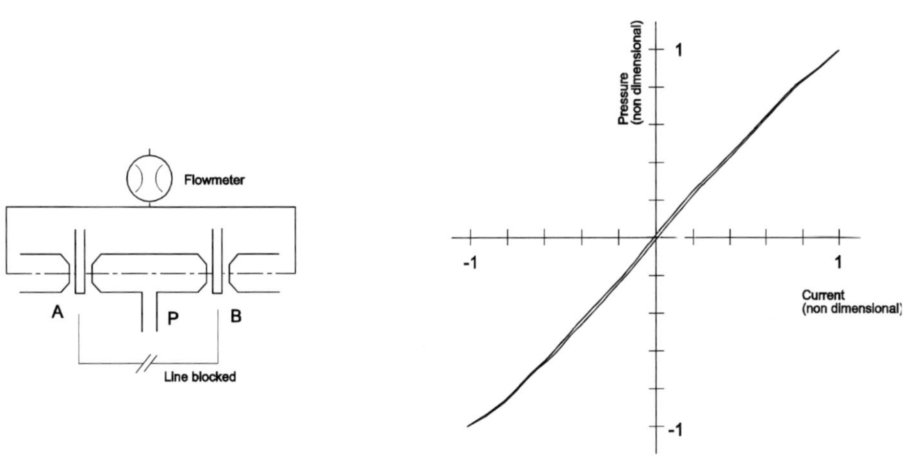

Figure 4.44. Characteristic curve pressure versus current for a four nozzle two flapper servovalve. Experimental.

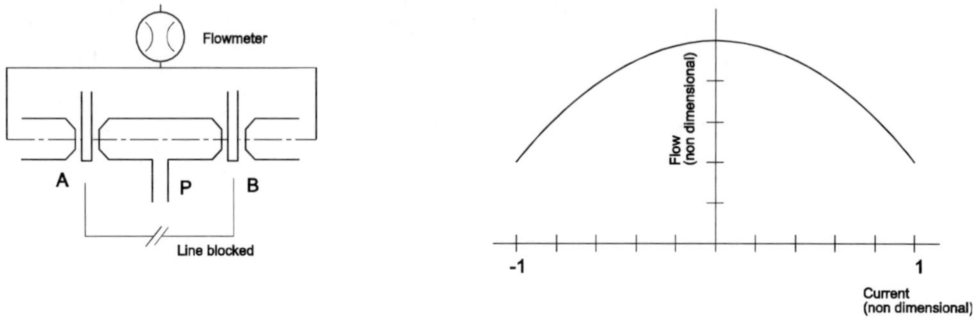

Figure 4.45. Characteristic curve leakage/tank flow versus current for a four nozzle two flapper servovalve. Experimental.

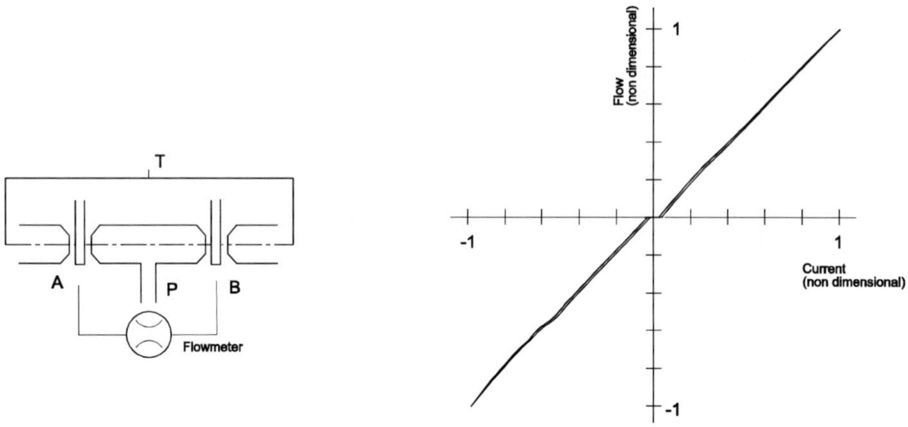

Figure 4.46. Characteristic curve flow between ports A and B versus servovalve intensity. Experimental.

It is very interesting to point out, from figures 4.44 and 4.46, that for the first stage servovalve under study, the gains pressure and flow versus intensity are rather linear, which is in reality the desired performance, it is also to be noticed that these curves present a very small hysteresis, again showing a very good servovalve performance. At this point it is interested to compare figures 4.13 and 4.46, which clearly show the high level performance of the servovalve studied.

Although not presented here, the servovalve dynamic curves, integrated in the Bode diagram, did also show a very high frequency response.

4.4.2.7. References

[1] Williams LJ. (1965). High performance of single-stage servovalve. *SAE Aerospace Fluid Power Systems and Equipment Conference.* Los Angeles. Moog Technical Bulletin N 106.

[2] Tchouprakov Y. (1979). *Comande Hydraulique et automatismes hydráuliques*. Moscou. Ed. Mir.

[3] Kassem SA; Arafa HA. (1982). Static and dynamic characteristics of four nozzle flapper valves. 10th Conference on fluid mechanics Czechoslovakia. 159-172.

[4] Arafa HA; Kassem SA; Osman TA. (1987). Performance of four nozzle flapper hydraulic servovalves. *Mechanisms Machines Theory*. V 22 N3; 243:251

[5] Kassem SA; Arafa HA. (1987). Design aspects of four nozzle hydraulic servovalves. *Journal of Machines Tools Manufacture*. V27 N4; 457-468.

[6] Bahr MK. (1988). Theoretical and experimental investigation of performance of four nozzle hydraulic servovalves. Thesis. *Department of Mechanical design and Production*. Cairo University.

[7] Elgamil MA. (1991). Investigation of performance of single stage hydraulic servovalve with four control gaps. Thesis. Department of Mechanical design and Production. Cairo University.

[8] Duggins RK. (1973). *Further studies of flow in a flapper valve*. 3rd International Fluid Power Symposim. Torino..

[9] Hayashi S; Matsui T; Ito T. (1975). Study of flow and thrust in nozzle flapper valves. *Journal of Fluids Engineering*. March. 39-50.

[10] Kirshner JM; Schmidlin AE. (1976). *Fluidic Sensors*. Fluidics Quarterly.

[11] Capdevila R. (1977). Desarrollo de una Servoválvula para control y regulación de caudal, con mínima histéresis y deriva de cero, por mando del distribuidor a partir de una señal de consigna eléctrica. Tesis. Departamento de Mecánica de Fluidos UPC Terrassa.

[12] Lebrun M; Scavarda A; Jutard A. (1978). *Simulation sur ordinateur d'un servovalve á deux étages.* Automatisme March-April.

[13] Hayashi S; Matsui T; Imai K. (1980). Stability and self sustained oscillations in nozzle flapper valve with pipe line. *Bulletin of the JSME* V23 N179.

[14] Watton J. (1980). Servovalve flapper nozzle dynamics with drain orifice damping. *The American Society of mechanical Engineers*. 84-WA/DSC-17.

[15] Nakada T. (1985). Range of control for electrohydraulic servovalves represented by the rate of flow and the frequency characteristics. *Fluid Control and Measurement*. V1 421-427.

[16] Lin SJ; Akers A. (1985). *A stand alone flapper nozzle servovalve*. Manuscript.

[17] Lin SJ; Akers A. (1988). The predicted performance of a flapper nozzle valve. American control conference. Atlanta June 15-17; V3 1945-1950.

[18] Lin SJ; Akers A. (1989). A dynamic model of the flapper nozzle component of an electrohydraulic servovalve. *Dynamic Systems Measurement and Control*. V11; 105-109.

[19] Akers A; Tsai ST. (1990). Lin SJ. The effect of configuration of the pilot stage on the performance of a two stage two spool pressure control servovalve. The American Society of mechanical Engineers 90-WA/FPST-12. Winter Annual Meeting. Texas November 25-30.

[20] Akers A; Lin SJ. (1990). Squeeze film damping of the motion of a control flapper nozzle. *Proc. Institution of Mechanical Engineers*. V204. 109-115. 1990.

[21] Akers A; Lin SJ. (1990). Dynamic properties of a single spool and two spool two stage servovalves. International Off-Highway and Powerplant congress and exposition. Milwakee Wisconsil September 10-13.

[22] Tsai ST; Akers A; Lin SJ. (1990). Dynamic analysis of a two stage two spool pressure control servovalve. American Control Conference. San Diego California. May 23-25.

[23] Lin SJ; Akers A. (1991). Dynamic analysis of a flapper nozzle valve. *Journal of Dynamic Systems Measurement and Control.* V113; 163-167.

[24] Tsai ST; Akers A; Lin SJ. (1991). Modelling and dynamic evaluation of a two stage two spool servovalve used for pressure control. *Journal of Dynamic Systems Measurement and Control.* V113; 909-713.

[25] Powell A. (1962). *Nature of feedback mechanism of some fluid flows producing sound.* 4th international Cong. On acoustics. Coopenhagen 1962.

[26] Blake WK; Powell A. (1986). The development of contemporary views of flow tone generation. 247-326. *Recent Advances in Aeroacoustics.* A Krothapalli; CA. Smith. Springer Verlag.

[27] Rockwell D; Naudascher E. (1978). Self sustained oscillations of flow past cavities. *Journal of fluid engineering*. 152-165.

[28] Heller H; Holmes D; Covert E. (1971). Flow induced pressure oscillations in shallow cavities. *J. sound and vibration* V.18 N4 1971: 545-553.

[29] Vogel H. (1920). Die Zungenpfeife als gekoppeltes System. *Analen der physik.* V 62; 247-282.

[30] Codina E; Bergada JM. (1992). Some irregular aspects of performance of hydraulic servoactuator for sparkmachining equipment. Proceedings of the international Fluid Power applications conference. USA. Vol 1: 321-331.

[31] BergadaJM; Codina E. (1994). Discharge coefficients for a four nozzle two flapper servovalve. *Proceedings of the 46th National conference on Fluid Power.* USA. Vol 1: 213-218.

[32] BergadaJM; Codina E. (1996). Main frequencies in the performance of a servovalve. *Proceedings of the 47th National conference on Fluid Power.* USA. 351-358.

[33] BergadaJM; Codina E. (2000). Flow features in a nozzle of a servovalve. *Proceedings of the 48th National conference on Fluid Power.* USA. 497-505.

[34] Watton J; Bergada JM. (1994). Progress towards an understanding of the pressure flow characteristics of aservovalve two flapper/double nozzle flow divider using CFD modelling. Four triennial international symposium on Fluid Control. Flucome 94 Tolouse, France. V1: 47-52.

[35] Bergada JM. (1996). Servoposicionador electro-oleohidráulico para una máquina de electroerosión. Doctoral Thesis ETSEIT-UPC.

[36] Lichtarowicz A. (1973). *Flow and force characteristics on flapper valves.* 3rd International Fluid Power symposium. Turin. Italy.

[37] Lichtarowicz A. (1965). Discharge coefficients for incompressible flow through long orifices. *Journal of Mech. Eng. Sci.* 7 N2: 210:219.

[38] Duggins RK. (1973). *Further studies of flow in a flapper valve.* 3rd International Fluid power symposium. Turin. Italy.

[39] Banieghbal MR; Pountney DC; Weston W. (1983). Experimental and numerical analysis of flow characteristics of hydraulic servovalves and orifices. *International conference on Optical techniques in process control.* The Hague. The Netherlands.

[40] Akers A. (1973). *Discharge coefficients for an annular orifice with a moving wall.* 3rd International Fluid Power symposium. Turin. Italy.

[41] McCloy D. Martin HR. (1980). *Control of Fluid Power. Ellis Horwood Series in Engineering Science.*

[42] Thomson WT. (1988). *Theory of vibration with applications.* Prentice Hall.

[43] Merrit HE. (1967). Hydraulic control systems. John Wiley.

4.5. NOMENCLATURE

$C_{d\,NF}$	Discharge coefficient when flow from nozzle to flapper.
$C_{d\,FN}$	Discharge coefficient when flow from flapper to nozzle.
d	Thread displacement. (m)
dA	Area variation. (m^2).
dh	Variation of height. (m^2).
ds	Surface differential. (m^2).
dx	Length differential along the conical seat. (m).
D	Nozzle diameter. (m).
Dh	Hydraulic diameter. (m).
F	Force on the conical spool. Force onto the flapper. (N).
g	Acceleration due to gravity. (m/s^2).
h	Generic distance between cone and seat. (m).
H	Distance between cone and seat. (m).
K_1 ; K_2 ; K_3	Constants.
l	Conical seat length. (m).
P	Pressure. (Pa).
Q	Volumetric flow. (m^3/s).
r	Generic radius. (m).
r_1	Generic internal radius. (m)
r_2	Generic external radius. (m).
$r_{2\,inlet}$	External radius at the inlet. (m).
$r_{2\,outlet}$	External radius at the outlet. (m).
Re	Reynolds number.
T	Temperature. (K).
V	Generic velocity. (m/s).
V_o	Valve inlet velocity. (m/s).
$\forall$	Volume. (m^3).
x	Generic length along the conical seat. Generic flapper displacement versus its centred position. (m).
x_0	Distance flapper nozzle when flapper centred. (m).
X_*	Distance flapper nozzle. (m).
α	Conical seat angle. (m).
μ	Dynamic viscosity. (Kg/(m s)).
ν	Kinematic viscosity. (m^2/s).
θ	Angle turned by the gap adjuster thread. (rad).
ρ	Fluid density. (Kg/m^3).
τ	Shear stress. (N/m^2).

Chapter 5

PUMPS AND MOTORS

5.1. INTRODUCTION

Efficiency improvement is a key issue in any machine. The fluid power industry relies on volumetric pups which need to pump high pressure fluid to a set of actuators located at different positions in a given device, the higher the pressure the smaller the actuators need to be, therefore allowing to reduce the weight of the machine, being this, a critical issue in any flying device. At the present, the pumps which are able to produce the highest fluid pressure are piston pumps, and among the different sort of piston pumps, axial piston pumps seem to be the most widely used, probably due to its high efficiency and reliability. Pumps and motors overall efficiency, is in reality the product of volumetric, mechanical and hydraulic efficiency, therefore a decrease in any of these efficiencies will bring an overall efficiency decrease. In this book chapter, shall be presented a deep study on the different axial piston pump moving parts, equations clarifying leakage and pressure distribution in all axial piston pump moving parts will be introduced, and therefore the dimensional parameters from which leakage depends will clearly be defined, as a result a tool to improve piston pump volumetric efficiency shall be established. To validate the equations presented, a comparison between results produced by the equations, by several CFD models of each axial piston pump moving parts and by several experimental measurements will be performed. Thanks to this comparison, the validity limits of the equations presented will be established.

Thanks to the theory developed and the different test rigs used, a better understanding of the slippers dynamic behavior and barrel dynamics was gathered, pressure distribution, forces and torques generated in the slipper-swash plate, barrel-port plate and piston barrel will be presented, comparisons between CFD, analytical equations and experimental results will validate the new theory produced.

One of the newest characteristics of the analytical, CFD and experimental development presented, is based on the performance of grooves being cut on slippers and pistons surfaces, the use of grooves is not fully extended, then each manufacturer decides whether shall or shall not be used for a given application. Nevertheless, a full understanding of its effect is not yet clarified, in the present chapter, the benefits and drawbacks of using grooves will be clearly established.

Several dynamic models are also included in the present book chapter, the first model will focus in understanding the barrel dynamics, some of the equations previously presented and validated will be included in the model, the barrel dynamic movement will be, thanks to this model and the experimental measurements performed, much better understood.

A second model presented, will again use the new leakage equations developed and join them to create a full dynamic model of the entire axial piston pump. The model will be able to predict the output flow and pressure ripple, comparisons between numerical and experimental results are used to validate the new model created. Please notice that a total of three different state of the art test rigs have been used to validate all the equations and models generated.

At the end of the book chapter shall be introduced some new trends on piston pumps and motors design, like new composite materials and the use of spherical slippers.

As a conclusion, the present book chapter is having sub chapters on: Flat and tilt slippers with grooves, the use of grooves on pistons, barrel dynamics, piston-slipper spherical journal and an overall pump pressure and flow ripple model. New analytical equations, CFD models and state of the art test rigs will be presented. The aim is to give a tool to better design axial piston pumps and improve its efficiency. All experimental work presented in this chapter was undertaken in the Prof John Watton Fluid Power Lab. at Cardiff University UK. A previous version of this book chapter having Prof. J Watton as co-author has already been published by Nova Science in 2012.

5.1.1. General Classification of Pumps and Motors

There exist two different sorts of hydraulic machinery, centrifugal and volumetric. The main characteristic of a centrifugal pump is that for a given turning speed output pressure decreases as input flow increases.

Table 5.1.1. Volumetric pump classification

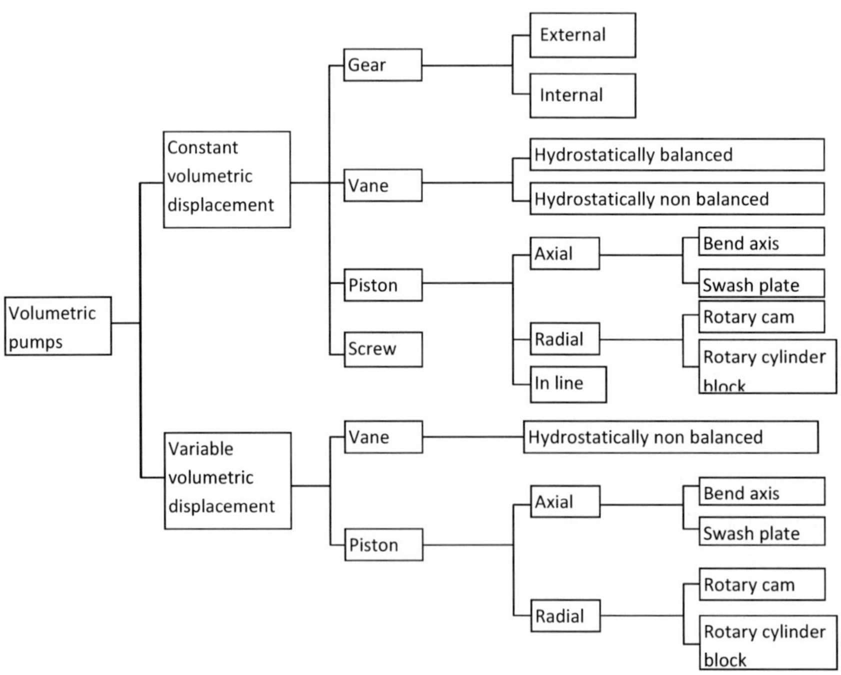

Volumetric machines on the other hand maintain the output flow nearly constant independently of the output pressure required. In reality, as output pressure increases, pump volumetric efficiency decreases, as a result output flow slightly decreases with the increase of output pressure. Due to their construction principle, a given volume of fluid is transported from the pump inlet to the pump outlet, volumetric pumps and motors are able to work at much higher pressures and with better efficiencies than centrifugal machines. This is why; fluid power industry relies on volumetric machinery.

Table 5.1.1 presents the different sort of volumetric pumps used in the fluid power industry. Notice that volumetric pumps can be subdivided into constant or variable volumetric displacement. Machines with variable volumetric displacement, allow adjusting the flow volume the machine transfers from inlet to outlet.

Among these classification, screw pumps are characterized for its low efficiency, low pressure and low pulsating pressure fluctuations. Gear pumps are widely used due to their construction simplicity and good performance, their total efficiency can reach 90%. The output flow is pulsating and they are capable of supplying pressures above 300 bar. Vane pumps produce a lower pulsation output flow than gear pumps, they are able to work at pressures slightly lower than gear pumps being the efficiency of both type of pumps, very similar. Piston pumps are the most efficient ones, they can supply a much higher pressure than any other pump, output flow is highly pulsating and the noise level is among the highest produced by any other sort of pump. Hydraulic motors fall in the main generic classification as hydraulic pumps, therefore gear, vane and piston motors do exist. Gear and vane motors are suitable for high and medium speed applications while piston motors are suitable for all sort of applications including low speed ones. As it has already been seen in pumps, gear motors, due to its construction have a constant volumetric displacement, vane and piston motors can have whether constant or variable volumetric displacement but in reality the vast majority of variable volumetric hydraulic machines are the piston ones.

Pump and motor descriptions, working principle, performance characteristics as well as determination of volumetric displacement of the different configurations, can be found in many books and shall not be repeated here. The lector is encouraged to study the references if further information in this direction is needed. In what follows, specific original research focused on axial piston pumps, shall be presented.

5.1.2. Axial Piston Pump under Research

As presented in table 5.1.1, an axial piston pump is a positive displacement machine which can be classified into two main categories, namely bend axis and swash plate piston pump. In the present chapter, the pump into consideration is a swash plate axial piston pump. It has an odd number of pistons arranged in a circular array within a housing which is commonly referred to as a cylindrical block, rotor or barrel. The cylinder block is driven to rotate about its axis of symmetry by a central shaft, aligned with the pumping pistons.

Axial piston pumps can be further classified into two categories, fixed displacement and variable displacement. In fixed displacement pumps, the stroke of the piston cannot be modified, on the other hand, piston stroke can be modified in variable displacement piston pumps. Figure 5.1.1 shows a cross section cut of a variable displacement axial piston pump, where its different parts are presented.

The study of axial piston pump moving parts is essential to evaluate volumetric, mechanical and hydraulic efficiencies, in fact, the overall pump efficiency and performance is directly linked with the fluid behavior understanding in all pump moving parts. From figure 5.1.1 it is to be noticed that relative movement appears between, pistons and barrel, slippers and swash plate, barrel and port plate and piston-slipper spherical journal.

The pump operating mechanism is as follows, as the cylinder block (barrel) rotates, the exposed ends of the pistons (slippers) are constrained to follow the surface of the swash plate plane. Since the swash plate plane is at an angle to the axis of rotation, the pistons must reciprocate axially as they proceed about the cylinder block axis. The axial motion of the pistons is sinusoidal. During the rising portion of the piston's reciprocating cycle, the piston moves towards the port plate, during this period, the fluid trapped between the buried end of the piston and the valve plate is vented to the pump's discharge port.

Whe a piston is positioned at the top reciprocating cycle (top death centre, TDC), the connection between the piston-cylinder chamber and the pump's discharge port is closed, shortly thereafter, piston-cylinder chamber is connected to pump's inlet port. The piston moves away from the port plate, thereby increasing the volume of piston-cylinder chamber, as this occurs, fluid enters the chamber from the pumps inlet to fill the void. This process continues until the piston reaches the bottom of the reciprocation cycle (bottom death centre, BDC). At BDC, the connection between the piston-cylinder chamber and the inlet port is closed, shortly thereafter, the chamber becomes open to the discharge port again and the pumping cycle starts over.

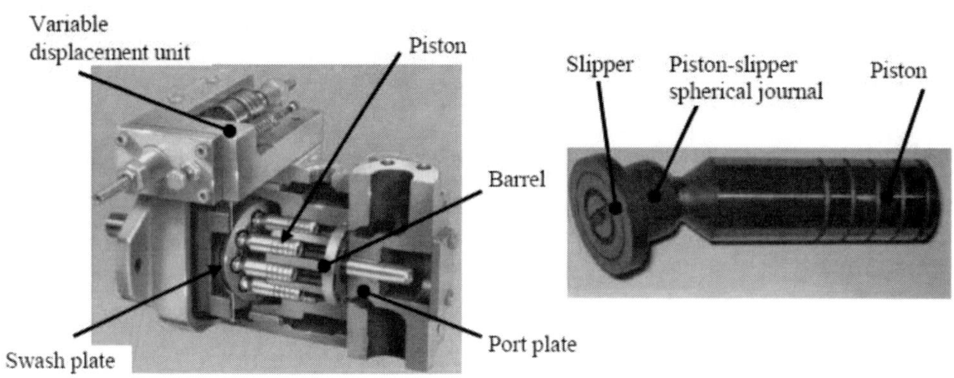

Figure 5.1.1. Axial piston pump main components.

5.2. Effect of Piston-Barrel Clearance and Grooves

5.2.1. Previous Research

Whenever a manufacturer designs a piston pump or motor to be used in high pressure applications, often comes across the question if grooves along the piston surface are needed, then depending on the manufacturer the pistons may or may not have grooves. Grooves are meant to stabilise the piston but the amount of grooves needed for a specific application and where should they be located along the piston length is at the moment very much linked with the designers' expertise. On the other hand it must be recalled that, among the most efficient pumps are to be found the piston ones, piston dynamics plays a fundamental role in two critical processes related to fluid flow in these pumps. The first is the flow leakage through the radial clearance, which may cause considerable reduction in the pump efficiency. The second process is the viscous friction associated, with the lubricant film in the radial clearance, eventually friction metal to metal might appear. Therefore the geometry of the pistons used affects the mechanical and volumetric efficiency of the pump and its long term performance. The present chapter clarifies the effect of the grooves being cut on piston surface and the necessity, or not, of their use.

The first studies about the groove balancing effect were conducted experimentally by Sweeney [1], who examined the pressure distribution in the piston-cylinder clearance and established a relationship between the leakage flow and the geometry of the clearance. Sadashivappa et al [2] examined also experimentally the pressure distribution in the clearance piston-cylinder and concluded that the eccentricity of the piston affected the performance of the piston by influencing the frictional and leakage aspects.

Some attempts have been pursued to find the flow and pressure distribution theoretically taking into account the effect of the grooves, Milani [3] applied the continuity equation to link the Poiseulle equation in each land, and considered a constant pressure in each groove. The same method was used by Borghi et al [4, 5], although they applied it to a single groove tapered spool. In both cases relative movement between piston and cylinder was not considered, yet eccentricity was taken into account. Blackburn et al [6] and Merrit [7] established an analytical formulation for the pressure distribution and forces in narrow clearances. They assumed that the pressure distribution in narrow gaps was not affected by peripheral flow rates; they made an easy estimation of the sticking phenomena effects.

In any case, the most precise way to find out the leakage and pressure distribution would be via using the two dimensional Reynolds equation of lubrication. The main difficulty here is that the equation needs to be integrated numerically. Such work although when grooves were not considered was undertaken by Ivantysynova [8, 9] which found the dynamic pressure distribution and leakage between piston and barrel considering piston tilt, piston displacement and heat transfer. Elastohydrodynamic friction was also considered.

Fang et al [10] carried out a numerical analysis in order to obtain the metal contact force between the piston and cylinder and he concluded that exist mixed lubrication between the piston and cylinder, being independent on pump operation conditions such as supply pressure or the rotation speed. Prata et al [11] performed a numerical simulation for a piston without grooves, by using finite volume method, considering both the axial and the radial piston motion and explained the effect of the operating conditions on the stability of the piston.

On the other hand, the study of the machine element surfaces with grooves and narrow gaps is more generic and has a mature foundation in literature. Berger et al [12] investigated the effect of the surface roughness and grooves on permeable wet clutches by using a finite element approach and considering the modified Reynolds and force balance equations. They concluded that friction and groove width significantly influence the engagement characteristics as torque, pressure and film thickness, on the other hand groove depth did not have a significant effect on engagement characteristics. Razzaque et al [13] applied a steady-state Reynolds type equation with inertia consideration to a coolant film entrapped between a grooved separator-friction plate pair of a multi disk wet clutch arrangement. Razzaque used finite difference technique to simulate pressure distribution and flow field for different groove shapes such as rounded, trapezoidal, and V-section at different angular orientations, and found that among the profiles studied, the rounded groove performed better under the leakage and force point of view, nevertheless the use of inclined grooves caused less viscous torque and, hence, less power loss.

Lipschitz et al [14] used Finite Difference Method to study a radial grooved thrust bearing operation and showed that rounded bottom grooves were superior to flat bottom grooves regarding the load carry capacity. Basu [15] justified the validity of the radial groove approximation when simulating parallel grooves in face seals. He used both FDM (Finite Difference Method) and FEM (Finite Element Method) and found that FDM was considerably faster. Kumar et al [16] investigated the effect of the groove in slipper-swash plate clearance, by doing finite volume formulation for a three dimensional Navier stokes equation in cylindrical coordinates. They demonstrated that the presence of the groove stabilized the pressure distribution in the clearance slipper swash plate. The grooves position was having a considerable effect on the force acting over the slipper.

An interesting amount of work has been undertaken until now, considering the geometric shape of the groves, friction parameters and its effect on the operating conditions in order to improve the piston performance, but despite all the work undertaken by previous researchers, there has never been studied, the effect of the number of grooves cut on the piston surface and specially the effect of modifying their position, including as well the piston tilt and its relative movement versus the cylinder. In this section, it is being investigated the piston performance by modifying the number of grooves and their position, pressure distribution in the clearance piston-cylinder, leakage force and torque acting over the piston will be discussed, also the locations where cavitation is likely to appear will be presented, discussing how to prevent cavitation from appearing via using grooves.

5.2.2. Mathematical Analysis

Figure 5.2.1 represents a picture of the initial configuration of the piston considered in this section and a two dimensional schematic diagram of it. It is important to notice from figure 5.2.1, that the piston into consideration has several grooves on the sliding surface, the aim of which is to increase stability, decrease friction and increase lateral forces.

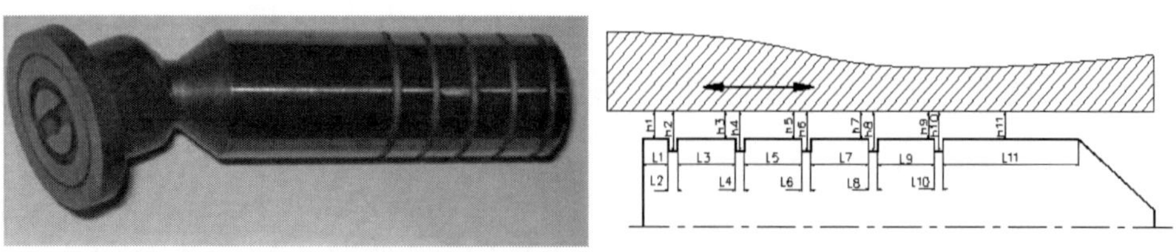

a) b)

Figure 5.2.1. Piston geometry [a] Piston considered in this project. [b] 2-D schematic diagram of the piston, with main dimensions.

In the present study, a direct method to find out the pressure distribution and leakage in the piston/cylinder gap will be described. The advantage of this new method is that the relative movement of the piston cylinder is taken into account, and also groove effect is considered. The disadvantage is that the eccentricity effect cannot be considered.

The equations about to be presented are based on the one-dimensional Reynolds equation of lubrication, the Couette-Poiseulle equation, and the continuity equation. The full description of the mathematical analysis is to be found in Bergada and Watton [17,18]. The assumptions considered are:

1. Laminar flow is being considered in all cases.
2. The flow is two-dimensional.
3. Relative movement between piston and barrel exists.
4. The gap piston cylinder is simulated as the gap between two flat plates.
5. No eccentricity is considered.
6. Each land and groove is modelled as a flat plate.

The piston main dimensions are seen figure 5.2.1.

$h_1 = h_3 = h_5 = h_7 = h_9 = h_{11} = 2.5$ microns.
$h_2 = h_4 = h_6 = h_8 = h_{10} = h_1 + 0.4$ mm.
$L_1 = 1.42$ mm
$L_{11} = 19.5$ mm
$L_2 = L_4 = L_6 = L_8 = L_{10} = 0.88$ mm
$L_3 = L_5 = L_7 = L_9 = 4$ mm

The one dimensional Reynolds equation of lubrication in Cartesian coordinates can be given as:

$$\frac{\partial}{\partial x}\left(\frac{h^3}{\mu}\frac{\partial p}{\partial x}\right)=0 \tag{5.2.1}$$

and its integration yields

$$P=\frac{A\mu}{h^3}x+B \tag{5.2.2}$$

Equation (5.2.2) gives the pressure distribution along the “x” axis, and the constants A and B must be found using the boundary conditions.

The Couette-Poiseulle flow between two flat plates [19] results in the following flow per unit depth:

$$\dot{\forall}=\frac{h\,u}{2}-\frac{\partial p}{\partial x}\frac{h^3}{12\mu} \tag{5.2.3}$$

Via substituting the first integration of equation (5.2.1) in (5.2.3) it is found that

$$\dot{\forall}=\frac{h\,u}{2}-\frac{A}{12} \tag{5.2.4}$$

Since equations (5.2.2) and (5.2.4) are applicable to any pair of flat plates, then for each flat plate shown in figure 5.2.1 there exist a pair of equations as follows.

$$P_1=\frac{A\,\mu}{h_1^3}x+B \tag{5.2.5}$$

$$\dot{\forall}_1=\frac{h_1\,u}{2}-\frac{A}{12} \tag{5.2.6}$$

range of applicability $0 \le x \le l_1$. For the last flat plate:

$$P_{11}=\frac{U\mu}{h_{11}^3}x+V \tag{5.2.7}$$

$$\dot{\forall}_{11}=\frac{h_{11}\,u}{2}-\frac{U}{12} \tag{5.2.8}$$

range of applicability $\left(\sum_{i=1}^{i=10} l_i\right) \le x \le \left(\sum_{i=1}^{i=11} l_i\right)$ (5.2.9)

The constants A….V have to be found using boundary conditions in both piston ends and in each pair of connected surfaces.

In this study the analysis results in 22 equations with 22 unknown constants.

In any connection of two surfaces:

$$x=\sum_{j=1}^{j=i} l_j\ ;\ P_i=P_{i+1};\ \dot{\forall}_i=\dot{\forall}_{i+1}\ ;\ 1 \le i \le 10 \tag{5.2.10}$$

If it is assumed that:

$l_2=l_4=l_6=l_8=l_{10}$; $l_3=l_5=l_7=l_9$;

$h_1=h_3=h_5=h_7=h_9=h_{11}$; $h_2=h_4=h_6=h_8=h_{10}$; (5.2.11)

The value of the constants will be:

$$A=\frac{P_{\tan k}-P_{Piston}-C_{A2}}{\dfrac{\mu}{h_{11}^3}\displaystyle\sum_{i=1}^{i=11} l_i+C_{A1}} \tag{5.2.12}$$

In the case under study:

A=E=I=M=Q=U; B = Ppiston; (5.2.13)

$$C_{A1} = \mu\left[\frac{1}{h_2^3} - \frac{1}{h_1^3}\right](l_2 + l_4 + l_6 + l_8 + l_{10}) \tag{5.2.14}$$

$$C_{A2} = 6u\mu\left[\frac{h_{10} - h_1}{h_{10}^3}(l_2 + l_4 + l_6 + l_8 + l_{10})\right] \tag{5.2.15}$$

According to the previous specifications:

$$C = K = G = O = S = 6u(h_2 - h_1) + A \tag{5.2.16}$$

$$D = A\mu\left[\frac{1}{h_1^3} - \frac{1}{h_2^3}\right]*(l_1) + 6u\mu\left[\frac{h_2 - h_1}{h_2^3}\right]*(-l_1) + B; \tag{5.2.17}$$

$$F = A\mu\left[\frac{1}{h_2^3} - \frac{1}{h_1^3}\right]*(l_2) + 6u\mu\left[\frac{h_2 - h_1}{h_2^3}\right]*(l_2) + B; \tag{5.2.18}$$

$$H = A\mu\left[\frac{1}{h_1^3} - \frac{1}{h_2^3}\right]*(l_1 + l_3) + 6u\mu\left[\frac{h_2 - h_1}{h_2^3}\right]*(-(l_1 + l_3)) + B; \tag{5.2.19}$$

$$J = A\mu\left[\frac{1}{h_2^3} - \frac{1}{h_1^3}\right]*(l_2 + l_4) + 6u\mu\left[\frac{h_2 - h_1}{h_2^3}\right]*(l_2 + l_4) + B \tag{5.2.20}$$

$$L = A\mu\left[\frac{1}{h_1^3} - \frac{1}{h_2^3}\right]*(l_1 + l_3 + l_5) + 6u\mu\left[\frac{h_2 - h_1}{h_2^3}\right]*(-(l_1 + l_3 + l_5)) + B; \tag{5.2.21}$$

$$N = A\mu\left[\frac{1}{h_2^3} - \frac{1}{h_1^3}\right]*(l_2 + l_4 + l_6) + 6u\mu\left[\frac{h_2 - h_1}{h_2^3}\right]*(l_2 + l_4 + l_6) + B; \tag{5.2.22}$$

$$P = A\mu\left[\frac{1}{h_1^3} - \frac{1}{h_2^3}\right]*(l_1 + l_3 + l_5 + l_7) + 6u\mu\left[\frac{h_2 - h_1}{h_2^3}\right]*(-(l_1 + l_3 + l_5 + l_7)) + B \tag{5.2.23}$$

$$R = A\mu\left[\frac{1}{h_2^3} - \frac{1}{h_1^3}\right]*(l_2 + l_4 + l_6 + l_8) + 6u\mu\left[\frac{h_2 - h_1}{h_2^3}\right]*(l_2 + l_4 + l_6 + l_8) + B; \tag{5.2.24}$$

$$T = A\mu\left[\frac{1}{h_1^3} - \frac{1}{h_2^3}\right]*(l_1 + l_3 + l_5 + l_7 + l_9) + 6u\mu\left[\frac{h_2 - h_1}{h_2^3}\right]*(-(l_1 + l_3 + l_5 + l_7 + l_9)) + B \tag{5.2.25}$$

$$V = A\mu\left[\frac{1}{h_2^3}-\frac{1}{h_1^3}\right]*(l_2+l_4+l_6+l_8+l_{10})+6u\mu\left[\frac{h_2-h_1}{h_2^3}\right]*(l_2+l_4+l_6+l_8+l_{10})+B; \tag{5.2.26}$$

With these set of equations, it is now possible to find the pressure distribution along the piston length, for a piston with five slots. In fact, the equations allow to investigate the pressure increase on each slot when different piston velocities are considered. See Bergada and Watton [17].

It is traditionally assumed that the leakage due to the gap piston cylinder is constant, and has a linear relationship with the pressure differential of the piston ends. In fact, if the previous results are considered it can clearly be seen that the leakage flow depends on the relative movement of the piston which could well be significant in practice. Since the piston velocity is sinusoidal the leakage will also be affected. The piston velocity can be given as:

$$u = -R_{sw}\ \tan\varepsilon\ \sin(\sigma t)\ \sigma \tag{5.2.27}$$

Substituting this equation into each leakage flow for real piston movement, results in the flow equation (5.2.28).

$$q_{piston-barrel} = \pi D_P\left[\frac{h_1 R_{sw}\tan\varepsilon\left(-\sin(\sigma t)\right)\sigma}{2}\right] - \frac{\pi D_P}{12}\left[\frac{P_{Tank}-P_{Piston}-6R_{sw}\ \tan\varepsilon\left(-\sin(\sigma t)\right)\sigma\mu\left[\frac{h_{10}-h_1}{h_{10}^3}\right](l_2+l_4+l_6+l_8+l_{10})}{\frac{\mu}{h_{11}^3}(l_1+l_2+l_3+.....+l_{11})+\mu\left[\left(\frac{1}{h_2^3}-\frac{1}{h_1^3}\right)(l_2+l_4+l_6+l_8+l_{10})\right]}\right] \tag{5.2.28}$$

Equation (5.2.28) assumes that the piston is always inside the barrel, but in reality this is not true since the piston length inside the barrel changes temporally, once the real piston length inside the barrel is taken into account from equation (5.2.27), the resulting piston-barrel dynamic leakage will be given as equation (5.2.29).

$$q_{piston-barrel} = \pi D_P\left[\frac{h_1 R_{sw}\tan\varepsilon\left(-\sin(\sigma t)\right)\sigma}{2}\right] - \frac{\pi D_P}{12}\left[\frac{P_{Tank}-P_{Piston}-6R_{sw}\ \tan\varepsilon\left(-\sin(\sigma t)\right)\sigma\mu\left[\frac{h_{10}-h_1}{h_{10}^3}\right](l_2+l_4+l_6+l_8+l_{10})}{\frac{\mu}{h_{11}^3}\left(l_1+l_2+l_3+.....+l_{11}-\frac{l_{11}}{2}-R_{sw}\ \tan\varepsilon\ \cos(\sigma t)\right)+\mu\left[\left(\frac{1}{h_2^3}-\frac{1}{h_1^3}\right)(l_2+l_4+l_6+l_8+l_{10})\right]}\right] \tag{5.2.29}$$

Equation which will give the temporal leakage piston barrel, for any clearance, pressure differential, and pump turning speed. At time $t = 0$, it has to be understood that the piston is at its bottom death centre.

5.2.3. 2-D CFD Approach

In order to check the quality of the equations previously derived, a two dimensional computer model was studied using Fluent CFD software package. For the model, the turbulent Navier Stokes and continuity equations were adopted, K-ε - RNG model of turbulence was used. Simulation was undertaken for inlet pressures of 4, 8, $16*10^6$ Pa, and for piston velocities of 1; 0.5; 0; –0.5 and 1 m/s. The fluid used for this model was water.

Figure 5.2.2 shows the grid generated to evaluate the pressure distribution along the piston barrel gap and the flow across it, the clearance piston barrel was considered to be of 2.54 microns. Five grid cells were used in the piston-cylinder, slipper-swash plate and piston-slipper spherical journal gaps.

Although not presented here, a perfect parabolic velocity distribution was found in the piston cylinder clearance, demonstrating that flow has to be laminar under all conditions studied. From the simulation was found that under static conditions, for a given pressure differential and clearance, the leakage flow between slipper and port plate is of an order of magnitude higher than the piston cylinder leakage.

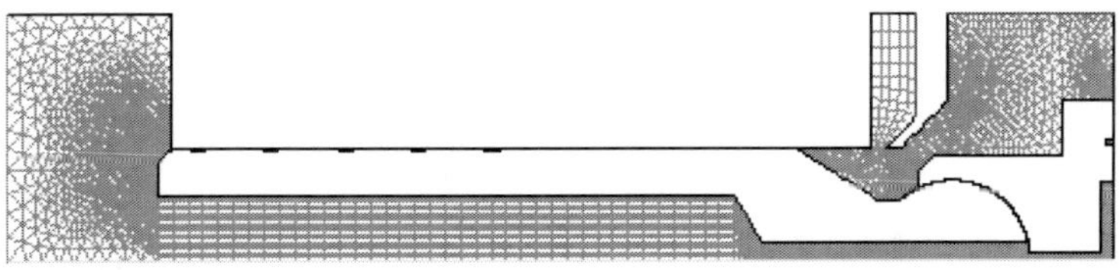

Figure 5.2.2. Two dimensional grid generated.

Figure 5.2.3 shows the piston-barrel leakage flow rate for a range of different pressure differentials and piston velocities, it can be seen that the CFD results and the analytical results from equation (5.28) have a good match, detailed comparisons revealing errors below 1% in almost all flows. Since the equations determined have the capability to give the pressure distribution along the gap, Figure 5.2.4 compares the results using the new set of equations and CFD analysis under static conditions and for three different pressure differentials 4, 8 and 16 MPa, clearance piston barrel being 2.54 microns. Notice the excellent agreement. In this figure has also been plotted the pressure distribution given by the former existing set of equations, when no groove is considered, see the clear difference in results.

5.2.4. Piston-Cylinder Numerical Model under Tilt Conditions

The piston model considered until now, is not able to grasp the effect of a tilt piston; this is why a numerical model using MATLAB was created to consider this situation. The numerical model about to be presented in this section considers the fluid flow as laminar, piston is tilted and piston grooves are considered. From the literature [8-15], it is noticed that Reynolds equation of lubrication is considered a good approach to investigate fluid flow in narrow gaps. In the present sub chapter, the Reynolds equation of lubrication (5.2.30) in Cartesian coordinates is applied to the piston-cylinder clearance. For a given generic piston location inside the cylinder, piston-cylinder clearance is variable and a function of the coordinate axis $\{\theta, L\}$, see figure 5.2.5, being its calculation vital for the simulation of the pressure field.

$$\left(\frac{2}{D_P}\right)^2 \frac{\partial}{\partial\theta}\left(\frac{h^3}{\mu}\frac{\partial p}{\partial\theta}\right)+\frac{\partial}{\partial L}\left(\frac{h^3}{\mu}\frac{\partial p}{\partial L}\right)=6\left(V_{S\theta}\frac{2}{D_P}\frac{\partial h}{\partial\theta}+V_{SL}\frac{\partial h}{\partial L}+2\frac{\partial h}{\partial t}\right) \quad (5.2.30)$$

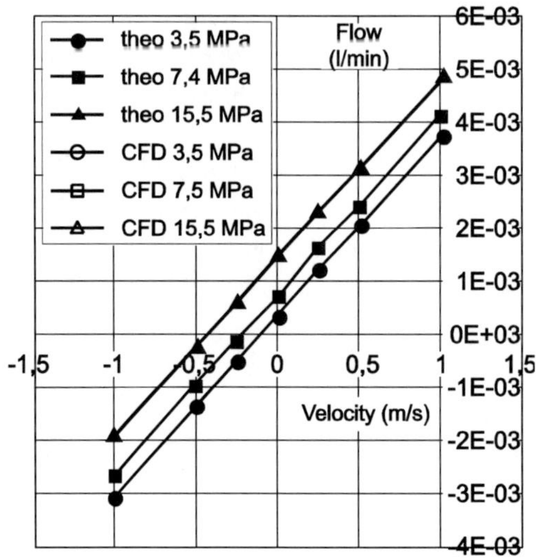

Figure 5.2.3. Comparison of flow rates between CFD and analytical solution. Water.

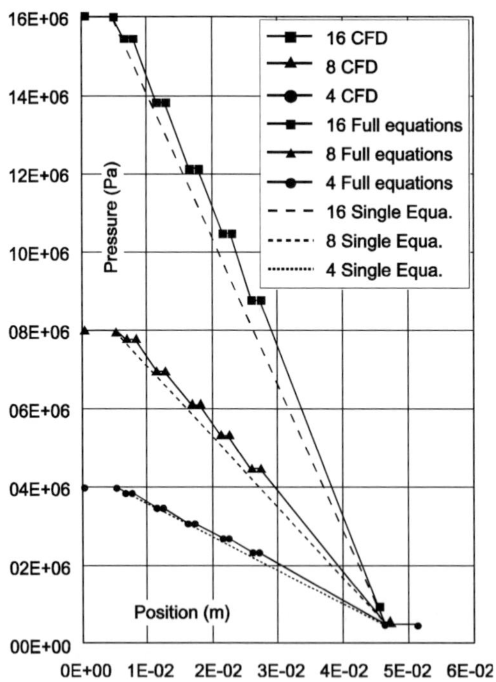

Figure 5.2.4. Pressure distribution along the piston-barrel clearance under different pressure differentials. Velocity = 0 m/s. Equations versus CFD. Water.

Figure 5.2.6 represents the cross sectional cut, perpendicular to the piston central axis, of the piston-cylinder assembly, the figure represents the clearance piston-cylinder for two different given lengths {L (0, L_t)}, where L is greater or smaller than L_l (position of the intersection of piston and cylinder axis).

The Length of the piston inside the barrel as a function of the piston-slipper position as the slipper slides around the swash plate (λ_{sw}) is given by equation (5.2.31).

$$L_t = L_0 + R_{sw} \tan \varepsilon \left(1 - \cos \lambda_{sw}\right) \tag{5.2.31}$$

To simulate the pressure distribution in the clearance piston-cylinder, first it is important to evaluate the clearance as a function of the known variables. Equation (5.2.32) represents the relationship between piston diameter (D_p), cylinder diameter (D_c), length of the piston inside barrel (L_t), minimum edge clearance (Ec_1 & Ec_2) and piston tilt from barrel axis (α). When knowing the values of D_p, D_c, L_t, Ec_1 and Ec_2, the tilt (α) can be evaluated numerically from equation (5.2.32).

Once the tilt (α) is known, the gap piston barrel at any point can be calculated from equations (2.33-2.36).

Equation (5.2.33) relates the piston minimum edge clearance Ec_2, piston tilt (α), piston diameter and cylinder diameter with the piston length between the origin of the coordinates system and the intersection piston axis with cylinder axis, see figure 5.2.5.

Equations (5.2.34) and (5.2.35) use the information calculated until this moment to find out the X and Y coordinates of the point (P_1) given by the intersection between the ellipse curve presented in figure 5.2.6 which represents the cylinder boundary, and a generic straight line which central position is the piston central axis, the straight line is defined as a function of a generic angle θ.

The intersection point (P_2) is the point between the same straight line and the piston diameter.

Once (P_1) and (P_2) are found, the straight distance between them, which represents the clearance piston-cylinder at a particular spatial coordinate, is given by equation (5.2.36).

$$Ec_1 + \left(L_t - \frac{D_p}{2} \tan \alpha \right) \sin \alpha + D_p \sec \alpha + Ec_2 - D_c = 0 \tag{5.2.32}$$

$$L_l \sin \alpha + \frac{D_p}{2} \sec \alpha + Ec_2 = \frac{D_c}{2} \tag{5.2.33}$$

$$X_1 = \frac{-\left(L_l - L\right) \tan \alpha \tan^2 \theta \pm \sqrt{\left(\frac{D_c \tan \theta}{2} \right)^2 + \left(\frac{D_c \cos \alpha}{2} \right)^2 - \left\{ \left(L_l - L\right) \sin \alpha \tan \theta \right\}^2}}{\cos^2 \alpha + \tan^2 \theta} \tag{5.2.34}$$

$$Y_1 = \tan \theta \left[X_1 + \left(L_l - L\right) \tan \alpha \right] \tag{5.2.35}$$

$$h=\sqrt{\left[X_1-\left\{\frac{D_P}{2}\cos\theta-(L_1-L)\tan\alpha\right\}\right]^2+\left[Y_1-\frac{D_P}{2}\sin\theta\right]^2} \tag{5.2.36}$$

The Reynolds equation of lubrication equation (5.2.30) has been integrated over a two dimensional staggered grid in theta and L direction via using the finite volume technique described by Patankar [20]. Dirichlet type pressure boundary conditions are specified at inlet and outlet boundary and a no slipping boundary condition is imposed on the walls, as defined in equations (5.2.37) and (5.2.38).

$$V_{S\theta}=0 \tag{5.2.37}$$

$$V_{SL}=R_{sw}\,\sigma\tan\varepsilon\sin\lambda_{sw} \tag{5.2.38}$$

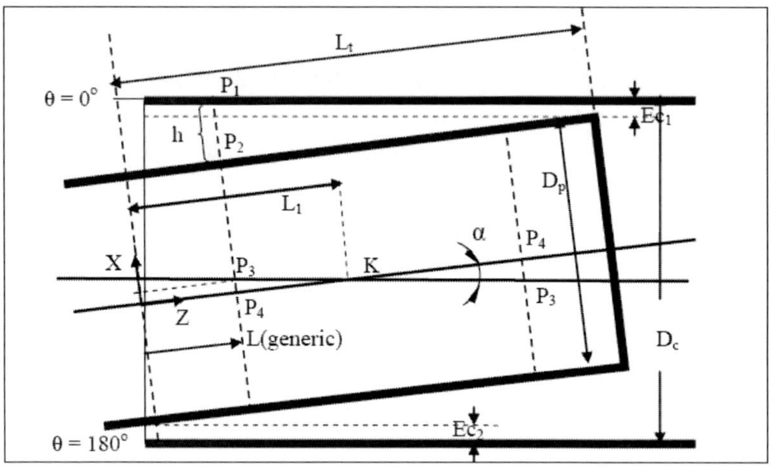

Figure 5.2.5. Tilt piston inside the cylinder, Lateral view.

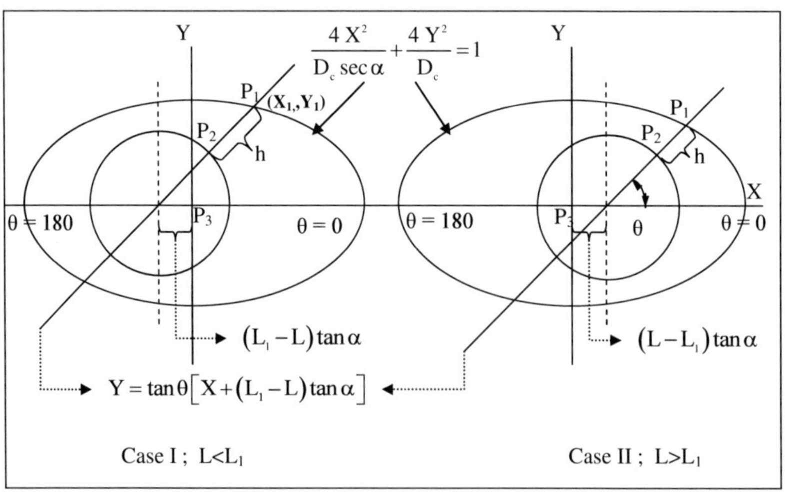

Figure 5.2.6. Piston cylinder clearance (Cross sectional view) with coordinate geometric equations used to calculate clearance.

A Couette - ***Poiseuille*** type velocity distribution profile is assumed at any point of the clearance piston-cylinder, then, once the pressure distribution in the clearance piston-cylinder will be determined, to calculate piston-cylinder leakage equation (5.2.39) will be used.

$$Q=\int_0^{2\pi}\int_0^{h}\left\{\frac{1}{2\mu}\frac{\partial p}{\partial L}\left(X^2\text{-}h\,X\right)+V_{SL}\frac{X}{h}\right\}\frac{D_P}{2}\,d\theta\,dX \tag{5.2.39}$$

The torque has been calculated with respect to both axis via using equations (5.2.40) and (5.2.41).

$$T_x=\int_0^{2\pi}\int_0^{L}P\frac{D_p}{2}L\sin\theta\;dl\,d\theta \tag{5.2.40}$$

$$T_y=\int_0^{2\pi}\int_0^{L}P\frac{D_p}{2}L\cos\theta\;dl\,d\theta \tag{5.2.41}$$

Using the methodology presented here, a set of computational tests were developed, all tests used 30 MPa pump outlet pressure, central clearances piston-cylinder were of 3, 10, 15 and 20 microns, pump turning speeds ranged from 200 rpm to 1000 rpm and a set of different piston tilts were also evaluated, three piston diameters of D_p (14.6mm), 1.5D_p (21.9mm) and 2D_p (29.2mm) were chosen.

For the numerical model created, a staggered type grid in both directions(θ, L) was chosen. Grid independency test has been performed on two different grid sizes (360-800) and (720-1600), results demonstrated that the less dense grid produce the same accuracy as the denser one, therefore the grid size of (360-800) was used for the entire simulation. Results from the simulation will be discussed next.

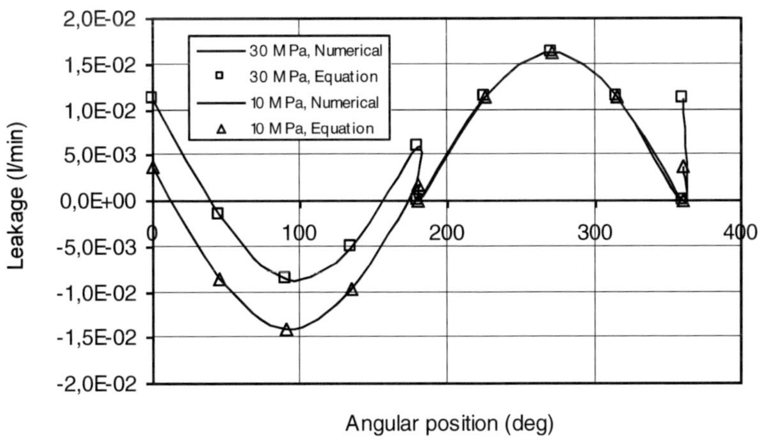

Figure 5.2.7. Leakage between piston cylinder clearance versus angular position at 1000 rpm pump turning speed, 10 microns central clearance, two different inlet pressures, comparison between numerical and analytical results. Fluid oil ISO 32.

5.2.5. Results. Piston without Grooves

Figure 5.2.7 introduces the comparison between the leakage obtained using equation (2.29) and the numerical model created in this sub chapter. The piston has no tilt, clearance piston-cylinder 10 microns, pump turning speed 1000 rpm and fluid used oil ISO 32. Notice that the agreement is very good. The leakage peaks at 0º; 180º and 360º are due to Poiseulle flow.

5.2.5.1. Results. The Effect of Grooves

Maybe the most interesting feature of the numerical model presented in this sub-chapter lays on the fact that piston tilt, number of grooves being cut on the piston surface, groove dimensions and position can be evaluated. To perform such evaluation a set of different groove configurations were studied. To understand the effect of groove positioning, eight different types of pistons, as shown in figure 5.2.8, were used to evaluate the piston performance. The nomenclature used is: G0 no groove, G1o one groove located at the outer edge, G1i one groove at inner edge, G2 two grooves, one at the inner edge and other at the outer edge, G5, five grooves placed at equiv distance from each other, $G1_{2i}$ one groove at the inner edge and located at the 2nd groove position, Gc_1 is the same configuration as $G1_{2i}$ with an extra groove located at the piston stroke length, Gc_2 is the same configuration as the $G1_{2i}$ with two extra grooves located at the piston stroke length. Piston stroke length is defined as the length of the piston which is moving in and out of the cylinder. All grooves cut on the piston surface, have a width of 0.8mm and a depth of 0.8mm. Nevertheless the grooves cut on the piston stroke length (see configurations Gc_1 and Gc_2 in figure 5.2.8) have a groove depth of 0.2mm and a width of 0.2mm.

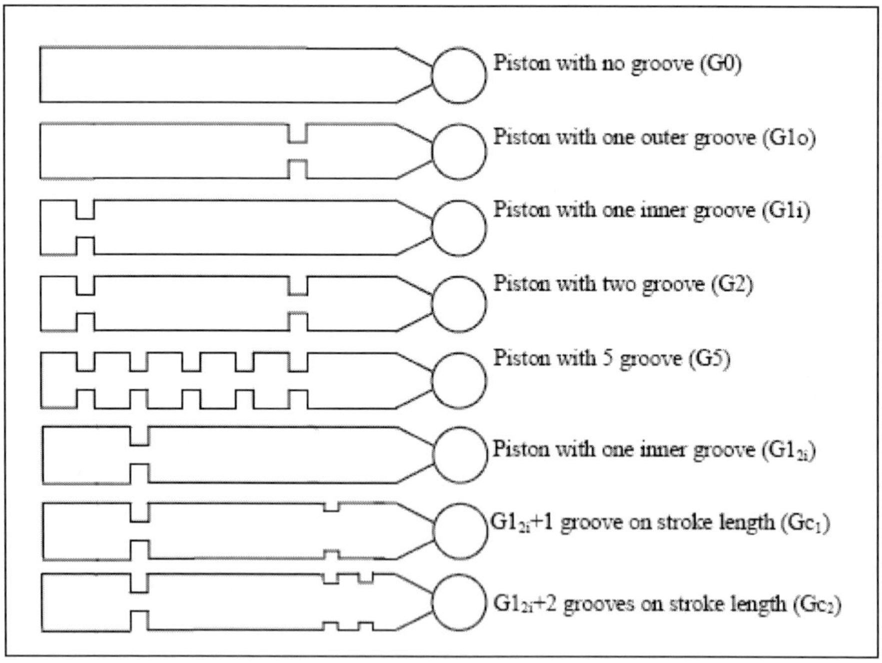

Figure 5.2.8. Eight different types of pistons studied, and for three piston diameters.

It must be recalled when viewing figure 5.2.8, that the five main grooves cut on the piston surface, remain inside the cylinder for all swash plate angular positions, just the one or two grooves cut on the piston stroke length will come in and out of the cylinder, depending on the swash plate angular position. These one/two extra grooves do not exist in the original manufactured piston shown in figure 5.2.1, and its use will be discussed in a further section.

5.2.5.2. Effect of Grooves on Piston-Barrel Pressure Distribution

Figure 5.2.9 represents the simulated pressure distribution in the piston-cylinder clearance for a piston without grooves and at different piston angular positions on swash plate, outlet pressure 30Mpa and 0.15Mpa tank pressure side, 1000 rpm pump turning speed, 10 microns central clearance and 5 microns piston eccentric displacement. Piston is connected to high pressure side from 0° to 180° swash plate angular positions and to the tank side (low pressure side) from 180° to 360° swash plate angular positions. Piston is moving in upward direction while connected to high pressure side and in downward direction when connected to the tank side.

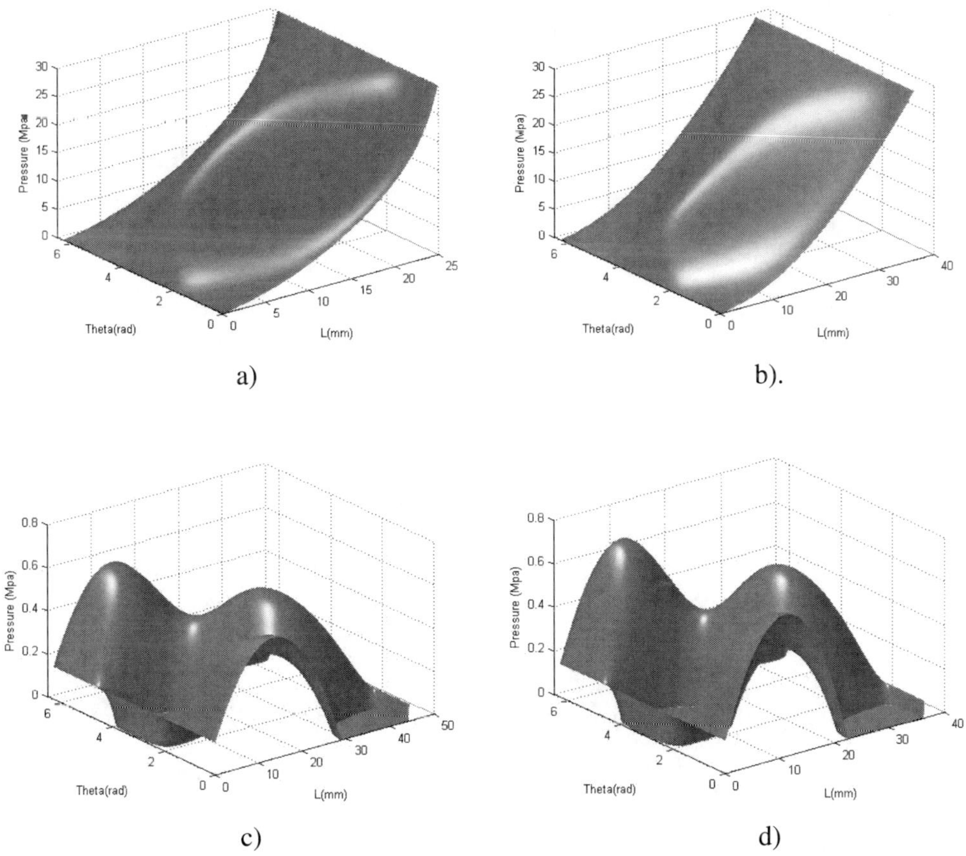

Figure 5.2.9. Pressure distribution between piston cylinder gaps at 1000 rpm rotation speed for different piston angular positions on swash plate, piston without grove, central clearance 10 microns, edge clearance 5 microns. Fluid oil ISO 32. [a] 0 degree, 30Mpa. [b] 90 degrees, 30Mpa. [c] 225 degrees, 0.15Mpa. [d] 270 degrees, 0.15Mpa.

It has to be noticed from figures 5.2.9c, d that negative pressure has been put to zero in order to see more clearly the area where cavitation is likely to appear.

It can be seen that cavitation appears when the piston is connected to tank and the area where cavitation appears is at its highest at 270° piston angular position, this happens because the piston velocity is negative and is at its maximum for this particular piston angular position.

Therefore during the cylinder incoming flow period it would be strongly desirable, to minimize the effect of cavitation in order to increase the life of the piston. The existence of cavitation is found in regions where piston-cylinder clearance is at its minimum, on both ends of the piston inside the cylinder.

Although not presented in the present section, the area under the influence of cavitation increases as the pump turning speed increases.

Another thing to be noticed from figure 5.2.9, is that the pressure peaks appearing in figure 5.2.9a, b produce a negative y-directional torque (see figures 5.2.5, 5.2.6), trying to restore the piston eccentric displacement, which arises from the differences in the friction of ball-cup and piston-cylinder pairs [21, 22].

Figure 5.2.10 presents the pressure distribution in the piston-cylinder clearance for different piston groove configurations at 30 Mpa outlet pressure, 1000 rpm and 90° piston angular position on swash plate.

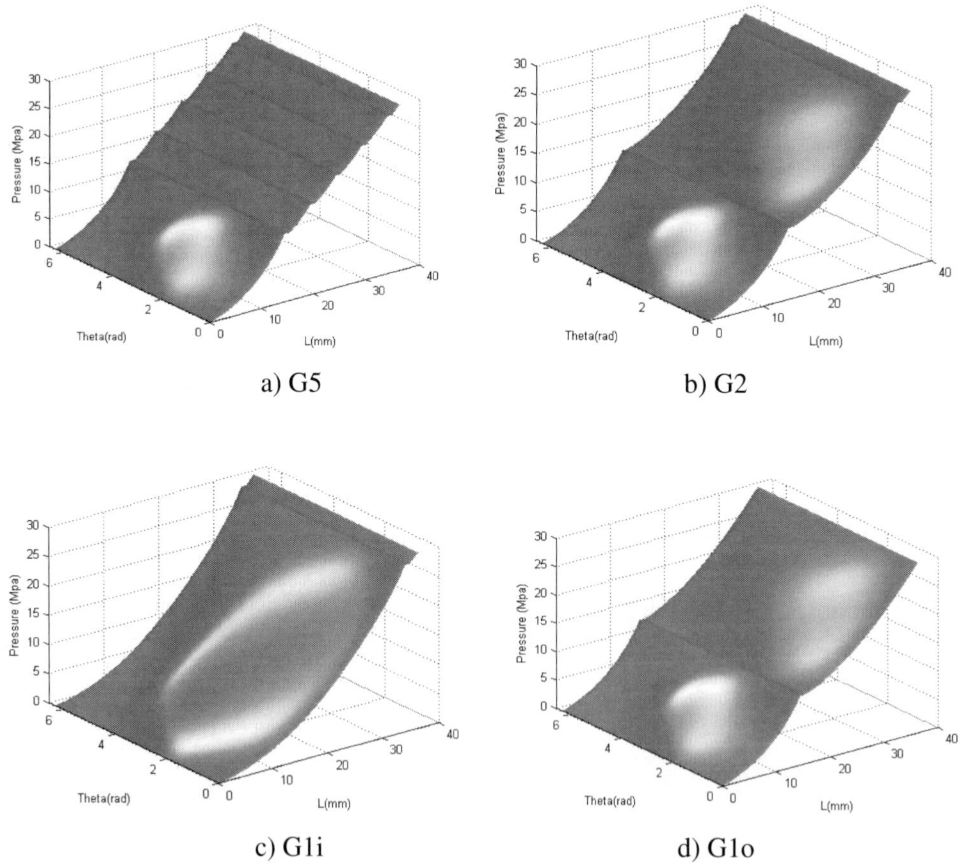

a) G5 b) G2

c) G1i d) G1o

Figure 5.2.10. (Continued)

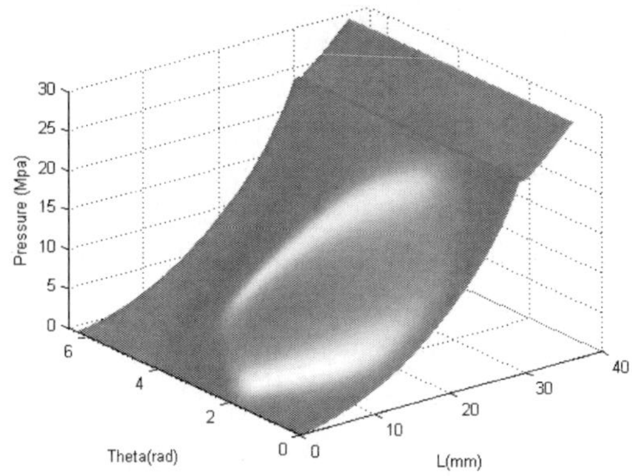

e) $G1_{2i}$

Figure 5.2.10. Pressure distribution in the piston-cylinder clearance at 30 MPa outlet pressure, 1000 rpm rotation speed, 90° piston angular positions on swash plate, 10 microns central clearance, 5 microns edge clearance, different pistons groove configurations. Fluid oil ISO 32.

This particular piston angular position has been chosen because the piston sliding velocity is at its maximum and therefore the pressure peaks are expected to be at its highest. It is clear from the figure that, grooves stabilize the pressure distribution along the angular direction of the piston. Such pressure distribution will result into more packed and stiff piston-cylinder system which will give a higher resistance to any movement created by external forces such as friction force. It can be noticed from figure 5.2.10b that when a groove is located at each piston side, the pressure distribution along the piston length is rather stable; such stability is nearly achieved when just a single groove is placed on the outer side of the piston, figure 5.2.10d. Therefore for stabilization purpose and pressure distribution point of view, it would be desirable to locate the grooves towards the edges of the piston rather than placing them in the centre.

Although not presented here, another important consideration to be noticed is, in presence of the grooves, the pressure distribution in piston-cylinder clearance is less dependent on the piston tilt. In reality, as the minimum edge clearance changes over time (tilt changes over time), for a piston without grooves the pressure distributions will be very much time dependent. On the other hand when considering the same piston dynamic movements and using a piston with grooves, the pressure distribution variation will be much less time dependent. Notice that in figure 5.2.10, the configuration G0, piston without groove is not presented since such configuration can be found in figure 5.2.10b.

5.2.5.3. Effect of Grooves on Piston-Barrel Leakage

Figure 5.2.11a, represents the leakage in the piston-cylinder assembly versus swash plate angular position for a non grooved piston and maintaining an edge clearance of 1.5 microns for all swash plate angular positions, being the maximum clearance when piston is centered of 3 microns. Two different pump turning speeds of 200 and 1000 rpm are considered. It can be seen that when piston is connected to the higher pressure side, the leakage is found to be negative for most of the cycle, (leakage flowing towards the cylinder chamber), this is due to the fact that Couette flow which is link with piston velocity, is higher that Poiseuille flow. Similar results were found in literature [8-9].

Figure 5.2.11b shows again the leakage at different swash plate angular positions for 10 microns piston-cylinder central clearance, 1000 rpm pump turning speed and different piston eccentric displacements (piston tilt), piston without grooves. It can be seen that piston tilt affects the leakage when piston moves from lower death centre to upper death centre (0°-180°), as tilt increases the leakage curves fall, increasing the leakage towards cylinder chamber, as a result it is expected the overall leakage of piston-cylinder for one full revolution, to be decreasing with the increase of piston tilt. Notice as well in figure 5.2.11b, the effect of piston tilt is just relevant when the piston is connected to the high pressure port, indicating that the tilt influences mostly the Poiseuille flow. The peaks at 180° and 360°, found in figures 5.2.11a, b are due to Poiseuille flow when the piston is at its upper and lower death centre.

It must be recalled that the overall leakage in a full cycle will be positive; the leakage flow direction is towards tank. The overall leakage can be found when integrating the temporal leakage presented in figure 5.2.11 as a function of angular position.

Figure 5.2.12a presents the overall leakage between piston-cylinder gap versus piston eccentric displacement for 10 microns central clearance and different piston groove configurations G0, G5, $G1_{2i}$, 30 MPa outlet pressure. It is noticed, as already established in figure 5.2.11b, that an increase in piston tilt decreases slightly the overall leakage, and such decrease is more relevant for pistons without grooves. As the number of groves cut on the piston increases, the piston-cylinder overall leakage tends to be constant and independent of piston tilt. Notice that just the inclusion of a single groove located at the 2nd groove position ($G1_{2i}$) brings a good stabilization of piston cylinder leakage at any piston tilt. Nevertheless, as the number of the grooves being cut on the piston surface increases the overall leakage increases.

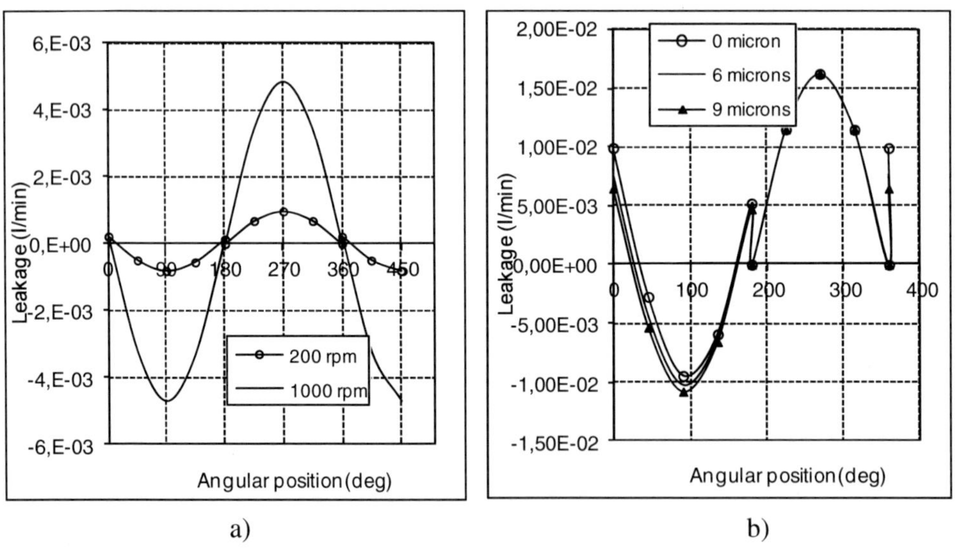

Figure 5.2.11. Leakage piston-cylinder clearance at 30MPa outlet pressure versus piston angular position on swash plate for non groove piston. Fluid oil ISO 32. [a] 1.5 microns edge clearance. 3 microns central clearance, 200-1000 rpm pump turning speed. [b] 10 microns central clearance, 1000 rpm pump turning speed, at different piston eccentric displacement.

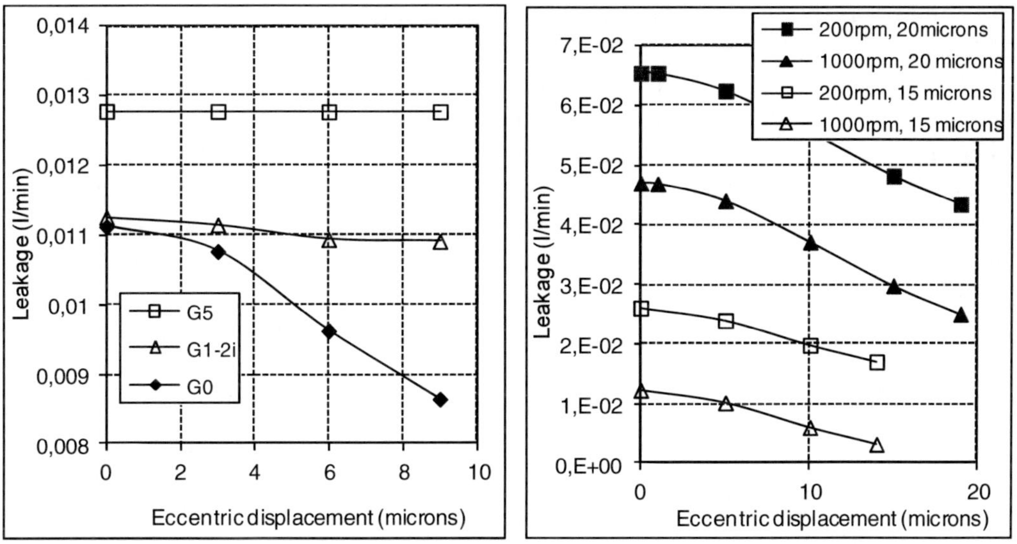

a) b)

Figure 5.2.12. Leakage piston cylinder versus piston eccentric displacement, 30 MPa outlet pressure. [a] overall leakage at 10 microns central clearance, different groove configurations. Fluid oil ISO 32. [b] Temporal leakage for non groove piston at 45° piston angular position on swash plate at two different clearance and turning speed.

To see more clearly the effect of piston eccentric displacement (tilt) on leakage, two given piston clearances of 15 and 20 microns and two turning speeds of 200 and 1000 rpm were evaluated, outlet pressure was 30 MPa and the angular position of the piston on the swash plate was 45 degrees, results are presented in figure 5.2.12b.

It can be seen that as the piston eccentric displacement increases, the leakage tends to decrease. The leakage decrease with piston eccentric displacement is higher for higher clearances.

One of the most important characteristic of figure 5.2.12b is, the leakage at this particular point (45° angular position of piston on swash plate), is positive and in figure 5.2.11b the same leakage at 10 microns central clearance was reported as negative.

This is due to the fact that as the clearances increase Poiseuille flow becomes more relevant than Couette flow, therefore the overall leakage towards tank will be much higher than for smaller clearances.

It is important to point out that in figure 5.2.12b, leakages have a higher value for low turning speeds and high clearances. In figure 5.2.12a it was explained that leakages had a higher value for higher turning speeds and in figure 5.2.12b it seems that the opposite is being said.

The explanation of this is that as clearance increases, (figure 5.2.11a is done at 3 microns central clearance and figure 5.2.12b is done at 15, 20 microns central clearance), the Poiseulle flow becomes more relevant than Couette flow, then even when the piston moves from the lower death centre towards the upper death centre the leakage piston-cylinder flows in direction to tank, its sign is positive, as pump turning speed increases the Couette flow gains relevance although Poiseulle flow is still dominant, the resulting flow will therefore be positive, but the magnitude will be smaller than the one found at high pump turning speeds.

5.2.5.4. Effect of the Grooves on Piston-Barrel Cavitation

As can be seen from figures 5.2.9c, d, in the absence of grooves, there is an important part of the piston, which is under the influence of cavitation; on the other hand in the presence of grooves, the effect of the cavitation tends to be reduced, as can be seen in figure 5.2.13. Figures 5.2.13a,b, have the same characteristics as figure 5.2.9d, 30 MPa outlet pressure, 10 microns central clearance, 5 microns edge clearance, 270 degrees of piston on the swash plate. The difference is that in figure 5.2.13a, the piston configuration G5 is used, while on figure 5.2.13b, configuration G2 is presented. It is noticed, the five groove configuration, G5, reduces more effectively the appearance of cavitation.

It is a standard procedure to avoid putting grooves on the piston stroke length, but according to the models developed, it would be desirable to put maybe a very shallow groove on piston stroke length, in order to be able to minimize the cavitation in this particular position. Figure 5.2.14 presents the pressure distribution between piston-cylinder clearance, for such piston configurations (Gc_1 and Gc_2) and under the same operating conditions as the ones used in figure 5.2.13. It is clear from figure 5.2.14a that putting 1 shallow groove on the piston stroke length reduces the appearance of cavitation; increasing the number of grooves to two, figure 5.2.14b, will produce a more stable pressure all around the angular position.

Despite the shape of the grooves has not been considered in this study, we are confident to say that using shallow V-shape grooves on the piston stroke length would bring very similar results to the ones presented in figure 5.2.14, such grooves would facilitate the incoming and outgoing of the piston into the cylinder.

It is important to notice from figures 5.2.10, 5.2.13 and 5.2.14 that, the position of the grooves on piston surface is very relevant, since it completely modifies the pressure distribution in the piston-cylinder clearance. The use of a groove located on the second groove position, is the configuration which appears to be reducing more effectively the appearance of cavitation, compare figure 5.2.13b and 5.2.14b. On the other hand, the grooves located on the central part of the piston have no effect regarding cavitation improvement. Further information regarding piston performance with grooves can be found in [23].

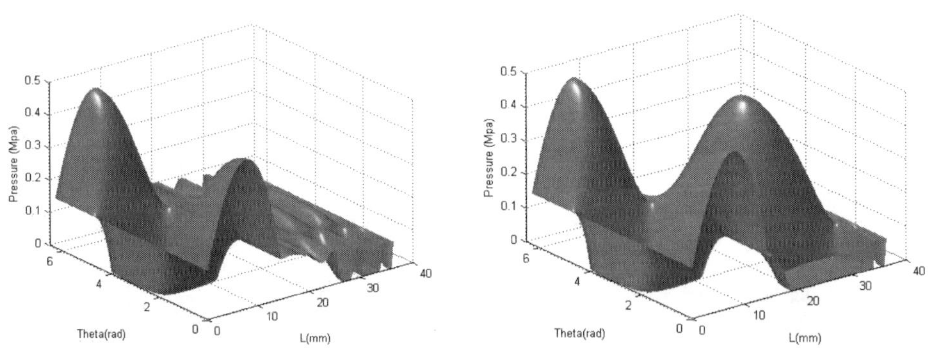

a) b)

Figure 5.2.13. Pressure distributions in piston cylinder clearance when piston is connected to tank side, 10 microns central clearance, 5 micron edge clearance, 270° piston angular position, 1000 rpm pump turning speed. Fluid oil ISO 32. [a] Configuration G5. [b] Configuration G2.

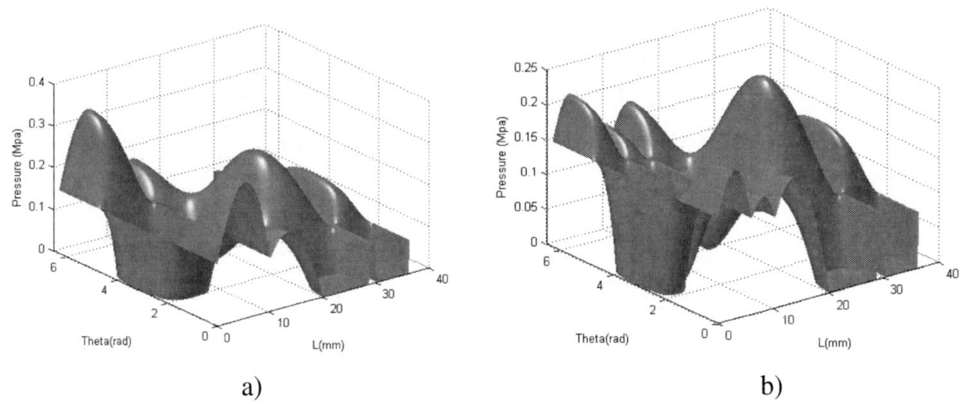

Figure 5.2.14. Pressure distributions in piston cylinder clearance when piston is connected to tank side, 10 microns central clearance, 5 micron edge clearance, 270 degrees piston angular position, 1000 rpm pump turning speed. Fluid oil ISO 32. [a] Configuration Gc_1. [b] Configuration Gc_2.

5.2.5.5. Effect of the Grooves on Total Piston Force and Y-Directional Torques

Initially, the different groove configurations will be studied when maintaining the piston perfectly aligned with the cylinder, under such conditions, it is understood that the force acting over the piston will be symmetrical, maintaining the piston always in the central position, the higher the force due to pressure distribution will be, the tighter the piston will be held in its central position, therefore the piston resistance to the external forces, such as friction forces will be maximum. Then the total force on piston surface can be seen as a measure of the system stiffness.

Figure 5.2.15 presents the percentage increase of the total force exerted by the pressure acting on piston surface, with respect to the non grooved piston configuration, the operating conditions are: 30 MPa inlet pressure, 3 microns central clearance, 400 rpm pump turning speed and when piston axis is parallel to cylinder axis (α=0). It can be seen that grooves placed towards the inner edge of the piston contribute to the force magnitude and produce 1.5% higher force on piston surface with respect to the non grooved piston. On the other hand, grooves placed towards the outer edge reduce the total force on the piston. Such statement, it is found to be independent of pump turning speed. It is important to clarify that figure 5.2.15 presents the percentage increase in force when the piston is connected to the higher pressure side (0°-180°) as the total force acting on the piston surface is much higher than the force when piston is connected to the tank side.

When the piston is tilt versus the cylinder axis, the force acting over the piston surface will create a restoring torque which can be seen as a stabilizing parameter. In figure 5.2.16, it is represented the y-directional torque acting on the piston versus swash plate angular position, for different turning speeds and groove configurations. It can be seen that, in most of the cases, the value of the torque is negative, which means that the torque acts over the piston in such a way that it tries to rotate it in clock wise direction to restore its central position. Although when grooves are placed near outer edge (G2, G1o), the y-direction torque does become positive, which means the pressure torque acts in such a way that it tries to tilt the piston even further, which is totally undesirable. Therefore configuration G2 and G1o are unacceptable under the restoring torque point of view.

From figure 5.2.16, configuration G5, which is the way the original piston was designed, produce a very small restoring torque, being the torque from 0° to 60° angular position less than 5Nm, nevertheless this configuration is the only one which creates a higher torque with the increase of pump turning speed. Configuration G1i and G0 produce good torque on low turning speeds but fail to do it on high turning speeds. Configuration $G1_{2i}$, creates a restoring torque which is pretty independent of pump turning speeds and will result into the most stable configuration.

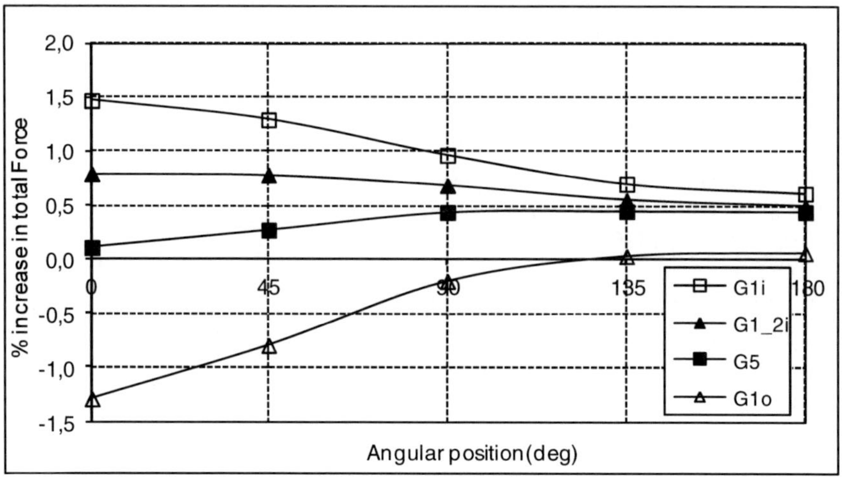

Figure 5.2.15. Percentage increase in total force with respect to no groove piston at 400 rpm, 30MPa outlet pressure, 3 microns central clearance, no eccentric displacement is considered.

5.2.5.6. Effect of the Piston Diameter on Leakage and Torque

Figure 5.2.17 presents the piston-cylinder clearance leakage for three different piston diameters, (1D = 14.6mm, 1.5D = 21.9mm, 2D = 29.2mm), three piston configurations G5, G0, $G1_{2i}$ and as a function of pump turning speed.

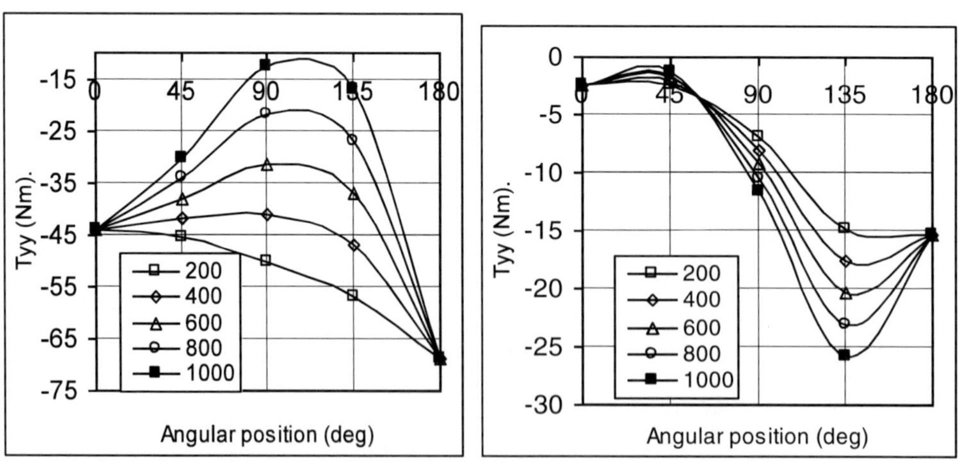

a) G0 b) G5

Figure 5.2.16. (Continued).

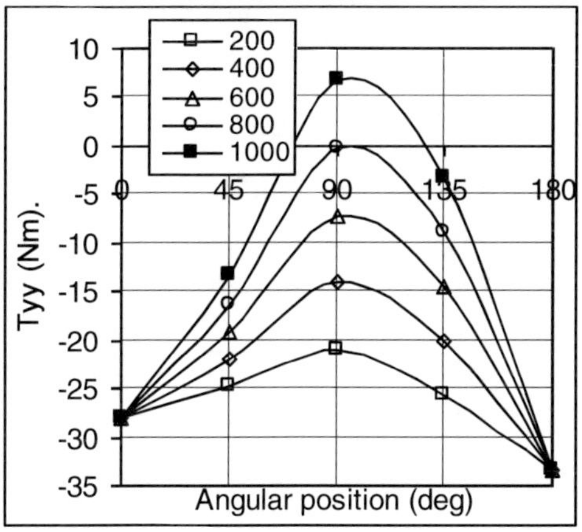

c) G2

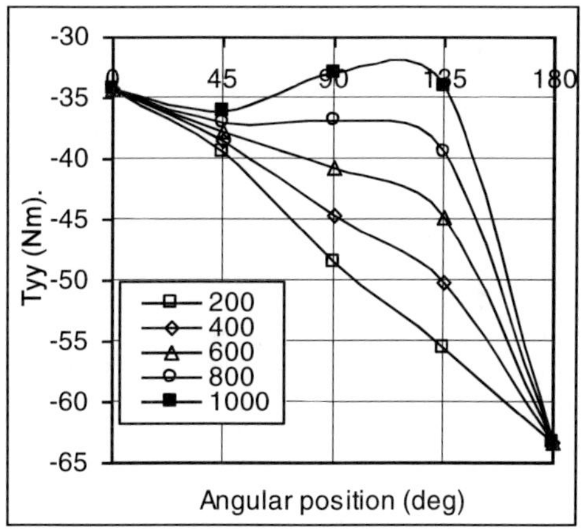

d) G1i

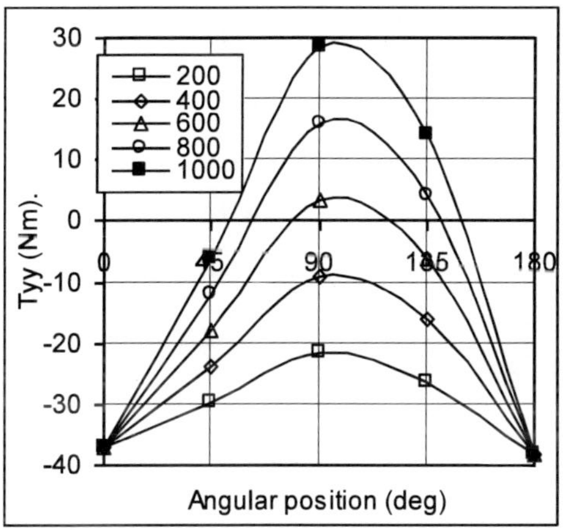

e) G1o.

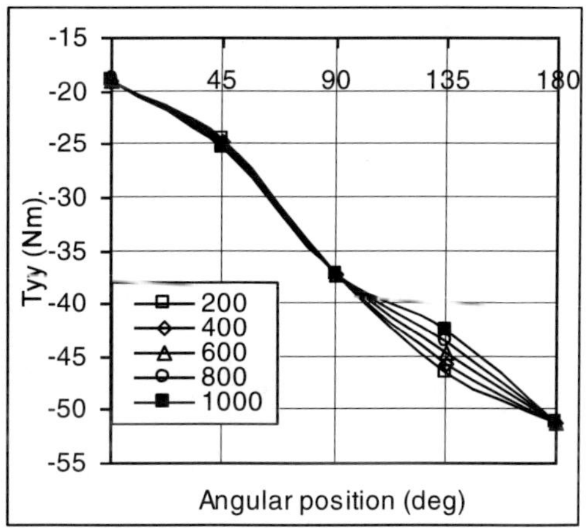

f) $G1_{2i}$.

Figure 5.2.16. Tyy with respect to piston swash plate angular position at different pump turning speed at 30 Mpa outlet pressure, 3 microns central clearance, 1.5 micron edge clearance. [a] No groove piston. [b] 5 groove piston. [c] 2 groove piston. [d] 1 inner groove piston. [e] 1 outer groove piston. [f] 1 inner groove located at 2nd groove position.

As expected, leakage tends to increase with diameter increase. An interesting thing to be noticed from figure 5.2.17 is that, the gradient of leakage increase with turning speed is highest for the piston with grooves; on the other hand such gradient is lowest for the piston without any grooves. The piston configuration $G1_{2i}$ lies in between.

Figure 5.2.18 represent the y-directional torque acting on the piston versus swash plate angular position, for 1000 pump turning speeds, two different groove configurations and three different piston diameters, 30 MPa outlet pressure.

It can be seen from figure 5.2.18 that the magnitude of the restoring torque, increases with the increase of piston diameter, such increase is maximum when using pistons with no grooves. From the results obtained, it can be concluded that, a piston with no grooves and with a bigger diameter, will produce the best restoring torque, which means, less contact metal to metal is to be expected and therefore the life of the machine is expected to be longer. Whenever the manufacturer requirements lead to a piston of small diameter, the configuration $G1_{2i}$ might be considered, since it produces a relatively low leakage (although a little higher than the non grooved pistons) and a higher restoring torque, which is independent of pump turning speed, see figure 5.2.16.

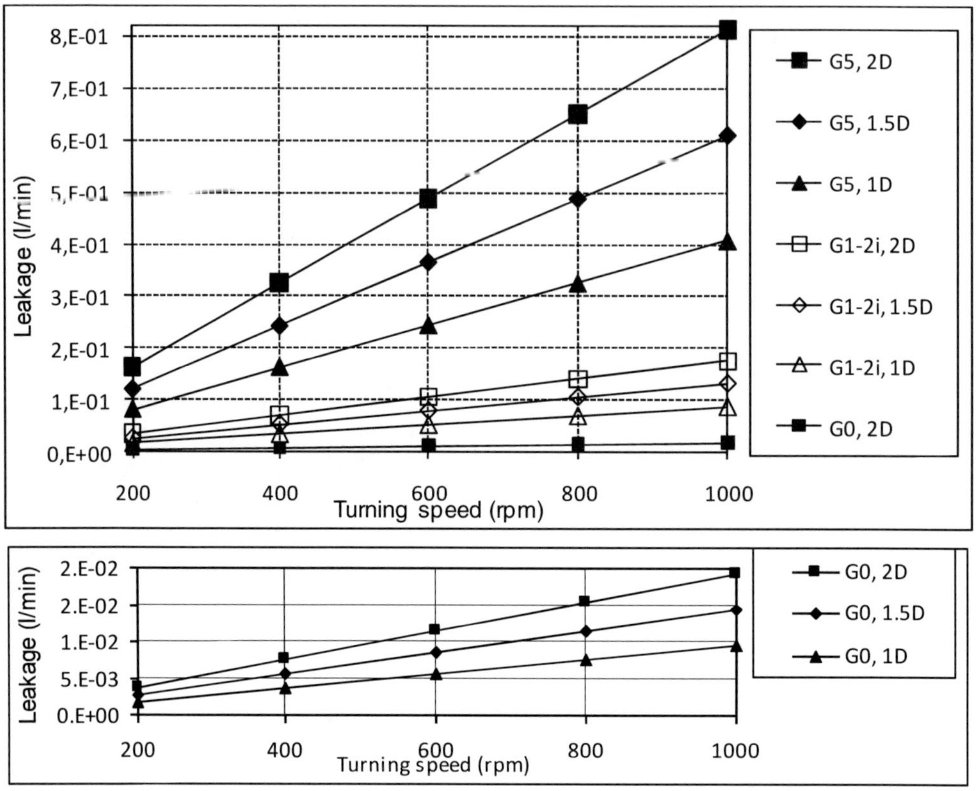

Figure 5.2.17. Leakage piston-cylinder clearance at 30MPa outlet pressure versus swash plate turning speed for three different type and piston and piston diameter, at 90° swash plat angular position. (1.5 microns edge clearance. 3 microns central clearance)

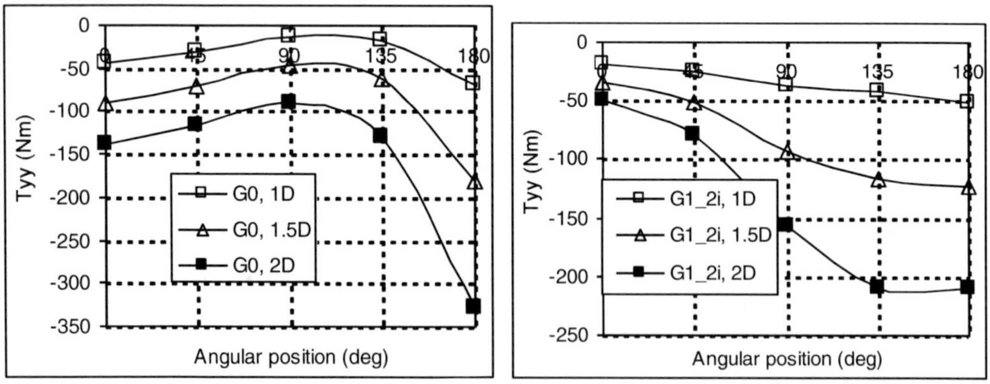

a) b)

Figure 5.2.18. Tyy with respect to piston swash plate angular position at 1000rpm pump turning speed, 30 Mpa outlet pressure, 3 microns central clearance, 1.5 micron edge clearance. [a] No groove piston. [b] 1 inner groove located at 2nd groove position.

5.2.6. Conclusion

1. Analytical equations able to give leakage flow and pressure distribution in the piston-cylinder clearance are being presented. The equations are applicable for laminar flow, piston without tilt and consider piston-cylinder relative movement.
2. A 2-D CFD model using the software Fluent has been used to validate the equations created.
3. A numerical model which includes piston tilt and piston cylinder relative movement has been created. This model allows studying the effect of grooves being cut on the piston surface.
4. It is demonstrated that just two grooves located respectively on both piston ends are sufficient to maintain piston dynamic equilibrium.
5. The use of grooves cut on the piston surface brings stability to the piston, since it increases piston stiffness.
6. The grooves located on the central part of the piston, appear not to be useful regarding the pressure stabilization, torque, force or even cavitation point of view.
7. To avoid cavitation, it is important to consider the inclusion of grooves at the piston stroke length and near to the piston pressure side. In fact, wherever it is expected cavitation to appear, the use of a groove will tent to prevent cavitation from appearing. The area where cavitation might appear, increases with the increase of pump turning speed.
8. Regarding the restoring torque point of view, configurations G2 and G1o are undesirable, configuration $G1_{2i}$ (one groove at the inner edge at second groove position) seems to be the best location from restoring torque point of view, since the magnitude of the restoring torque produced by the pressure distribution is quite high and rather independent of pump turning speed.
9. For a piston without grooves, leakage piston-cylinder decreases as the piston tilt increases, the decrease is more relevant at higher clearances. For such a piston, pressure distribution in the piston-cylinder clearance is very much time dependent.
10. As the number of grooves being cut on piston surface increases, the effect of piston tilt becomes less relevant regarding the piston cylinder overall leakage. As already established in previous literature, the increase in number of grooves will bring an increase in leakage.
11. Grooves are good to prevent cavitation, but from leakage and restoring torque point of view their presence is useful only when piston diameter is small. On the other hand, in case of a piston with a relatively bigger diameter, non groove piston produces low leakage and higher restoring torque.
12. For a smaller diameter piston, among the configurations studied, $G1_{2i}$ is the one bringing the best performance because it produces higher restoring torque at all pump turning speeds, leakage is fairly independent of piston tilt and although overall leakage is slightly higher that the non groove configuration, it is much lower than the configuration G5.
13. As an overall conclusion, it can be said that, it is desirable to produce either pistons with relatively big diameters and without grooves on its surface, or pistons with smaller diameters and a single groove positioned near the inner part of the piston.

5.2.7. References

[1] Sweeney D.C. (1951). Preliminary investigations of hydraulic lock. *Engineering*. 172, 513-16.

[2] Sadashivappa K., Singaperumal M., Narayanasamy K. (2001). Piston eccentricity and friction force measurement in a hydraulic cylinder in dynamic conditions considering the form deviations on a piston". *Mechatronics* 11, 251-66.

[3] Milani M. (2001). Design hydraulic locking balancing grooves. *Proceedings Institution of Mechanical Engineers Part I*. 453-465.

[4] Borghi M, Cantore G., Milani M. and Paoluzzi P. (1998). Numerical analysis of the lateral forces acting on spools of hydraulic components. FPST5, *Fluid Power Systems and Technology, ASME*.

[5] Borghi M. (2001). Hydraulic locking-in spool-type valves: tapered clearances analysis. *Proceedings Institution of Mechanical Engineers, Part I*. 157-168.

[6] Blackburn J.F., Reethof G. and Shearer J. L. (1960). *Fluid power control.* New York. MIT Press – John wiley.

[7] Merrit H. E. (1967). *Hydraulic control systems.* New York. John Wiley.

[8] Ivantysynova M; Huang C. (2002). Investigation of the Flow in Displacement Machines Considering Elastohydrodynamic Effect. *Proceedings of the 5th JFPS International Symposium on Fluid Power,* November 13, Nara, Japan Vol. 1, 219-229.

[9] Ivantysynova M; Lasaar R. (2004). An Investigation into Micro – and Macrogeometric Design of Piston Cylinder Assembly of Swash plate machines. *International Journal of Fluid Power* 5:1, 23-36.

[10] Fang, Y. and Shirakashi, M. (1995). Mixed lubrication characteristics between the piston and cylinder in hydraulic piston pump motor. *J. of Tribology Transactions of ASME*, 117:1, 80-85.

[11] Prata AT, Fernandes JRS, Fagotti F. (2000). Dynamic analysis of piston secondary motion for small reciprocating compressors, *J. of Tribology Transactions of ASME*. 122:4, 752-760.

[12] Berger E. J., Sadeghi F., Krousqrill C.M. (1996). Finite element modeling of engagement of rough and grooved wet clutches. *J. of Tribology Transactions of ASME*, 118:1, 137-146.

[13] Razzaque M. M., Kato T. (1999). Effects of Groove Orientation on Hydrodynamic Behavior of Wet Clutch Coolant Films. *Transactions of the ASME*. 121, 56-61.

[14] Lipschitz, A., Basu, P., and Johnson, R. P. (1991). A Bi-Directional Gas Thrust Bearing. *STLE Tribology Transactions*, 34: 1, 9-16.

[15] Basu, P. (1992). Analysis of a Radial Groove Gas Face Seal. *STLE Tribology* Transactions. 35:1, 11-20.

[16] Kumar, S., Bergada, JM., Watton, J. (2009). Axial piston pump grooved slipper analysis by CFD simulation of three dimensional NVS equation in cylindrical coordinates. *Computer & Fluids* 38, 648-663.

[17] Bergada JM; Watton J. (2003). A New Approach Towards the Understanding of the flow in Small clearances applicable to Hydraulic Pump Pistons With Pressure Balancing Grooves. *7th International Symposium on Fluid Control, Measurement and Visualization.* Flucome, Sorrento Italy. August 25-28. 1-8.

[18] Bergada JM; Kumar S; Davies DLl; Watton J. (2012). A complete analysis of axial piston pump leakage and output flow ripples. *Applied Mathematical Modeling* 36, 1731-1751.
[19] Pnueli D., Gutfinger C. (1992). *Fluid Mechanics*. Cambridge university press.
[20] Patankar, SV. (1980). *Numerical Heat Transfer and Fluid Flow*. Taylor & Francis Group: Hemisphere Publishing Corporation.
[21] Hooke C.J. Kakoullis Y.P. (1978). The lubrication of slippers on axial piston pumps. *5th International Fluid Power Symposium*. Durham, England. B2-(13-26)
[22] Hooke C.J., Kakoullis Y.P. (1981). The effects of centrifugal load and ball friction on the lubrication of slippers in axial piston pumps. *6th International Fluid Power Symposium,* Cambridge, England. 179-191.
[23] Kumar S; Bergada JM. (2013). The effect of piston grooves performance in an axial piston pump. CFD analysis. *Int. Journal of mechanical sciences*. 66, 168-179.

5.3. Slipper Performance, Effect of Grooves on Slipper Surface

Slippers have been extensively studied, since it has been traditionally assumed that leakage slipper-swash plate was higher than leakage in any other pump clearance and therefore pump volumetric efficiency was very much dependent on slipper-swash plate performance. In what follows a deep study of slipper behavior shall be presented, in section 5.6 slipper-swash plate leakage will be compared with other piston pump leakages finding out how correct the traditional assumptions are.

5.3.1. Previous Research on Slippers

The importance of understanding slippers behavior is made relevant when is considered that most of the leakage in piston pumps and motors happens to be through slippers. Good performance of the machine is directly linked with smooth slipper/swash plate running, being necessary to avoid metal to metal contact and excessive film thickness. Therefore, volumetric, hydraulic and mechanic efficiencies in piston pumps and motors will be affected by slipper performance.

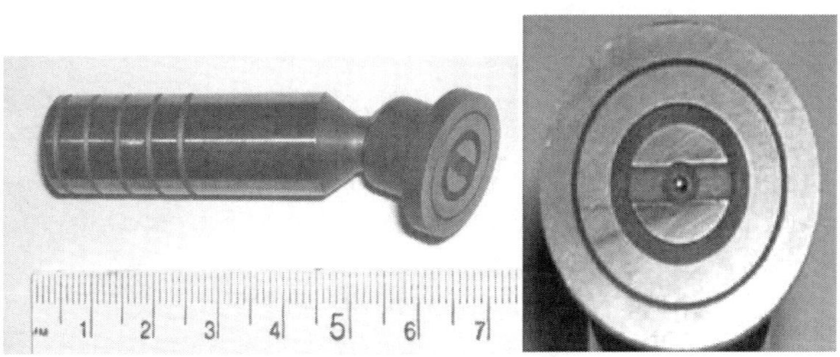

Figure 5.3.1. Piston and slipper assembly [Courtesy Oilgear Towler UK Ltd].

In the majority of the researches presented until now, the effect of the different pressure balancing grooves cut on pistons and slippers has been neglected, and although the groove effect on the flow and the pressure distribution is not expected to give a completely different pattern from previous knowledge using single-land slippers, the introduction of a groove brings a far more complicated mathematical approach when aiming to fully understand its behavior.

The main piston and slipper assembly used in this study is shown in figure 5.3.1, and is one of nine pistons from a pump which maximum volumetric displacement is 0,031 dm^3 /rev. It will be seen that the slipper design uses two full lands, an alternative being to machine additional slots across the second land to balance the groove and outlet pressure. The approach selected seems to be the corporate design philosophy of the particular pump manufacturer.

There have been many publications in this general subject area over the past 40 years, many concerned with improving the slipper performance of piston pumps and motors. Fisher [1] studied the case of a slipper with single land on a rotating plate, in both cases, when the slipper was parallel and tilted with respect to the swash plate and the load capacity, restoring moment, and flow characteristics were studied. Fisher demonstrated that if a flat slipper tilts slightly so that the minimum clearance occurs at the rear; the hydrodynamic loads generated tend to return the slipper to the non tilted position. Fisher concluded that when the ratio of the angle of tilt to the angle at which the slipper would just touch the plate is higher than 0.675 then slipper equilibrium would be impossible since the load plus the dynamic force cannot be balanced by the hydrostatic force.

Böinghoff, [2] performed a deep study on slippers. He studied theoretically the static and dynamic forces and torques acting on a single piston, via analyzing carefully the slipper performance as it rotates around the swash plate, he also took into account the torque generated on the spherical bearing. Large quantities of experimental results where also generated, in which torque and leakage were evaluated for different position angles and turning speed. The effect of oil viscosity on the torques created was also taken into account. Pump leakage was studied for different swash plate angles and turning speeds. Leakage was found to be smaller at low speeds < 5 rad/s and low swash plate angles and increased with turning sped. He also studied experimentally the influence of slippers with different lands, focusing on torque and leakage at different turning speeds. It must be pointed out that although the slipper studied had four lands, just one of them can be considered as full land, the rest were vented. He found that torque remained pretty much constant with turning speed when 1 or 2 lands were used, and torque was quickly increasing with speed when using four lands. Leakage was found to be lower when decreasing the number of lands and for speeds higher than 10 rad/s.

Hooke [3] showed that a degree of non-flatness was essential to ensure the successful operation of the slipper and the non-flatness must have a convex profile. He concluded that the lift contribution due to spin had an effect of second order. The centripetal forces resulting from the speed of the pump had a tendency to tilt the slipper outwards thus reducing the clearance on the inside of the slipper path. He also pointed out that the friction on the piston ball played a major role in determining the behavior of the slipper. In a further paper [4] Hooke studied more carefully the couples created by the slipper ball, finding that the major source of variation between slippers did not arise from differences on surface profile, but from differences in the friction in the ball-cup and piston-cylinder pairs. He concluded that ball-cup friction increased with pressure, and contact metal to metal may appear when lubrication was deficient.

Iboshi and Yamaguchi [5-7], working with single land slippers, found a set of equations based on the Reynolds equation of lubrication which gave the flow and the main moments acting on the slipper by taking into account the slipper displacement velocity and tilt. They found that there was a limit of fluid film lubrication for the specific supply pressure and rotational speed. They also defined a diagram checking the conditions under which metal to metal contact on the slipper may appear. It was pointed out that the friction of the spherical bearing affects significantly the tilt angles, and the rotational speed affects the central clearance of the slipper plate. Experimentally they found that the slipper plate clearance, under steady rotational conditions, was fluctuating.

Hooke et al [8] studied more carefully the effect of non-flatness and the inlet orifice on the performance of the slipper. He gave a very good explanation of the equations used and the mathematical process to find them, finding the moments along the two main axes of the slipper. He found out that 2-5% of the load was being supported by hydrodynamic forces and tilt was necessary to produce the desired hydrodynamic lift. It was also found that the increase of the film thickness with reduction of slipper non-flatness was very small. In all geometrical conditions studied, it was found that slippers with no inlet orifices had larger clearances than slippers with orifices. However starvation effects and cavitation may appear. In [9] Hooke and Li focused on the lubrication of over clamped slippers, the clamping ratio being defined as the relation between the hydrostatic lift acting on the slipper and the piston load. Typical over clamped ratios ranged between 1-10%. He noticed that to have successful slipper lubrication, the plate where the slipper slides must be well supplied with fluid. The tilt was found to be proportional to the non-flatness magnitude divided by the square root of the slipper central clearance. In this paper the Reynolds equation of lubrication for tilted slippers was integrated numerically by finite difference method. In [10] Hooke and Li analyzed carefully the three different tilting couples acting on slipper, finding that the tilting couple due to friction at the slipper running face is much smaller than the ones created at the piston-cylinder, piston-slipper interfaces and the centrifugal one. All slippers tested had a single land. The slippers were found to operate relatively flat, clearances were highly dependent on the offset loads and the minimum clearance was found to be not particularly sensitive to the type of non flatness magnitude.

Takahashi et al [11] studied the unsteady laminar incompressible flow between two parallel disks with the fluid source at the centre of the disks. Both the flow rate and the gap between disks were varied arbitrarily with time and independently of each other. The two dimensional Navier-Stokes equations were solved via spectral method. The theory presented gave light to the study of the complicated characteristics of the inertial forces.

Li et al [12] studied the lubrication of composite slippers on water based fluids. It was found out that the slipper plate clearance was smaller than when using hydraulic oil and it was essential that the surfaces of the slipper and plate should be highly polished in order to accomplish a successful slipper operation. Even for the best material combinations, problems were encountered when the system was run at high fluid pressures and low running speeds. When turning at speeds lower than 300 rpm, slipper plate metal to metal contact was found. The slipper plate clearance increased when increasing the slipper surface.

Koc et al [13] focused their work on checking whether under clamped flat slippers could operate successfully or whether a convex surface was required. A good understanding of the three couples acting on the slipper, previously defined by Hooke [4, 10], was essential. They took into account the work done by Kobayashi et al [14] on the measurements of the ball friction. They concluded that polishing of the running face of the slipper to a slightly convex form appeared to be essential for successful operation under all conditions. It was also found that the insertion of an inlet orifice at the centre of the slippers result in an increase of the central clearance, though tending to destabilize the slippers. Notice that the insertion of an inlet orifice seams to give opposite effects in references [8] and [13]. It must be bared in mind that in reference [8] the slipper used was having conical lands, while in reference [13] the sliding surface is slightly convex. The size of the central orifice in under clamped slippers appeared to be most critical for a successful operation. Harris et al [15, 16], created a mathematical dynamic model for slipper pads, in which lift and tilt could be predicted, the model was able to handle the effect of the possible contact with the swash plate. The simulation shows that slipper tilt is much higher at suction that at delivery and at delivery tilts increases with pump speed.

In [17, 18] Koc and Hooke studied more carefully the effects of orifice size, finding that the under clamped slippers and slippers with larger orifice sizes run with relatively larger central clearances and tilt more than those of over clamped slippers with no orifice. Slippers with no orifice had greatest resistance to tilting couples and the largest minimum film thickness. One of the major effects of the orifices was to greatly reduce the slipper resistance to tilting couples. They pointed out that the use of two lands, an inner and outer land, brought more stability to the slipper. They also indicated that when a slipper incorporates a second land, the space between lands needs to be vented to avoid the generation of excessive hydrostatic lift, allowing the flow trapped between lands to escape. The direction and magnitude of the tilt was found to be directly dependent on the offsets imposed.

Tsua et al [19], analyzed in detail the slipper dynamics in a piston pump. As other authors before [1, 5], Tsua used the Reynolds equation of lubrication considering slipper spin and tangential velocity over the pump axis and integrated this equations by using Newmark β method. Pressure distribution found from the numerical scheme was later used to find out the force and torques over the slipper. Wieczoreck, U. Ivantysynova [20, 21] developed a package called CASPAR which uses the two dimensional equation of lubrication and the energy equation in differential form. Transient cylinder pressure has been computed by considering leakage in piston, slipper, and port plate. In addition, the clearance and tilt of the slipper was shown to vary over one revolution of the pump and a single land slipper plate was used in the theoretical and experimental analysis.

Manring [22, 23] analyzed the slipper by using classic lubrication equation based on pressure and volumetric flow rates, but the slipper he took into account had no groove. He found out that the minimum fluid film thickness between the bearing and the thrust surface is of the order of the surface roughness. Therefore, metal to metal contact might be possible. Kazama [24-26] formulated a time dependent mathematical model for slipper-swash plate model for the use of tap water under mixed and fluid film lubrication condition by considering the surface roughness and the revolution radius. He found that the radius of revolution of the pad influences the bearing performance because of the hydrodynamics wedge effect and the minimum power loss happened when the balance ratio become close to unity. The performance of slippers with grooves was reported by [2, 8, 18, 27], where it was found that a groove brought stability to the slipper dynamics. In all these cases, the second land was vented and therefore the pressure on the groove was reported to be atmospheric. As a result the groove itself was not creating lift. It was also reported that for a given central clearance, reducing the number of lands give a reduction in leakage. It has to be noticed that in the present case the groove is not vented and therefore as it will later be demonstrated the second land and the groove will create hydrostatic and hydrodynamic lift. Analytical solution for slippers with multiple lands was outlined in [28, 29] and more clearly defined in [30], although the effect of tangential velocity was not considered. Reynolds equation of lubrication is applied to the slipper swash plate gap by considering the flow only in radial direction, which turn out to be accurate for flat slipper but while considering tilt slipper, flow tend to move into angular direction too, as a result this analysis does not produce very good result for higher tilts. The equations developed in the full mathematical analysis of the slipper with groove are complex enough not to be solved analytically without further approximation. Another possible way to tackle such complex equations while retaining its accuracy is implementing a computational technique. There has been some previous efforts made in [11, 18 - 21, 24, 26, 31 - 34] to analyze the slipper through various computational techniques. Some of these works used spectral method [11, 19, 24] other have used finite difference method [18] through Reynolds equations.

Brajdic [31] analyzed the low friction pad bearing in two dimensions Cartesian coordinates system taking into account the compressibility of fluid. He showed the development of the fluid recirculation behavior within the pocket. Helene et al [32] also investigated a hybrid journal bearing in two dimensional Cartesian coordinate system. She also took into account the turbulent flow conditional (Re up to 5000) by implementing k-ε model with logarithmic wall functions and pointed out that turbulent pressure pattern is less affected by recirculation zones. Braun et al [33] analyzed the effect of pocket depth by applying two dimensional Navier stokes equation to the slipper pocket gap and pointed out that the deep pockets show a lesser degree of coupling between the pocket flow and the clearance flow than the shallow pockets. Niels and Santos [34] formulated a numerical model based on Reynolds equation to minimize the friction in tilting pad and showed that a large amount of energy can be saved by using low length to width ration of the cavity. If we focus on the problem a bit more conceptually rather than technically, the problem we are tackling here is similar in behaviour of three dimensional open cavities in cylindrical coordinates. The literature available for pressure/flow simulation in cavities is quite vast and the SIMPLE (semi implicit method for pressure linked equation) family algorithm [35] has been widely applied to this kind of simulation. Most of the work has been done for rectangular cavities [36-42] where Cartesian coordinates were applied. Literature available on cavities in cylindrical coordinates [43-45] is much less common. The cavities analyzed in [43-45] are 2-dimensional and the analysis done in these papers is focusing in analyzing the heat transfer. Although the analysis performed in [44] considers the effect on flow performance when changing the sealing gaps, still the flow is axis-symmetric and therefore the cavity is considered as two dimensional. In the present study, the flow does not have any kind of symmetry, as a result a complete three dimensional analysis needs to be considered. Despite the fact that exist some literature available in Cartesian coordinates, where dimensions and shape of cavities as well as clearances between plates were analyzed [39-42], no evidence has been found of a flow involving the complexities considered in the present study. For example, a 2d simulation in curvilinear coordinates using the stream function method was done in [39] where the Vorticity in triangular, circular and rectangular cavities were studied; the conclusion of the study was that for a given Reynolds number, triangular shape cavity created the smallest leakage, and for Reynolds numbers smaller than 100, the vortex created in all different cavities was positioned at the width centre of the cavity. The effect of upstream boundary layer thickness and the effect of the cavity dimensions on three dimensional rectangular cavities were studied in [40], it was found that the flow became increasingly unstable as the upstream boundary layer thickness decreased. Rectangular three dimension flow inside a cavity was also studied in [41], the paper focused on studding the Vorticity created inside the groove, they concluded that the corner Vorticity tended to increase flow transport. This paper also presented a graph of particle tracer explaining the vortex decay along the groove. In [42] the vortex created inside a rectangular cavity was studied when the lid was submitted to a sinusoidal displacement at different frequencies. It is also interesting to point out that all the previous studies presented regarded the flow as incompressible. Despite the amount of work developed on slippers, no evidence has been found of any research focused on finding the leakage, pressure distribution, force and torque created by a slipper with a non vented groove and considering spin and tangential velocity. Then such requirements can only be analyzed if three dimensional Navier Stokes equations in cylindrical coordinates are applied in the gap slipper swash/plate. In the present section this problem will be analyzed.

5.3.2. Flat Slipper with Grooves, Static Equations

As a first approach, in this sub chapter, equations able to determine the pressure distribution, force and leakage in the slipper-swash plate clearance will be presented, a generalization of the equations for any number of grooves being cut on the slipper surface will also be provided. All grooves are non-vented. The beauty of the new set of equations to be presented is that despite their simplicity, they bring a deep understanding of the slipper behaviour. Regarding the analytical study, the following assumptions are appropriate:

- Flow will be considered laminar and incompressible in all cases
- Static conditions for the slipper and the plate are considered
- Flow will be radially dominant
- The slipper is considered as a rigid body, no mechanical deformation is considered

Reynolds equation of lubrication applied to the slipper-swash plate clearance when the slipper moves tangentially with a velocity “U”, spins with an angular velocity “ω” and has a tilt which depends on the slipper angular position“θ” and the slipper radius “r” is given in cylindrical coordinates according to [46], chapter three, as equation (5.3.1).

$$\frac{1}{r^2}\frac{\partial}{\partial\theta}\left(h^3\frac{\partial p}{\partial\theta}\right)+\frac{1}{r}\frac{\partial}{\partial r}\left(r\,h^3\frac{\partial p}{\partial r}\right)=6\mu\left(U\cos\theta\frac{\partial h}{\partial r}-\frac{U\sin\theta}{r}\frac{\partial h}{\partial\theta}+\omega\frac{\partial h}{\partial\theta}\right) \tag{5.3.1}$$

In this equation, the slipper angular position is represented by “θ” which has a value between 0 and 360 degrees, the slipper tilt is considered by the terms $\frac{\partial h}{\partial r}$ and $\frac{\partial h}{\partial \theta}$.

When considering constant viscosity, slipper without tilt and no relative movement between slipper and plate, the equation becomes:

$$\frac{\partial}{\partial r}\left(\frac{r\ h^3}{\mu}\frac{\partial p}{\partial r}\right)=0 \tag{5.3.2}$$

Its integration yields:

$$P=\frac{C_1\,\mu}{h^3}\ln r+C_2 \tag{5.3.3}$$

C_1 and C_2 are constants which have to be found from the boundary conditions.

The equations representing velocity profile and flow rate between two cylindrical flat plates separated by a very small gap and for a pressure differential between the inner and outer radius are given by:

$$u=-\frac{1}{\mu}\frac{dp}{dr}\frac{y}{2}(h-y) \tag{5.3.4}$$

$$Q=\int_0^h u\,2\pi\ r\,dy=-\frac{\pi\,r}{\mu}\frac{dp}{dr}\frac{h^3}{6} \tag{5.3.5}$$

Substituting the first derivative of equation (3.2) into equation (3.5) yields:

$$Q=-\pi\frac{C_1}{6} \tag{5.3.6}$$

Equations (5.3.3) and (5.3.6) give the pressure distribution and radial flow between the gap of two cylindrical flat plates. To find the constants C_1 and C_2, knowledge of two boundary conditions is required:

$r = r_i;\ p = p_i.$ (5.3.7)

$r = r_j;\ p = p_j.$

Equations (5.3.3) and (5.3.6) can be applied to any number of consecutive cylindrical flat plates, understanding that the flow will be laminar at all points and having in mind that for every plate two new constants will appear. For the case under study, a slipper with a central pocked, two lands and a groove separating them, Figure 5.3.2a, can be established.

Slipper central pocket

$$p_1=C_1\frac{\mu}{h_0^3}\ln r+C_2 \text{ range of applicability } r_0<r<r_1 \tag{5.3.8}$$

$$Q_1 = -\frac{\pi C_1}{6} \tag{5.3.9}$$

First Land.

$$p_2 = C_3 \frac{\mu}{h_1^3} \ln r + C_4 \text{ range of applicability } r_1 < r < r_2 \tag{5.3.10}$$

$$Q_2 = -\frac{\pi C_3}{6} \tag{5.3.11}$$

Groove

$$p_3 = C_5 \frac{\mu}{h_2^3} \ln r + C_6 \text{ range of applicability } r_2 < r < r_3 \tag{5.3.12}$$

$$Q_3 = -\frac{\pi C_5}{6} \tag{5.3.13}$$

Second Land

$$p_4 = C_7 \frac{\mu}{h_3^3} \ln r + C_8 \text{ range of applicability } r_3 < r < r_4 \tag{5.3.14}$$

$$Q_4 = -\frac{\pi C_7}{6} \tag{5.3.15}$$

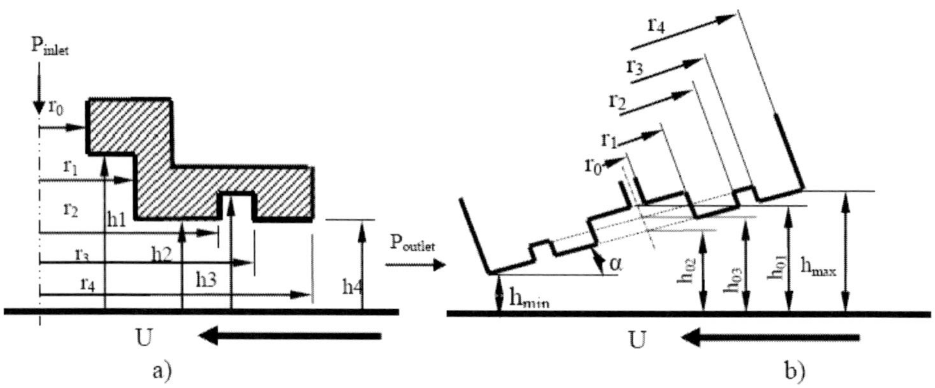

Figure 5.3.2. Diagram of the flat/tilt slipper under study with two main lands. a) Flat, b) Tilt.

The boundary conditions are:

$r = r_0$ $p_1 = p_{inlet}$ (5.3.16)
$r = r_1$ $p_1 = p_2$ $Q_1 = Q_2$
$r = r_2$ $p_2 = p_3$ $Q_2 = Q_3$
$r = r_3$ $p_3 = p_4$ $Q_3 = Q_4$
$r = r_4$ $p_4 = p_{outlet}$.

It needs to be considered that Reynolds equation of lubrication must be used under laminar conditions.

On the slipper first and second lands, the distance slipper plate, for a flat slipper, is constant and very narrow, usually around 5 to 15 microns, the fluid velocity is rather high, the Reynolds number is considered to be laminar.

When the fluid enters the slipper, it faces the slipper central pocked, which depth for the case studied is 1.4 mm, the flow when the slipper is held perfectly parallel to the plate (flat slipper) has to be considered radial, and the velocity will be very small, the Reynolds number will be much smaller than the one found in the slipper first and second lands, as a conclusion, the assumption of laminar flow is perfectly valid in the slipper pocked.

Reynolds equation of lubrication it is absolutely applicable in the slipper pocked and under the conditions established. The same phenomenon happens in the slipper groove, which depth is 0.8 mm.

The assumption of flow in radial direction, it is perfectly true for a slipper held perfectly parallel to the plate, and under static conditions, under these conditions the flow is perfectly symmetric.

The assumption of velocity parabolic profile, typical of laminar flow, it is perfectly correct in the slipper central pocked, groove, first and second lands.

Once the constants are found and substituted in equations (5.3.8)-(5.3.15), then the equations describing the pressure distribution across the central pocked each slipper land and the slipper groove can be determined, also the leakage flow between the slipper and plate will be characterized.

For the present case of a slipper with a central pocked, first land, groove and second land, total number of flat plates (total lands), n = 4, the equations giving pressure distribution leakage and force are:

The pressure distribution at each slipper land for the present case n = 4 is given by the following equations:

$$p_1 = p_{inlet} - \frac{(p_{inlet} - p_{outlet})}{\frac{1}{h_1^3}\ln\left(\frac{r_1}{r_0}\right) + \frac{1}{h_2^3}\ln\left(\frac{r_2}{r_1}\right) + \frac{1}{h_3^3}\ln\left(\frac{r_3}{r_2}\right) + \frac{1}{h_4^3}\ln\left(\frac{r_4}{r_3}\right)} \frac{1}{h_1^3}\ln\left(\frac{r}{r_0}\right) \tag{5.3.17}$$

Range of applicability $r_0 < r < r_1$

$$p_2 = p_{inlet} - \frac{(p_{inlet} - p_{outlet})}{\frac{1}{h_1^3}\ln\left(\frac{r_1}{r_0}\right) + \frac{1}{h_2^3}\ln\left(\frac{r_2}{r_1}\right) + \frac{1}{h_3^3}\ln\left(\frac{r_3}{r_2}\right) + \frac{1}{h_4^3}\ln\left(\frac{r_4}{r_3}\right)} \left[\frac{1}{h_1^3}\ln\left(\frac{r_1}{r_0}\right) + \frac{1}{h_2^3}\ln\left(\frac{r}{r_1}\right)\right] \tag{5.3.18}$$

Range of applicability $r_1 < r < r_2$

$$p_3 = p_{inlet} - \frac{(p_{inlet} - p_{outlet})}{\frac{1}{h_1^3}\ln\left(\frac{r_1}{r_0}\right) + \frac{1}{h_2^3}\ln\left(\frac{r_2}{r_1}\right) + \frac{1}{h_3^3}\ln\left(\frac{r_3}{r_2}\right) + \frac{1}{h_4^3}\ln\left(\frac{r_4}{r_3}\right)} \left[\frac{1}{h_1^3}\ln\left(\frac{r_1}{r_0}\right) + \frac{1}{h_2^3}\ln\left(\frac{r_2}{r_1}\right) + \frac{1}{h_3^3}\ln\left(\frac{r}{r_2}\right)\right] \tag{5.3.19}$$

Range of applicability $r_2 < r < r_3$

$$p_4 = p_{inlet} - \frac{(p_{inlet} - p_{outlet})}{\frac{1}{h_1^3}\ln\left(\frac{r_1}{r_0}\right) + \frac{1}{h_2^3}\ln\left(\frac{r_2}{r_1}\right) + \frac{1}{h_3^3}\ln\left(\frac{r_3}{r_2}\right) + \frac{1}{h_4^3}\ln\left(\frac{r_4}{r_3}\right)}\left[\frac{1}{h_1^3}\ln\left(\frac{r_1}{r_0}\right) + \frac{1}{h_2^3}\ln\left(\frac{r_2}{r_1}\right) + \frac{1}{h_3^3}\ln\left(\frac{r_3}{r_2}\right) + \frac{1}{h_4^3}\ln\left(\frac{r}{r_3}\right)\right] \quad (5.3.20)$$

Range of applicability $r_3 < r < r_4$

The leakage flow equation for the actual slipper n = 4, will take the form:

$$Q = \frac{\pi}{6\mu}\frac{(p_{inlet} - p_{outlet})}{\frac{1}{h_1^3}\ln\left(\frac{r_1}{r_0}\right) + \frac{1}{h_2^3}\ln\left(\frac{r_2}{r_1}\right) + \frac{1}{h_3^3}\ln\left(\frac{r_3}{r_2}\right) + \frac{1}{h_4^3}\ln\left(\frac{r_4}{r_3}\right)} \quad (5.3.21)$$

The lift force can be found by integrating the radial pressure. Since the slipper under study has an inner pocket and two lands separated by a groove, the integral has to be split into four parts as follows:

$$F_{lift} = \int_{r_0}^{r_1} P_1(r)2\pi r\,dr + \int_{r_1}^{r_2} P_2(r)\,2\pi r\,dr + \int_{r_2}^{r_3} P_3(r)2\pi r\,dr + \int_{r_3}^{r_4} P_4(r)2\pi r\,dr \quad (5.3.22)$$

where for n = 4, $P_1(r)$, $P_2(r)$, $P_3(r)$ and $P_4(r)$ are given by the equations (5.3.17) to (5.3.20).

As a result, the equation giving the lift force on the slipper face as a function of the slipper dimensions and the inlet pressure, for the actual slipper under study, number of lands n = 4 is:

$$F_{lift} = P_{inlet}\,\pi(r_4^2 - r_0^2) - C\pi\, r_4^2\left(\frac{1}{h_1^3}\ln\left(\frac{r_1}{r_0}\right) + \frac{1}{h_2^3}\ln\left(\frac{r_2}{r_1}\right) + \frac{1}{h_3^3}\ln\left(\frac{r_3}{r_2}\right) + \frac{1}{h_4^3}\ln\left(\frac{r_4}{r_3}\right)\right)$$
$$+ C\pi\left(\frac{1}{h_1^3}\frac{r_1^2 - r_0^2}{2} + \frac{1}{h_2^3}\frac{r_2^2 - r_1^2}{2} + \frac{1}{h_3^3}\frac{r_3^2 - r_2^2}{2} + \frac{1}{h_4^3}\frac{r_4^2 - r_3^2}{2}\right) \quad (5.3.23)$$

where the constant C takes the form.

$$C = \frac{p_{inlet} - p_{outlet}}{\frac{1}{h_1^3}\ln\left(\frac{r_1}{r_0}\right) + \frac{1}{h_2^3}\ln\left(\frac{r_2}{r_1}\right) + \frac{1}{h_3^3}\ln\left(\frac{r_3}{r_2}\right) + \frac{1}{h_4^3}\ln\left(\frac{r_4}{r_3}\right)} \quad (5.3.24)$$

These equations can be generalized for a slipper with any number of grooves. The generic equations giving the leakage flow and pressure distribution for a generic number of (total lands) "n" are presented next. The equation which gives the leakage flow between a slipper and plate for a slipper with a generic number of (total lands) "n", which include the slipper pocked, and the groove or grooves, takes the form. Notice that when talking about (total lands) the first land is in reality the slipper central pocked.

$$Q=\frac{\pi}{6\mu}\frac{(p_{inlet}-p_{outlet})}{\sum_{i=1}^{i=n}\frac{1}{h_i^3}\ln\frac{r_i}{r_{i-1}}} \tag{5.3.25}$$

The generic pressure distribution for a slipper with any number of (total lands) "n" will be:

For the slipper pocked: $r_0 < r < r_1$.

$$p_1=p_{inlet}-\frac{(p_{inlet}-p_{outlet})}{\sum_{i=1}^{i=n}\frac{1}{h_i^3}\ln\left(\frac{r_i}{r_{i-1}}\right)}\left[\frac{1}{h_1^3}\ln\left(\frac{r}{r_0}\right)\right] \tag{5.3.26}$$

For the rest of the lands, including the groove: $2 \le j \le n$;

$$p_j=p_{inlet}-\frac{(p_{inlet}-p_{outlet})}{\sum_{i=1}^{i=n}\frac{1}{h_i^3}\ln\left(\frac{r_i}{r_{i-1}}\right)}\left[\frac{1}{h_j^3}\ln\left(\frac{r}{r_{j-1}}\right)+\sum_{k=1}^{k=j-1}\frac{1}{h_k^3}\ln\left(\frac{r_k}{r_{k-1}}\right)\right] \tag{5.3.27}$$

The generic lift force equation for a slipper with any number of (total lands) is now developed as follows:

$$F_{lift}=P_{inlet}\,\pi(r_n^2-r_0^2)\;-C\pi\;r_n^2\sum_{i=1}^{i=n}\frac{1}{h_i^3}\ln\left(\frac{r_i}{r_{i-1}}\right)+C\pi\sum_{i=1}^{i=n}\frac{1}{h_i^3}\frac{r_i^2-r_{i-1}^2}{2} \tag{5.3.28}$$

where the generic constant C will be:

$$C=\frac{P_{inlet}\;-\;P_{outlet}}{\sum_{i=1}^{i=n}\frac{1}{h_i^3}\;\ln\left(\frac{r_i}{r_{i-1}}\right)} \tag{5.3.29}$$

It is now analytically possible to determine the condition for maximum lift using the previously derived set of equations.

5.3.3. Tilted Slipper with Grooves, Static Analytical Equations

A crucial aspect of the method proposed is based on linking the first direct integration of the Reynolds equation with the flow leakage differential equation and until now no attempt has been made to explain the non vented two land slipper behaviour both mathematically and experimentally. The method proposed allows the behaviour of the slipper grooves and second land to be defined without the necessity of having the second land vented. The basis of the theory was outlined in Bergada and Watton [28, 29] and Watton [47] for flat slipper and in [30, 48] for tilted slipper. Consider figure 5.3.3:

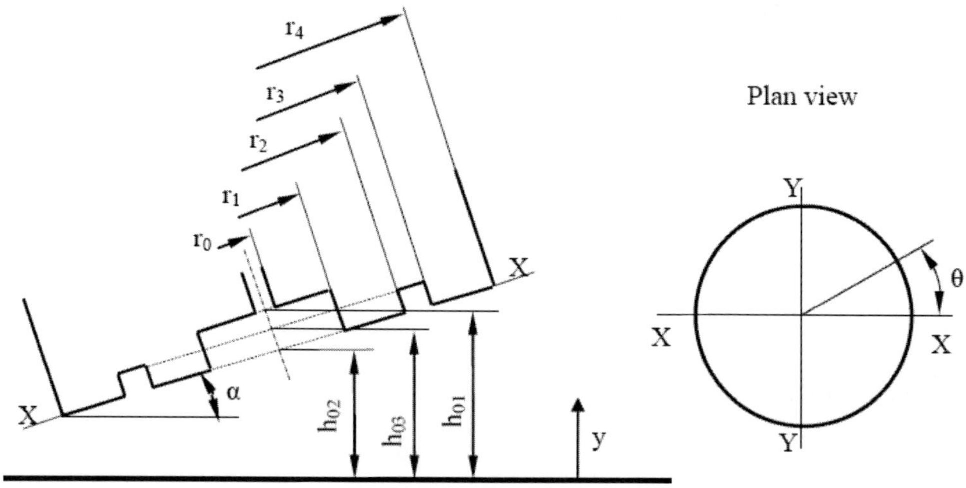

Figure 5.3.3. Slipper main parameters.

The following assumptions are then made:

- Flow will be considered laminar.
- The slipper plate clearance is not uniform; the slipper is tilted.
- Steady conditions are considered.
- Slipper spin is taken into account.
- Flow will be considered as radial.
- Slipper pocket, groove and slipper lands are flat.
- The only relative movement between slipper/swash plate is slipper spin.

Reynolds equation applicable to this study is given as:

$$\frac{\partial}{\partial r}\left(r h^3 \frac{\partial p}{\partial r} \right) = 6 \mu \omega r \frac{\partial h}{\partial \theta} \tag{5.3.30}$$

The film thickness in the clearance is given by:

$$h = h_0 + \alpha\, r_m \cos\theta \tag{5.3.31}$$

The average radius between land ends is used, and the film thickness is:

$$\frac{\partial h}{\partial \theta} = -\alpha\, r_m \sin\theta \tag{5.3.32}$$

The first integration of the differential equation (3.30) will then give:

$$\frac{\partial p}{\partial r} = \frac{-3\mu\omega\alpha\, r_m \sin\theta\ r}{(h_0 + \alpha r_m \cos\theta)^3} + \frac{k_1}{r\,(h_0 + \alpha r_m \cos\theta)^3} \tag{5.3.33}$$

The second integration gives:

$$p = \frac{-3\mu\omega\alpha\, r_m \sin\theta\ r^2}{2\,(h_0 + \alpha r_m \cos\theta)^3} + \frac{k_1}{(h_0 + \alpha r_m \cos\theta)^3}\ln(r) + k_2 \tag{5.3.34}$$

The slipper leakage through a generic radius will be:

$$Q_{leakage} = \int_0^{2\pi}\int_0^{h} u\, r\, dy\, d\theta \tag{5.3.35}$$

Assuming Poiseulle flow, the velocity distribution is given by:

$$u = \frac{1}{\mu}\frac{dp}{dr}\frac{y}{2}(y-h) \tag{5.3.36}$$

Then:

$$Q_{leakage} = \int_0^{2\pi}\int_0^{h} \frac{1}{\mu}\frac{dp}{dr}\frac{y}{2}(y-h)\, r\, dy\, d\theta \tag{5.3.37}$$

Substituting the pressure distribution versus radius, equation (5.3.33), into equation (5.3.37) and after some integration and rearrangement gives:

$$Q_{leakage} = \int_0^{2\pi} -\frac{1}{12\mu}\left[-3\mu\omega\alpha\, r_m \sin\theta\, r^2 + k_1\right] d\theta \tag{5.3.38}$$

It must be remembered at this point that a second integration cannot be performed since the unknown constant k_1 depends on the angular position θ. Nevertheless for a tilted slipper with several lands as shown in figure 5.3.3, and assuming that the flow and pressure distribution in the slipper pocket and groove behave in the same way as in a conventional land, then equations (5.3.34) and (5.3.38) can be applied to each slipper land obtaining:

Slipper pocket: $r_0 < r < r_1$

$$p_1 = \frac{-3\mu\omega\alpha\, r_{m1} \sin\theta\ r^2}{2\,(h_{01} + \alpha r_{m1} \cos\theta)^3} + \frac{k_1}{(h_{01} + \alpha r_{m1} \cos\theta)^3}\ln(r) + k_2 \tag{5.3.39}$$

$$Q_{leakage\,1} = \int_0^{2\pi} -\frac{1}{12\mu}\left[-3\mu\omega\alpha\, r_{m1} \sin\theta\, r^2 + k_1\right] d\theta \tag{5.3.40}$$

$$r_{m1} = \frac{r_1 + r_0}{2} \tag{5.3.41}$$

First land: $r_1<r<r_2$

$$p_2 = \frac{-3\mu\omega\alpha\, r_{m2} \sin\theta\; r^2}{2\left(h_{02}+\alpha\, r_{m2}\cos\theta\right)^3} + \frac{k_3}{\left(h_{02}+\alpha\, r_{m2}\cos\theta\right)^3}\ln(r) + k_4 \quad (5.3.42)$$

$$Q_{leakage\,2} = \int_0^{2\pi} -\frac{1}{12\mu}\left[-3\mu\omega\alpha\, r_{m2}\sin\theta\, r^2 + k_3\right]d\theta \quad (5.3.43)$$

$$r_{m2} = \frac{r_2 + r_1}{2} \quad (5.3.44)$$

Slipper groove: $r_2<r<r_3$

$$p_3 = \frac{-3\mu\omega\alpha\, r_{m3} \sin\theta\; r^2}{2\left(h_{03}+\alpha\, r_{m3}\cos\theta\right)^3} + \frac{k_5}{\left(h_{03}+\alpha\, r_{m3}\cos\theta\right)^3}\ln(r) + k_6 \quad (5.3.45)$$

$$Q_{leakage\,3} = \int_0^{2\pi} -\frac{1}{12\mu}\left[-3\mu\omega\alpha\, r_{m3}\sin\theta\, r^2 + k_5\right]d\theta \quad (5.3.46)$$

$$r_{m3} = \frac{r_3 + r_2}{2} \quad (5.3.47)$$

Second land: $r_3<r<r_4$

$$p_4 = \frac{-3\mu\omega\alpha\, r_{m4} \sin\theta\; r^2}{2\left(h_{04}+\alpha\, r_{m4}\cos\theta\right)^3} + \frac{k_7}{\left(h_{04}+\alpha\, r_{m4}\cos\theta\right)^3}\ln(r) + k_8 \quad (5.3.48)$$

$$Q_{leakage\,4} = \int_0^{2\pi} -\frac{1}{12\mu}\left[-3\mu\omega\alpha\, r_{m4}\sin\theta\, r^2 + k_7\right]d\theta \quad (5.3.49)$$

$$r_{m4} = \frac{r_4 + r_3}{2} \quad (5.3.50)$$

The boundary conditions necessary to determine the constants will be:

$r = r_0$ $p_1 = p_{inlet}$ (5.3.51)
$r = r_1$ $p_1 = p_2$ $Q_{leakage\,1} = Q_{leakage\,2}$
$r = r_2$ $p_2 = p_3$ $Q_{leakage\,2} = Q_{leakage\,3}$
$r = r_3$ $p_3 = p_4$ $Q_{leakage\,3} = Q_{leakage\,4}$
$r = r_4$ $p_4 = p_{outlet}$

After appropriate mathematical development, the value of the constants is found.The expressions for the different constants are given as a function of the first constant k_1

$$k_1 = \frac{p_{tank} - p_{inlet} - \dfrac{3\mu\omega\alpha\sin\theta}{2}\left[\dfrac{r_{m1}\left(r_0^2 - r_1^2\right)}{\left(h_{01} + \alpha\, r_{m1}\cos\theta\right)^3} + \dfrac{r_{m2}\left(r_1^2 - r_2^2\right)}{\left(h_{02} + \alpha\, r_{m2}\cos\theta\right)^3} + \dfrac{r_{m3}\left(r_2^2 - r_3^2\right)}{\left(h_{03} + \alpha\, r_{m3}\cos\theta\right)^3} + \dfrac{r_{m4}\left(r_3^2 - r_4^2\right)}{\left(h_{04} + \alpha\, r_{m4}\cos\theta\right)^3}\right]}{\dfrac{\ln\left(\dfrac{r_1}{r_0}\right)}{\left(h_{01} + \alpha\, r_{m1}\cos\theta\right)^3} + \dfrac{\ln\left(\dfrac{r_2}{r_1}\right)}{\left(h_{02} + \alpha\, r_{m2}\cos\theta\right)^3} + \dfrac{\ln\left(\dfrac{r_3}{r_2}\right)}{\left(h_{03} + \alpha\, r_{m3}\cos\theta\right)^3} + \dfrac{\ln\left(\dfrac{r_4}{r_3}\right)}{\left(h_{04} + \alpha\, r_{m4}\cos\theta\right)^3}}$$

$$- \frac{3\mu\omega\alpha\sin\theta\left[\dfrac{r_1^2\left(r_{m2} - r_{m1}\right)\ln\left(\dfrac{r_2}{r_1}\right)}{\left(h_{02} + \alpha\, r_{m2}\cos\theta\right)^3} + \dfrac{\left[r_1^2\left(r_{m2} - r_{m1}\right) + r_2^2\left(r_{m3} - r_{m2}\right)\right]\ln\left(\dfrac{r_3}{r_2}\right)}{\left(h_{03} + \alpha\, r_{m3}\cos\theta\right)^3} + \dfrac{\left[r_1^2\left(r_{m2} - r_{m1}\right) + r_2^2\left(r_{m3} - r_{m2}\right) + r_3^2\left(r_{m4} - r_{m3}\right)\right]\ln\left(\dfrac{r_4}{r_3}\right)}{\left(h_{04} + \alpha\, r_{m4}\cos\theta\right)^3}\right]}{\dfrac{\ln\left(\dfrac{r_1}{r_0}\right)}{\left(h_{01} + \alpha\, r_{m1}\cos\theta\right)^3} + \dfrac{\ln\left(\dfrac{r_2}{r_1}\right)}{\left(h_{02} + \alpha\, r_{m2}\cos\theta\right)^3} + \dfrac{\ln\left(\dfrac{r_3}{r_2}\right)}{\left(h_{03} + \alpha\, r_{m3}\cos\theta\right)^3} + \dfrac{\ln\left(\dfrac{r_4}{r_3}\right)}{\left(h_{04} + \alpha\, r_{m4}\cos\theta\right)^3}}$$

(5.3.52)

$$k_3 = 3\,\mu\,\omega\,\alpha\,\sin\theta\; r_1^2\left[r_{m2} - r_{m1}\right] + k_1 \tag{5.3.53}$$

$$k_5 = 3\,\mu\,\omega\,\alpha\,\sin\theta\left[r_1^2\left[r_{m2} - r_{m1}\right] + r_2^2\left[r_{m3} - r_{m2}\right]\right] + k_1 \tag{5.3.54}$$

$$k_7 = 3\,\mu\,\omega\,\alpha\,\sin\theta\left[r_1^2\left[r_{m2} - r_{m1}\right] + r_2^2\left[r_{m3} - r_{m2}\right] + r_3^2\left[r_{m4} - r_{m3}\right]\right] + k_1 \tag{5.3.55}$$

$$k_2 = p_{inlet} + \frac{3\mu\omega\alpha\, r_{m1}\sin\theta}{\left(h_{01} + \alpha r_{m1}\cos\theta\right)^3}\frac{r_0^2}{2} - \frac{k_1}{\left(h_{01} + \alpha r_{m1}\cos\theta\right)^3}\ln r_0 \tag{5.3.56}$$

$$k_4 = p_{inlet} + k_1\left[\frac{\ln\left(\dfrac{r_1}{r_0}\right)}{\left(h_{01} + \alpha\, r_{m1}\cos\theta\right)^3} - \frac{\ln r_1}{\left(h_{02} + \alpha\, r_{m2}\cos\theta\right)^3}\right] + \frac{3\mu\omega\alpha\,\sin\theta}{2}\left[\frac{r_{m1}\left[r_0^2 - r_1^2\right]}{\left(h_{01} + \alpha\, r_{m1}\cos\theta\right)^3} + \frac{r_{m2}\, r_1^2}{\left(h_{02} + \alpha\, r_{m2}\cos\theta\right)^3}\right]$$

$$- \frac{3\mu\omega\alpha\,\sin\theta\, r_1^2\left[r_{m2} - r_{m1}\right]}{\left(h_{02} + \alpha r_{m2}\cos\theta\right)^3}\ln r_1$$

(5.3.57)

$$k_6 = p_{inlet} + k_1\left[\frac{\ln\left(\dfrac{r_1}{r_0}\right)}{\left(h_{01} + \alpha\, r_{m1}\cos\theta\right)^3} + \frac{\ln\left(\dfrac{r_2}{r_1}\right)}{\left(h_{02} + \alpha\, r_{m2}\cos\theta\right)^3} - \frac{\ln r_2}{\left(h_{03} + \alpha\, r_{m3}\cos\theta\right)^3}\right]$$

$$+ \frac{3\mu\omega\alpha\,\sin\theta}{2}\left[\frac{r_{m1}\left[r_0^2 - r_1^2\right]}{\left(h_{01} + \alpha\, r_{m1}\cos\theta\right)^3} + \frac{r_{m2}\left[r_1^2 - r_2^2\right]}{\left(h_{02} + \alpha\, r_{m2}\cos\theta\right)^3} + \frac{r_{m3}\, r_2^2}{\left(h_{03} + \alpha\, r_{m3}\cos\theta\right)^3}\right]$$

$$+ 3\mu\omega\alpha\,\sin\theta\left[\frac{r_1^2\left[r_{m2} - r_{m1}\right]\ln\left(\dfrac{r_2}{r_1}\right)}{\left(h_{02} + \alpha\, r_{m2}\cos\theta\right)^3} - \frac{r_1^2\left[r_{m2} - r_{m1}\right] + r_2^2\left[r_{m3} - r_{m2}\right]}{\left(h_{03} + \alpha r_{m3}\cos\theta\right)^3}\ln r_2\right]$$

(5.3.58)

$$k_8 = p_{inlet} + k_1\left[\frac{\ln\left(\frac{r_1}{r_0}\right)}{\left(h_{01}+\alpha r_{m1}\cos\theta\right)^3}+\frac{\ln\left(\frac{r_2}{r_1}\right)}{\left(h_{02}+\alpha r_{m2}\cos\theta\right)^3}+\frac{\ln\left(\frac{r_3}{r_2}\right)}{\left(h_{03}+\alpha r_{m3}\cos\theta\right)^3}-\frac{\ln r_3}{\left(h_{04}+\alpha r_{m4}\cos\theta\right)^3}\right]$$

$$+\frac{3\mu\omega\alpha\sin\theta}{2}\left[\frac{r_{m1}\left[r_0^2-r_1^2\right]}{\left(h_{01}+\alpha r_{m1}\cos\theta\right)^3}+\frac{r_{m2}\left[r_1^2-r_2^2\right]}{\left(h_{02}+\alpha r_{m2}\cos\theta\right)^3}+\frac{r_{m3}\left[r_2^2-r_3^2\right]}{\left(h_{03}+\alpha r_{m3}\cos\theta\right)^3}+\frac{r_{m4}\,r_3^2}{\left(h_{04}+\alpha r_{m4}\cos\theta\right)^3}\right]$$

$$+3\mu\omega\alpha\sin\theta\left[\frac{r_1^2\left[r_{m2}-r_{m1}\right]\ln\left(\frac{r_2}{r_1}\right)}{\left(h_{02}+\alpha r_{m2}\cos\theta\right)^3}+\frac{r_1^2\left[r_{m2}-r_{m1}\right]+r_2^2\left[r_{m3}-r_{m2}\right]}{\left(h_{03}+\alpha r_{m3}\cos\theta\right)^3}\ln\left(\frac{r_3}{r_2}\right)-\frac{r_1^2\left[r_{m2}-r_{m1}\right]+r_2^2\left[r_{m3}-r_{m2}\right]+r_3^2\left[r_{m4}-r_{m3}\right]}{\left(h_{04}+\alpha r_{m4}\cos\theta\right)^3}\ln r_3\right]$$

(5.3.59)

when substituting the different constants into the equations for the pressure distribution at different lands, equations (5.3.39), (5.3.42), (5.3.45) and (5.3.48), explicit equations for the pressure distribution at each land are obtained and take the following form.

$$p_1 = p_{inlet} + \frac{k_1\ \ln\left(\frac{r}{r_0}\right)}{\left(h_{01}+\alpha r_{m1}\cos\theta\right)^3}+\frac{3\mu\omega\alpha r_{m1}\sin\theta}{\left(h_{01}+\alpha r_{m1}\cos\theta\right)^3}\left[\frac{r_0^2}{2}-\frac{r^2}{2}\right] \qquad (5.3.60)$$

$$p_2 = p_{inlet} + k_1\left[\frac{\ln\left(\frac{r_1}{r_0}\right)}{\left(h_{01}+\alpha r_{m1}\cos\theta\right)^3}+\frac{\ln\left(\frac{r}{r_1}\right)}{\left(h_{02}+\alpha r_{m2}\cos\theta\right)^3}\right]+\frac{3\mu\omega\alpha\sin\theta}{2}\left[\frac{r_{m1}\left[r_0^2-r_1^2\right]}{\left(h_{01}+\alpha r_{m1}\cos\theta\right)^3}+\frac{r_{m2}\left[r_1^2-r^2\right]}{\left(h_{02}+\alpha r_{m2}\cos\theta\right)^3}\right]$$

$$+\frac{3\mu\omega\alpha\sin\theta\, r_1^2\left[r_{m2}-r_{m1}\right]}{\left(h_{02}+\alpha r_{m2}\cos\theta\right)^3}\ln\left(\frac{r}{r_1}\right)$$

(5.3.61)

$$p_3 = p_{inlet} + k_1\left[\frac{\ln\left(\frac{r_1}{r_0}\right)}{\left(h_{01}+\alpha r_{m1}\cos\theta\right)^3}+\frac{\ln\left(\frac{r_2}{r_1}\right)}{\left(h_{02}+\alpha r_{m2}\cos\theta\right)^3}+\frac{\ln\left(\frac{r}{r_2}\right)}{\left(h_{03}+\alpha r_{m3}\cos\theta\right)^3}\right]$$

$$+\frac{3\mu\omega\alpha\sin\theta}{2}\left[\frac{r_{m1}\left[r_0^2-r_1^2\right]}{\left(h_{01}+\alpha r_{m1}\cos\theta\right)^3}+\frac{r_{m2}\left[r_1^2-r_2^2\right]}{\left(h_{02}+\alpha r_{m2}\cos\theta\right)^3}+\frac{r_{m3}\left[r_2^2-r^2\right]}{\left(h_{03}+\alpha r_{m3}\cos\theta\right)^3}\right] \qquad (5.3.62)$$

$$+3\mu\omega\alpha\sin\theta\left[\frac{r_1^2\left[r_{m2}-r_{m1}\right]\ln\left(\frac{r_2}{r_1}\right)}{\left(h_{02}+\alpha r_{m2}\cos\theta\right)^3}+\frac{r_1^2\left[r_{m2}-r_{m1}\right]+r_2^2\left[r_{m3}-r_{m2}\right]}{\left(h_{03}+\alpha r_{m3}\cos\theta\right)^3}\ln\left(\frac{r}{r_2}\right)\right]$$

$$p_4 = p_{inlet} + k_1\left[\frac{\ln\left(\frac{r_1}{r_0}\right)}{\left(h_{01}+\alpha r_{m1}\cos\theta\right)^3}+\frac{\ln\left(\frac{r_2}{r_1}\right)}{\left(h_{02}+\alpha r_{m2}\cos\theta\right)^3}+\frac{\ln\left(\frac{r_3}{r_2}\right)}{\left(h_{03}+\alpha r_{m3}\cos\theta\right)^3}+\frac{\ln\left(\frac{r}{r_3}\right)}{\left(h_{04}+\alpha r_{m4}\cos\theta\right)^3}\right]$$

$$+\frac{3\mu\omega\alpha\sin\theta}{2}\left[\frac{r_{m1}\left[r_0^2-r_1^2\right]}{\left(h_{01}+\alpha r_{m1}\cos\theta\right)^3}+\frac{r_{m2}\left[r_1^2-r_2^2\right]}{\left(h_{02}+\alpha r_{m2}\cos\theta\right)^3}+\frac{r_{m3}\left[r_2^2-r_3^2\right]}{\left(h_{03}+\alpha r_{m3}\cos\theta\right)^3}+\frac{r_{m4}\left[r_3^2-r^2\right]}{\left(h_{04}+\alpha r_{m4}\cos\theta\right)^3}\right] \qquad (5.3.63)$$

$$+3\mu\omega\alpha\sin\theta\left[\frac{r_1^2\left[r_{m2}-r_{m1}\right]\ln\left(\frac{r_2}{r_1}\right)}{\left(h_{02}+\alpha r_{m2}\cos\theta\right)^3}+\frac{r_1^2\left[r_{m2}-r_{m1}\right]+r_2^2\left[r_{m3}-r_{m2}\right]}{\left(h_{03}+\alpha r_{m3}\cos\theta\right)^3}\ln\left(\frac{r_3}{r_2}\right)+\frac{r_1^2\left[r_{m2}-r_{m1}\right]+r_2^2\left[r_{m3}-r_{m2}\right]+r_3^2\left[r_{m4}-r_{m3}\right]}{\left(h_{04}+\alpha r_{m4}\cos\theta\right)^3}\ln\left(\frac{r}{r_3}\right)\right]$$

A key, and new, feature of the analysis is that it allows a generalised set of equations to be developed for the constants and from these generalised equations the value of any constant for a slipper with any number of lands can be determined. Since all constants are given as a function of the parameter k_1 the generic equation for this parameter will be presented first. It must be remembered that $1 \le i \le n$ where n is the number of lands, including grooves and the slipper pocket. Notice that the minimum value of "n" for a single land slipper is 2, for the case under study n = 4. The generic equation for the constant k_1 will be:

$$k_1 = \frac{p_{\tan k} - p_{inlet} - \dfrac{3\mu\omega\alpha \sin\theta}{2}\left[\displaystyle\sum_{i=1}^{i=n} \frac{r_{mi}\left(r_{i-1}^2 - r_i^2\right)}{\left(h_{0i} + \alpha r_{mi}\cos\theta\right)^3}\right] - 3\mu\omega\alpha \sin\theta\left[\displaystyle\sum_{i=1}^{i=(n-1)} \frac{\ln\left(\dfrac{r_{i+1}}{r_i}\right)\left[\displaystyle\sum_{j=1}^{j=i} r_j^2\left(r_{m(j+1)} - r_{mj}\right)\right]}{\left(h_{0(i+1)} + \alpha r_{m(i+1)}\cos\theta\right)^3}\right]}{\displaystyle\sum_{i=1}^{i=n} \frac{\ln\left(\dfrac{r_i}{r_{(i-1)}}\right)}{\left(h_{0i} + \alpha r_{mi}\cos\theta\right)^3}} \qquad (5.3.64)$$

The generic equation for the odd constants k_3, k_5, k_7.....will be:

$$k_L = 3\mu\omega\alpha\, \mathrm{Sin}\theta\left(\sum_{j=1}^{j=\left(\frac{L-1}{2}\right)} \left[r_j^2\left[r_{m(j+1)} - r_{mj}\right]\right]\right) + k_1 \qquad (5.3.65)$$

The value of "L" will have to be odd and between $3 \le L \le (2n-1)$. The generic equation for the even constants k_2, k_4, k_6, k_8 ....will be:

$$k_M = P_{inlet} + \left\{\sum_{i=1}^{i=\left(\frac{M-2}{2}\right)}\left(\frac{\ln\left(\dfrac{r_i}{r_{(i-1)}}\right)}{\left(h_{0i} + \alpha r_{mi}\cos\theta\right)^3}\right) - \frac{\ln r_{\left(\frac{M-2}{2}\right)}}{\left(h_{0(M/2)} + \alpha r_{m(M/2)}\cos\theta\right)^3}\right\}k_1 +$$
$$+\frac{3\mu\omega\alpha\sin\theta}{2}\left\{\sum_{i=1}^{i=\left(\frac{M-2}{2}\right)}\left(\frac{r_{mi}\left(r_{i-1}^2 - r_i^2\right)}{\left(h_{0i} + \alpha r_{mi}\cos\theta\right)^3}\right) + \frac{r_{m(M/2)}\, r_{\left(\frac{M-2}{2}\right)}^2}{\left(h_{0(M/2)} + \alpha r_{m(M/2)}\cos\theta\right)^3}\right\} +$$
$$+3\mu\omega\alpha\sin\theta\left\{\sum_{i=1}^{i=\left(\frac{M-4}{2}\right)}\left(\frac{\left[\displaystyle\sum_{j=1}^{j=i}\left[r_j^2\left(r_{m(j+1)} - r_{mj}\right)\right]\right]\ln\left(\dfrac{r_{i+1}}{r_i}\right)}{\left(h_{0(i+1)} + \alpha r_{m(i+1)}\cos\theta\right)^3}\right) - \frac{\displaystyle\sum_{j=1}^{j=\left(\frac{M-2}{2}\right)}\left(r_j^2\left(r_{m(j+1)} - r_{mj}\right)\right)}{\left(h_{0(M/2)} + \alpha r_{m(M/2)}\cos\theta\right)^3}\ln r_{\left(\frac{M-2}{2}\right)}\right\}$$

(5.3.66)

The values of the variable "M" in equation (5.3.66) must be even, and in the range $2 \le M \le 2n$. When producing the explicit equations based on the generic equations given here it must be remembered that in a summation $\sum_{\text{lower lim it}}^{\text{upper lim it}}$ then if the upper limit has a value smaller than the lower limit it means that the value of the entire term is zero. A generic equation capable of generating the pressure distribution equations for a slipper with any number of lands, would take the form.

$$p_i = p_{inlet} + k_1 \left[\frac{\ln\left(\frac{r}{r_{i-1}}\right)}{\left(h_{0i} + \alpha r_{mi}\cos\theta\right)^3} + \sum_{j=2}^{j=i} \frac{\ln\left(\frac{r_{j-1}}{r_{j-2}}\right)}{\left(h_{0(j-1)} + \alpha r_{m(j-1)}\cos\theta\right)^3} \right] +$$

$$\frac{3\mu\omega\alpha\sin\theta}{2}\left[\frac{r_{mi}\left[r_{i-1}^2 - r^2\right]}{\left(h_{0i} + \alpha r_{mi}\cos\theta\right)^3} + \sum_{j=2}^{j=i} \frac{r_{m(j-1)}\left[r_{j-2}^2 - r_{j-1}^2\right]}{\left(h_{0(j-1)} + \alpha r_{m(j-1)}\cos\theta\right)^3} \right] + \tag{5.3.67}$$

$$3\mu\omega\alpha\sin\theta \left[\frac{\left[\sum_{j=2}^{j=i} r_{j-1}^2\left[r_{mj} - r_{m(j-1)}\right]\right]}{\left(h_{0i} + \alpha r_{mi}\cos\theta\right)^3} \ln\left(\frac{r}{r_{i-1}}\right) + \sum_{j=3}^{j=i} \left[\frac{\ln\left(\frac{r_{j-1}}{r_{j-2}}\right)\left[\sum_{k=3}^{k=i} r_{k-2}^2\left[r_{m(k-1)} - r_{m(k-2)}\right]\right]}{\left(h_{0(j-1)} + \alpha r_{m(j-1)}\cos\theta\right)^3} \right] \right]$$

When substituting the constants into the leakage equations, (5.3.40), (5.3.43), (5.3.46) and (5.3.49), the same equation for the leakage flow is found:

$$Q_{leakage} = -\int_0^{2\pi} \frac{k_1}{12\,\mu}\, d\theta \tag{5.3.68}$$

Due to the complexity of the integral, equation (3.68) must be integrated numerically to determine the leakage.

The force over the tilted slipper can be now much carefully studied, since the effect of slipper turning speed and slipper tilt are being taken into account. The force over the slipper can be given as:

$$F = \int_0^{2\pi}\int_{r_0}^{r_1} p_1\, r\, dr\, d\theta + \int_0^{2\pi}\int_{r_1}^{r_2} p_2\, r\, dr\, d\theta + \int_0^{2\pi}\int_{r_2}^{r_3} p_3\, r\, dr\, d\theta + \int_0^{2\pi}\int_{r_3}^{r_4} p_4\, r\, dr\, d\theta \tag{5.3.69}$$

when substituting the values of the pressure distribution and integrating partially, the total force over the slipper can be found, having the form:

$$F = p_{inlet}\,\pi(r_4^2 - r_0^2) +$$

$$\int_0^{2\pi} \frac{k_1\, r_4^2}{2} \left[\frac{\ln\left(\frac{r_1}{r_0}\right)}{(h_{01} + \alpha r_{m1}\cos\theta)^3} + \frac{\ln\left(\frac{r_2}{r_1}\right)}{(h_{02} + \alpha r_{m2}\cos\theta)^3} + \frac{\ln\left(\frac{r_3}{r_2}\right)}{(h_{03} + \alpha r_{m3}\cos\theta)^3} + \frac{\ln\left(\frac{r_4}{r_3}\right)}{(h_{04} + \alpha r_{m4}\cos\theta)^3} \right] d\theta + \tag{5.3.70}$$

$$\int_0^{2\pi} \frac{k_1}{4} \left[\frac{\left(r_0^2 - r_1^2\right)}{(h_{01} + \alpha r_{m1}\cos\theta)^3} + \frac{\left(r_1^2 - r_2^2\right)}{(h_{02} + \alpha r_{m2}\cos\theta)^3} + \frac{\left(r_2^2 - r_3^2\right)}{(h_{03} + \alpha r_{m3}\cos\theta)^3} + \frac{\left(r_3^2 - r_4^2\right)}{(h_{04} + \alpha r_{m4}\cos\theta)^3} \right] d\theta;$$

Due to its complexity, the integrals defined in equation (5.3.70) will be integrated numerically.

The generic force equation for the tilted slipper with any number of lands will be.

$$F = p_{inlet}\pi(r_n^2 - r_0^2) + \int_0^{2\pi} \frac{k_1 r_n^2}{2}\left[\sum_{i=1}^{i=n} \frac{\ln\left(\frac{r_i}{r_{i-1}}\right)}{\left(h_{0i} + \alpha r_{mi}\cos\theta\right)^3}\right] d\theta + \int_0^{2\pi} \frac{k_1}{4}\left[\sum_{i=1}^{i=n} \frac{(r_{i-1}^2 - r_i^2)}{\left(h_{0i} + \alpha r_{mi}\cos\theta\right)^3}\right] d\theta; \qquad (5.3.71)$$

Due to its complexity, the integrals defined in equation (5.3.71) will also be integrated numerically.

What may be most interesting is to find out the torque created by the non symmetric pressure distribution over the two main axis of the slipper. As defined by Fisher [1] among others, such torque will tend to maintain the slipper in its non tilted position.

The torque versus the main slipper central axis will be given by the equations:

$$M_{xx} = \int_0^{2\pi}\int_{r_0}^{r_1} p_1 r^2 \sin\theta \, dr\, d\theta + \int_0^{2\pi}\int_{r_1}^{r_2} p_2 r^2 \sin\theta \, dr\, d\theta + \int_0^{2\pi}\int_{r_2}^{r_3} p_3 r^2 \sin\theta \, dr\, d\theta + \int_0^{2\pi}\int_{r_3}^{r_4} p_4 r^2 \sin\theta \, dr\, d\theta \qquad (5.3.72)$$

$$M_{yy} = \int_0^{2\pi}\int_{r_0}^{r_1} p_1 r^2 \cos\theta \, dr\, d\theta + \int_0^{2\pi}\int_{r_1}^{r_2} p_2 r^2 \cos\theta \, dr\, d\theta + \int_0^{2\pi}\int_{r_2}^{r_3} p_3 r^2 \cos\theta \, dr\, d\theta + \int_0^{2\pi}\int_{r_3}^{r_4} p_4 r^2 \cos\theta \, dr\, d\theta \qquad (5.3.73)$$

when substituting the pressure distribution in each integral of the previous two equations it is obtained.

$$M_{XX} = \int_0^{2\pi} \frac{k_1 r_4^3 \sin\theta}{3}\left[\frac{\ln\left(\frac{r_1}{r_0}\right)}{(h_{01} + \alpha r_{m1}\cos\theta)^3} + \frac{\ln\left(\frac{r_2}{r_1}\right)}{(h_{02} + \alpha r_{m2}\cos\theta)^3} + \frac{\ln\left(\frac{r_3}{r_2}\right)}{(h_{03} + \alpha r_{m3}\cos\theta)^3} + \frac{\ln\left(\frac{r_4}{r_3}\right)}{(h_{04} + \alpha r_{m4}\cos\theta)^3}\right] d\theta +$$

$$\int_0^{2\pi} \frac{k_1 \sin\theta}{9}\left[\frac{\left(r_0^3 - r_1^3\right)}{(h_{01} + \alpha r_{m1}\cos\theta)^3} + \frac{\left(r_1^3 - r_2^3\right)}{(h_{02} + \alpha r_{m2}\cos\theta)^3} + \frac{\left(r_2^3 - r_3^3\right)}{(h_{03} + \alpha r_{m3}\cos\theta)^3} + \frac{\left(r_3^3 - r_4^3\right)}{(h_{04} + \alpha r_{m4}\cos\theta)^3}\right] d\theta +$$

$$\int_0^{2\pi} \frac{\sin^2\theta\ \mu\omega\alpha}{(h_{01} + \alpha r_{m1}\cos\theta)^3}\left[\frac{r_{m1} r_4^3 (r_0^2 - r_1^2)}{2} + \frac{r_{m1}\left(r_1^5 - r_0^5\right)}{5}\right] d\theta +$$

$$\int_0^{2\pi} \frac{\sin^2\theta\ \mu\omega\alpha}{(h_{02} + \alpha r_{m2}\cos\theta)^3}\left[\frac{r_{m2} r_4^3 (r_1^2 - r_2^2)}{2} + \frac{r_{m2}\left(r_2^5 - r_1^5\right)}{5} + r_1^2 (r_{m2} - r_{m1})\left\{r_4^3 \ln\left(\frac{r_2}{r_1}\right) + \frac{\left(r_1^3 - r_2^3\right)}{3}\right\}\right] d\theta +$$

$$\int_0^{2\pi} \frac{\sin^2\theta\ \mu\omega\alpha}{(h_{03} + \alpha r_{m3}\cos\theta)^3}\left[\frac{r_{m3} r_4^3 (r_2^2 - r_3^2)}{2} + \frac{r_{m3}\left(r_3^5 - r_2^5\right)}{5} + \left[r_1^2 (r_{m2} - r_{m1}) + r_2^2\left(r_{m3} - r_{m2}\right)\right]\left\{r_4^3 \ln\left(\frac{r_3}{r_2}\right) + \frac{\left(r_2^3 - r_3^3\right)}{3}\right\}\right] d\theta +$$

$$\int_0^{2\pi} \frac{\sin^2\theta\ \mu\omega\alpha}{(h_{04} + \alpha r_{m4}\cos\theta)^3}\left[\frac{r_{m4} r_4^3 \left(r_3^2 - r_4^2\right)}{2} + \frac{r_{m4}\left(r_4^5 - r_3^5\right)}{5} + \left[r_1^2 (r_{m2} - r_{m1}) + r_2^2\left(r_{m3} - r_{m2}\right) + r_3^2\left(r_{m4} - r_{m3}\right)\right]\left\{r_4^3 \ln\left(\frac{r_4}{r_3}\right) + \frac{\left(r_3^3 - r_4^3\right)}{3}\right\}\right] d\theta;$$

(5.3.74)

$$M_{YY} = \int_0^{2\pi} \frac{k_1 r_4^3 \cos\theta}{3}\left[\frac{\ln\left(\frac{r_1}{r_0}\right)}{(h_{01} + \alpha r_{m1}\cos\theta)^3} + \frac{\ln\left(\frac{r_2}{r_1}\right)}{(h_{02} + \alpha r_{m2}\cos\theta)^3} + \frac{\ln\left(\frac{r_3}{r_2}\right)}{(h_{03} + \alpha r_{m3}\cos\theta)^3} + \frac{\ln\left(\frac{r_4}{r_3}\right)}{(h_{04} + \alpha r_{m4}\cos\theta)^3}\right] d\theta + \qquad (5.3.75)$$

$$\int_0^{2\pi} \frac{k_1 \cos\theta}{9}\left[\frac{\left(r_0^3 - r_1^3\right)}{(h_{01} + \alpha r_{m1}\cos\theta)^3} + \frac{\left(r_1^3 - r_2^3\right)}{(h_{02} + \alpha r_{m2}\cos\theta)^3} + \frac{\left(r_2^3 - r_3^3\right)}{(h_{03} + \alpha r_{m3}\cos\theta)^3} + \frac{\left(r_3^3 - r_4^3\right)}{(h_{04} + \alpha r_{m4}\cos\theta)^3}\right] d\theta;$$

As in the previous equations, the remaining integrals need to be solved numerically. The torque equations for a generic number of lands are presented next.

$$M_{XX}=\int_0^{2\pi}\frac{k_1 r_n^3\sin\theta}{3}\left[\sum_{i=1}^{i=n}\frac{\ln\left(\frac{r_i}{r_{i-1}}\right)}{(h_{0i}+\alpha r_{mi}\cos\theta)^3}\right]d\theta+\int_0^{2\pi}\frac{k_1\sin\theta}{9}\left[\sum_{i=1}^{i=n}\frac{(r_{i-1}^3-r_i^3)}{(h_{0i}+\alpha r_{mi}\cos\theta)^3}\right]d\theta+$$
$$\int_0^{2\pi}\sin^2\theta\mu\omega\alpha\left[\sum_{i=1}^{i=n}\left[\frac{r_{mi}r_n^3(r_{i-1}^2-r_i^2)}{2}+\frac{r_{mi}(r_i^5-r_{i-1}^5)}{5}\right]\frac{d\theta}{(h_{0i}+\alpha r_{mi}\cos\theta)^3}\right]+ \quad (5.3.76)$$
$$\int_0^{2\pi}\sin^2\theta\,\mu\omega\alpha\left[\sum_{i=2}^{i=n}\left[\frac{r_n^3\ln\left(\frac{r_i}{r_{i-1}}\right)+\frac{(r_{i-1}^3-r_i^3)}{3}}{(h_{0i}+\alpha r_{mi}\cos\theta)^3}\left[\sum_{j=2}^{j=i}r_{j-1}^2(r_{mj}-r_{m(j-1)})\right]\right]\right]d\theta;$$

Following the same procedure, the torque versus the y axis will be given as:

$$M_{YY}=\int_0^{2\pi}\frac{k_1 r_n^3\cos\theta}{3}\left[\sum_{i=1}^{i=n}\frac{\ln\left(\frac{r_i}{r_{i-1}}\right)}{(h_{0i}+\alpha r_{mi}\cos\theta)^3}\right]d\theta+\int_0^{2\pi}\frac{k_1\cos\theta}{9}\left[\sum_{i=1}^{i=n}\frac{(r_{i-1}^3-r_i^3)}{(h_{0i}+\alpha r_{mi}\cos\theta)^3}\right]d\theta; \quad (5.3.77)$$

Notice that the integral type that should appear in equation (5.3.77) is $\int_0^{2\pi}A\frac{\sin\theta\cos\theta}{(b+c\cos\theta)^3}d\theta=0$

As in the previous cases the remaining integrals will have to be solved numerically.

It is to be noticed that the equations developed for tilted slipper, whenever tilt is removed become the equations for flat slipper, equations given in the previous sub-chapter.

5.3.4. CFD Model of Flat Slipper under Static Conditions

The validity of some of the equations presented has been assessed using a 3D Computational Fluid Dynamics model. In order to do this, a model of the central part of the slipper scale 2:1 was developed, and a 5-degree section of the slipper was only necessary due to the absence of slipper tilt. It is impossible to convey the overall grid image due to the large number of cells used, but figure 5.3.4 shows the grid developed at just one small part at the entry from the groove to the second land. It will be noticed that 5 cells are used across the clearance. The total number of cells exceeded 1.6 million for the 5 degree portion of the slipper, and a Sun Enterprise 4500 computer with 6Gb ram memory and 12 processors was used, a typical run time being 24 hours when using one processor. For the model, the turbulent Navier-Stokes and continuity equations were solved using the K-ε RNG model of turbulence, although the resulting flow was laminar as expected. Several simulations were undertaken for a range inlet pressures between 1 and 16MPa, and in all cases the reservoir pressure was set to 100Pa. Distance slipper plate 2.54 microns.

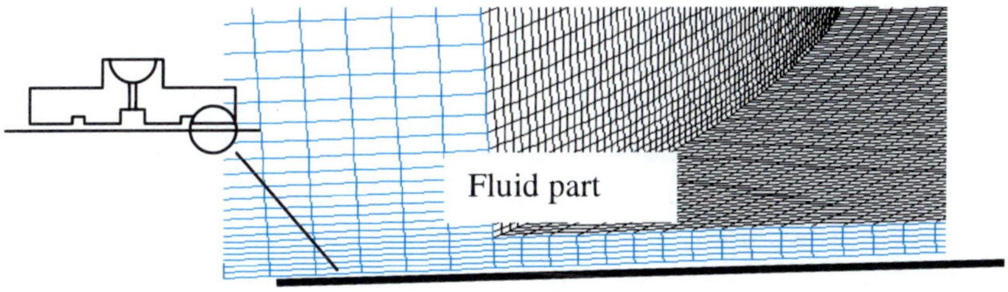

Figure 5.3.4. Partial grid of 5 degree portion of the slipper clearance groove→ second land.

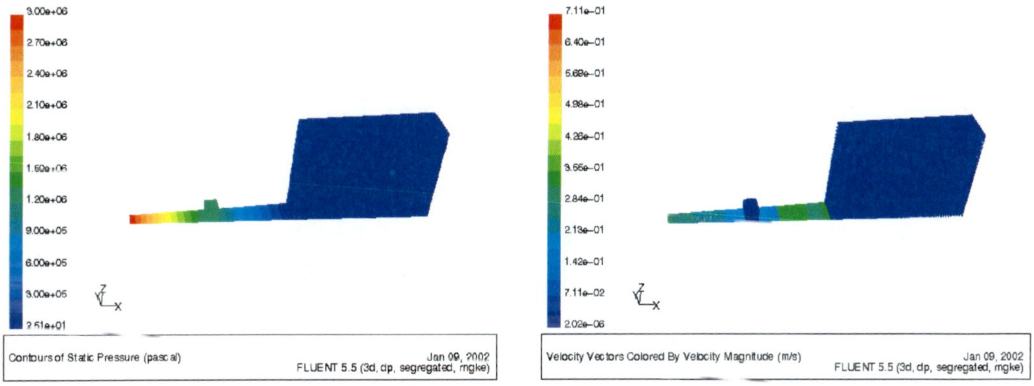

Figure 5.3.5. Pressure and velocity distribution along the slipper radius. Pressure differential 3 MPa distance slipper plate 2,54 microns.

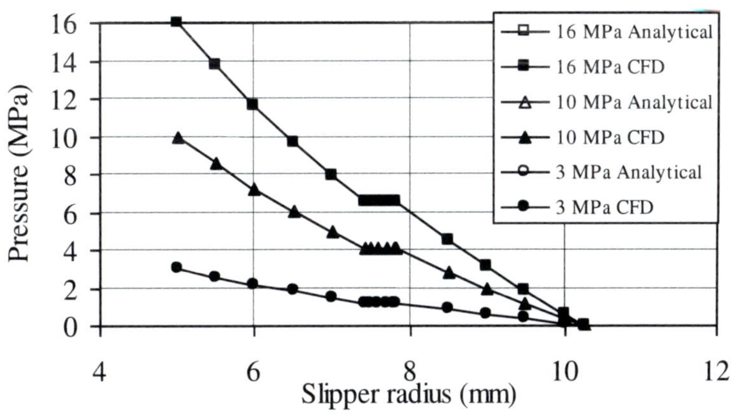

Figure 5.3.6. Pressure distribution comparison between the theoretical equations and the CFD predictions, slipper clearance 2,54 microns. Slipper scale 2:1.

In reality, two other models of the slipper scale 2:1 including the groove, were developed using the Fluent 6.1 CFD package. The first model considered the slipper without tilt, under static conditions and for a clearance of 10 microns, maintaining the groove dimensions, the groove was positioned in three different radial locations. A second model considered the slipper with a central clearance of 10 microns and two very small tilts of 0.0014 and 0.0028 degrees, which corresponds to 1 and 2 microns tilt over the slipper diameter. Such small tilts are the ones expected to be found in practice. For all cases studied, the effect of plate turning speed in the range 0 – 1250 rpm, was considered. A single inlet pressure of 15 MPa was used for these tests. In the present section just the static case for flat slipper is being used to compare with the theoretical static equations previously presented. From the simulation undertaken and under static conditions, it can be stated that for the small tilts studied the pressure distribution along the slipper diameter remains very much the same as the one found for the flat slipper case. Pressure and velocity distribution along the slipper radius can be seen in figures 5.3.5 and 5.3.6. Notice that results produced by the equations and the CFD ones have a perfect agreement.

5.3.5. Flat Slipper with Grooves, Static and Dynamic Numerical Model

Figure 5.3.7 shows a schematic drawing of the slipper and swash plate clearance, the two slipper-relative movements, spin and tangential velocity are also outlined. The shaded area presents the computational domain $\Omega\{\Omega : R^3\ (r,\theta,z)\}$ with shown boundary $\Gamma\{\Gamma_{in}, \Gamma_{out}\}$. Notice that the combination of tangential velocity and spin acting over the slipper will create a flow field below the slipper which can not be considered symmetric in angular direction, even though the clearance between slipper-plate is constant at all points. Therefore it becomes necessary to consider a 3-dimentional cylindrical computational domain.

The continuity equation and conservative form of Navier Stokes in cylindrical coordinates is given in Equations (5.3.78) and (5.3.79) respectively. Where $\phi\{V_r, V_\theta, V_z\}$ is considered as flux vector and the value of S_ϕ is given in Table 5.3.1 for different ϕ.

$$\frac{1}{r}\frac{\partial(r.V_r)}{\partial r} + \frac{1}{r}\frac{\partial V_\theta}{\partial \theta} + \frac{\partial V_z}{\partial z} = 0 \tag{5.3.78}$$

$$\rho\frac{\partial\phi}{\partial t} + \rho\left[\frac{1}{r}\frac{\partial}{\partial r}(r.V_r.\phi) + \frac{1}{r}\frac{\partial}{\partial\theta}(V_\theta.\phi) + \frac{\partial}{\partial z}(V_z.\phi)\right] = \mu\left[\frac{\partial}{\partial r}\left(\frac{1}{r}\frac{\partial(r.\phi)}{\partial r}\right) + \frac{1}{r^2}\frac{\partial^2\phi}{\partial\theta^2} + \frac{\partial^2\phi}{\partial z^2}\right] + S_\phi \tag{5.3.79}$$

A no slipping boundary condition is imposed in all walls. The pressure at inlet and outlet boundaries is specified and the flow variables at inlet and outlet boundary are specified by zero normal derivatives at each sub iteration step.

$$P|_{\Gamma_{in}} = P_{in};\ P|_{\Gamma_{out}} = P_{out};\ \left.\frac{\partial(r.V_r)}{\partial r}\right|_{\Gamma_{in} or\ \Gamma_{out}} = 0;\ \left.\frac{\partial V_\theta}{\partial r}\right|_{\Gamma_{in} or\ \Gamma_{out}} = 0;\ V_z|_{\Gamma_{in}} = 0;\ \left.\frac{\partial V_z}{\partial r}\right|_{\Gamma_{out}} = 0; \tag{5.3.80}$$

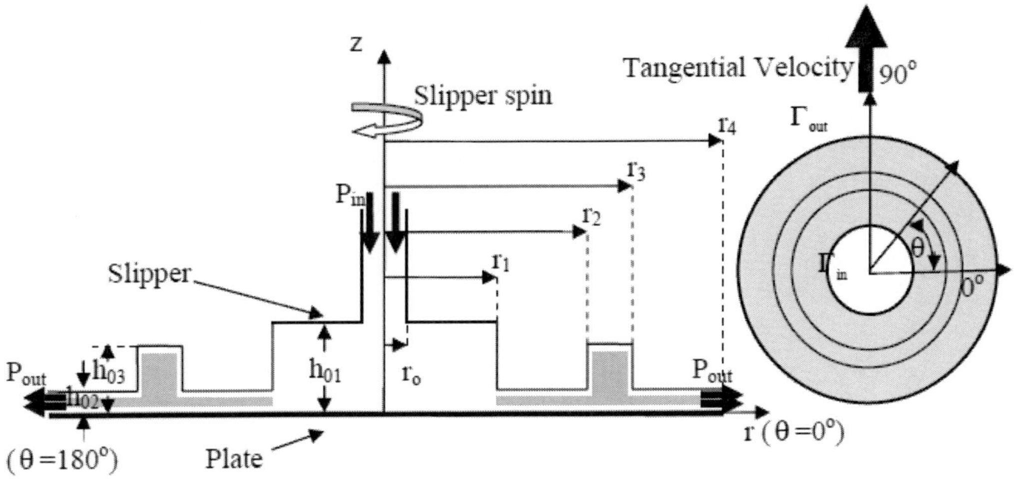

Figure 5.3.7. Schematic diagram of the slipper-swash plate computational domain.

Table 5.3.1. S_ϕ for different ϕ

ϕ	S_ϕ
V_r	$-\frac{\partial P}{\partial r}-\frac{2\mu}{r^2}\frac{\partial V_\theta}{\partial \theta}-\frac{\mu V_r}{r^2}+\frac{V_\theta^2 \cdot \rho}{r}$
V_θ	$-\frac{1}{r}\frac{\partial P}{\partial \theta}+\frac{2\mu}{r^2}\frac{\partial V_r}{\partial \theta}-\frac{\mu V_\theta}{r^2}-\frac{\rho V_r V_\theta}{r}$
V_z	$-\frac{\partial P}{\partial z}$

5.3.6. Numerical Solution Technique

Equations (5.3.78) and (5.3.79) are discretized by control volume formulation over a staggered grid as defined in [35, 49]. Four different control volumes for r-momentum, θ-momentum, z momentum and continuity are presented in Figure 5.3.8 with dependent neighbour variables. As can be seen velocities and pressure are staggered and therefore are the corresponding control volumes. In figure 5.3.8 all the neighbour velocities and effective pressures for each corresponding control volume are shown in terms of grid coordinates. The grid size is uniform in r and θ direction but increases in the z direction when moving down towards groove bottom. Since in “z” direction the grid is not uniform, and the flow is considered incompressible, it was decided to put velocity grid lines in the centre of pressure grid lines, therefore velocity will represent the best approximation over the entire control volume when applied to momentum equation.

To maintain the positivity of the coefficients for ensuring stability of the method, the nonlinear source term of θ-momentum equation is linearized according to Equation (5.3.81) with negative gradient as described in [35]. A power law scheme is used as a shape function between two points. After discretization, the momentum equation can be written as Equations (5.3.82-5.3.84).

$$\left(\frac{\rho V_r}{r}\right)\phi = \left(\overline{\frac{\rho V_\theta}{r}\max[-V_r,0]}\right) + \left(\frac{\rho}{r}\max[V_r,0]\right)\phi \qquad \text{where } \phi = V_\theta \tag{5.3.81}$$

$$\frac{a_{r.p}}{\alpha_v}V^*_{r.p} = \sum_{nb} a_{r.nb}V^*_{r.nb} + b_r + (P^*_p - P^*_E).r.d\theta.dz + \frac{(1-\alpha_v)}{\alpha_v}a_{r.p}V^*_{r.p.old} \tag{5.3.82}$$

$$\frac{a_{\theta.p}}{\alpha_v}V^*_{\theta.p} = \sum_{nb} a_{\theta.nb}V^*_{\theta.nb} + b_\theta + (P^*_N - P^*_S).dr.dz + \frac{(1-\alpha_v)}{\alpha_v}a_{\theta.p}V^*_{\theta.p.old} \tag{5.3.83}$$

$$\frac{a_{z.p}}{\alpha_v}V^*_{z.p} = \sum_{nb} a_{z.nb}V^*_{z.nb} + b_z + (P^*_B - P^*_T).r.d\theta.dr + \frac{(1-\alpha_v)}{\alpha_v}a_{z.p}V^*_{z.p.old} \tag{5.3.84}$$

By using this imperfect velocity field, calculated from equations (5.3.82-5.3.84), a pressure correction formula (5.3.85) is developed from continuity equation (5.3.78) to improve the imperfect velocity field into a prefect velocity field which satisfy mass conservation as described by Patankar et al [35], this method is named SIMPLE (Semi-Implicit Method for Pressure-Linked Equation) based algorithm.

$$A_p P'_p = \sum_{nb} A_{nb} P'_{nb} + B_p \tag{5.3.85}$$

When improving the guessed pressure using the calculated correction from equation (5.3.85), an under-relaxation factor α_p is implemented, as shown in equation (5.3.86).

$$P = P^* + \alpha_p P' \tag{5.3.86}$$

Although the value of the under-relaxation factor suggested by Patankar et al [35] is 0.8, often, this value can be as low as a 0.1 as discussed by Anderson [50]. In the present work a very low value of α_p (0.01) is used to achieve convergence at the beginning of the simulation. After a few iterations, the rapidly varying component of error becomes negligible and error becomes a smooth function of spatial coordinates, therefore the value of the under-relaxation factor (α_p) is updated to 0.5 to increase the rate of convergence. When solving equations (5.3.82-5.3.84) an under-relaxation technique as suggested by Patankar et al [35] and Anderson [50] was used, an under-relaxation factor α_V is implemented during successive iterations. The value of under-relaxation factor α_V used in the present work is 0.1 which is a value suggested by Anderson [50].

By using the above specified computational method and under-relaxation factors, first, several numerical test were performed at different grid sizes, to ensure a grid independent solution. Then pressure distribution, force, torque and leakage were tested for a set of different pressure boundary conditions (from 3 to 13 MPa) and also for a different slipper-swash plate gaps, (from 6.5 to 20 microns). The effect of turning speed on pressure distribution, force, torque, leakage and groove Vorticity was also studied by modifying turning speed from 200 to 1000 rpm. Finally the effect of groove dimensions and position on leakage, pressure distribution force and torque was also analyzed by modifying the groove width (1 mm and 2 mm), groove depth (from 0.1 mm to 1.2mm) and its radial position. Computational results were compared with experimental ones, showing a good agreement. The scale of the slipper studied is 2:1 in comparison with the slipper presented in figure 5.3.1. Slipper dimensions are to be found in the next section, experimental test rig. Further information regarding the numerical procedure used is to be found in [51, 52]

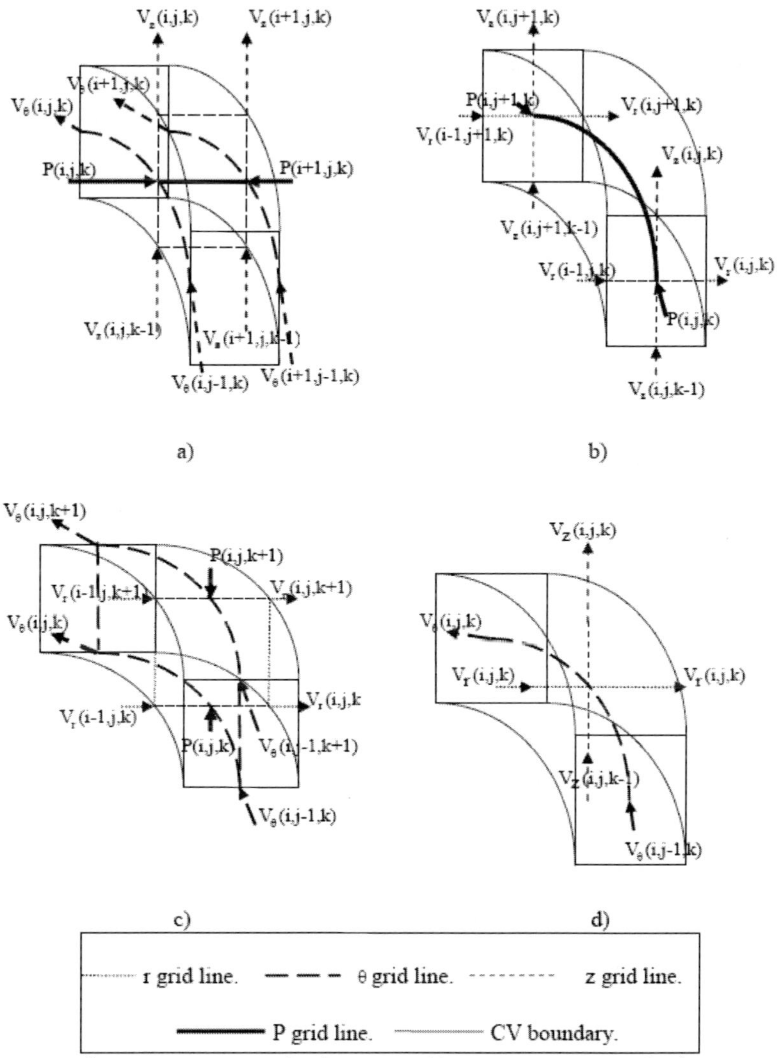

a) r-momentum control volume. b) θ -momentum control volume.
c) z-momentum control volume. d) Continuity control volume.

Figure 5.3.8. Four different control volumes with their neighbour dependent variables.

5.3.7. Experimental Test Rigs

In order to experimentally validate the equations developed, the CFD and numerical models created, the test rig represented in figure 5.3.9 was constructed.

Three position transducers, having a measurement accuracy of 0.5 microns, were attached to the slipper at 120° intervals. These sensors require a non-ferrous measuring face for optimum performance and therefore the housing assembly was manufactured from aluminium while the slipper assembly was manufactured from stainless steel. The slipper is held in position using four screws and the required slipper orientation is achieved by turning four additional positioning screws. Using this method the slipper can be positioned completely parallel to the swash plate or to any desired tilt. Four holes, 0.5 mm diameter and at every 90°, were drilled at the centre of slipper groove allowing measurement of the pressure inside the groove at its four cardinal positions. Static tests were performed with a slipper central clearance ranging from 5 to 30 microns and for a range of slipper tilts at each central clearance. In all positions, a set of different pressures were applied to the slipper, these being from 1 to 15MPa. The slipper was built with a scale 2:1 when compared with the slipper that initiated this project, shown in figure 5.3.1. This was done to be able to physically locate both the position sensors and the pressure measuring points. However, the size of the test rig slipper is not unlike those that exist in larger pumps such as the set of pistons and slippers supplied by Oilgear Towler UK to the authors for reference. The test rig slipper dimensions scale 2:1are defined in table 5.3.2:

Table 5.3.2. Slipper dimensions for the two different test rigs used

Parameters.	Test rig 1. Slipper scale 2:1	Test rig 2. Slipper scale 1:1
Orifice radius r_1.	1mm	0.5 mm
Inner land inside radius r_2.	10.15 mm	5 mm
Inner land outside radius r_3.	14.7 mm	7.4 mm
Groove width.	1 mm	0.4 mm
Outside radius r_4.	20,5 mm	10.2 mm
Film thickness $h_2 = h_4$.	Modifiable 5 to 35 μm	Modifiable 5 to 35 μm
Slipper pocked depth h_1.	h_2 + 1.4 mm	h_2 + 0.65 mm
Groove depth h_3.	h_2 + 0.8 mm	h_2 + 0.4 mm

Since the test rig allows rotation of the swash plate, a second set of tests were performed to study the effect of tangential velocity on slipper plate leakage and slipper groove pressure distribution. The variable-speed tests were performed for a single central clearance of 15 microns and for a single tilt of 0.035°. A set of swash plate turning speeds were studied in the range 0 to 1350 rpm, the maximum turning speed corresponding to a tangential velocity on the slipper main axis of 13 m/s.

During dynamic experiments it was observed that slipper/plate clearance changes over time, due to temporal turning plate run out. Disk run out was found to be dependent on slipper input pressure and disk turning speed and independent of slipper/plate clearance. Therefore in reality, dynamic measurements are referenced to average central clearances. Figure 5.3.10 presents disk run out at two pressures and turning speeds, from where it is noticed that at low pressures disk run out is highly dependent on turning speed, but at high pressures such dependency falls to a minimum. When studying slipper dynamics, the slipper/plate clearance needs to be modified after calculating the weighted average of disk run out.

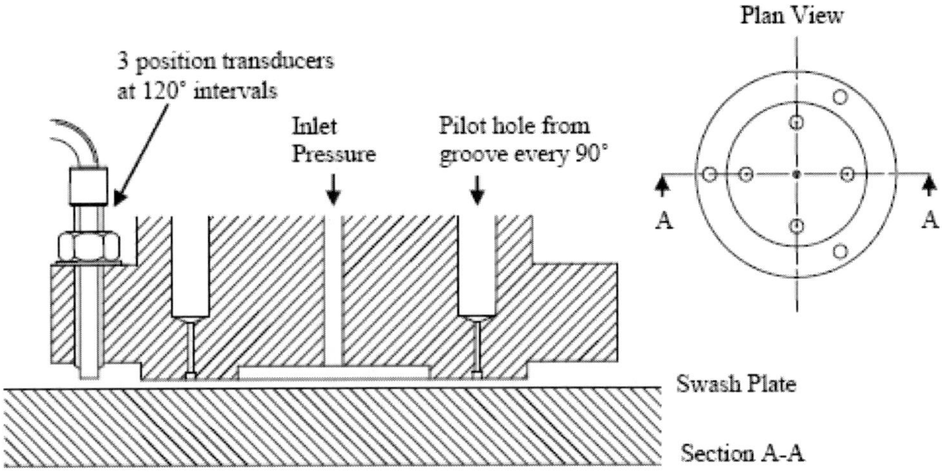

a) Cross section of the slipper area

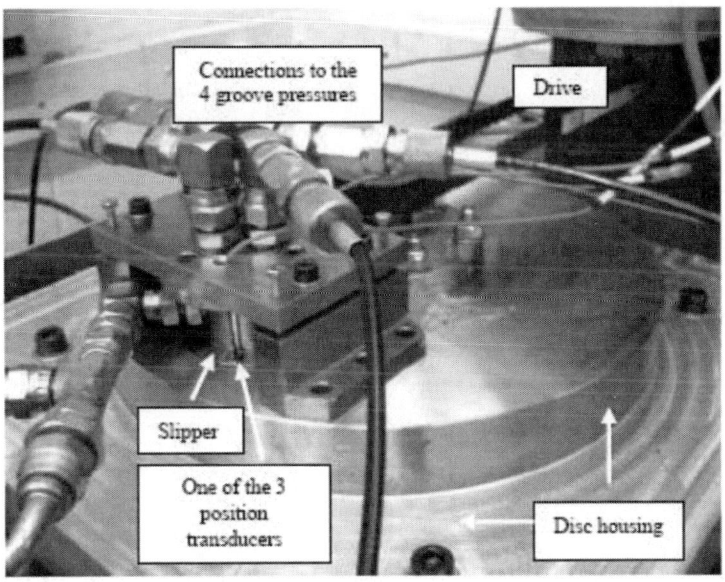

b) Slipper, disc housing and drive system

Figure 5.3.9. Experimental test rig 1, slipper scale 2:1.

The test conditions considered should be put into context with those that would probably exist for the test slipper when used in a real pump application. Slipper force measurement is not considered in this chapter, but a calculation can be made using existing lubrication theory [53, 54]. It is assumed that:

- the maximum pump pressure would be 35 MPa
- the ISO32 fluid viscosity $\mu = 0.032 Ns/m^2$

- the pocket pressure is marginally different from the pump pressure
- the swash plate angle is 20°
- the pressure distribution across the slipper is approximated by an equivalent logarithmic decay passing through the centre of the groove

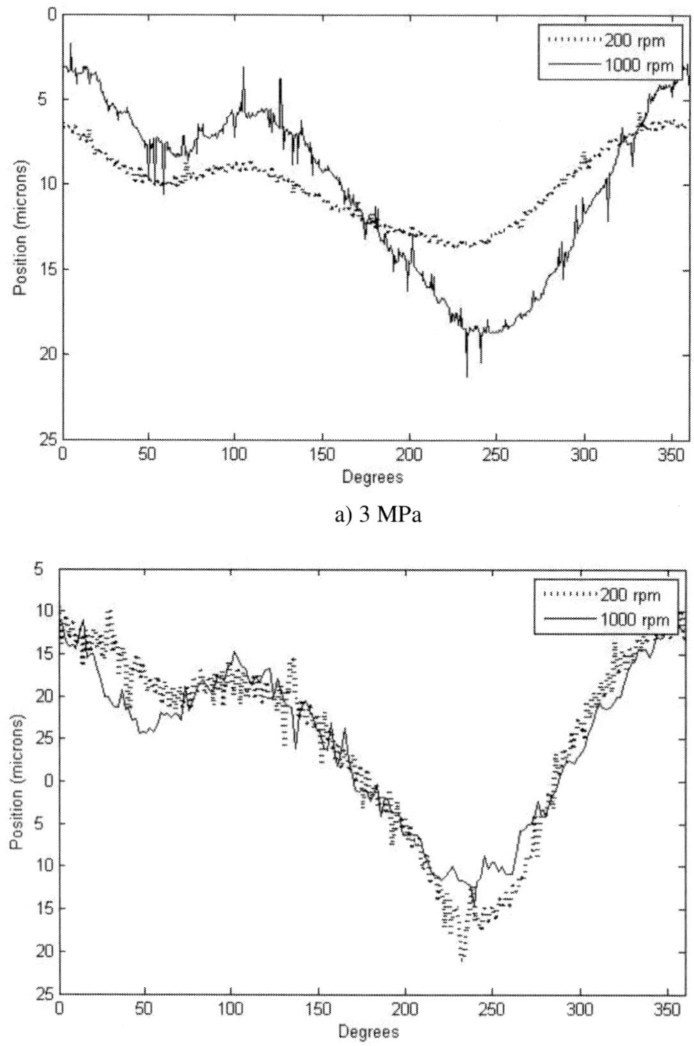

a) 3 MPa

b) 13 MPa

Figure 5.3.10. Disk run out at several pressures and turning speeds.

The maximum hydrostatic force generated on the slipper is then 23kN. The force balance across the slipper and piston is then determined by the pump manufacturer, perfect force balance occurring for a piston diameter of 28.9mm. If an additional hydrodynamic force is required then this will not be greater than typically 5% of the hydrostatic force, and is based upon well-established design knowledge. There is no explicit theory for determining the hydrodynamic lift for a circular slipper, but a good approximation can be made by using square plain bearing theory with side leakage effects taken into account [54]. It will be further assumed that:

- a square plain bearing of equivalent area, 36.3mm x 36.3mm, applies
- the bearing central clearance h_o = 10micron
- a side leakage compensating factor of 0.44 applies [54]
- a tangential velocity of 13m/s still applies

The bearing tilt is then calculated to be equivalent to a 0.26 microns increase from the trailing edge to the leading edge. This gives a square bearing tilt angle of 0.00041° and therefore smaller by a factor of 12.2 compared with the minimum non-zero value of 0.005° that was set in the tests. Even if all the slipper lift force was created by hydrodynamic effects, a condition that would not normally occur with a non-blocked slipper orifice, the tilt angle would still only be 0.0079°.

The first test rig allowed measurements of leakage across the slipper and pressure inside the groove, but pressure decay along the lands could not be measured. To overcome this difficulty, a second and much simpler test rig was built as shown in Figure 5.3.11

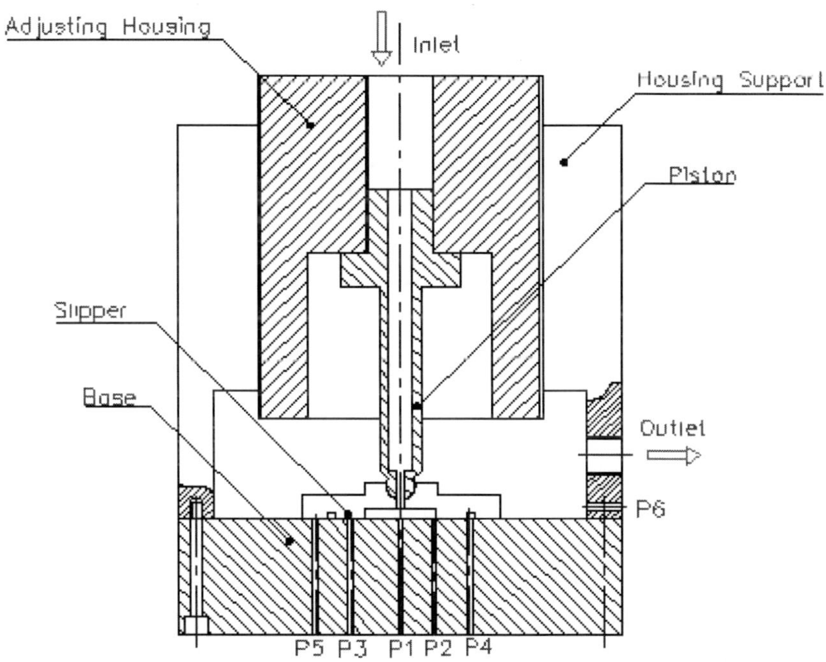

Figure 5.3.11. Test rig 2, slipper scale 1:1.

This second test rig housed the actual piston/slipper assembly shown in Figure 5.3.1 and was only capable of static testing for the constant clearance condition, slipper dimensions being defined in table 5.3.2. A micrometer gauge thread was machined on the adjuster allowing known clearances to be set and a range of no-load clearances from 0 to 35microns in 5micron steps were studied and up to a maximum inlet pressure of 16MPa. However before the results are compared with theory it is essential to determine the actual clearance as pressure is applied. This is due to the small yet significant compression between the adjusting housing and the housing support fine thread, the net result being that the actual clearance increases with applied pressure. This compression was measured with a precision position transducer mounted to the bed plate holding the test unit and it was found that the compression increased to 4 microns as the pressure increased to 16MPa. The accuracy of the displacement transducer used to measure the relative displacement between the adjusting housing and the housing support was determined as ± 0.25 microns. Pressure tapings in the base unit then allowed the pressure distribution to be measured across one axis of the slipper, including the groove, using calibrated test Bourdon gauges.

5.3.8. Results

5.3.8.1. Leakage and Pressure Distribution for a Non Tilted Static Slipper

To experimentally evaluate the leakage slipper/plate under static conditions, test rig 1 was used, since slipper position could be established accurately using the position transducers. For the case of no tilt, comparisons between experimental and analytical results, given by equation (5.3.21) or the generic (5.3.25), for a set of inlet pressures and clearances, are to be found in Figure 5.3.12 where it is noticed that a good agreement between measurement and theory is obtained.

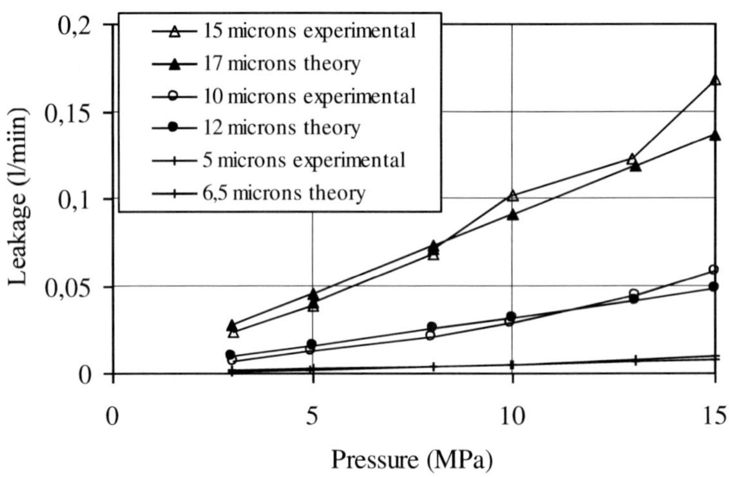

Figure 5.3.12. Comparison between theoretical and measured flow rates, test rig 1.

It is important to point out that there will be an error in the apparent clearance and the true clearance which varies from point to point due to disk and slipper surface roughness. Surface roughness measurements were presented in [55] from where it can be stated that the variation in surface finish is typically 1micron for both materials.

Test rig 1 also allows measurement of the pressure inside the groove at four points. As the slipper had no tilt and no relative movement was considered, the pressure at all four measurement points of the groove was the same. Figure 5.3.13a shows the analytical radial pressure distribution below the groove for three different inlet pressures using equations

(5.3.17-5.3.20) or the generic ones (5.3.26-5.3.27). It is noticed that according to the theory the pressure distribution does not depend on the clearance yet the experimental results show that there is a dependency, as demonstrated in Figure 5.3.13b.

In Figure 5.3.13a, the comparison between analytical and CFD results at 15 MPa is presented, showing an excellent agreement, and therefore indicating that the theory presented give the same information as a CFD model. On the other hand, the pressure inside the groove was found experimentally to be changing with inlet pressure and central clearance, as can be seen in Figure 5.3.13b.

The average pressure tends to decrease as central clearance increases, indicating that as central clearance increases, the force over the slipper will decrease.

The comparisons between analytical and experimental results, Figure 5.3.13b, show some discrepancies at high pressures and high clearances, which can be understood when it is noticed that the equations proposed consider the shear stresses between the fluid and the slipper/plate boundaries for the ideal case; this is, when the slipper and plate surfaces are perfectly smooth. As demonstrated in [55], in reality these surfaces have a measurable roughness. Therefore the shear stresses occurring between the metal surface and the fluid in the clearance between the slipper and plate will be higher than the theoretical ones.

The consequence is that the pressure drop at each slipper land will be higher than the theoretical values presented, and this leads to a lower average pressure inside the groove than the one expected in theory. Shear stresses increase with the velocity gradient, which is higher for higher flow, and this is why the experimental/theoretical discrepancy is higher at higher clearances and pressures.

It is important to point out that the clearance slipper swash plate is usually around 5 – 10 microns, where the agreement between experimentation and theory is much higher.

Using test rig 2 the pressure was measured at the centre of each slipper land, in two points inside the slipper pocked and inside the groove for a set of different inlet pressures and clearances of 15, 25, 35 microns.

Figure 5.3.14 compares the average of all the measurements at each inlet pressure with the analytical results, given by the equations (5.3.17-5.3.20). The agreement is very good although it is noticed that at high pressures there is some disagreement, the explanation of which was given when Figure 5.3.13 was discussed.

These comparisons raise further issues regarding the measurement of pressure for practical piston/slipper assemblies. The pressure tappings were created by drilling ostensibly 0.3mm diameter holes, and the pressure drop over this distance is 1.2MPa for an inlet pressure of 16MPa.

The exact location of the pressure tappings with respect to the slipper cannot be precisely measured for test units of this scale.

In addition any variation in the set clearance or the induction of tilt, during testing cannot also be determined.

The net result is that it is proposed for this test rig that the experimental error for pressure measurement is ±0.6MPa at the highest inlet pressure used of 16MPa. Figure 5.3.14 shows that the comparison between theory and measurement is good for the inner land but with experimental measurements lower than predicted for the outer land, particularly at the highest pressure.

A displacement error of 0.3mm for the pressure tapping position in this region would explain the difference.

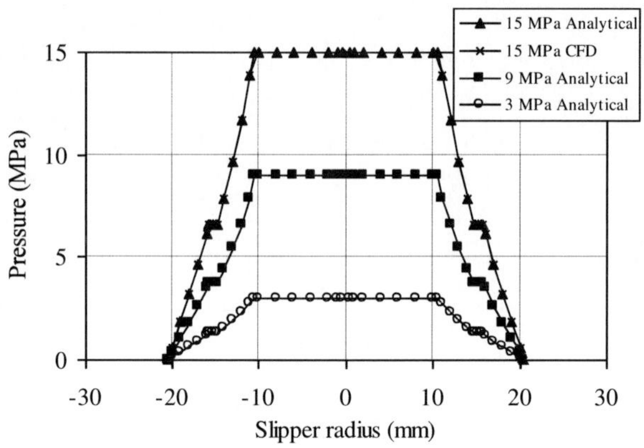

a) Theoretical and CFD radial distribution on pressure.

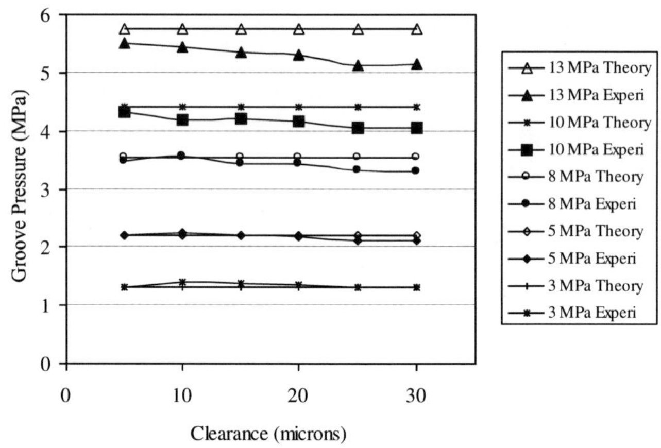

b) Groove pressure variation

Figure 5.3.13. Pressure distribution and average groove pressure, test rig 1. (scale 2:1).

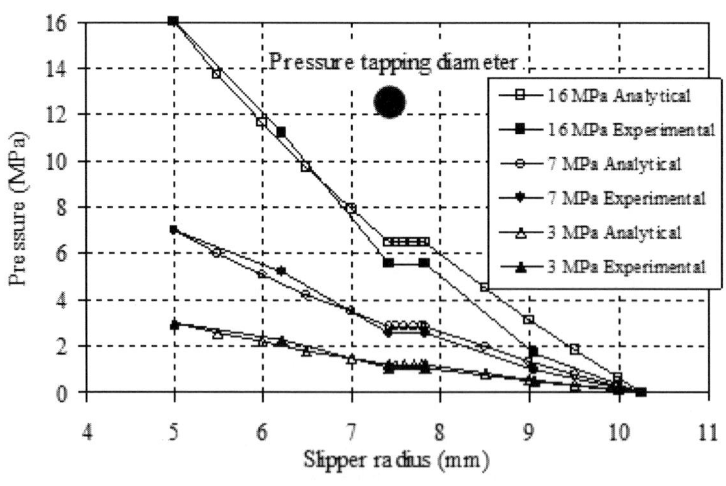

Figure 5.3.14. Measured pressure distributions for a set of clearances, test rig 2 (scale 1:1).

5.3.8.2. Influence of Groove Position on Non Tilted Static Slipper Leakage and Force

When designing a slipper, it should be realised that a priory, the smaller the slipper the larger the pump mechanical efficiency, then higher dimension means higher weigh and therefore higher amount of energy is needed to move the piston/slipper assembly, nevertheless, as defined in [56-59] the frictional power loss linked with the slipper dimensions, must be considered when aiming to fully analyse slipper efficiency. Also the slipper should create enough lift to compensate for the force acting at the opposite end of the piston while maintaining a small oil film between the slipper and the swash plate. It then needs to be recalled that the thicker the oil film the lower the pump volumetric efficiency. In order to increase slipper stability while running around the swash plate, some manufacturers have decided to use grooves. Notice that when the slipper slides around the swash plate, during about 150 degrees the pressure on the top of the piston is high, and according to [15,16,20] the slipper runs nearly flat, but when the piston faces the tank kidney port, the slipper tilt increases sharply. Then at each revolution, the slipper needs to accommodate from a very high tilt condition to a nearly flat position, good slipper stability it is required under such conditions. Slipper stability it is also required when running nearly flat around the swash plate, since metal to metal contact should be avoided. In some cases the grooves are vented with the intention of reducing slipper spin, and this will allow hydrodynamic lift but not hydrostatic lift. However, the use of non vented grooves allows the entire slipper including the groove to create both hydrostatic and hydrodynamic lift. This means that the slipper external diameter can be smaller, maintaining a higher mechanical efficiency. If just the slipper lift characteristics are to be taken into account, the use of a non grooved slipper with a bigger central pocket would be desirable, yet slipper stability might be compromised. If a bigger central pocket is used, the remaining slipper land becomes smaller, wear is more rapid which would create higher leakages and thus a decrease in volumetric efficiency. This is why a compromise has to be reached between achieving a lift via increasing the central pocket diameter and compromising slipper stability, or achieving the same lift using a smaller central pocket diameter and inserting a non vented groove. This aspect of the analysis presented is considered to be very relevant to pump manufacturers.

A further advantage of the equations proposed is that slipper performance can now be evaluated for different groove positions. Equations (5.3.21, 5.3.23) or the generic ones (5.3.25, 5.3.28) are used to calculate the leakage flow rate and slipper lift, given the pressures at the slipper central radius r_0 and external radius r_4. Some results for lift and flow rate are shown in Figure 5.3.15 for variations in groove inner r_2 and outer radius r_3 and considering $r_1 \le r_2 \le r_3$. These figures demonstrate that groove length and position for given slipper dimensions will drastically change the slipper performance. This could not have been deduced from previous work and illustrates a particular design feature of the analytical approach presented. Figures 5.3.15a, b, demonstrate that for a specific radius r_2, the modification of r_3 in the range selected will create an increase or decrease in lift while the leakage flow rate will always suffer an increase, the minimum leakage will always occur for a non grooved slipper. For the specific boundary conditions set, there is a unique relationship between r_2 and r_3 that will give maximum lift. Notice that according to the force diagram, Figure 5.3.15a, the best groove to create maximum lift would be the one covering the entire slipper land and almost reaching the slipper external diameter, in other words, the best groove to achieve maximum lift, is the one which extends the slipper pocked to nearly the external slipper diameter. However, an increase in force via using such a groove would bring a huge increase of leakage.

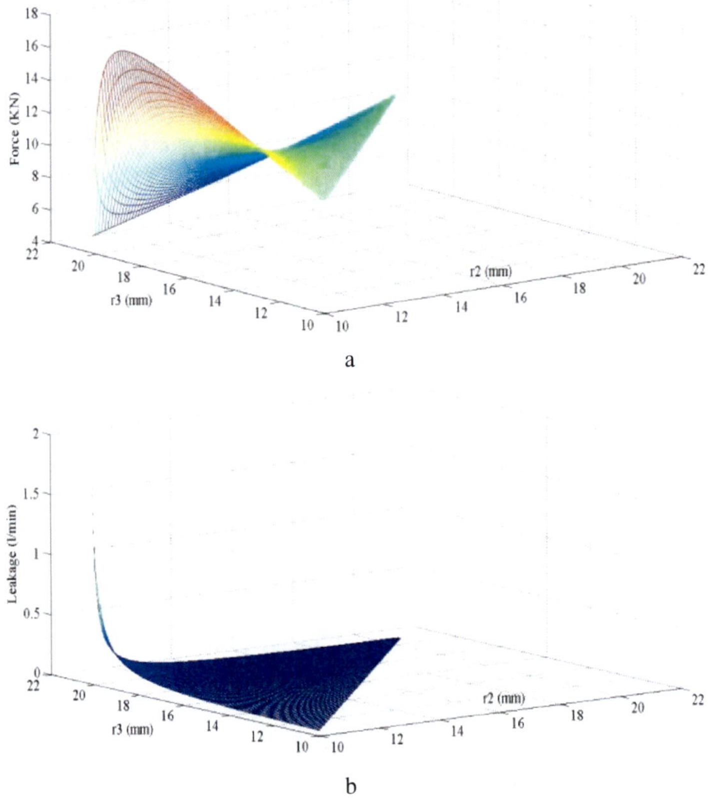

Figure 5.3.15. Force and leakage over the original single groove slipper (scale 2:1) when modifying groove dimensions r_3 and r_4. Groove depth is maintained constant at 0.8 mm. $h_2 = h_4 = 10$ microns. Inlet pressure 15MPa, applicable to test rig 1. a) Force; b) Leakage.

In order to further clarify the effect of a groove on a non tilted slipper, Figure 5.3.16 presents the force and leakage variation as a percentage of the flat slipper without a groove, for a set of groove position central radii while maintaining constant groove dimensions. Slipper scale 2:1. It clearly demonstrates that the inclusion of a groove may increase or decrease the force compared with a non grooved slipper, but the leakage will always increase. Such percentage variation is independent of the slipper/plate clearance and the inlet pressure. A comparison between the leakage obtained using the theory presented and the CFD model has also been undertaken, showing a very good agreement.

Figure 5.3.17 presents the effect on leakage in percentage versus the non grooved slipper as groove depth is modified. It is clearly shown that for a given groove width and groove position, leakage increases with the increase of groove depth, although when depth overcomes a certain value (0.2mm in the present study), leakage will be nearly constant and independent on groove depth increase. The explanation of such behaviour is found latter when analyzing vorticity inside groove. Notice that at very low groove depths, slipper plate clearance slightly modifies leakage increase.

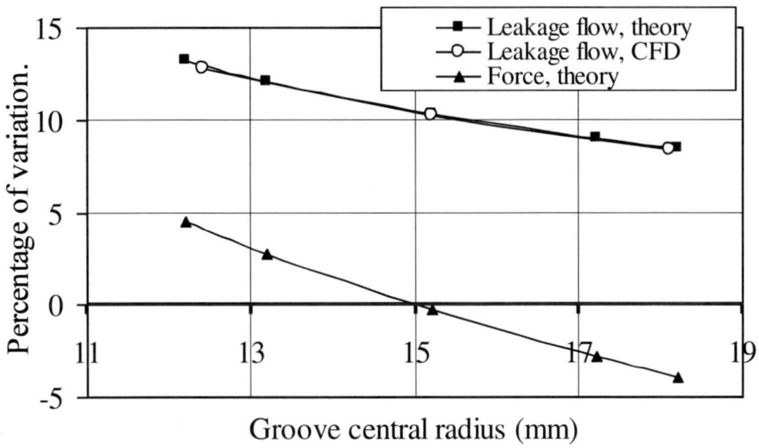

Figure 5.3.16. Leakage flow and force variation, as a percentage of the non grooved case. The groove size remains constant. Slipper scale 2:1.

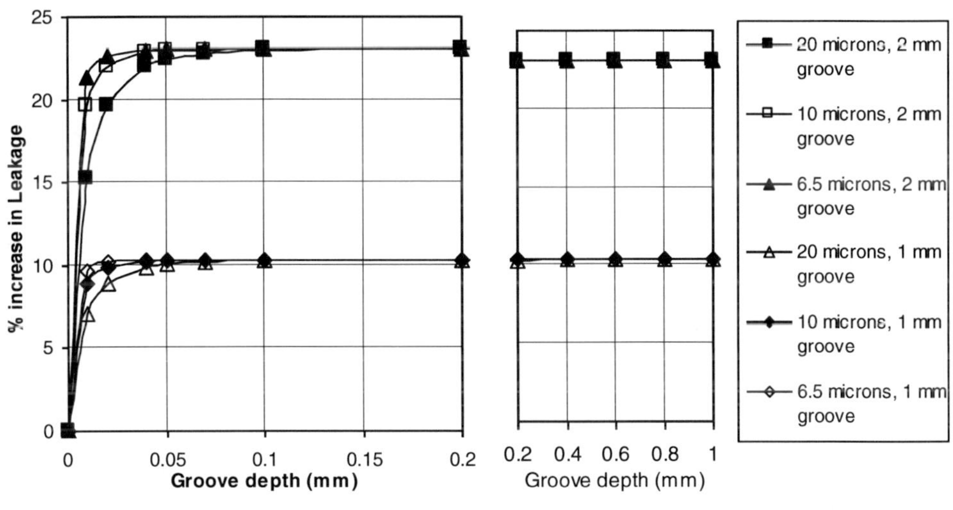

Figure 5.3.17. Leakage increase in percentage versus non grooved slipper, for two different withds (1, 2 mm), inlet pressure 10 MPa and different clearance. Groove centered (CFD).

5.3.8.3. Non Tilted Dynamic Slipper

As in reality the slipper turns around the swash plate and also spins with respect to its own axis, to understand its dynamic behaviour it will be necessary to consider such movement. When using the test rig presented in figure 5.3.9 under dynamic conditions, it has to be considered the disk run out. Then, the clearance slipper/plate will need to be corrected by the disk run out and during experimentation it was observed that disk run out depends on slipper inlet pressure and disk turning speed. The disk run out has been measured using the position transducer located at the slipper leading edge and figure 5.3.10 presents results at 3 MPa and 13 MPa and for two rotational speeds of 200 rpm and 1000 rpm. It is noticed that at

3 MPa, the effect of rotational speed modifies substantially the disk run out; while at 13 MPa the disk run out is rather independent of turning speed. The maximum run out is found at low pressures and high speeds. Disk run out (for the cases studied) was found to be independent on the film thickness. During experimentation it was found that slipper plate clearance was increasing with the increase of inlet pressure.

Figure 5.3.18 shows a diagram of the two slipper rotations considered and existent in reality, with respect to slipper axis and with respect to the swash plate axis, the different angular positions considered over the slipper are also presented. The turning with respect to swash plate axis is much higher than the spin and therefore the flow is expected to be deeply affected by its tangential velocity associated.

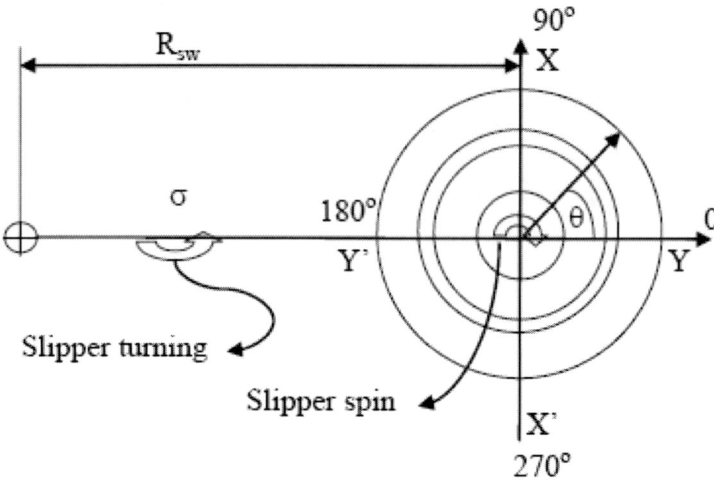

Figure 5.3.18. Diagram of slipper turning speeds with respect to swash plat axis. R_{sw} = 92 mm.

5.3.8.3.1. Non Tilted Dynamic Slipper, Pressure, Force and Torque, Experimental and Numerical Results

In figure 5.3.13a, it was presented the pressure distribution for several inlet pressures under static conditions, it has been noticed that for the cases studied, pressure distribution below the slipper was independent on slipper/plate clearance, therefore, it can be expected that under dynamic conditions, disk run out is not going to affect very much the pressure distribution below the slipper. On the other hand, leakage slipper plate it is expected to be highly dependent on disk run out.

The experimental test rig 1 was able to evaluate the pressure inside the slipper groove in four points separated 90 degrees from each other, and under all static and dynamic conditions pressure inside the groove was found to be approximately constant on turning speed and just dependent on inlet pressure.

In figure 5.3.19 the average simulated groove pressures for a set of turning speeds and slipper inlet pressures is compared with the experimental ones.

The results show a good concordance specially at low and medium slipper inlet pressures, nevertheless, such agreement decreases at high pressures, such discrepancies are well understood when taken into account that as inlet pressure increase leakage across the slipper also increases and shear stresses increase with leakage, the NVS equations consider shear stresses for a perfectly smooth surface, but in reality the slipper and plate surfaces have some roughness, as a result pressure decay in slipper first land is higher in reality than the one found numerically, specially at high pressure differentials.

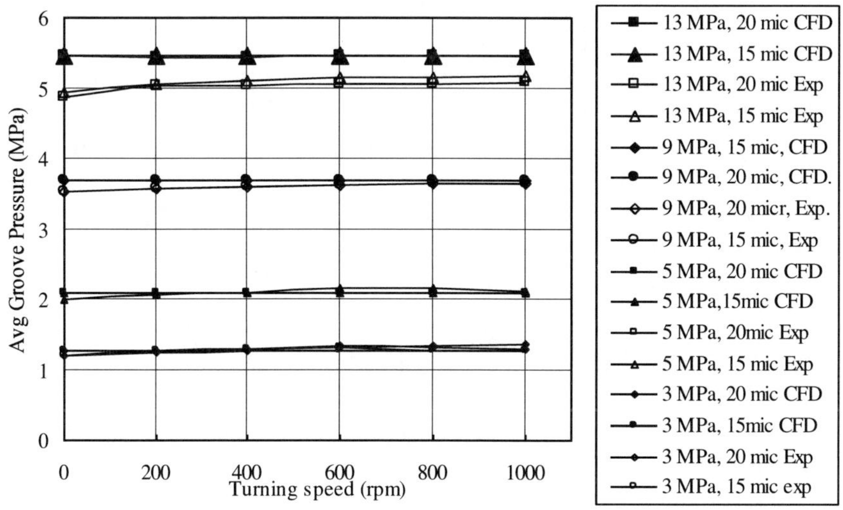

Figure 5.3.19. Averaged groove pressure at different inlet pressure, central clearance and turning speed (comparison between CFD and experiment).

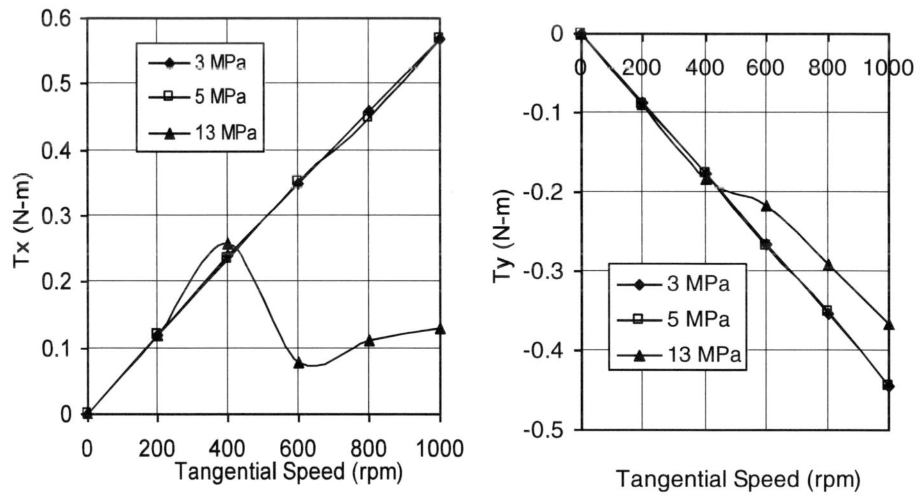

a) Torque Vs x-axis. b) Torque Vs y-axis.

Figure 5.3.20. X and Y axis numerical torque at different turning speeds and inlet pressures. Clearance 20 microns.

Pressure distribution below the non tilted slipper under dynamic conditions, it is expected to be rather similar to the one under static conditions. Nevertheless, as pump turning speed increases, pressure differential below the slipper becomes angular dependent, the pressure inside the groove is not constant anymore, experimentally and via numerical simulation, it has been found that groove pressure differential is approximately 0.3MPa at a pump turning speed of 1000 rpm. It is interesting to point out that at high turning speeds, the pressure measured experimentally at the four groove measuring points was fluctuating and therefore just the average pressure could be gathered. As a conclusion, it can be said that pressure distribution below the actual non tilted grooved slipper is not much affected by tangential velocity; as a result force will remain nearly the same as in static conditions. For the cases studied the variation found was in the range of 0.02 % (3 Mpa, 1000 rpm) to 0.08 % (13 MPa, 1000 rpm).

Figure 5.3.20 presents for a clearance of 20 microns, the torque at different turning speeds and pressures. It can be seen that, at low inlet pressures, torque is independent of pressure, but as the pressure increases the torque becomes dependent on pressure and turning speed. The explanation of this phenomena can be found when noticing that the magnitude of flow due to pressure difference (Poiseulle flow) is at some point of the same order of magnitude as the flow due to slipper movement (Couette flow), there is then a mutual adjustment between these two flows, as a result, the direction and magnitude of flow change at different angular position as a function of turning speed and pressure, pressure distribution below the slipper and shear stresses are also affected by this effect. On the other hand, at low inlet pressure, the magnitude of Poiseulle flow is low with respect to Couette flow and flow pattern is manly determined by Couette flow. Nevertheless, the torques acting over a flat slipper, are of a fraction of a Newton meter.

5.3.8.3.2. Non Tilted Dynamic Slipper, Effect on Slipper/Swash Plate Leakage, Experimental and Numerical Results

When measuring experimentally the slipper/plate leakage in dynamic conditions, needs to be considered the disk run out, since leakage strongly depends on central clearance between slipper and plate. Therefore for each turning speed and inlet pressure, the average clearance will have to be calculated to find out the real clearance slipper/plate. Figure 5.3.21b represents the measured average plate run out for a set of inlet pressures and turning speeds. It seems that for a given pressure, the increase of turning speed slightly decreases the disk run out, but it is needed to point out that as the pressure increase the clearance will also increase, due to plate displacement when submitted under pressure. The clearance slipper/plate was measured statically and for a pressure of 3 MPa. Such distance was measured at the highest point of the disk run out, to make sure that this distance was the minimum possible between slipper and plate for the corresponding pressure. When the pressure was increased, the clearance slipper/plate also increased and this plate displacement was measured via measuring the change in amplitude of the disk run out versus the amplitude found at 3 MPa. Figure 5.3.21a represents the plate displacement at different pressures.

To compare the experimentally measured leakage with the simulated one, the clearances need to be modified with plate run out and plate displacement, according to figure 5.3.21. Figure 5.3.22 represents the comparison between experimentally measured and numerically simulated leakage at modified clearances, being the initial static clearance of 15 microns. The modified clearance at each point is given by equation (5.3.87). The comparison shows a very good agreement.

$$\text{Modefied clearance} = \text{initial clearance} + \text{plate run out} + \text{plate displacement} \qquad (5.3.87)$$

Figure 5.3.23 presents the leakage calculated from numerical simulation, at fix gap and different turning speeds. It can be seen that, for a flat slipper at a fix clearance, turning speed does not affect the leakage. This is due to the fact that when the slipper turns around the pump swash plate, in half part of the slipper (0 to 180^0) flow is supported by the surface movement and on the other half part flow is opposed by surface movement and since the slipper is flat, these effect cancel out each other and as a result the effective leakage remain constant with turning speed. Therefore it can be concluded that, in dynamic conditions leakage variation is due to plate run out and plate displacement, not because of turning speed.

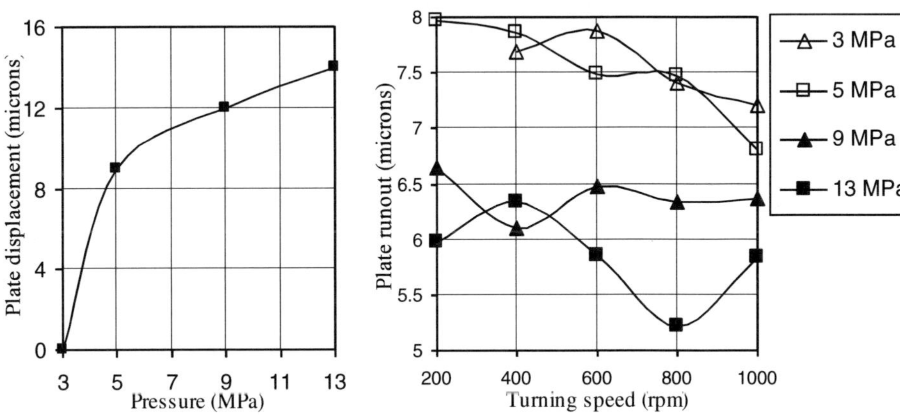

Plate displacement. b) Plate run out.

Figure 5.3.21. Experimental plate run out and plate displacement at different turning speeds and inlet pressures.

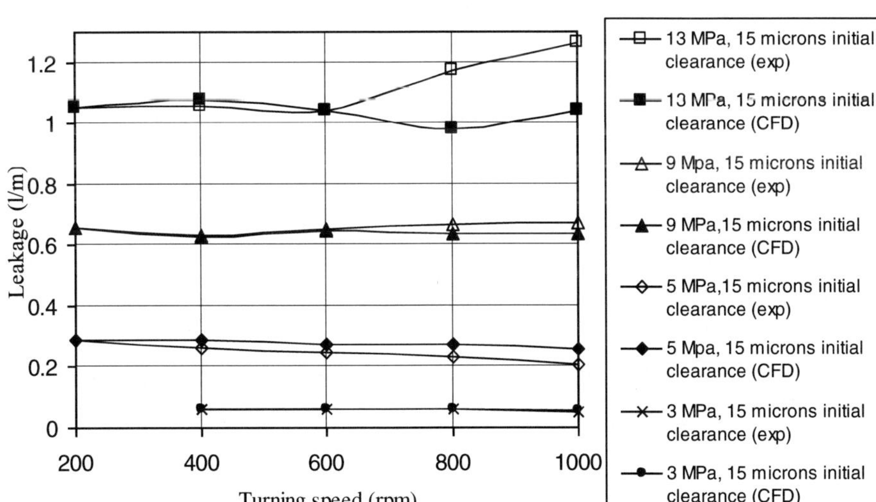

Figure 5.3.22. Comparison between measured and simulated leakage as a function of pump turning speed and for different inlet pressures. 15 microns initial clearance.

5.3.8.3.3. Non Tilted Slipper, Vorticity Inside the Groove Under Static and Dynamic Conditions

It is assumed that a groove maintains a constant pressure all along, to understand this assumption it is necessary to study the momentum transfer of the fluid particles inside the groove, and such momentum transfer entirely depends on the Vorticity created inside the slipper groove. In the present section, Vorticity is analyzed in both static and dynamic conditions.

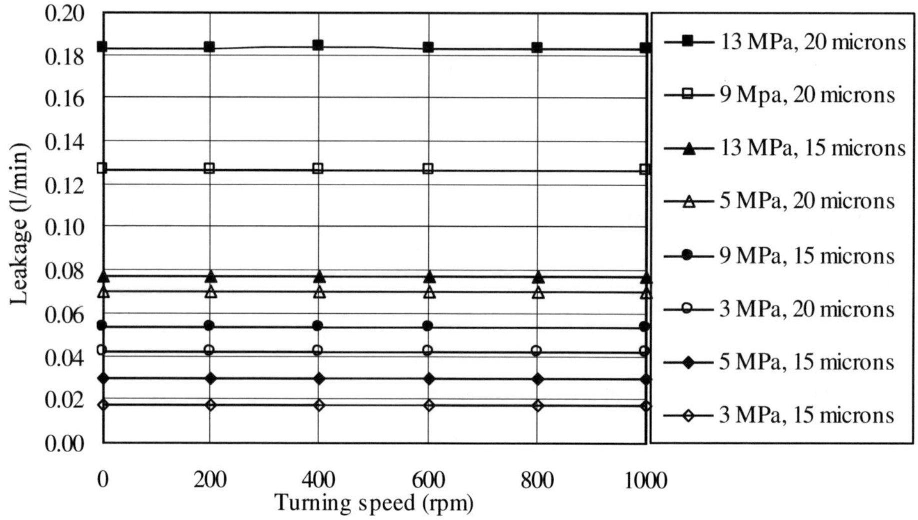

Figure 5.3.23. Slipper-Swash plate leakage versus turning speed at different inlet pressures and central clearances. Numerical results.

Under static conditions, the flow movement in the clearance slipper-plate is created by the pressure difference across the slipper and the vorticity exist only in r-z plane (see figure 5.3.7), then, there is no need to transfer momentum along angular direction.

Figure 5.3.24 represents the streamlines inside the groove in the r-z plane at different groove depths. It can be noticed, there exist an interaction between the external flow and the recirculating one inside the groove and exist a big vortex at the groove bottom. Although not stated in figure 5.3.24, it is found that the position of this vortex is independent from inlet pressure and clearance, being the groove depth the most relevant parameter. It is noticed from figure 5.3.24 that there exists a saddle point at 0.2 mm in z direction versus the slipper face.

The position of this saddle point remains constant regardless of groove depth or inlet pressure. Since the flow across the slipper and across the groove has to be the same, and for a given slipper/plate clearance the flow depends on the distance between the slipper face and the saddle point, and such distance remains constant for groove depths higher than 0.2 mm, it can be concluded that for the groove studied the leakage will remain constant if groove depths are higher than 0.2mm, understanding that pressure and slipper plate clearance remain constant. On the other hand, for groove depths lower than 0.2mm the leakage will sharply decrease, as presented in figure 5.3.17.

In section 5.3.8.3.1, it has been demonstrated experimentally and via numerical simulation that pressure inside the non tilted slipper groove remains constant in angular direction regardless of slipper tangential speed. The explanation of why pressure remains constant under dynamic conditions needs to be found in the extremely quick momentum interchange between fluid particles inside the groove. Such momentum interchange is enhanced by the action of recirculating fluid inside the groove which will now depend on slipper tangential velocity among other parameters.

The velocity on the slipper surface, at a given radius (r) and angular position (θ), due to slipper rotation with respect to swash plate axis is given by equations (5.3.88-5.3.89).

$$V_r = R_{sw}\ \sigma \sin(\theta) \tag{5.3.88}$$

$$V_\theta = R_{sw}\ \sigma \cos(\theta) + r\,\sigma \tag{5.3.89}$$

When analyzing the Vorticity in dynamic conditions, the flow inside the groove, which was recirculating in r-z plane in static conditions, is expected in dynamic conditions to recirculate in angular direction as well due to surface movement. Figure 5.3.25 represents a three dimensional streamlines flow pattern in the slipper-swash plate groove at 13 MPa and 200 rpm rotation speed. It can be seen that in dynamic conditions, there exist two vortexes inside the groove, a primary vortex at groove bottom and a secondary vortex near groove entrance.

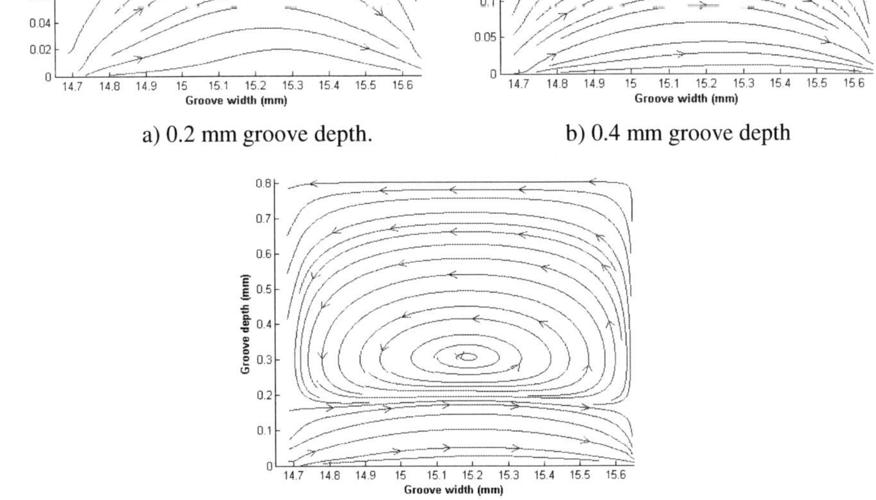

Figure 5.3.24. Streamlines in r-z plane in side groove at different groove depth in static conditions. 10 MPa, 20 microns clearance (CFD).

Primary vortex is created by the interaction between the flow coming into the groove and the no slipping boundary conditions at slipper walls. The primary vortex is displaced along the θ direction by the groove bottom surface movement defined in equation (5.3.89). The rotation of the primary vortex is directed by the radial component of slipper surface movement, given by equation (5.3.88).

Since the radial movement of the groove bottom surface is a sine wave, as a result, the primary vortex rotation will also change as a sine wave, therefore the vortex will rotate anticlockwise from 0° to 180° and clockwise from 180° to 360° angular position.

This primary vortex is the most relevant one since exist at all angular positions and play the most important role in stabilizing the pressure. Despite the fact that the secondary vortex is created by the mutual adjustment between the external flow and the primary vortex, the effect of the incoming flow is of higher relevance. As a result, the structure of the secondary vortex depends on the direction of slipper/plate flow, which depends upon inlet pressure and tangential speed (operating condition). Therefore the vortex dimensions and turning speed will be completely different at different angular positions. In some conditions the vortex might also disappear.

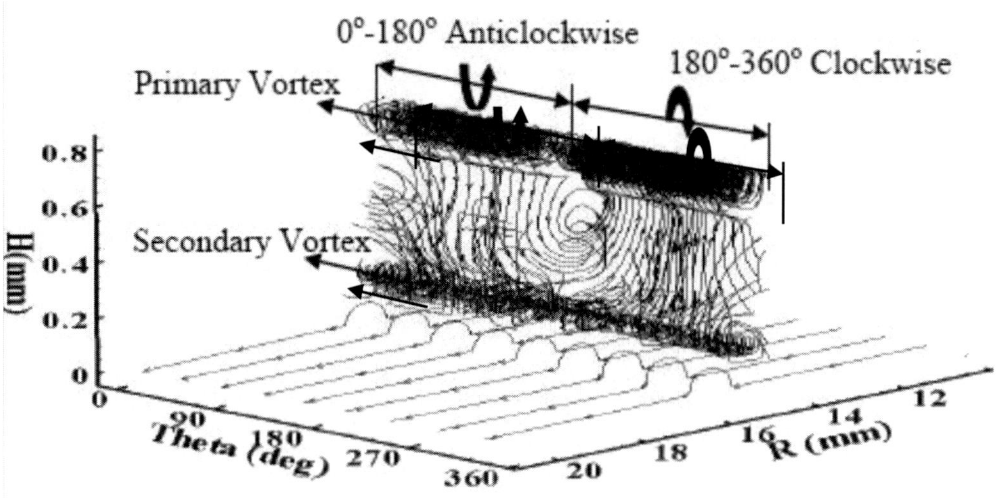

Figure 5.3.25. Streamlines plot at 13 MPa, 200 rpm and 20 microns clearance.

Before further analyzing the vortex, it is important to understand the direction and relative magnitude of the slipper/plate flow (radial flow) as a function of angular position at different inlet pressures and turning speeds.

The structure of vortexes is determined by the magnitude and direction of the shear stresses which depend on velocity gradient.

Figure 5.3.26 represents the normalized slipper/plate flow (radial flow) as a function of angular position at different tangential speeds (200 – 1000 rpm) and inlet pressures (3 and 13MPa). It can be seen from figure 5.3.26 that depending upon inlet pressure and tangential speed, the direction of slipper/plate flow changes from positive (outward) to negative (inward) at different angular positions.

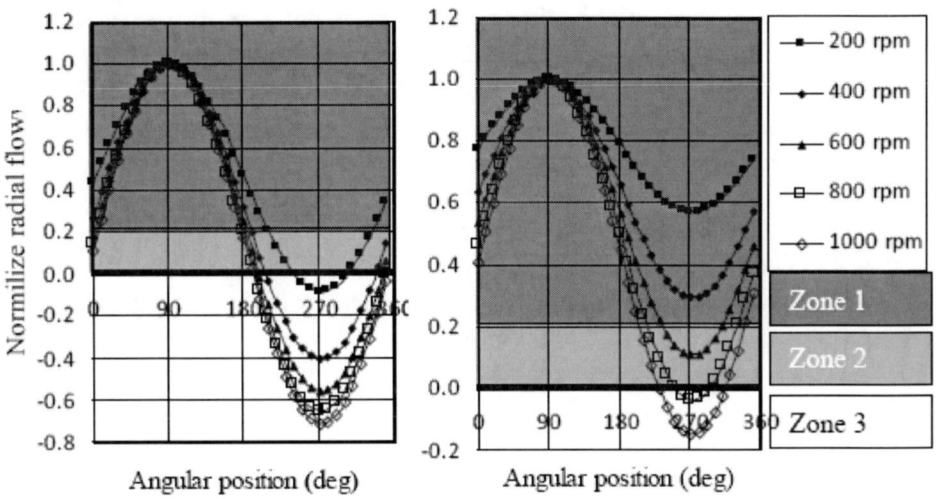

a) 3 MPa, 20 microns clearance. b) 13 MPa, 20 microns clearance.

Figure 5.3.26. Normilize radial flow at different angular position of slipper.

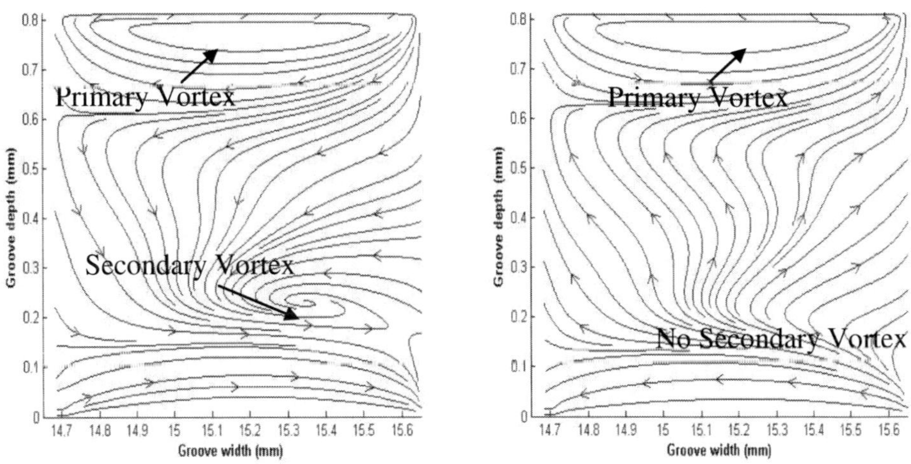

a) 90° angular position (Zone 1). b) 270° angular position. (Zone 3)

Figure 5.3.27. Streamlines plot in r-z plane at different angular position for 13 MPa inlet pressure, 1000 rpm rotation and 20 microns clearance.

By taking into account the mutual adjustment between Poiseulle and Couette flow, the slipper can be divided into three flow zones (see figure 5.3.26):

1. Zone 1 [Net radial flow = Poiseulle flow + Couette flow] – In this zone, the flow is radially outward. Poiseulle and Couette flows might be whether having the same direction (slipper leading edge) or opposite directions (slipper trailing edge), in any case the magnitude of Poiseulle flow at each angular position is much higher than the magnitude of Couette flow. Under such conditions, the velocity gradient at the groove face is at its highest, therefore the secondary vortex rotation speed will also be maximum. As a result it is expected that the momentum transfer between particles

at different angular positions will also be higher. In this zone the secondary vortex is helping the primary vortex to stabilize the pressure inside the groove.

2. Zone 2 [Net radial flow = Poiseulle flow - Couette flow; & (Poiseulle flow > Couette flow)] – In this zone, Poiseulle and Couette flows have the same order of magnitude and opposite directions, as a result, the net flow is very small. The secondary vortex will be very weak and tending to disappear.
3. Zone 3 [Net radial flow = Poiseulle flow - Couette flow; & (Poiseulle flow < Couette flow)] – In this zone, Couette flow is radially inward and the magnitude of Couette flow is slightly higher than Poiseulle flow. Notice for example from figure 5.3.26b, at 13 Mpa and 1000 rpm, that the magnitude of inward flow is about 15 % of the magnitude of outward flow of zone 1. Therefore inward net flow velocity in zone 3 is very weak; as a result the velocity gradient is not big enough to create a secondary vortex.

As a conclusion figure 5.3.26 can be used to quickly visualize the existence of vortexes at different slipper groove angular positions for a set of inlet pressures and turning speeds. In zone 1 both vortexes exist and in zones 2 and 3 the secondary vortex is whether nonexistent or very weak. Figure 5.3.27a,b present a 2-D streamlines plot corresponding to zone 1 (90°) and zone 3 (270°) at 13Mpa and 1000 rpm, such figures corroborate the statement previously defined.

5.3.8.4. Tilt Slipper, Static Performance

5.3.8.4.1. Tilt Static Slipper Leakage

Leakage at every slipper angular position, under static conditions, can be studied using equation (5.3.68), once the numerical integration is done. This results in figure 5.3.28 where can clearly be seen that as tilt increases the difference between the front and back slipper leakage also increases, please notice that the tilts used are much greater than those that occur in practice.

What is most remarkable is that for the range of slipper spin speed values studied, the total leakage does not depend upon the slipper turning speed; it just depends upon the pressure differential, the slipper central gap and tilt. Figure 5.3.29 represents the leakage given as a percentage increase plotted relative to the slipper non tilted position. It is evident that leakage increases with slipper tilt, but for a given central clearance such an increase does not depend on the pressure differential applied to the slipper.

Using test rig 1 a set of leakage measurements for three central clearances of 10, 15 and 20 microns, and for a range of slipper tilts and inlet pressures were performed. Some of the experimental results, for a central clearance of 15 microns, are shown in figure 5.3.30. It can clearly be noticed that as tilt increases then leakage also increases and leakage increases as pressure increase as expected. Although not presented in this chapter, it is also very interesting to point out that at a central clearance of 10 microns, and for any given pressure, the leakage seemed to first increase and then decrease with slipper tilt. The explanation of this particular behaviour is to be found when realizing that at some clearances and tilts the flow at the entrance of the slipper first land changes from reattached to separated, reducing the flow section and therefore reducing the leakage flow.

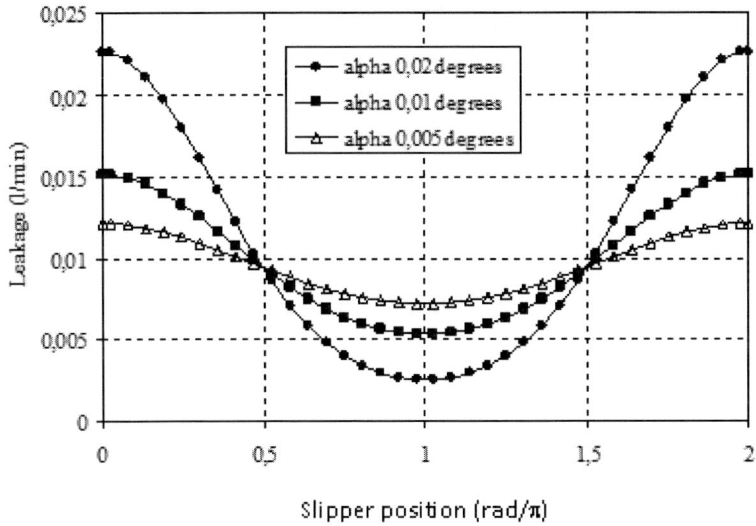

Figure 5.3.28. Leakage flow, (analytical), central distance $h_{02} = 15$ microns, $P_{inlet} = 10$ MPa.

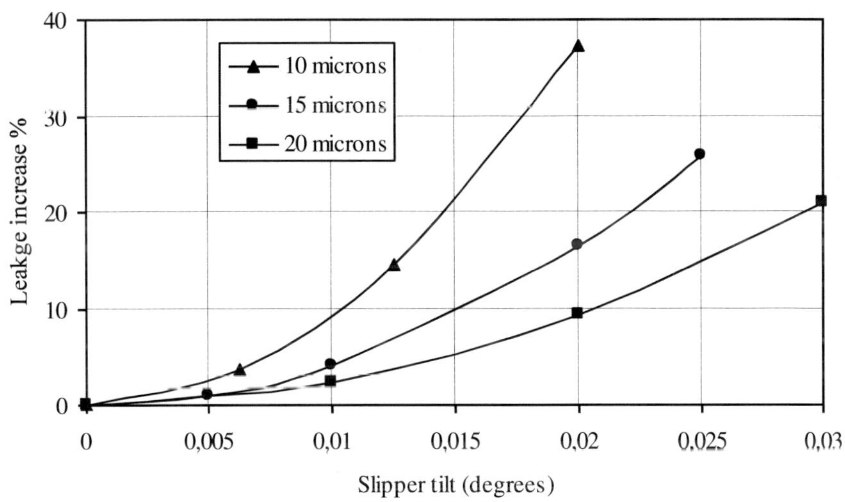

Figure 5.3.29. Slipper leakage percentage increase versus non tilt slipper. Analytical.

According to the theory presented the leakage increase, given as a percentage of non tilted slipper leakage, should be independent on inlet pressure, as shown in figure 5.3.29. When leakage in figure 5.3.30 is represented as a percentage of the non tilted slipper leakage, it can be seen that for a given central clearance, all the different curves can be brought together. Therefore figure 5.3.31 presents the trend curve for all the central clearances studied which are compared with the theoretical predictions.

It can be seen that a good agreement is found, especially at the very low tilts which exist in practice. From these results it can be stated that leakage percentage increase versus a non tilted slipper is mostly independent on the inlet pressure. Nevertheless it has been found experimentally that as the inlet pressure increases the percentage increase trend line curve tends to slightly increase beyond the predictions.

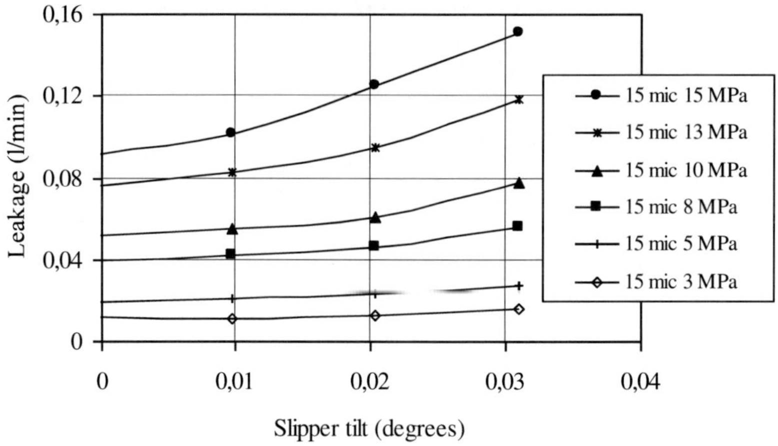

Figure 5.3.30 Experimental leakage as a function of slipper tilt and inlet pressure. Central clearance 15 microns.

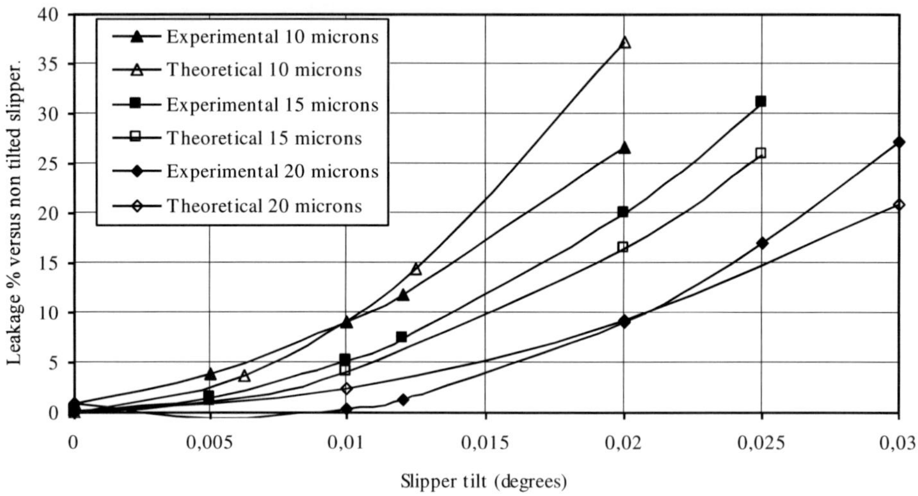

Figure 5.3.31. Comparison experimental and theoretical leakages, given as a percentage of the non tilted slipper leakage.

Slipper leakage was measured experimentally by using the test rig 1. A comparison between the leakage obtained via computer numerical simulation and experimentally is presented under static conditions and for a set of inlet pressures and clearances in figure 5.3.32. While dealing with such a tiny clearance, roughness plays an important role in determining the actual slipper-plate clearance. Surface roughness measurements clarified that average variation in surface finish is typically 1 micron for both materials. Then the measured transducer clearance needs to be modified by the surface roughness in order to get the true clearance between slipper and plate, as defined in equation (5.3. 90).

True clearance = Measured clearance+2*Average roughness of the surface (5.3.90)

From figure 5.3.32, it is noticed that leakage slightly increases with slipper tilt. The reason behind such increment can be understood when noticed that as tilt increases the overall flow resistance created by the slipper slightly decreases. When comparing the experimental and numerical results, it is seen a very good agreement.

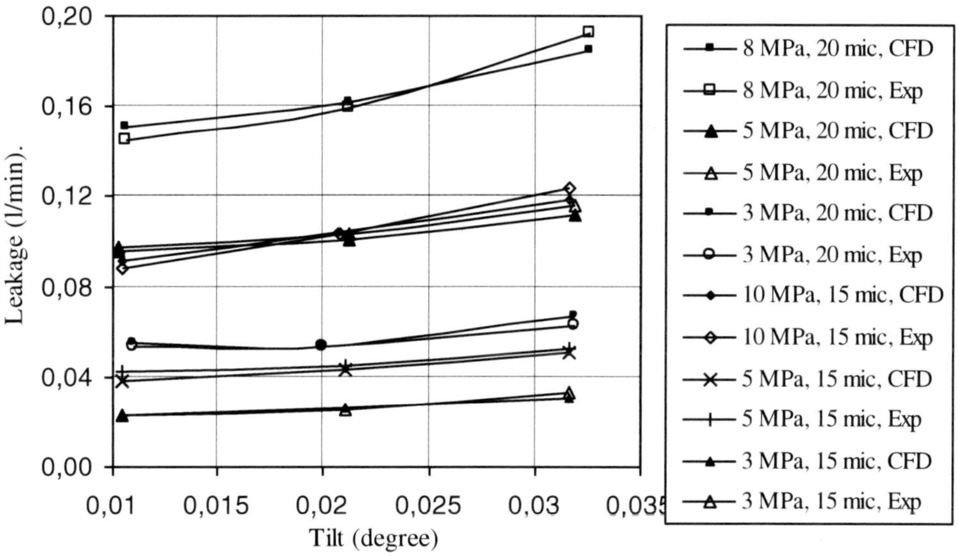

Figure 5.3.32. Slipper leakage with tilt at different inlet pressures and central clearances (Comparison between numerical and experimental).

5.3.8.4.2. Tilt Static Slipper. Pressure Distribution

Regarding the pressure distribution, the equations presented (5.3.60-5.3.63) or the generic one (5.3.67), are capable of predicting the pressure at all points below the slipper, as it is represented in figure 5.3.33. It has to be said that due to the consideration of radial flow, the theoretical pressure differential inside the slipper groove is slightly higher than what has been found experimentally. In fact the experiments have revealed that the pressure inside the groove is mostly constant for the set of tilts and central clearances studied. For a given central clearance the groove pressure, although constant at all four pick up points, tends to decrease as tilt increases. This is shown in figure 5.3.34.

Also represented in figure 5.3.34 is the theoretical pressure variation. In agreement with the theory, the pressure inside the groove does change with angular position. The pressure at angle $\theta = 0$ is computed, and represents the analytical minimum pressure on the slipper groove.

Theoretically the pressure inside the groove increases for a tilted slipper as the slipper clearance decreases, and the question arises as to which of the range of theoretical pressures is likely to appear experimentally. Thanks to the experimentation undertaken it can be said that the minimum theoretical pressure is the most likely to appear in reality. A well-designed groove geometry allows flow from the theoretical groove high pressure points to move almost instantaneously towards the groove theoretical low pressure points, thus equalising the pressure within the groove.

It can then be concluded that for the groove studied, a rate of momentum exchange exists between fluid particles at the top of the groove. Although not presented here, it has been observed that for smaller central clearances the pressure decay with slipper tilt inside the groove is higher. A very good agreement between theory and experimentation is found under all conditions studied.

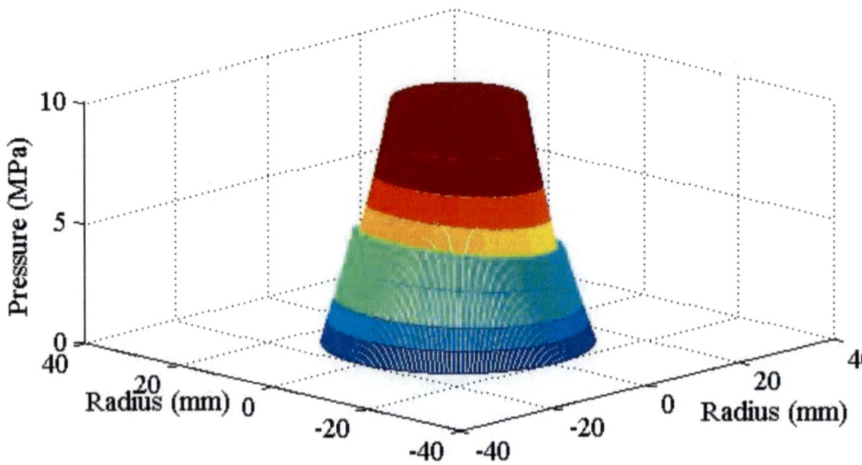

Figure 5.3.33. Theoretical pressure distribution below the slipper. h_{02}=15 microns; α =0,01 deg; ω = 25,12 rad/s; 10 MPa.

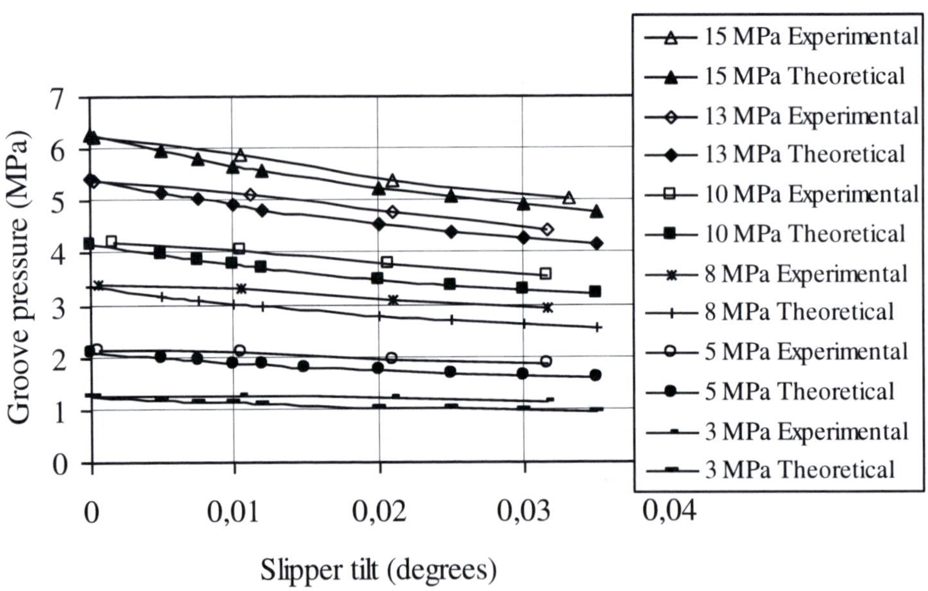

Figure 5.3.34. Groove pressure decrease with tilt. Comparison experimental and analytical results. 15 microns central clearance.

For the slipper with groove studied, and when working under expected operating conditions, the pressure inside the groove is maintained constant. However it has been found experimentally that as the tilt and slipper inlet pressure increases, some pressure differential inside the groove can be expected. To illustrate this point figure 5.3.35 is presented, showing that when the slipper is operating outside the normal working conditions, the flow circulation around the groove and therefore the momentum exchange around the groove is not enough to maintain constant pressure. Figure 5.3.35 also demonstrates that if the groove depth is decreased then a much bigger pressure differential inside the groove is to be expected. Notice that an increase of inlet pressure also creates a higher pressure differential inside the groove.

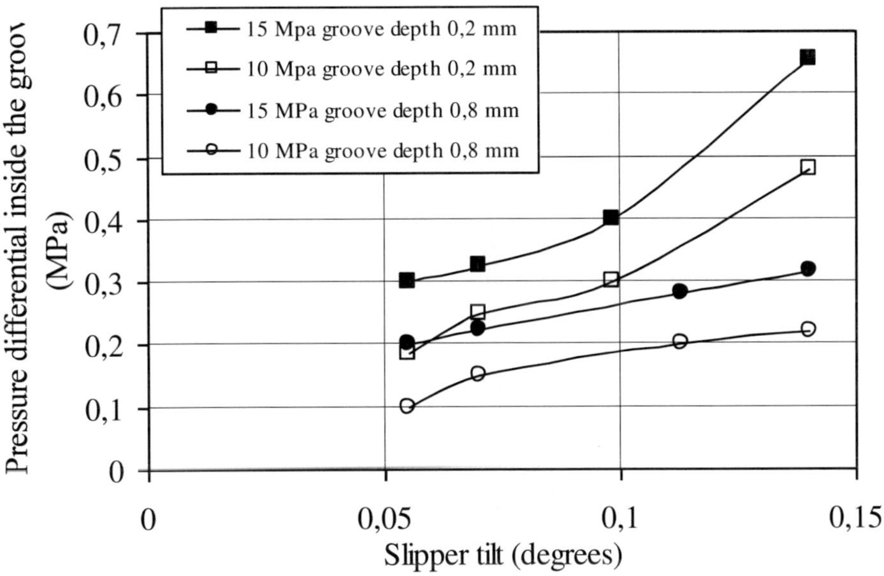

Figure 5.3.35 Operating conditions under which pressure differential inside the groove can be expected. Experimental.

5.3.8.4.3. Tilt Static Slipper, Vorticity Inside the Groove

The vorticity inside the groove when slipper is placed parallel to the swash plate was explained in Kumar et al [51]. Vorticity in tilted static conditions is found to be far more complex than for the flat slipper case. The flow inside the groove is highly angular and depends on central clearance, slipper tilt and input pressure. Figure 5.3.36 present the three dimensional stream line plot at 30 microns central clearance, 0.03 degree tilt and 5 MPa inlet pressure. To better understand the vortexes and its evolution, figure 5.3.37 present the 2D stream lines plots corresponding to figure 5.3.36. As can be seen, there exist three vortexes, one at the entrance of the groove (Entrance Vortex), a second at the groove exit (Exit Vortex) and a third at the groove bottom (Bottom Vortex).

It can be noticed from figure 5.3.36 that the top two vortexes exist throughout the angular positions ($0^o – 360^o$). The entrance vortex tends to move towards the outer radius and towards the bottom of the groove, when moving from 180° angular position to 0° angular position. The vortex displacement can be understood by the fact that when moving from 180° towards 0° angular position, higher amount of flow tends to enter inside the groove, then the available slipper/plate gap is higher, such flow increase push the entrance vortex towards the groove bottom and towards a higher radius position, while tending to increase its diameter.

It has also been noticed that an increase of pressure and or tilt, produce an increase of the entrance vortex turning speed. Regarding the entrance vortex diameter along the groove angular position, from the numerical simulation performed, it is stated that the higher the inlet pressure the bigger the vortex diameter will be. An increase of tilt brings a decrease on the vortex diameter, especially at slipper 180°, since around this angular position, the leakage flow will be at its minimum. For slipper angular position between 160°-0°, the entrance vortex diameter increase with the increase of tilt, then for the cases studied, under these angular positions the leakage flow increases with increase of tilt.

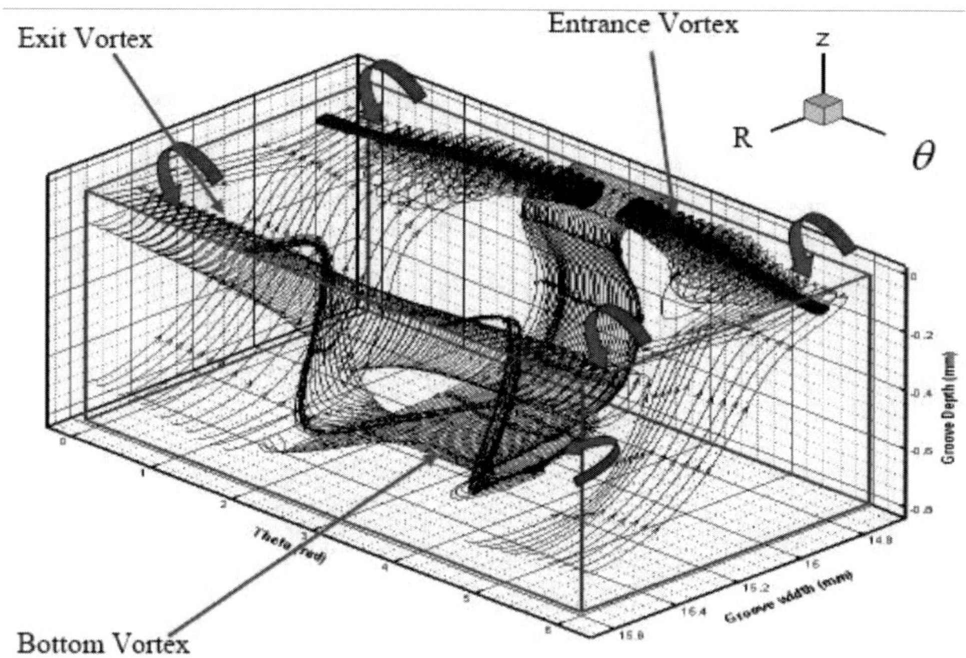

Figure 5.3.36. 3-D stream line plot inside the groove at 5 MPa inlet pressure, 0.03 degree tilt and 30 microns central clearance.

When studying the exit vortex, the first thing to be noticed is that for the cases studies, such vortex maintains its shape rather constant along its 360°, regardless of slipper tilt and inlet pressure. Regarding the exact vortex variation with tilt and pressure, it can be seen from figure 5.3.37a, that a tilt increase brings a small decrease in vortex diameter, especially at 180 degrees, while a pressure increase create a negligible effect on the exit vortex.

The evolution of exit and entrance vortexes is fully linked with the evolution of the bottom vortex. The bottom vortex exists along an angular position when huge momentum transfer between particles is needed. This is why at low tilts and low pressures the bottom vortex length is smaller than at higher tilts and pressures. The conclusion is, that the bottom vortex job is to maintain a constant pressure along the groove bottom, this is why the vortex transfers momentum to a longer distances when needed, this is for higher tilts and or higher pressures. For the cases studied the effect of tilt on the bottom vortex is more relevant than the effect of the inlet pressure.

A very interesting point regarding the bottom vortex is the movement of the vortex central core. At slipper 180°, leakage flow comes into the groove, pushing the bottom vortex towards the groove inner radius; the vortex at this angular position is rather tiny and close to the groove bottom. As the bottom vortex moves along the groove angular positions, the vortex central core moves from the groove inner radius towards the groove outer radius, creating a horse shoe shape. As soon as the vortex reaches the groove outer radius, changes its direction in 90° and the flow leaves the groove. Regarding the bottom vortex dimensions, at 180° it is noticed, the vortex is small, as the vortex moves in angular direction, its dimension first increases and then, just before changing direction and leaving the groove, the vortex abruptly decreases its diameter and disappears.

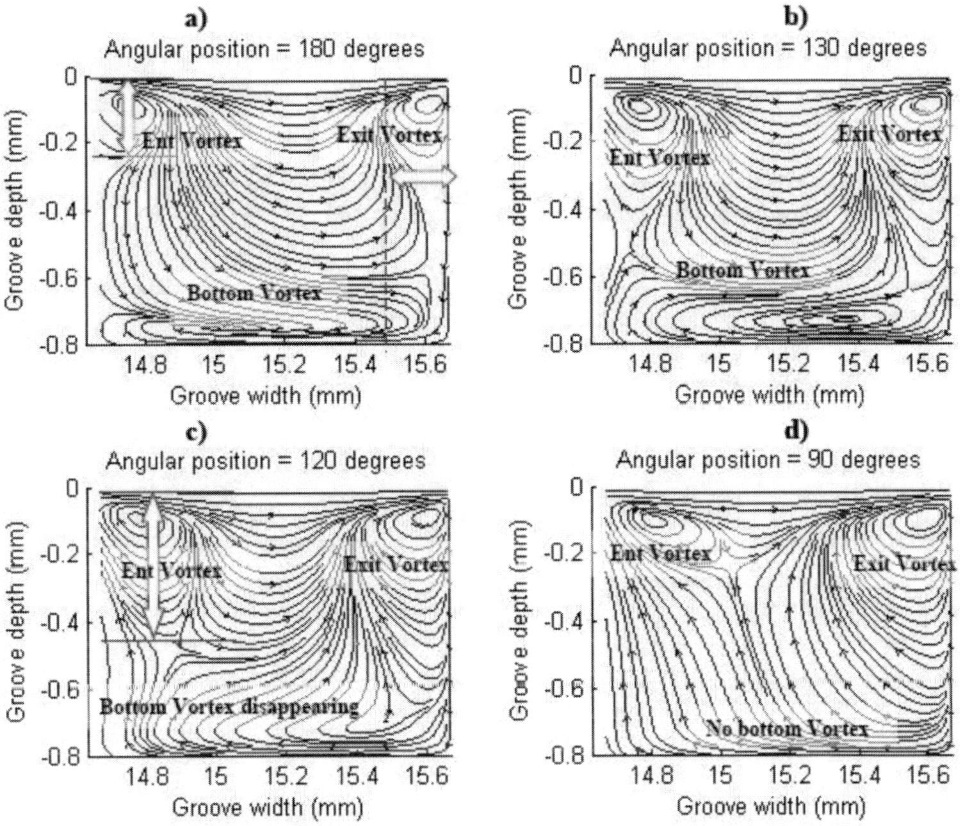

Figure 5.3.37. 2-D stream line plot inside the groove at 5 MPa inlet pressure, 0.03 degree tilt and 30 microns central clearance.

It must be recalled that a good vortex understanding is decisive to understand how a groove behaves and the benefits of its behaviour. For the present case, and according to figures 5.3.36; 5.3.37, it can be stated that the groove pressure is maintained constant all along, thanks to the existence of the entrance and exit vortexes. Nevertheless, at the slipper trailing edge, a much higher momentum interchange between particles is needed, and this is why, a third vortex, the bottom vortex, appears during a certain angular position. Such angular position vortex increases with the increase of slipper inlet pressure and the increase of slipper tilt.

5.3.8.5. Tilt Slipper, Dynamic Performance

Slippers are designed to run almost parallel to the swash plate. This means that lift is created mostly hydrostatically, hydrodynamic lift being just an small percentage (around 5%) of the total lift. In this section the effect of tangential velocity on tilted slippers with a groove will be discussed. Leakage and average pressure distribution inside the groove shall be presented as a function of tilt and tangential velocity. Figure 5.3.38 presents the measured average pressure inside the slipper groove for a set of inlet pressures and turning speeds, the film thickness has been assessed by taking into account the weighed average of the disk runout and the disk mean axial displacement.

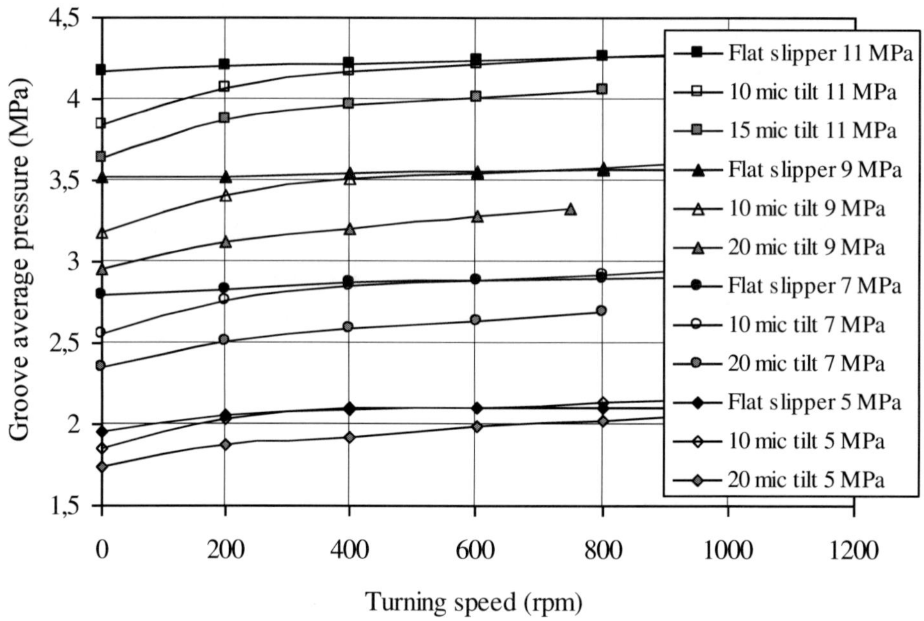

Figure 5.3.38. Measured average pressure inside the groove for several inlet pressures and initial static slipper/plate clearances and tilts.

The results show that for the non-tilted slipper case, the pressure at the four cardinal points of the groove remains the same and this pressure slightly increases with turning speed, demonstrating that the lift force will remain constant with turning speed. Also during experimental work it was found that as the clearance increases, the average groove pressure slightly decreases. Such an effect is well explained when considering that an increase of the film thickness creates an increase of flow and the pressure decay along the slipper first land depends directly on the shear stresses on the slipper face, which increase with the flow. Figure 5.3.38 also presents the effect on the groove average pressure, with slipper tilt, where it is demonstrated that as tilt increases the average pressure inside the groove decreases. The average pressure will quickly increase with the increase of turning speed, demonstrating that for slippers with tilt the increase of turning speed will bring an increase of lift. It is also interesting to realize that the results presented in Figure 5.3.38 are very much dependent of the clearance, except for the non-tilted slipper case.

It is very important to point out that the effect of tangential velocity increases the pressure difference inside the groove between the leading and the trailing edge of the slipper.

This increase in pressure difference, although small, will be higher for higher clearances, as figure 5.3.39 presents, demonstrating that at high clearances, the actual groove depth is not enough to maintain a constant pressure along its path.

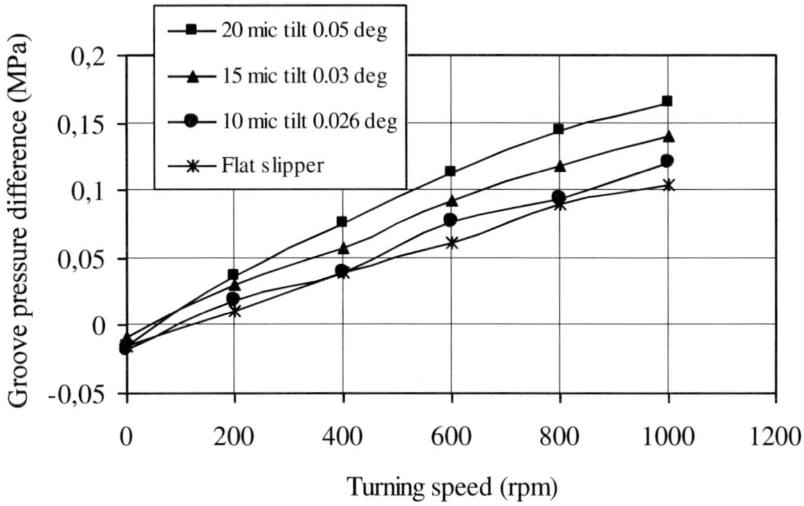

Figure 5.3.39. Measured pressure difference between the trailing and the leading edge of the slipper groove, as a function of slipper tilt and turning speed.

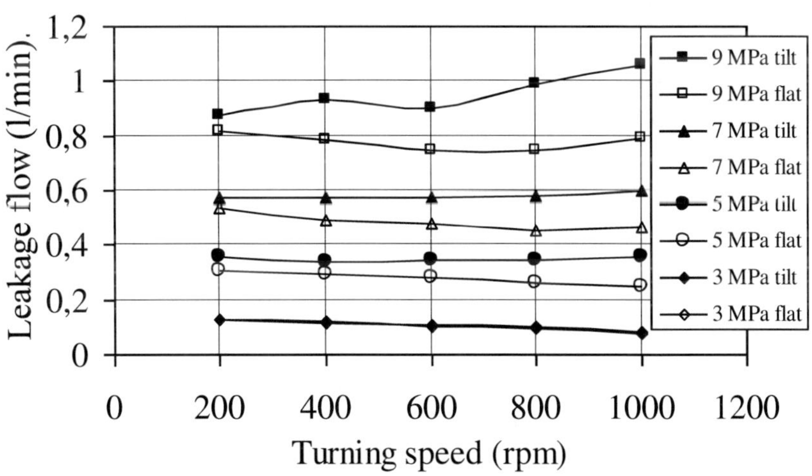

Figure 5.3.40. Comparison between flat and tilt slipper performance with turning speed. Initial static central clearance 15 microns, 0,03 degrees tilt. Experimental.

Figure 5.3.40 presents the leakage variation with rotational speed for a central initial static clearance of 15 microns and with a tilt of 0.03 degrees. The results are compared with the ones obtained for the non-tilted slipper at the same initial static clearance. It clearly shows that the leakage obtained with a tilted slipper is always higher than the one obtained for the non-tilted case. Since it has been earlier demonstrated that the leakage for the non-tilted slipper remains constant with turning speed, figure 5.3.40 demonstrates that the effect of turning speed on a tilt slipper, tends to increase the leakage flow rate. Such an increase appears to be more relevant at higher pressures. This effect has been experimentally observed in all the tests performed, yet, it was noticed that at small clearances, the heat generated by the test rig was being transferred to the fluid thereby decreasing the viscosity and therefore increasing the overall leakage flow. At high clearances nevertheless, the flow passing through the test rig, was big enough to dissipate the heat without suffering a relevant temperature increase. This is why the graph presented in figure 5.3.40 is for an initial static central clearance of 15 microns, its equivalent average dynamic central clearance, once axial displacement and plate runout was considered, being 21 microns.

5.3.9. Conclusion

1. A new set of equations and tests are presented capable of directly evaluating leakage flow rate, the hydrostatic pressure distribution and lift on a grooved slipper having an ostensibly constant clearance. In practice, experimental measurements must consider:
   - surface roughness
   - pressure tapping point diameter and its relative position between slipper and base
   - test rig small displacement under pressure.

The hydrostatic theoretical characteristics of a grooved slipper have been validated experimentally.

The equations have been generalised to be used for a slipper with any number of lands. Results were achieved for slipper tilts far beyond those that would exist in practice, had the test slipper geometry been used in a pump application.

2. It is demonstrated that the equations can be used to optimise the slipper design and clarifies the effect on the slipper force and leakage when groove position and dimensions are modified.
3. Lift is higher when the groove is located along the inner land and decreases as the groove move towards the external radius. However, leakage increases as the groove moves towards the slipper pocket. The inclusion of a groove in a slipper will result in an increase of leakage flow rate.
4. For a slipper held parallel to the plate, is has been demonstrated via numerical analysis and experimentally that the leakage flow rate will remain constant and therefore independent of turning speed. For the case of tilted slippers, the experiments have demonstrated that the increase of plate turning speed will bring a small increase in leakage flow rate.
5. For both a non-tilted and tilted slipper, the pressure difference between the trailing and leading edge of the slipper will increase with turning speed. For the tilted slipper case, the average pressure inside the groove sharply increases with turning speed and such an increase is almost negligible for the flat slipper case. It is therefore to be expected that the lift force onto a tilt slipper will increase as turning speed increases, while it will remain rather constant for the non-tilted slipper case.

6. A particular feature of the design equations presented is that they can be used to determine the groove geometry for optimum lift at a specified leakage flow rate. A methodology has been established to design grooved systems; therefore a door to use the same methodology for other applications is opened.
7. It has been experimentally demonstrated that the well-chosen groove depth resulted in a constant pressure around the groove and therefore a groove needs to be properly designed to avoid a pressure differential effect.
8. In static conditions, it is found that the normalized pressure inside the groove is independent of inlet pressure, force acting on the slipper and leakage are a linear function of pressure. Leakage strongly depends on clearance slipper/plate while slipper pressure distribution is for the cases studied, independent of clearance.
9. Under dynamics conditions, the tangential speed has negligible effect on the force acting over the slipper. It creates nevertheless a small torque respect to the two slipper main axis. At higher speed, there exists a noticeable pressure differential inside the groove. Leakage is independent on turning speed.
10. Vorticity inside the non tilted slipper groove has been studied to analyze the momentum transfer inside the groove. In general two forced vortexes appear inside the groove. The primary one located at the groove bottom is the most responsible for maintaining the pressure along the groove in angular direction. This vortex exist under all working condition, is created by mutual adjustment between slipper/plate flow and no slipping condition on slipper groove wall. A secondary vortex is also near the groove face. It existence is due to interaction between slipper/plate flow and primary vortex. This secondary vortex exists only in the region of higher velocity gradient.
11. Under tilted static conditions, pressure is found to be very stable along the angular direction in presence of the groove. The maximum pressure differential across the slipper radius, for inlet pressure 10 MPa, 0.042° tilt and 15μm central clearance, decreases from 0.3 MPa to 0.03 MPa due to the presence of a groove.
12. As the slipper tilt is considered along the X-axis, the torque with respect to X-axis is found to be zero. On the other hand there exists Y directional torque. The magnitude of the Y torque is found to be increasing with the increase of tilt.
13. Slipper leakage is found to be a strong function of clearance as it was found in Kumar et al [51] for flat slipper. In fact, slipper leakage is a function of the clearance to the power 3, see Bergada et al [30; 55] . Slipper leakage increases with the increase of tilt.
14. Under tilt conditions, it is found, there exist three vortexes inside the groove, two at groove top edges and one at the bottom of the groove. The existence of the bottom vortex depends on tilt and inlet pressure. At higher tilt and higher pressure, the angular length of the bottom vortex increases. The bottom vortex appears in the locations where a huge momentum interchange between particles is needed. The two small vortexes appearing at the groove top edges remain rather constant in shape along the slipper groove, tilt and inlet pressure have a second order effect on the top vortexes.

5.3.10. References

[1] Fisher, M.J. (1962). A theoretical determination of some characteristics of a tilted hydrostatic slipper bearing. *B.H.R.A.* Rep. RR 728 April 1962.

[2] Böinghoff, O. (1977). Untersuchen zum Reibungsverhalten der Gleitschuhe in Schrägscheiben-Axialkolbenmascinen. *VDI-Forschungsheft* 584. VDI-Verlag. 1-46.

[3] Hooke C.J. , Kakoullis Y.P. (1978). The lubrication of slippers on axial piston pumps. *5th International Fluid Power Symposium September,* B2-(13-26) Durham, England.

[4] Hooke C.J. , Kakoullis Y.P. (1981). The effects of centrifugal load and ball friction on the lubrication of slippers in axial piston pumps. *6th International Fluid Power Symposium*, 179-191, Cambridge, England.

[5] Iboshi N., Yamaguchi A. (1982). Characteristics of a slipper Bearing for swash plate type axial piston pumps and motors, theoretical analysis. *Bulletin of the JSME*, 25:210, 1921-1930.

[6] Iboshi N., Yamaguchi A. (1983). Characteristics of a slipper Bearing for swash plate type axial piston pumps and motors, experiment. *Bulletin of the JSME*, 26:219, 1583-1589.

[7] Iboshi N. (1986). Characteristics of a slipper Bearing for swash plate type axial piston pumps and motors, Design method for a slipper with a minimum Power loss in fluid lubrication. *Bulletin of the JSME*, 29:254.

[8] Hooke C.J., Kakoullis Y.P. (1983). The effects of non flatness on the performance of slippers in axial piston pumps. *Proceedings of the Institution of Mechanical Engineers*, 197 C, 239-247.

[9] Hooke C.J., Li K.Y. (1988). The lubrication of overclamped slippers in axial piston pumps centrally loaded behaviour. *Proceedings of the Institution of Mechanical Engineers* 202: C4, 287-293.

[10] Hooke C.J., Li K.Y. (1989). The lubrication of slippers in axial piston pumps and motors. The effect of tilting couples. Proceedings of the Institution of Mechanical Engineers, 203:C, 343-350.

[11] Takahashi K. Ishizawa S. (1989).Viscous flow between parallel disks with time varying gap width and central fluid source. *JHPS International Symposium on Fluid Power*, Tokyo, 407-414.

[12] Li K.Y., Hooke C.J. (1991). A note on the lubrication of composite slippers in water based axial piston pumps and motors. *Wear*, 147, 431-437.

[13] Koc. E., Hooke C.J., Li K.Y. (1992). Slipper balance in axial piston pumps and motors. Trans ASME, *Journal of Tribology,* 114, 766-772.

[14] Kobayashi, S., Hirose, M., Hatsue, J., Ikeya M. (1988). Friction characteristics of a ball joint in the swashplate type axial piston motor. *Proc Eighth International Symposium on Fluid Powe*r, Birmingham, England. J2-565-592,

[15] Harris RM. Edge KA. And Tilley DG. (1993). Predicting the behaviour of slipper pads in swash plate-type axial piston pumps. ASME Winter Annual Meeting. New Orleans, Louisiana. November 28-December 3, 1-9.

[16] Harris RM. Edge KA. And Tilley DG. (1996). Predicting the behaviour of slipper pads in swash plate-type axial piston pumps. *J. Dyn. Syst. Meas. Control* 114: 766-772.

[17] Koc E., Hooke C.J. (1996). Investigation into the effects of orifice size, offset and oveclamp ratio on the lubrication of slipper bearings. *Tribology International*, 29:4, 299-305.

[18] Koc E. Hooke C.J. (1997). Considerations in the design of partially hydrostatic slipper bearings. *Tribology International*, 30:11, 815-823.

[19] Tsuta, T. Iwamoto, T. Umeda T. (1999). Combined dynamic response analysis of a piston-slipper system and lubricants in hydraulic piston pump. Emerging Technologies in Fluids, Structures and Fluid/Structure Interactions. *ASME*.396, 187-194.

[20] Wieczoreck, U. Ivantysynova M. (2000). CASPAR-A computer aided design tool for axial piston machines. *Proceedings of the Power Transmission Motion and Control International Workshop*, PTMC2000, Bath, UK. 113-126.

[21] Wieczoreck U and Ivantysynova M. (2002). Computer aided optimization of bearing and sealing gaps in hydrostatic machines-the simulation tool CASPAR. *International Journal of fluid Power* 3:1, 7-20.

[22] Crabtree AB, Manring ND, Johnson RE. (2005). Pressure measurements for translating hydrostatic trust bearings. *International Journal of Fluid Power* 6:3.

[23] Johnson RE, Manring ND. (2005). Translating circular thrust bearings. *J. Fluid Mech.* 530, 197-212.

[24] Kazama T., Yamaguchi A. (1993). Application of a mixed lubrication model for hydrostatic equipment. *Tribology transactions of ASME*. 115, 686-91.

[25] Kazama T., Yamaguchi A. Fujiwara M. (2002). Motion of Eccentrically and dynamically loaded hydrostatic thrust bearing in mixed lubrication. *Proceedings of the 5th JFPS International.*

[26] Kazama T. (2004). Numerical simulation of a slipper model for water hydraulic pumps/motors in mixed lubrication. *Proceedings of the 6th JFPS International.*

[27] Kakoulis YP. (1977). Slipper lubrication in axial piston pumps. M.Sc. Thesis University of Birmingham.

[28] Bergada JM and Watton J. (2002). A direct leakage flow rate calculation method for axial pump grooved pistons and slippers, and its evaluation for a 5/95 fluid application. 5th JFPS international Symposium on fluid power. Nara, Japan.

[29] Bergada JM and Watton J. (2002). Axial Piston pump slipper balance with multiple lands. *ASME International Mechanical Engineering Congress and exposition*, New Orleans Louisiana. 2 No 39338.

[30] Bergada JM, Haynes JM, Watton J. (2008). Leakage and groove pressure of an axial piston pump slipper with multiple lands. *Tribol Transactions*. 5:4, 469-82.

[31] Brajdic-Mitideri P, Gosman A. D, Loannides E, Spikes H. A. (2005). CFD Analysis of a low friction pocketed pad bearing. *Journal of Tribology, ASME*. 127, 803-12.

[32] Helene M., Arghir M., Frene J. (2003). Numerical study of the pressure pattern in a two dimensional hybrid journal bearing recess, laminar and turbulent flow results. *Journal of tribology – ASME*. 125: 283-90.

[33] Braun M.J., Dzodzo M. (1995). Effect of the feedline and the hydrostatic pocket depth on the flow pattern and pressure distribution. *Tribology transactions of ASME*. 117, 224-32.

[34] Niels H., Santos F. (2008). Reducing friction in tilting pad bearing by the use of enclosed recesses. ASME, *Journal of Tribology*. 130.

[35] Patankar, Suhas V. (1980). Numerical Heat Transfer and Fluid Flow. Taylor & Francis Group: *Hemisphere Publishing Corporation.*

[36] Estrada C.A, Alvarez G, and Hinojosa J.F. (2005). Three-dimensional numerical simulation of the natural convection in an open tilted cubic cavity. *Revista Mexicana De Fisica*, 52:2, 111-119.

[37] Zeng M. and Tao W.Q. (2003). A comparison study of the convergence characteristics and robustness for four variants of SIMPLE-family at fine grid. *Engineering Computations* 20:3, 320-340.

[38] Shi, X. Khodadadi J.M. (2002). Laminar fluid flow and heat transfer in a lid driven cavity due to thin film. Journal of heat transfer, *ASME.* 124: 1056-1063.

[39] Chen C.L., Cheng C.H. (2006). Numerical study of flow and thermal behaviour of lid driven flow in cavities of small aspect ratio. *International journal for numerical methods in fluids*. 52, 785-799.

[40] Yao H., Cooper R.K, Raghunathan S. (2004). Numerical Simulation Incompressible Laminar Flow Over Three Dimensional Rectangular Cavities. *Journal of Fluid Engineering* 126, 919-927.

[41] Ching, T. P. Hwang, R. R. and Sheu, W.H. (1997). On End-Wall Corner Vortices in a Lid-Driven Cavity. *ASME Journal of Fluid Eng*. 119, 201-204.

[42] Iwatsu, R., Hyun J.M. and Kuwahara K. (1993). Numerical Simulation of Three Dimensional Flow in a Cubic Cavity with an Oscillating Lid. *ASME-J. Fluid Eng*. 115, 680-686.

[43] Tasnim S.H., Mahmud S. and Das P.K. (2002). Effect of aspect ratio and eccentricity on heat transfer from a cylinder in a cavity. *International journal of Numerical Method for Heat & Fluid Flow.*12:7, 855-869.

[44] Luan Z., Khonsari M.M. (2006). Numerical simulation of the flow field around the ring of mechanical seals. *Journal of Tribology, ASME.*128, 559-565.

[45] Molki M., Faghri M. (1999). Interaction between a buoyancy-driven flow and an array of annular cavities. *Sadhana academy proceedings in engineering science*. 19, 705-721.

[46] Cameron A. (1966). The principles of lubrication. Longman.

[47] Watton J. (2007). Modelling Monitoring and Diagnostic *Techniques for Fluid Power Systems.* Springer.

[48] Bergada JM, Watton J. (2005). Force and flow through hydrostatic slippers with grooves. The 8th International Symposium on Flow Control, Measurement and Visualization, FLUCOME 2005, Chengdu, China, Paper 240.

[49] Harlow FH, Welch JE. (1965). Numerical calculation of time-dependent viscous incompressible flow of fluid with free surface. *Physics of Fluids.* 8:12, 2182.

[50] Anderson JD. (1995). Computational fluid dynamics. *The basic with applications.* McGraw-Hill, Inc.

[51] Kumar S, Bergada JM, Watton J. (2009). Axial piston pump grooved slipper analysis by CFD simulation of three dimensional NVS equation in cylindrical coordinates. *Computer & Fluids.* 38:3, 648-663.

[52] Kumar S. (2010). CFD analysis of an axial piston pump. PhD Thesis. ETSEIAT-UPC.

[53] Konami S and Nishiumi T. (1999). Hydraulic Control Systems (in Japanese). Published by TDU.

[54] Freeman P. (1962). Lubrication and friction. Pitman.

[55] Bergada JM, Watton J, Haynes JM, Davies DLl. (2010). The hydrostatic/hydrodynamic behaviour of an axial piston pump slipper with multiple lands. *Meccanica* 45, 585-602.
[56] Huanlong L; Jian K; Guozhi W; Lanying Y. (2006). Research on the lubrication characteristics of water hydraulic slipper friction pairs. *J. Mech. Eng. Sci.* 220, 1559-1567.
[57] Canbulut F; Sinanoglu C; Yildirim S; Koç E. (2004). Design of neural network model for analysing hydrostatic circular recessed bearings with axial piston pump slipper. *Ind. Lubr. Tribol.* 56:5, 288-299.
[58] Canbulut F; Koç E; Sinanoglu C. (2009). Design of artificial neural networks for slipper analysis of axial piston pumps. *Ind. Lubr. Tribol.*61:2, 67-77.
[59] Canbulut F; Sinanoglu C; Koç E. (2009). Experimental analysis of frictional power loss of hydrostatic slipper bearings. *Ind. Lubr. Tribol.* 61:3, 123-131.

5.4. Barrel-Port Plate Performance

It is known that an axial piston barrel experiences small oscillations due to the forces acting over it. Cavitation also occurs in many cases, sometimes damaging the plate and barrel sliding surfaces and therefore reducing the volumetric and overall efficiency of the pump. More importantly, the resulting failure of the pump is often a critical issue in modern industrial applications. Piston pumps and motors are not fully understood in analytical detail, since problems related to cavitation, mixed friction and barrel dynamics, among others, are yet to be resolved via explicit methods. This book chapter attempts to bridge this gap by bringing together purely analytical solutions, with numerical validation, in connection with an important area of barrel/port plate leakage flow and associated torque dynamics. The barrel complex fluctuation will be in the present work experimentally evaluated and the clearance between barrel plate and port plate will be analyzed. The present work demonstrates the importance of properly designing the barrel-plate sliding surface, since pump efficiency is highly dependent on it.

5.4.1. Previous Research

Some of the most relevant research related to piston pump barrel dynamics and leakage barrel-plate are next outlined.

Helgestad et al [1] studied theoretically and experimentally the effect of using silencing grooves on the temporal pressure and leakage fluctuation in one piston cycle. Triangular and rectangular silencing grooves versus port plate ‘ideal timing’ and standard port plate were compared. For a range of operating conditions, the choice of triangular entry grooves was deduced to be the most appropriate. Martin and Taylor [2] analysed in detail the start and finish angles for the pressure and tank grooves to have ideal timing. As in [1] graphs are presented to understand the temporal pressure and flow in a single piston, but leakage flow was not considered. The results showed that triangular silencing grooves were more appropriate in all cases except when the pump parameters are fixed; in such case ideal timing main grooves were preferable. Edge et al [3] presented an improved analysis able to evaluate piston temporal pressure and flow, the improvement being based on taking into account the rate of change of momentum of the fluid during port opening. As in previously reported work, triangular silencing grooves were shown to be most appropriate for a piston pump operating over a wide range of working conditions. With regard to cavitation erosion, they defined the most severe region to be at the end of the inlet port and at the start of the delivery port.

Jacazio and Vatta [4] studied the pressure, hydrodynamic force and leakage between the barrel and plate. The study used Reynolds equation of lubrication, integrating it when considering pressure decay in the radial direction and including rotational speed. They found equations for the pressure distribution and lift force which showed the dependency of these parameters with rotational speed. Yamaguchi [5] demonstrated that a port plate with hydrostatic pads allows fluid film lubrication over a wide range of operating conditions. When analysing the barrel dynamics he took into account the spring effect of the shaft and by changing some physical parameters he determined the most likely cases for metal to metal contact between barrel and valve plate to occur. Yamaguchi [6] experimentally studied the barrel and plate dynamics, using position transducers, and used 4 different plates for experimentation, three of them with a groove, one without a groove and no outer pad. He found that the gap between the barrel and plate oscillates, the oscillation having a large peak and an intermediate smaller peak. For any kind of fluid used, it was found that the film thickness and amplitude increased with increasing inlet pressure.

Matsumoto and Ikeya [7] experimentally studied the friction, leakage and oil film thickness between the port plate and cylinder for low speeds. They found that the friction force was almost constant with rotational speed, but strongly depended on supply pressure and static force balance. In a further paper, Matsumoto and Ikeya [8] focussed more carefully on the leakage characteristics between the cylinder block and plate, again for low speed conditions. The results showed that the fluctuation of the tilt angle of the barrel and the azimuth of minimum oil film thickness depended mainly on the high pressure side number of pistons. Kobayashi and Matsumoto [9] studied the leakage and oil film thickness fluctuation between a port plate and barrel. They integrated numerically the Reynolds equation of lubrication, taking into account the pressure distribution in both the radial and the tangential direction. The flow, barrel tilt and barrel/port plate clearance versus angular position were determined at very low rotational speeds. Weidong and Zhanlin [10] studied the temporal leakage flow between a barrel and plate and between piston and barrel, and considered separately the leakage from each barrel groove and the effect of the inlet groove. Barrel tilt was not taken into consideration. Yamaguchi [11] gives an overview of the different problems found when considering tribological aspects of pumps. When assessing the plate and cylinder block performance, he pointed out the effect of the leakage for different fluid viscosities when the port plate has or has not a hydrodynamic groove. It was found that the use of a groove stabilizes the leakage for different fluid viscosities [12].

Manring [13] evaluated the forces acting on a cylinder block and its torque over the cylinder main axis. He considered the pressure distribution at the pump outlet as constant and the decay along the barrel lands as logarithmic, independent of the barrel tilt and turning speed. In a further study [14] he also investigated various port plate timing geometries within an axial piston pump. It was found that a constant area timing groove design had the advantage of minimizing the required discharge area of the timing groove, the linearly varying timing groove design having the advantage of utilizing the shortest timing groove length, and the quadratic timing groove design had no particular advantages over the other two. Zeiger and Akers [15] considered the dynamic equations of the swash plate which were linked with piston chamber pressures. They defined first the temporal piston chamber pressure, taking into account the area variation at the inlet and outlet groove entrance. The torque over the swash plate was dynamically and statically evaluated, finding that the torque average changed mainly with the swash plate angle, turning speed and outlet pressure. They compared simulation and experimental results finding a good correlation, although leakage was not evaluated. In a further study [16] they presented a model consisting of a second order differential equation of the swash plate motion and two first-order equations describing the flow continuity into the pump discharge chamber and into the swash plate control actuator.

One of the first studies focussing on the understanding of the operating torques on a pump swash plate was undertaken by Inoue et al [17,18]. They found theoretically that the exciting torque acting on the swash plate had a saw tooth shape. They also measured the torque on the swash plate finding that it had two peaks while the exciting one had a single peak. They defined the second peak as the one appearing when the system reached its natural frequency. Manring and Johnson [19] defined the dynamic equations of the swash plate in an axial piston pump, such equations having regard to the effect of the two actuators which maintain the swash plate in position. Wicke et al [20] simulated the dynamic behaviour of an axial piston pump using the program *bathfp*. They focussed the study on understanding the influence of swash plate angle variation on the piston forces and the yoke moment around the turning axis They found that an increase of swash plate angle increased the risk of cavitation in the cylinder chamber at the beginning of the suction port, and also decreased the time averaged yoke moment and increased peak to peak variations. In the paper by Manring [21], he further analyzed the dynamic torque acting on the swash plate. As in a previous study he did not consider the swash plate inertia and damping. He noticed that piston and slipper inertia tends to destabilize the swash plate position, although the most important term which created torque onto the swash plate was due to piston pressure. Gilardino et al [22] defined the dynamic equations which gives the torque onto the swash plate and including the torque created by the displacement control cylinders.

In Ivantysynova et al [23] a new method of prediction of the swash plate torque based on the software CASPAR is presented and which calculates the non isothermal gap flow and pressure distribution across all piston pump gaps. The study defined a direct link between the dynamic torque acting on the swash plate and the small groove dimension located at the entrance and exit of the valve plate main groove. Manring [24] studied the forces acting on the swash plate in an axial piston pump and took into account "secondary swash-plate angle" as well as the primary swash plate angle. He demonstrated that the use of a secondary swash plate angle will require a control and containment device that is capable of exerting a thrust load in the swash plate horizontal axis direction. In a further study [25] he examined the control and containment forces for a cradle-mounted, axial-actuated swash plate, showing that an axial-actuated swash plate tends to keep the swash-plate well seated within the cradle during all operating conditions. Bahr et al [26] used the swash plate dynamic equations, found in previous papers, to create a dynamic model of a pressure compensated swash plate axial piston pump with a conical cylinder block. They implemented the equations of the compensating unit to create a full model of the pump. The equations were integrated using Matlab Simulink, finding that the lateral moment acting on the swash plate fluctuates in a periodic fashion and contains nine harmonics and a negative mean value.

One of the most prolific researchers on piston pumps, which has published a large amount of papers in the last 10 years, is Ivantysynova et al [27-36]. Research regarding leakage in all piston pump gaps, forces and torques acting on slippers and swash plate, piston dynamics, plate surface temperature prediction, and pump design innovations, among other piston pump topics, are to be found in her papers. The latest research being developed in piston pumps focuses in reducing noise, several PhD's [37-39] and papers [40; 41] are to be found among the top quality work recently produced. It therefore seams that reducing noise and increasing pump efficiency are hot topics at the moment.

It is nevertheless important to remember that topics like using new materials on the sliding surfaces, to decrease friction and therefore increase hydraulic efficiency [42; 43], and piston pump barrel dynamics or pump dynamics [44-46], still need further development.

From all the studies undertook in the past nearly 40 years, it can be stated that the performance of silencing grooves used in axial piston pumps barrel-port plate sliding surfaces and considering their effect on pressure ripple, leakage, noise generation, dynamic forces and torques acting over the barrel-port plate, was studied among others by [1-3; 5; 37; 38; 40; 41; 44]. The clearance and leakage between the barrel and port plate has been studied experimentally by [6-8; 45], analytical research in this area has been presented in [4; 5; 9-13; 23; 27; 28; 32; 45], particularly innovative CFD research which included pressure distribution, thermal effects and the effect of micro structured wave surface in the barrel-port plate sliding surface has been undertaken by [29-31; 33; 34]. The piston pressure-flow dynamics was presented in [15-20; 28; 35; 36; 44], torques and forces acting on the swash plate were studied in [16-19; 21-26; 37; 39], friction barrel plate was analysed by [7; 30; 34; 42; 43]. Despite the amount of papers published on axial piston pumps and the huge knowledge gathered, it appears there is still further research to be done in order to better understand the barrel-port plate film thickness and the barrel dynamics associated. This chapter considers these issues with the intention of establishing more detailed experimental data and validation of design equations that may be used to improve axial piston pumps overall efficiency.

5.4.2. Mathematical Analysis

The equations giving leakage barrel-port plate, pressure distribution, force and torques acting on the barrel and port plate are to be presented next. Figure 5.4.1 represents the barrel and port plate face of an axial piston pump, one of the pistons being drawn for clarification. The port plate transfers the flow rate from the external connecting ports via two large kidney-shaped slots machined in the port plate inner face, one at the pump inlet and the other at the pump outlet.

These port plate slots, often called grooves, are shown in figure 5.4.1 where the main dimensions and the central axes are also shown. Notice that a timing groove is placed at the entrance of the main groove on the pressurised side.

The entrance to each piston in the barrel is via an associated small kidney shaped port referred to later as the 'piston groove', that is, there are 9 piston grooves machined on the face of the barrel. The sign convention is that the positive side of the 'X' axis is towards the left side of the 'Y' axis, and the barrel is slightly tilted with respect to the port plate with a tilt angle 'α'. The port plate is secured to the main body of the pump with four bolts and the barrel is pushed towards the port plate by a spring located at the bottom of the barrel, (not shown in figure 5.4.1).

This fixing mechanism therefore carries an additional load induced by the torque created by the pressure differential across the pump when in operation. Since laminar flow exists then taking tilt and rotation into account, assuming the flow moves in a radial direction, then Reynolds equation of lubrication takes the following polar coordinate form:

$$\frac{\partial}{\partial r}\left(r\,h^3\frac{\partial p}{\partial r}\right)=6\mu\sigma r\frac{\partial h}{\partial \theta} \tag{5.4.1}$$

This equation will be applied to four different lands, what it is called the internal and external land on the main port plate groove and the timing groove, see figure 5.4.1 where the internal and external lands on both sides of the main groove are clearly stated.

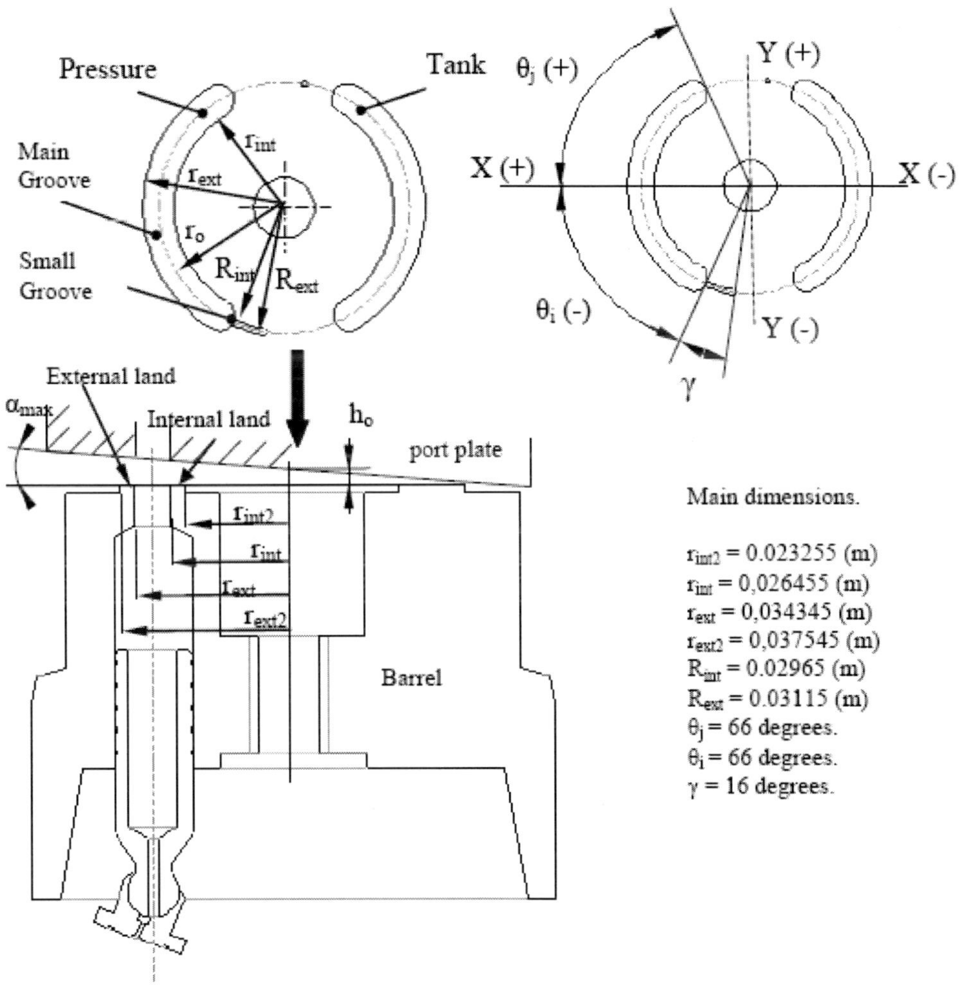

Figure 5.4.1. Barrel/port plate configuration.

For any generic land it will be assumed:

$$h = h_0 + \alpha\, r_m \cos\theta \tag{5.4.2}$$

where r_m is the average radius of each particular land.

Derivation of equation (5.4.2) versus θ will give:

$$\frac{\partial h}{\partial \theta} = -\alpha\, r_m \sin\theta \tag{5.4.3}$$

Substituting equations (5.4.3) and (5.4.2) in (5.4.1) and after the first integration it is found:

$$\frac{dp}{dr} = -\frac{3\mu\sigma\alpha r_m \sin\theta\, r}{\left(h_0 + \alpha r_m \cos\theta\right)^3} + \frac{c_1}{r\left(h_0 + \alpha r_m \cos\theta\right)^3} \tag{5.4.4}$$

After the second integration:

$$p = -\frac{3\mu\sigma\alpha r_m \sin\theta}{\left(h_0 + \alpha r_m \cos\theta\right)^3}\frac{r^2}{2} + \frac{c_1}{\left(h_0 + \alpha r_m \cos\theta\right)^3}\ln r + c_2 \tag{5.4.5}$$

This equation can be applied to any generic land, for each case the constants of integration will be found via boundary conditions

5.4.2.1. Pressure Distribution and Leakage between Barrel and Port Plate. Main Groove Effect

From figure 5.4.1, the boundary conditions for the external or internal land will be:

External land.

$$r = r_{ext} \quad p = p_{int} \tag{5.4.6}$$

$$r = r_{ext2} \quad p = p_{ext} = p_{tank}$$

$$r_{m\ ext} = \frac{r_{ext} + r_{ext2}}{2}$$

Internal land.

$$r = r_{int} \quad p = p_{int} \tag{5.4.7}$$

$$r = r_{int2} \quad p = p_{ext} = p_{tank}$$

$$r_{m\ int} = \frac{r_{int} + r_{int2}}{2}$$

When applying the boundary conditions for the external land it is found:

$$p_{int} = -\frac{3\mu\sigma\alpha r_{mext}\sin\theta}{\left(h_0 + \alpha r_{mext}\cos\theta\right)^3}\frac{r_{ext}^2}{2} + \frac{c_1}{\left(h_0 + \alpha r_{mext}\cos\theta\right)^3}\ln r_{ext} + c_2 \tag{5.4.8}$$

$$p_{ext} = -\frac{3\mu\sigma\alpha r_{mext}\sin\theta}{\left(h_0 + \alpha r_{mext}\cos\theta\right)^3}\frac{r_{ext2}^2}{2} + \frac{c_1}{\left(h_0 + \alpha r_{mext}\cos\theta\right)^3}\ln r_{ext2} + c_2 \tag{5.4.9}$$

and for the internal land:

$$p_{int} = -\frac{3\mu\sigma\alpha r_{m\,int}\sin\theta}{\left(h_0 + \alpha r_{m\,int}\cos\theta\right)^3}\frac{r_{int}^2}{2} + \frac{c_3}{\left(h_0 + \alpha r_{m\,int}\cos\theta\right)^3}\ln r_{int} + c_4 \tag{5.4.10}$$

$$p_{ext} = -\frac{3\mu\sigma\alpha r_{m\,int}\sin\theta}{\left(h_0 + \alpha r_{m\,int}\cos\theta\right)^3}\frac{r_{int2}^2}{2} + \frac{c_3}{\left(h_0 + \alpha r_{m\,int}\cos\theta\right)^3}\ln r_{int2} + c_4 \tag{5.4.11}$$

From equations (5.4.8) and (5.4.9) the value of the constants C_1 and C_2 can be found, equations (5.4.10) and (5.4.11) will be used to find the constant C_3 and C_4, the result is:

$$c_1 = \left[p_{int} - p_{ext} - \frac{3\mu\sigma\alpha r_{m\,ext}\sin\theta}{\left(h_0 + \alpha r_{m\,ext}\cos\theta\right)^3}\frac{\left(r_{ext2}^2 - r_{ext}^2\right)}{2}\right]\frac{\left(h_0 + \alpha r_{m\,ext}\cos\theta\right)^3}{\ln\left(\frac{r_{ext}}{r_{ext2}}\right)} \tag{5.4.12}$$

$$c_3 = \left[p_{int} - p_{ext} - \frac{3\mu\sigma\alpha r_{mint}\sin\theta}{\left(h_0 + \alpha r_{mint}\cos\theta\right)^3}\frac{\left(r_{int2}^2 - r_{int}^2\right)}{2}\right]\frac{\left(h_0 + \alpha r_{mint}\cos\theta\right)^3}{\ln\left(\frac{r_{int}}{r_{int2}}\right)} \tag{5.4.13}$$

$$c_2 = p_{int}\left(1 - \frac{\ln r_{ext}}{\ln\left(\frac{r_{ext}}{r_{ext2}}\right)}\right) + p_{ext}\frac{\ln r_{ext}}{\ln\left(\frac{r_{ext}}{r_{ext2}}\right)} + \frac{3\mu\sigma\alpha r_{m\,ext}\sin\theta}{\left(h_0 + \alpha r_{m\,ext}\cos\theta\right)^3 2}\left[r_{ext}^2 + \frac{\ln r_{ext}}{\ln\frac{r_{ext}}{r_{ext2}}}\left(r_{ext2}^2 - r_{ext}^2\right)\right] \tag{5.4.14}$$

$$c_4 = p_{int}\left(1 - \frac{\ln r_{int}}{\ln\left(\frac{r_{int}}{r_{int2}}\right)}\right) + p_{ext}\frac{\ln r_{int}}{\ln\left(\frac{r_{int}}{r_{int2}}\right)} + \frac{3\mu\sigma\alpha r_{m\,int}\sin\theta}{\left(h_0 + \alpha r_{m\,int}\cos\theta\right)^3 2}\left[r_{int}^2 + \frac{\ln r_{int}}{\ln\frac{r_{int}}{r_{int2}}}\left(r_{int2}^2 - r_{int}^2\right)\right] \tag{5.4.15}$$

The pressure distribution for the external land $r_{exter} < r < r_{exter2}$ after substituting the constants C_1 and C_2 in equation (5.4.5) will be:

$$p_{ext\ land} = p_{int}\left(1+\frac{\ln\frac{r}{r_{ext}}}{\ln\frac{r_{ext}}{r_{ext2}}}\right)+p_{ext}\frac{\ln\frac{r_{ext}}{r}}{\ln\frac{r_{ext}}{r_{ext2}}}+\frac{3\mu\sigma\alpha r_{m\,ext}\sin\theta}{\left(h_0+\alpha r_{m\,ext}\cos\theta\right)^3 2}\left[\left(r_{ext}^2-r^2\right)+\frac{\left(r_{ext2}^2-r_{ext}^2\right)\ln\left(\frac{r_{ext}}{r}\right)}{\ln\left(\frac{r_{ext}}{r_{ext2}}\right)}\right] \tag{5.4.16}$$

For the internal land $r_{int2} < r < r_{int}$, when substituting C_3 and C_4 in an equation homologous to equation (5.4.5), the pressure distribution will be:

$$p_{int\ land} = p_{int}\left(1+\frac{\ln\frac{r}{r_{int}}}{\ln\frac{r_{int}}{r_{int2}}}\right)+p_{ext}\frac{\ln\frac{r_{int}}{r}}{\ln\frac{r_{int}}{r_{int2}}}+\frac{3\mu\sigma\alpha r_{m\,int}\sin\theta}{\left(h_0+\alpha r_{m\,int}\cos\theta\right)^3 2}\left[\left(r_{int}^2-r^2\right)+\frac{\left(r_{int2}^2-r_{int}^2\right)\ln\left(\frac{r_{int}}{r}\right)}{\ln\left(\frac{r_{int}}{r_{int2}}\right)}\right] \tag{5.4.17}$$

For all the equations, the relation between α and θ is for $\theta = 0$; then $\alpha = \alpha_{maximum}$. Therefore according to figure 5.4.1, the main groove will exist between $-\theta_i < \theta < \theta_j$.

Once the equations giving pressure distribution has been found, a logical next step would be to determine the leakage. The total leakage due to the main groove has to be expressed as:

$$Q_{leakage} = \int_{-\theta_i}^{\theta_j}\int_0^h v_e\, r\, dy\, d\theta + \int_{-\theta_i}^{\theta_j}\int_0^h v_i\, r\, dy\, d\theta = Q_{ext} + Q_{int} \tag{5.4.18}$$

the velocity distribution according to Poiseulle's law can be given as: For the external land.

$$v_e = \frac{1}{\mu}\frac{dp}{dr}\frac{y}{2}(y-h) \tag{5.4.19}$$

where:

$$h = h_0 + \alpha r_{m\ ext}\cos\theta \tag{5.4.20}$$

And for the internal land:

$$v_i = \frac{1}{\mu}\frac{dp}{dr}\frac{y}{2}(y-h) \tag{5.4.21}$$

where:

$$h = h_0 + \alpha r_{m\ int}\cos\theta \tag{5.4.22}$$

The pressure distribution versus radius from the first integration of equation (5.4.1) will be: For the external land.

$$\frac{dp}{dr} = -\frac{3\mu\sigma\alpha r_{m\,ext} \sin\theta r}{\left(h_0 + \alpha r_{m\,ext} \cos\theta\right)^3} + \frac{c_1}{r\left(h_0 + \alpha r_{m\,ext} \cos\theta\right)^3} \tag{5.4.23}$$

And for the internal land

$$\frac{dp}{dr} = \left[-\frac{3\mu\sigma\alpha r_{m\,int} \sin\theta r}{\left(h_0 + \alpha r_{m\,int} \cos\theta\right)^3} + \frac{c_3}{r\left(h_0 + \alpha r_{m\,int} \cos\theta\right)^3} \right] * (-1) \tag{5.4.24}$$

It is necessary at this point to state that for the internal land the pressure decreases as the radius decreases, therefore its sign has to be changed to produce the required effect.

When substituting equation (5.4.19) into the first integral of equation (5.4.18) and considering also the relations defined in (5.4.20) and (5.4.23), after performing one of the two integrations, the leakage at the external land will be given as:

$$Q_{ext} = \int_{-\theta i}^{\theta j} -\frac{\left(h_0 + \alpha r_{m\,ext} \cos\theta\right)^3 r}{12\,\mu} \left[-\frac{3\mu\sigma\alpha r_{m\,ext} \sin\theta\, r}{\left(h_0 + \alpha r_{m\,ext} \cos\theta\right)^3} + \frac{c_1}{r\left(h_0 + \alpha r_{m\,ext} \cos\theta\right)^3} \right] d\theta \tag{5.4.25}$$

For a symmetrical groove, or in other words, when $|\theta_j| = |-\theta_i|$ and after some integration, the external flow is given as:

$$Q_{ext} = \int_{-\theta_i}^{\theta_j} \frac{(p_{ext} - p_{int})}{12\,\mu \ln\left(\dfrac{r_{ext}}{r_{ext2}}\right)} \left(h_0 + \alpha r_{m\,ext} \cos\theta\right)^3 d\theta \tag{5.4.26}$$

Once the final integration is performed it is obtained:

$$Q_{ext} = \frac{(p_{ext} - p_{int})}{12\,\mu \ln\left(\dfrac{r_{ext}}{r_{ext2}}\right)} \left[\begin{array}{l} h_0^3 \left[\theta\right]_{-\theta_i}^{\theta_j} + 3 h_0^2\, \alpha\, r_{m\,ext} \left[\sin\theta\right]_{-\theta_i}^{\theta_j} + \\ 3 h_0\, \alpha^2\, r_{m\,ext}^2 \left[\dfrac{1}{4}\sin(2\theta) + \dfrac{\theta}{2}\right]_{-\theta_i}^{\theta_j} + \alpha^3\, r_{m\,ext}^3 \left[\dfrac{1}{12}\sin(3\theta) + \dfrac{3}{4}\sin\theta\right]_{-\theta_i}^{\theta_j} \end{array} \right] \tag{5.4.27}$$

Operating similarly, when equations (5.4.21), (5.4.22) and (5.4.24) are substituted in the second integral of equation (5.4.18) the leakage across the internal land will be given as:

$$Q_{int} = \int_{-\theta i}^{\theta j} \frac{\left(h_0 + \alpha r_{m\,int} \cos\theta\right)^3 r}{12\,\mu} \left[-\frac{3\mu\sigma\alpha r_{m\,int} \sin\theta\, r}{\left(h_0 + \alpha r_{m\,int} \cos\theta\right)^3} + \frac{c_3}{r\left(h_0 + \alpha r_{m\,int} \cos\theta\right)^3} \right] d\theta \tag{5.4.28}$$

when $|\theta_j| = |-\theta_i|$; and after some minor integrations, the internal flow will be:

$$Q_{int} = \int_{-\theta_i}^{\theta_j} -\frac{(p_{ext} - p_{int})}{12\,\mu \ln\left(\frac{r_{int}}{r_{int\,2}}\right)} \left(h_0 + \alpha\, r_{m\,int} \cos\theta\right)^3 d\theta \tag{5.4.29}$$

After integration, the internal flow due to the main groove will be given by:

$$Q_{int} = -\frac{(p_{ext} - p_{int})}{12\,\mu \ln\left(\frac{r_{int}}{r_{int\,2}}\right)} \left[\begin{array}{l} h_0^3\,[\theta]_{-\theta_i}^{\theta_j} + 3\,h_0^2\,\alpha\, r_{m\,int}\,[\sin\theta]_{-\theta_i}^{\theta_j} + \\ 3\,h_0\,\alpha^2\, r_{m\,int}^2 \left[\frac{1}{4}\sin(2\theta) + \frac{\theta}{2}\right]_{-\theta_i}^{\theta_j} + \alpha^3\, r_{m\,int}^3 \left[\frac{1}{12}\sin(3\theta) + \frac{3}{4}\sin\theta\right]_{-\theta_i}^{\theta_j} \end{array}\right] \tag{5.4.30}$$

The total leakage for the barrel-plate will be the addition of the leakage due to the main port plate grove and the leakage due to the timing groove. For the main groove, the leakage will be the addition of leakages given by equations (5.4.27) and (5.4.30). The leakage will depend on the geometry, internal and external pressures, tilt, and the central clearance.

5.4.2.2. Barrel/Port Plate, Pressure Distribution and Leakage. Effect of the Entrance Timing Groove

As for the main port plate groove, the equations for the timing groove will be based on the Reynolds equations of lubrication equation (5.4.1). The equations giving the pressure distribution along the internal and external lands next to the timing groove are similar to the ones already found for the main groove, the main differences when solving the differential equation in this case being the boundary conditions and the limits of integration. The boundary conditions when focusing on the small groove, see figure 5.4.1, will be:

For the external land:

$$r = R_{ext};\ p = p_{int} \tag{5.4.31}$$

$$r = r_{ext\,2};\ p = p_{ext} = p_{tank}$$

$$R_{m\,ext} = \frac{R_{ext} + r_{ext2}}{2}$$

For the internal land:

$$r = R_{int};\ p = p_{int} \tag{5.4.32}$$

$$r = r_{int2};\ p = p_{ext} = p_{tank}$$

$$R_{m\,int} = \frac{R_{int} + r_{int\,2}}{2}$$

The limits of integration would be from $-\theta$ to $-(\theta+\gamma)$. Following the same procedure as in the main groove and taking into account that the constants C_1, C_2, C_3 and C_4 will be having the same form although it is necessary for this case to change r_{int} by R_{int} and r_{ext} by R_{ext}, see figure 5.4.1, it is found:

For the external land, $R_{ext} < r < r_{ext2}$

$$p_{ext\,land_{sg}} = p_{int}\left(1+\frac{\ln\frac{r}{R_{ext}}}{\ln\frac{R_{ext}}{r_{ext2}}}\right)+p_{ext}\frac{\ln\frac{R_{ext}}{r}}{\ln\frac{R_{ext}}{r_{ext2}}}+\frac{3\mu\sigma\alpha R_{mext}\sin\theta}{\left(h_0+\alpha R_{mext}\cos\theta\right)^3 2}\left[\left(R_{ext}^2-r^2\right)+\frac{\left(r_{ext2}^2-R_{ext}^2\right)\ln\left(\frac{R_{ext}}{r}\right)}{\ln\left(\frac{R_{ext}}{r_{ext2}}\right)}\right] \quad (5.4.33)$$

For the internal land: $r_{int2} < r < R_{int}$

$$p_{int\,land_{sg}} = p_{int}\left(1+\frac{\ln\frac{r}{R_{int}}}{\ln\frac{R_{int}}{r_{int2}}}\right)+p_{ext}\frac{\ln\frac{R_{int}}{r}}{\ln\frac{R_{int}}{r_{int2}}}+\frac{3\mu\sigma\alpha R_{m\,int}\sin\theta}{\left(h_0+\alpha R_{m\,int}\cos\theta\right)^3 2}\left[\left(R_{int}^2-r^2\right)+\frac{\left(r_{int2}^2-R_{int}^2\right)\ln\left(\frac{R_{int}}{r}\right)}{\ln\left(\frac{R_{int}}{r_{int2}}\right)}\right] \quad (5.4.34)$$

The leakage associated to the small groove would be given as:

$$Q_{leakage_{sg}} = \int_{-(\theta_i+\gamma)}^{-\theta_i}\int_0^h v_e\, r\, dy\, d\theta + \int_{-(\theta_i+\gamma)}^{-\theta_i}\int_0^h v_i\, r\, dy\, d\theta = Q_{ext_{sg}} + Q_{int_{sg}} \quad (5.4.35)$$

where the velocities v_e, v_i, the generic gap depth h and the pressure variation with radius, will have the same generic form as in the main groove case, equations (5.4.19) to (5.4.24), being in this case necessary to substitute in these equations, $r_{m\,ext}$ by $R_{m\,ext}$ and $r_{m\,int}$ by $R_{m\,int}$. Since the limits of integration are now non-symmetrical, some of the terms for the main groove that were zero, now do exist. Therefore, following the procedure established in the main groove case, the internal and external land leakage will be:

$$Q_{int_{sg}} = \frac{3\omega\alpha R_{m\,int} R_{int}^2}{12}\left[\cos(-\theta_i)-\cos(-(\theta_i+\gamma))\right]+\frac{3\omega\sigma R_{m\,int}}{12\ln\left(\frac{R_{int}}{r_{int2}}\right)}\frac{(r_{int2}^2-R_{int}^2)}{2}\left[\cos(-\theta_i)-\cos(-(\theta_i+\gamma))\right]$$

$$-\frac{p_{ext}-p_{int}}{12\mu\ln\left(\frac{R_{int}}{r_{int2}}\right)}\left[\begin{array}{l} h_0^3\left[-\theta_i-(-(\theta_i+\gamma))\right]+3h_0^2\alpha R_{m\,int}\left[\sin(-\theta_i)-\sin(-(\theta_i+\gamma))\right]+ \\ 3h_0\alpha^2 R_{m\,int}^2\left[\frac{1}{4}\sin(2*(-\theta_i))+\frac{-\theta_i}{2}-\frac{1}{4}\sin(2*(-(\theta_i+\gamma)))-\frac{-(\theta_i+\gamma)}{2}\right]+ \\ \alpha^3 R_{m\,int}^3\left[\frac{1}{12}\sin(3*(-\theta_i))+\frac{3}{4}\sin(-\theta_i)-\frac{1}{12}\sin(3*(-(\theta_i+\gamma)))-\frac{3}{4}\sin(-(\theta_i+\gamma))\right] \end{array}\right] \quad (5.4.36)$$

$$Q_{ext_{sg}} = \frac{3\sigma\alpha R_{m\,ext} R_{ext}^2}{12}\left[\cos(-\theta_i)-\cos(-(\theta_i+\gamma))\right]-\frac{3\sigma\alpha R_{m\,ext}}{12\ln\left(\frac{R_{ext}}{r_{ext2}}\right)}\frac{(r_{ext2}^2-R_{ext}^2)}{2}\left[\cos(-\theta_i)-\cos(-(\theta_i+\gamma))\right]$$

$$+\frac{p_{ext}-p_{int}}{12\mu\ln\left(\frac{R_{ext}}{r_{ext2}}\right)}\left[\begin{array}{l} h_0^3\left[-\theta_i-(-(\theta_i+\gamma))\right]+3h_0^2\alpha R_{m\,ext}\left[\sin(-\theta_i)-\sin(-(\theta_i+\gamma))\right]+ \\ 3h_0\alpha^2 R_{m\,ext}^2\left[\frac{1}{4}\sin(2*(-\theta_i))+\frac{-\theta_i}{2}-\frac{1}{4}\sin(2*(-(\theta_i+\gamma)))-\frac{-(\theta_i+\gamma)}{2}\right]+ \\ \alpha^3 R_{m\,ext}^3\left[\frac{1}{12}\sin(3*(-\theta_i))+\frac{3}{4}\sin(-\theta_i)-\frac{1}{12}\sin(3*(-(\theta_i+\gamma)))-\frac{3}{4}\sin(-(\theta_i+\gamma))\right] \end{array}\right] \quad (5.4.37)$$

It is interesting to notice that the leakage due to the small entrance groove depends on the barrel rotational speed, this effect appearing due to the groove asymmetry.

5.4.2.3. Force and Torque on the Barrel Due to the Pressure Distribution. Main Groove Effect

The force between the cylinders block and the pump plate due to the main groove is given as:

$$F=\int_{-\theta_i}^{\theta_j}\int_{r_{int}}^{r_{ext}} P_{int}\; r\, d\theta\, dr+\int_{-\theta_i}^{\theta_j}\int_{r_{int\,2}}^{r_{int}} P_{int\ land}\; r\, d\theta\, dr+\int_{-\theta_i}^{\theta_j}\int_{r_{ext}}^{r_{ext\,2}} P_{ext\ land}\; r\, d\theta\, dr \qquad (5.4.38)$$

Assuming that the pressure inside the groove is constant (although it is time dependant), the first term of the integration will be:

$$F=P_{int}\,\frac{r_{ext}^2-r_{int}^2}{2}(\theta_j-(-\theta_i))+\int_{-\theta_i}^{\theta_j}\int_{r_{int2}}^{r_{int}} P_{int\ land}\; r\, d\theta\, dr+\int_{-\theta_i}^{\theta_j}\int_{r_{ext}}^{r_{ext\,2}} P_{ext\ land}\; r\, d\theta\, dr \qquad (5.4.39)$$

The external and internal land pressures are given by equations (5.4.16) and (5.4.17) and after some integrations and rearrangement it is found:

$$F=P_{int}\,\frac{(\theta_j-(-\theta_i))}{4}\left[\frac{\left(r_{ext}^2-r_{ext2}^2\right)}{\ln\left(\dfrac{r_{ext}}{r_{ext2}}\right)}-\frac{\left(r_{int}^2-r_{inr2}^2\right)}{\ln\left(\dfrac{r_{int}}{r_{int\,2}}\right)}\right]+P_{ext}\,\frac{\left(\theta_j-(-\theta_i)\right)}{2}\left[r_{ext\,2}^2-r_{int\,2}^2\right]+$$

$$P_{ext}\,\frac{(\theta_j-(-\theta_i))}{4}\left[\frac{\left(r_{int}^2-r_{int\,2}^2\right)}{\ln\left(\dfrac{r_{int}}{r_{int\,2}}\right)}-\frac{\left(r_{ext}^2-r_{ext\,2}^2\right)}{\ln\left(\dfrac{r_{ext}}{r_{ext\,2}}\right)}\right]+\int_{-\theta_i}^{\theta_j}\frac{3\mu\sigma\alpha\, r_{m\,int}\sin\theta}{(h_0+\alpha\, r_{m\,int}\cos\theta)^3}d\theta\,*$$

$$\left\{\left(r_{int}^2-r_{int\,2}^2\right)^2\left[\frac{1}{8}-\frac{\ln(r_{int})}{4\ln\left(\dfrac{r_{int}}{r_{int\,2}}\right)}-\frac{1}{8}\frac{1}{\ln\left(\dfrac{r_{int}}{r_{int\,2}}\right)}\right]+\frac{r_{int\,2}^2-r_{int}^2}{4}\frac{1}{\ln\left(\dfrac{r_{int}}{r_{int\,2}}\right)}\left(r_{int\,2}^2\ln(r_{int\,2})-r_{int}^2\ln(r_{int})\right)\right\}+$$

$$\int_{-\theta_i}^{\theta_j}\frac{3\mu\sigma\alpha\, r_{m\,ext}\sin\theta}{(h_0+\alpha\, r_{m\,ext}\cos\theta)^3}d\theta\,*$$

$$\left\{\left(r_{ext2}^2-r_{ext}^2\right)^2\left[-\frac{1}{8}+\frac{\ln(r_{ext})}{4\ln\left(\dfrac{r_{ext}}{r_{ext2}}\right)}+\frac{1}{8}\frac{1}{\ln\left(\dfrac{r_{ext}}{r_{ext2}}\right)}\right]-\frac{r_{ext2}^2-r_{ext}^2}{4}\frac{1}{\ln\left(\dfrac{r_{ext}}{r_{ext2}}\right)}\left(r_{ext2}^2\ln(r_{ext2})-r_{ext}^2\ln(r_{ext})\right)\right\} \qquad (5.4.40)$$

The remaining integrals need to be solved numerically.

Although, as reported by Jacazzio and Vatta [4], the force depends on the turning speed ω, in reality the terms given by the integrals are much smaller than the first terms of equation (5.4.40). Such integration terms would play a much bigger role when using non symmetrical slots. In the case under study the groove is symmetrical and integrations of the type $\int_{-\theta_i}^{\theta_j} \frac{\sin\theta}{(h_0 + \alpha r_m \cos\theta)^3} d\theta$ will be equal to zero. Therefore the force on the barrel due to the main port plate groove effect will just depend on the geometry and the internal pressure:

$$F = P_{int} \frac{(\theta_j - (-\theta_i))}{4} \left[\frac{\left(r_{ext}^2 - r_{ext2}^2\right)}{\ln\left(\frac{r_{ext}}{r_{ext2}}\right)} - \frac{\left(r_{int}^2 - r_{inr2}^2\right)}{\ln\left(\frac{r_{int}}{r_{int2}}\right)} \right] + P_{ext} \frac{\left(\theta_j - (-\theta_i)\right)}{2} \left[r_{ext2}^2 - r_{int2}^2 \right] +$$

$$P_{ext} \frac{(\theta_j - (-\theta_i))}{4} \left[\frac{\left(r_{int}^2 - r_{int2}^2\right)}{\ln\left(\frac{r_{int}}{r_{int2}}\right)} - \frac{\left(r_{ext}^2 - r_{ext2}^2\right)}{\ln\left(\frac{r_{ext}}{r_{ext2}}\right)} \right] \tag{5.4.41}$$

According to equation (5.4.41) a linear variation between force and internal pressure is expected.

The torque over both axis created by the non uniform pressure distribution along the main port plate groove and the lands associated will have the following general form:

$$M_{XX} = \int_{-\theta_i}^{\theta_j} \int_{r_{int}}^{r_{ext}} P_{int}\, r^2 \sin\theta\, dr\, d\theta + \int_{-\theta_i}^{\theta_j} \int_{r_{int2}}^{r_{int}} P_{int\ land}\, r^2 \sin\theta\, dr\, d\theta + \int_{-\theta_i}^{\theta_j} \int_{r_{ext}}^{r_{ext2}} P_{ext\ land}\, r^2 \sin\theta\, dr\, d\theta \tag{5.4.42}$$

$$M_{YY} = \int_{-\theta_i}^{\theta_j} \int_{r_{int}}^{r_{ext}} P_{int}\, r^2 \cos\theta\, dr\, d\theta + \int_{-\theta_i}^{\theta_j} \int_{r_{int2}}^{r_{int}} P_{int\ land}\, r^2 \cos\theta\, dr\, d\theta + \int_{-\theta_i}^{\theta_j} \int_{r_{ext}}^{r_{ext2}} P_{ext\ land}\, r^2 \cos\theta\, dr\, d\theta \tag{5.4.43}$$

As when studying the force, the external and internal land pressure distribution is given by equations (5.4.16) and (5.4.17). Since the case studied is for a symmetrical groove, $|\theta_i| = |\theta_j|$, the first integral of equation (5.4.42) amongst others, has a zero value, therefore equation (5.4.42) after substituting the equations for the pressure distribution will look like.

$$M_{XX} = \int_{-\theta_i}^{\theta_j} \int_{r_{int2}}^{r_{int}} \left[p_{int} + p_{int} \frac{\ln\left(\frac{r}{r_{int}}\right)}{\ln\left(\frac{r_{int}}{r_{int2}}\right)} + p_{ext} \frac{\ln\left(\frac{r_{int}}{r}\right)}{\ln\left(\frac{r_{int}}{r_{int2}}\right)} + \frac{3\mu\sigma\alpha r_{m\,int} \sin\theta}{\left(h_0 + \alpha\, r_{m\,int} \cos\theta\right)^3} \left[\frac{r_{int}^2 - r^2}{2} + \frac{r_{int2}^2 - r_{int}^2}{2} \frac{\ln\left(\frac{r_{int}}{r}\right)}{\ln\left(\frac{r_{int}}{r_{int2}}\right)} \right] \right] r^2 \sin\theta\, dr\, d\theta$$

$$+ \int_{-\theta_i}^{\theta_j} \int_{r_{ext}}^{r_{ext2}} \left[p_{int} + p_{int} \frac{\ln\left(\frac{r}{r_{ext}}\right)}{\ln\left(\frac{r_{ext}}{r_{ext2}}\right)} + p_{ext} \frac{\ln\left(\frac{r_{ext}}{r}\right)}{\ln\left(\frac{r_{ext}}{r_{ext2}}\right)} + \frac{3\mu\sigma\alpha r_{m\,ext} \sin\theta}{\left(h_o + \alpha\, r_{m\,ext} \cos\theta\right)^3} \left[\frac{r_{ext}^2 - r^2}{2} + \frac{r_{ext2}^2 - r_{ext}^2}{2} \frac{\ln\left(\frac{r_{ext}}{r}\right)}{\ln\left(\frac{r_{ext}}{r_{ext2}}\right)} \right] \right] r^2 \sin\theta\, dr\, d\theta \tag{5.4.44}$$

Once integrated versus the radius and after some rearrangement, the torque over the "x" axis due to the main port plate groove will look like.

$$M_{XX} = \int_{-\theta_i}^{\theta_j} \frac{3\mu\sigma\alpha r_{m\,int}\sin^2\theta\, d\theta}{\left(h_0 + \alpha r_{m\,int}\cos\theta\right)^3} *$$

$$\left[\frac{r_{int}^2}{6}\left(r_{int}^3 - r_{int2}^3\right) - \frac{\left(r_{int}^5 - r_{int2}^5\right)}{10} + \frac{\left(r_{int2}^2 - r_{int}^2\right)}{6}\left(r_{int}^3 - r_{int2}^3\right)\frac{\ln\left(r_{int}\right)}{\ln\left(\frac{r_{int}}{r_{int2}}\right)} - \frac{\left(r_{int2}^2 - r_{int}^2\right)}{2\ln\left(\frac{r_{int}}{r_{int2}}\right)}\left[\frac{\left(r_{int2}^3 - r_{int}^3\right)}{9} + \frac{r_{int}^3\ln\left(r_{int}\right) - r_{int2}^3\ln\left(r_{int2}\right)}{3}\right]\right] + \qquad (5.4.45)$$

$$+\int_{-\theta_i}^{\theta_j} \frac{3\mu\sigma\alpha r_{m\,ext}\sin^2\theta\, d\theta}{\left(h_0 + \alpha r_{m\,ext}\cos\theta\right)^3} *$$

$$\left[\frac{r_{ext}^2}{6}\left(r_{ext2}^3 - r_{ext}^3\right) - \frac{\left(r_{ext2}^5 - r_{ext}^5\right)}{10} + \frac{\left(r_{ext2}^2 - r_{ext}^2\right)}{6}\left(r_{ext2}^3 - r_{ext}^3\right)\frac{\ln\left(r_{ext}\right)}{\ln\left(\frac{r_{ext}}{r_{ext2}}\right)} - \frac{\left(r_{ext2}^2 - r_{ext}^2\right)}{2\ln\left(\frac{r_{ext}}{r_{ext2}}\right)}\left[\frac{\left(r_{ext}^3 - r_{ext2}^3\right)}{9} + \frac{r_{ext2}^3\ln\left(r_{ext2}\right) - r_{ext}^3\ln\left(r_{ext}\right)}{3}\right]\right]$$

The remaining integrals need to be done numerically.

The torque over the "Y" axis when substituting the equations giving the pressure distribution (5.4.16) and (5.4.17) in equation (5.4.43) is given by:

$$M_{YY} = p_{int}\frac{r_{ext}^3 - r_{int}^3}{3}\left[\sin\theta\right]_{-\theta_i}^{\theta_j} +$$

$$+\int_{-\theta_i}^{\theta_j}\int_{r_{int2}}^{r_{int}}\left[p_{int} + p_{int}\frac{\ln\left(\frac{r}{r_{int}}\right)}{\ln\left(\frac{r_{int}}{r_{int2}}\right)} + p_{ext}\frac{\ln\left(\frac{r_{int}}{r}\right)}{\ln\left(\frac{r_{int}}{r_{int2}}\right)} + \frac{3\mu\sigma\alpha r_{m\,int}\sin\theta}{\left(h_0 + \alpha r_{m\,int}\cos\theta\right)^3}\left[\frac{r_{int}^2 - r^2}{2} + \frac{r_{int2}^2 - r_{int}^2}{2}\frac{\ln\left(\frac{r_{int}}{r}\right)}{\ln\left(\frac{r_{int}}{r_{int2}}\right)}\right]\right] r^2\cos\theta\, dr\, d\theta$$

$$+\int_{-\theta_i}^{\theta_j}\int_{r_{ext}}^{r_{ext2}}\left[p_{int} + p_{int}\frac{\ln\left(\frac{r}{r_{ext}}\right)}{\ln\left(\frac{r_{ext}}{r_{ext2}}\right)} + p_{ext}\frac{\ln\left(\frac{r_{ext}}{r}\right)}{\ln\left(\frac{r_{ext}}{r_{ext2}}\right)} + \frac{3\mu\sigma\alpha r_{m\,ext}\sin\theta}{\left(h_0 + \alpha r_{m\,ext}\cos\theta\right)^3}\left[\frac{r_{ext}^2 - r^2}{2} + \frac{r_{ext2}^2 - r_{ext}^2}{2}\frac{\ln\left(\frac{r_{ext}}{r}\right)}{\ln\left(\frac{r_{ext}}{r_{ext2}}\right)}\right]\right] r^2\cos\theta\, dr\, d\theta$$

(5.4.46)

When performing the rest of the integrations, taking into account that the groove under consideration is symmetrical and after rearrangement it is found:

$$M_{YY} = \frac{p_{int}\left[\sin\theta\right]_{-\theta_i}^{\theta_j}}{9}\left[\frac{\left(r_{int2}^3 - r_{int}^3\right)}{\ln\left(\frac{r_{int}}{r_{int2}}\right)} + \frac{\left(r_{ext}^3 - r_{ext2}^3\right)}{\ln\left(\frac{r_{ext}}{r_{ext2}}\right)}\right] + \frac{p_{ext}\left[\sin\theta\right]_{-\theta_i}^{\theta_j}}{9}\left[-\frac{\left(r_{int2}^3 - r_{int}^3\right)}{\ln\left(\frac{r_{int}}{r_{int2}}\right)} - \frac{\left(r_{ext}^3 - r_{ext2}^3\right)}{\ln\left(\frac{r_{ext}}{r_{ext2}}\right)}\right] + \frac{p_{ext}\left[\sin\theta\right]_{-\theta_i}^{\theta_j}}{3}\left[r_{ext2}^3 - r_{int2}^3\right]$$

(5.4.47)

It is noticed when checking the torque equations, that the torque over the "X" axis depends on the pump turning speed and plate tilt, which means that for the symmetrical groove case studied here such a torque will be zero if any of these parameters is zero. On the other hand, the torque over the "Y" axis is independent of the tilt and pump turning speed, and just depends on the geometry and the internal pressure.

5.4.2.4. Force and Torque Caused by the Action of the Timing Groove

The force over the barrel due to the timing groove can be given according to the following general equation.

$$F_{sg} = \int_{-(\theta_i+\gamma)}^{-\theta_i}\int_{R_{int}}^{R_{ext}} P_{int}\, r\, d\theta\, dr + \int_{-(\theta_i+\gamma)}^{-\theta_i}\int_{r_{int2}}^{R_{int}} P_{int\ land_{sg}}\, r\, d\theta\, dr + \int_{-(\theta_i+\gamma)}^{-\theta_i}\int_{R_{ext}}^{r_{ext2}} P_{ext\ land_{sg}}\, r\, d\theta\, dr \qquad (5.4.48)$$

When substituting equations (5.4.33) and (5.4.34) in (5.4.48), after integration and rearrangement, it is found:

$$F_{sg} = P_{int}\frac{(-\theta_i-(-(\theta_i+\gamma)))}{4}\left[\frac{\left(R_{ext}^2-r_{ext2}^2\right)}{\ln\left(\frac{R_{ext}}{r_{ext2}}\right)}-\frac{\left(R_{int}^2-r_{inr2}^2\right)}{\ln\left(\frac{R_{int}}{r_{int2}}\right)}\right]+P_{ext}\frac{(-\theta_i-(-(\theta_i+\gamma)))}{2}\left[r_{ext2}^2-r_{int2}^2\right]+$$

$$P_{ext}\frac{(-\theta_i-(-(\theta_i+\gamma)))}{4}\left[\frac{\left(R_{int}^2-r_{int2}^2\right)}{\ln\left(\frac{R_{int}}{r_{int2}}\right)}-\frac{\left(R_{ext}^2-r_{ext2}^2\right)}{\ln\left(\frac{R_{ext}}{r_{ext2}}\right)}\right]+\int_{-(\theta_i+\gamma)}^{-\theta_i}\frac{3\mu\sigma\alpha R_{mint}\sin\theta}{(h_0+\alpha R_{mint}\cos\theta)^3}d\theta\,*$$

$$\left\{\left(R_{int}^2-r_{int2}^2\right)^2\left[\frac{1}{8}-\frac{\ln(R_{int})}{4\ln\left(\frac{R_{int}}{r_{int2}}\right)}-\frac{1}{8}\frac{1}{\ln\left(\frac{R_{int}}{r_{int2}}\right)}\right]+\frac{r_{int2}^2-R_{int}^2}{4}\frac{1}{\ln\left(\frac{R_{int}}{r_{int2}}\right)}\left(r_{int2}^2\ln(r_{int2})-R_{int}^2\ln(R_{int})\right)\right\}+$$

$$\int_{-(\theta_i+\gamma)}^{-\theta_i}\frac{3\mu\sigma\alpha R_{mext}\sin\theta}{(h_0+\alpha R_{mext}\cos\theta)^3}d\theta\,*$$

$$\left\{\left(r_{ext2}^2-R_{ext}^2\right)^2\left[-\frac{1}{8}+\frac{\ln(R_{ext})}{4\ln\left(\frac{R_{ext}}{r_{ext2}}\right)}+\frac{1}{8}\frac{1}{\ln\left(\frac{R_{ext}}{r_{ext2}}\right)}\right]-\frac{r_{ext2}^2-R_{ext}^2}{4}\frac{1}{\ln\left(\frac{R_{ext}}{r_{ext2}}\right)}\left(r_{ext2}^2\ln(r_{ext2})-R_{ext}^2\ln(R_{ext})\right)\right\}$$

(5.4.49)

The remaining integrals need to be solved numerically.

The torque over the "X" and "Y" axes created by the timing groove will be given as:

$$M_{XX_{sg}} = \int_{-(\theta_i+\gamma)}^{-\theta_i}\int_{R_{int}}^{R_{ext}} P_{int}\, r^2\sin\theta\, dr\, d\theta + \int_{-(\theta_i+\gamma)}^{-\theta_i}\int_{r_{int2}}^{R_{int}} P_{int\ land_{sg}}\, r^2\sin\theta\, dr\, d\theta + \int_{-(\theta_i+\gamma)}^{-\theta_i}\int_{R_{ext}}^{r_{ext2}} P_{ext\ land_{sg}}\, r^2\sin\theta\, dr\, d\theta$$

(5.4.50)

$$M_{YY_{sg}} = \int_{-(\theta_i+\gamma)}^{-\theta_i}\int_{R_{int}}^{R_{ext}} P_{int}\, r^2\cos\theta\, dr\, d\theta + \int_{-(\theta_i+\gamma)}^{-\theta_i}\int_{r_{int2}}^{R_{int}} P_{int\ land_{sg}}\, r^2\cos\theta\, dr\, d\theta + \int_{-(\theta_i+\gamma)}^{-\theta_i}\int_{R_{ext}}^{r_{ext2}} P_{ext\ land_{sg}}\, r^2\cos\theta\, dr\, d\theta$$

(5.4.51)

When substituting the pressure distribution in each land, equations (5.4.33) and (5.4.34), in the torque equations, following the procedure established for the main port plate groove and after some development it is found:

$$M_{XX_{sg}}=\frac{p_{ext}}{3}(r_{ext2}^3-r_{int2}^3)*\left[-\cos\theta\right]_{-(\theta_i+\gamma)}^{-\theta_i}+\frac{(p_{int}-p_{ext})}{9}\left[\frac{r_{int2}^3-R_{int}^3}{\ln\left(\frac{R_{int}}{r_{int2}}\right)}+\frac{R_{ext}^3-r_{ext2}^3}{\ln\left(\frac{R_{ext}}{r_{ext2}}\right)}\right]*\left[-\cos\theta\right]_{-(\theta_i+\gamma)}^{-\theta_i}+$$

$$\int_{-(\theta_i+\gamma)}^{-\theta_i}\frac{3\mu\sigma\alpha R_{mint}\sin^2\theta\, d\theta}{\left(h_0+\alpha R_{mint}\cos\theta\right)^3}*\left[\frac{\left(R_{int}^5-r_{int2}^5\right)}{15}-\frac{\left(r_{int2}^2-R_{int}^2\right)\left(r_{int2}^3-R_{int}^3\right)}{18\ln\left(\frac{R_{int}}{r_{int2}}\right)}\right]+$$

$$+\int_{-(\theta_i+\gamma)}^{-\theta_i}\frac{3\mu\sigma\alpha R_{mext}\sin^2\theta\, d\theta}{\left(h_0+\alpha R_{mext}\cos\theta\right)^3}*\left[\frac{\left(r_{ext2}^5-R_{ext}^5\right)}{15}-\frac{\left(r_{ext2}^2-R_{ext}^2\right)\left(R_{ext}^3-r_{ext2}^3\right)}{18\ln\left(\frac{R_{ext}}{r_{ext2}}\right)}\right]$$

(5.4.52)

$$M_{YY_{sg}}=\frac{p_{ext}}{3}(r_{ext2}^3-r_{int2}^3)*\left[\sin\theta\right]_{-(\theta_i+\gamma)}^{-\theta_i}+\frac{(p_{int}-p_{ext})}{9}\left[\frac{r_{int2}^3-R_{int}^3}{\ln\left(\frac{R_{int}}{r_{int2}}\right)}+\frac{R_{ext}^3-r_{ext2}^3}{\ln\left(\frac{R_{ext}}{r_{ext2}}\right)}\right]*\left[\sin\theta\right]_{-(\theta_i+\gamma)}^{-\theta_i}+$$

$$\frac{3\mu\sigma\alpha R_{mint}}{h_0^3}\left[\frac{\left(R_{int}^5-r_{int2}^5\right)}{15}-\frac{\left(r_{int2}^2-R_{int}^2\right)\left(r_{int2}^3-R_{int}^3\right)}{18\ln\left(\frac{R_{int}}{r_{int2}}\right)}\right]*\left[\frac{1}{\left(\frac{\alpha R_{mint}}{h_0}\right)^2}\left[\frac{1}{1+\frac{\alpha R_{mint}}{h_0}\cos\theta}-\frac{1}{2\left(1+\frac{\alpha R_{mint}}{h_0}\cos\theta\right)^2}\right]_{-(\theta_i+\gamma)}^{-\theta_i}\right]+$$

$$\frac{3\mu\sigma\alpha R_{mext}}{h_0^3}\left[\frac{\left(r_{ext2}^5-R_{ext}^5\right)}{15}-\frac{\left(r_{ext2}^2-R_{ext}^2\right)\left(R_{ext}^3-r_{ext2}^3\right)}{18\ln\left(\frac{R_{ext}}{r_{ext2}}\right)}\right]*\left[\frac{1}{\left(\frac{\alpha R_{mext}}{h_0}\right)^2}\left[\frac{1}{1+\frac{\alpha R_{mext}}{h_0}\cos\theta}-\frac{1}{2\left(1+\frac{\alpha R_{mext}}{h_0}\cos\theta\right)^2}\right]_{-(\theta_i+\gamma)}^{-\theta_i}\right]$$

(5.4.53)

The integrals in all the remaining equations, need to be integrated numerically. The overall leakage, pressure distribution, force and torque versus the barrel main axis, will be the addition of the equations of the main groove and the small groove for each case. Special attention has to be made when regarding the torque, since different signs will mean different torque directions. The already developed equations need to be implemented by the effect of the pressure inside the cylinders as explained next.

5.4.2.5. The Effect of Cylinder Pressure

To complete the analysis, it is now necessary to include the effect of the pressure inside each cylinder chamber. Such a summation of pressure forces will create both forces and torques that will act in an opposite direction than the ones already presented. The first thing to be noticed is that pressure inside the cylinders chamber changes with time. Following previous work and a consideration of fluid volumes in this study, compressibility effects are negligible and therefore to find the temporal pressure the following equation is sufficient:

$$P_{piston} = P_{int} + \frac{\rho}{2}\left(\frac{Q}{Cd\ A_{flow}}\right)^2 = P_{int} + \frac{\rho}{2}\left(\frac{V_{piston}\ A_{cylinder}}{Cd\ A_{flow}}\right)^2 \quad (5.4.54)$$

It has been assumed that since the leakage flow between piston and cylinder is of a much lower order of magnitude than the piston flow, its effect can be neglected when calculating pressure. Zero reference time is defined when one of the pistons is at its bottom dead centre. The piston velocity can be calculated as follows:

$$V_{SL} = \sigma\ R_{sw}\ \tan\varepsilon\ \sin(\sigma\ t) \quad (5.4.55)$$

For the case under study, $\varepsilon_{max} = 20°$, $\sigma = 1440$ rpm, $R_{sw} = 0.03434$ m, $\rho = 875$ kg/m^3

There is little useful data regarding detailed discharge coefficients inside piston pump cylinders, and it will be assumed that an average discharge coefficient value of 0.75 is reasonable. When all values are substituted into equation (4.54) it is noticed that the maximum pressure inside the cylinder is only around 0.003MPa higher than the pressure outside, taking into account that the outside pressure can be typically 10-35MPa. As the barrel rotates the force and torque created by the pressure inside the cylinders will vary with rotation angle, hence time. It is necessary to point out that for 40 degrees of rotation, 28 degrees embraces 5 pistons that are connected to pump pressure, while during the other 12 degrees just four pistons are connected to pump pressure port.

5.4.3. Barrel Port Plate, Numerical Simulation

The developed equations are complex, some requiring additional numerical integrations, and are human error-prone. Therefore to validate the solutions developed, a numerical solution of the original differential equations has also been undertaken specifically for this study of flow through the gap between the barrel and the port plate. Regarding the mesh selection, it is important to point out that the mesh created in the θ direction along the main groove lands has a step size much bigger than the step size needed in the θ direction along the timing groove lands. This is due to the fact that the land for the barrel plate in the radial direction associated with the main groove is much smaller than the one associated with the timing groove. Therefore although the flow will be mostly radial in both cases, the effect of barrel rotation will be more intense along the small groove lands. Nevertheless, it was found that the radial direction pressure drop in both cases was much bigger than the tangential pressure drop. To increase the accuracy of the results at any cell, the average pressure of all four grid points was used. The program was written and executed using the software MATLAB. Data may be directly entered into the code which then automatically plots the 3-D pressure distribution.

5.4.4. Experimental Test Rig and Measuring Procedure

Figures 5.4.2 and 5.4.3 present the pump used for the experimentation; it is a nine piston axial design being its maximum volumetric displacement of 24.1 cc/rev and giving a maximum flow of about 35 l/min at its maximum swash plate angle of 20º. Figure 5.4.2a) presents the pump internal view, in figure 5.4.2b) it is seen the pump external view and the position of the transducers used, figure 5.4.2c) presents the pump cross section where two of the three position transducers used to perform the measurements are to be seen. Pump axis, pump spring, swash plate and the bearings located at both axis ends are also presented in figure 5.4.2c).

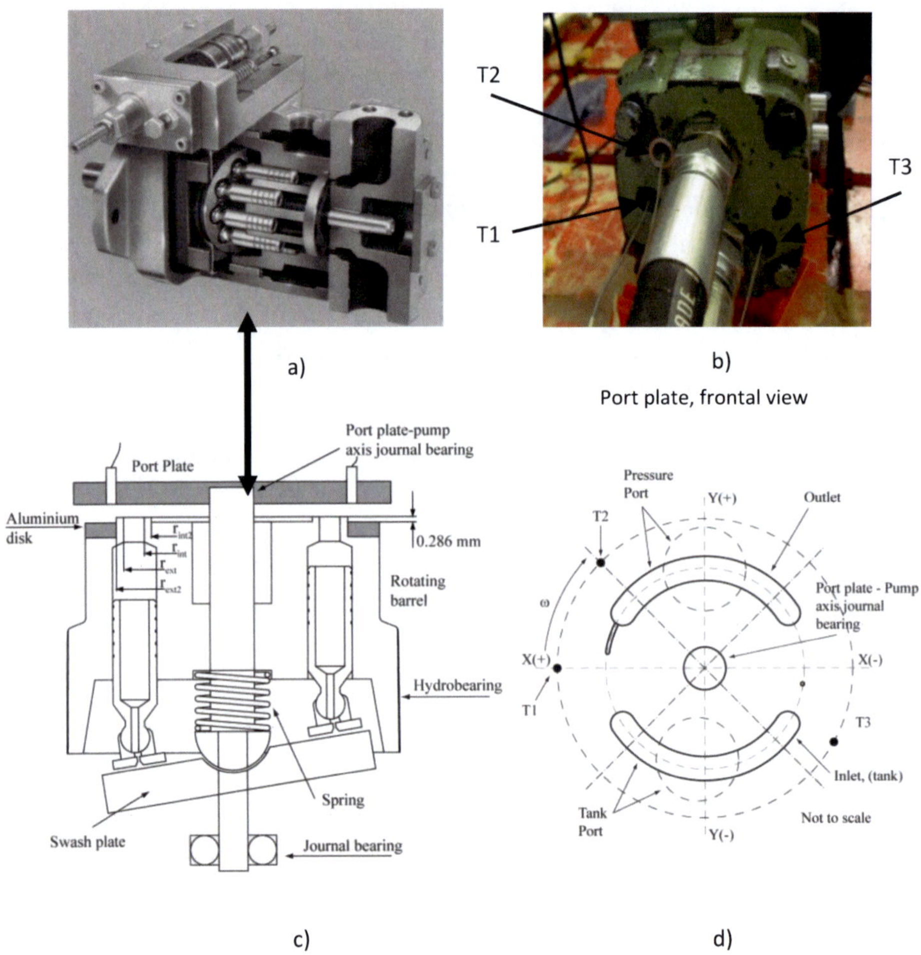

Figure 5.4.2. Pump under study and location of the three displacement transducers.

Figure 5.4.2d) defines the axes used to measure the exact position of the transducers presented in figure 5.4.2b), it must be noticed that the coordinate axis presented in figure 5.4.2d, are not the same as the axis defined in figure 5.4.1, also the pump rotational speed direction is given in the same figure 5.4.2d). The lower dash circle presented in this figure represents the pump inlet, the upper one represents the pump outlet and the big central one is approximately the diameter where the position transducers were located. On the pump port plate, figure 5.4.2b), three Micro-Epsilon unshielded inductive position transducers, model NCDT 3010 were allocated, the maximum measuring range for each transducer was of 0.5 mm and each of them was capable of measuring to an accuracy of 0.1μm. Such transducers offer a unique thermal stability of 0.05% FSO, being its temperature operation range of -50 to 150 C. The exact position of the transducers related to the XY axis represented in figure 5.4.2d) was:

Transducer 1: $X_1 = 47.96$ mm, $Y_1 = 0.285$ mm
Transducer 2: $X_2 = 33.68$ mm, $Y_2 = 34.155$ mm
Transducer 3: $X_3 = -39.93$ mm; $Y_3 = -27.34$ mm

The transducers were calibrated one by one in a specific calibration test rig. Transducer's calibration showed an excellent linearity, the calibration equation of each transducer being:

Transducer 1; $d_1 = 0.0487399\, v + 0.0296410$ mm

Transducer 2; $d_2 = 0.0452037\, v + 0.0328352$ mm

Transducer 3; $d_3 = 0.0476972\, v + 0.0322051$ mm

where v is the measured voltage [V] and d_i is the distance [mm].

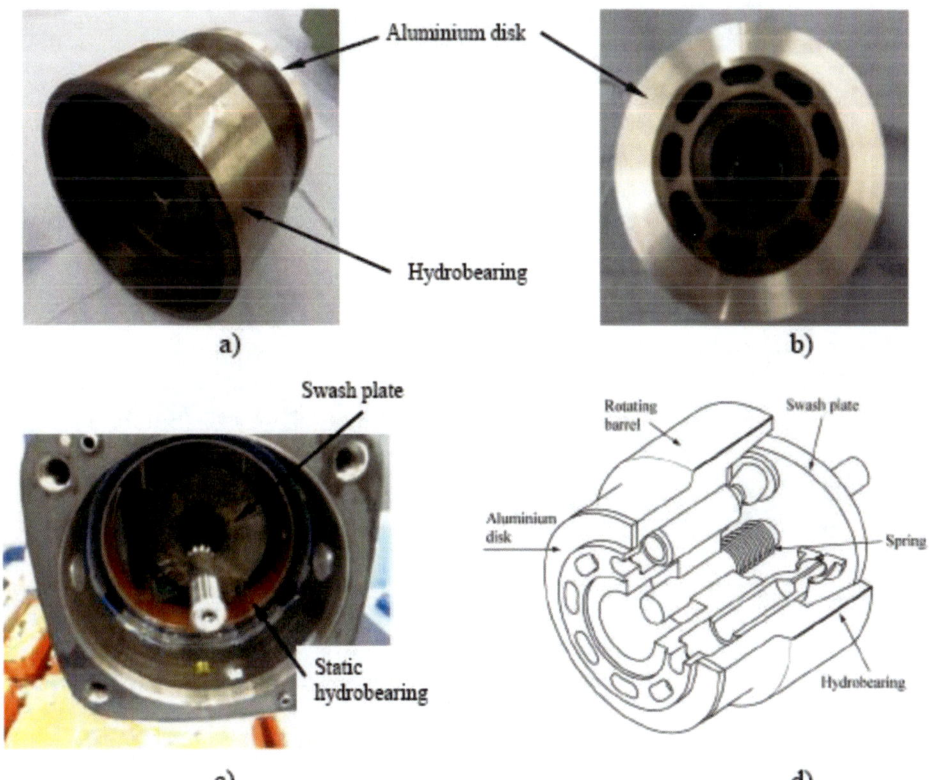

Figure 5.4.3. Pump under study internal view.

After being calibrated, the transducers were screwed into the fixed port plate and facing the inner part of the pump, the rotating barrel. To find the transducers zero position, the port plate with the transducers inserted in it was placed over an aluminium plate perfectly flat, the readings given by the transducers when facing such plate gave the transducers zero position. Once having done it, the next step was to fix the port plate to the pump. Due to the preferred operation of each transducer, which needs to point at a non magnetic material, a thin aluminium plate was bonded to the barrel end, as shown in figures 5.4.2c), 5.4.3a) and 5.4.3b).

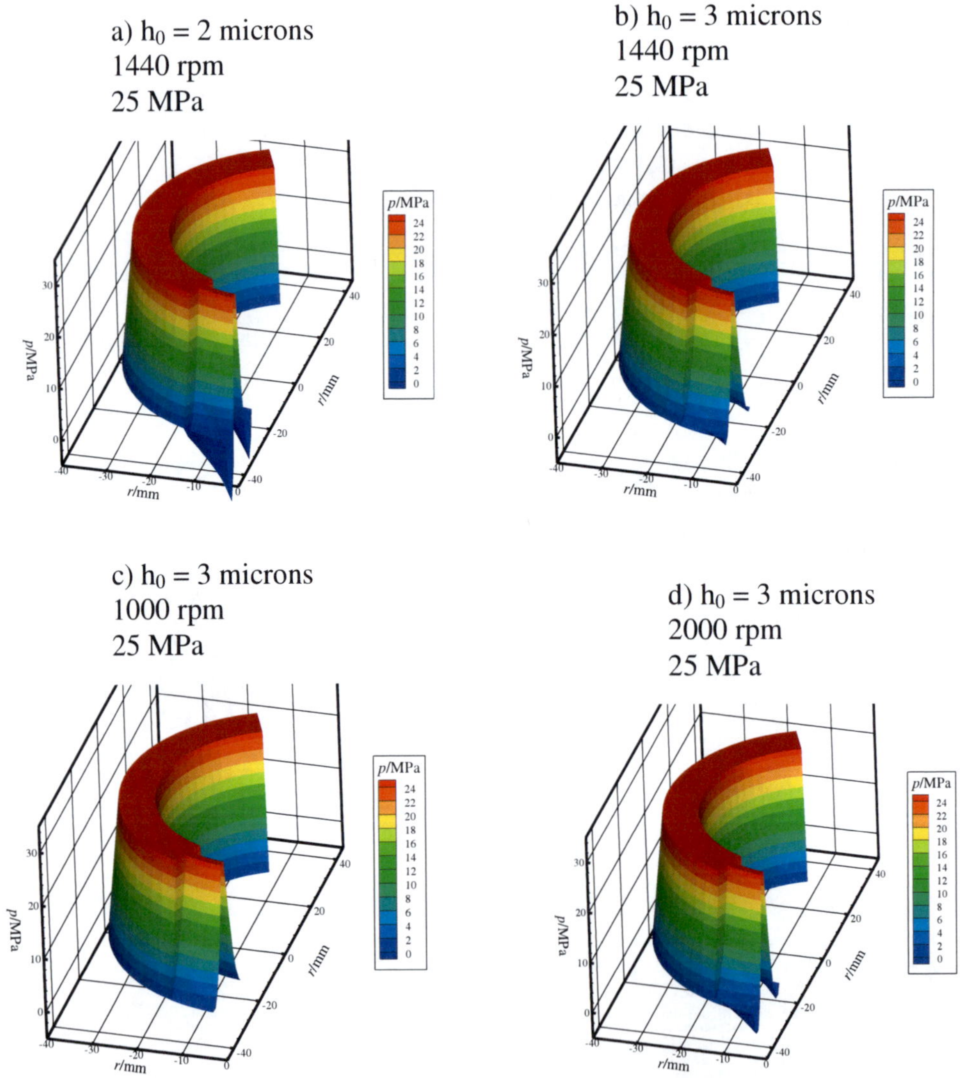

Figure 5.4.4. Theoretical pressure distribution along the main and small grooves for a set of different parameters. Maximum tilt.

In order to show more precisely the modifications made on the barrel, figure 5.4.3 has been introduced. Figure 5.4.3a) presents the axial piston pump barrel with the aluminium ring inserted on it. Figure 5.4.3b) shows the frontal view of the aluminium disk used. Figure

5.4.3c) presents the pump internal view, where the pump axis, swash plate and the so called by the manufacturer, hydrobearing, used to maintain the barrel aligned with the pump axis, are to be seen. It is to be noticed that the barrel external diameter turns around the static hydrobearing fixed to the pump case; the clearance between them is, according to the manufacturer of around 10 μm being the static hydrobearing part, made of a plastic material similar to PTFE. Such a pump design, assures that once the barrel is inserted into the pump, the barrel can only be tilted a fraction of a degree versus its central position. Figure 5.4.3d) introduces a three dimensional view of the barrel, where the barrel spring, aluminium disk and barrel hydrobearing are to be seen. The barrel and port plate are made of cast iron and the sliding surfaces are carbonitrided to a typical depth of 10μm. The sliding surfaces roughness is of 0.4Ra. The static run-out of the aluminium plate was of 7 microns, the average distance between the barrel face and the aluminium disk being 0.286 mm as indicated in figure 5.4.2c). Two sets of test were undertaken, at swash plate angles of 10° and 20°, and for each case two oil temperatures 28°C and 45°C were studied.

At each test, measurements were taken at pressures of 2.5, 5, 7.5, 10, 12.5, 15, 17.5, 19.5 MPa. Since the pump used has a pressure compensating unit, the maximum available pressure for the tests performed at 10 degrees swash plate angle and 45°C oil temperatures was 15MPa. The measurements presented in the following sections, shows that the barrel position fluctuates, being the fluctuation highly dependent on pump inlet pressure and oil temperature. For the purpose of performance analysis the temporal barrel position and the barrel-plate oil film thickness will be expressed as a temporal average. The fluid used to perform the experiments was hydraulic oil ISO 32, being its main physical characteristics fully defined in the ISO standards.

5.4.5. Results

5.4.5.1. Numerical and Analytical Results

5.4.5.1.1. Pressure Distribution.

The theoretical pressure distribution according to equations (5.4.16), (5.4.17), (5.4.33) and (5.4.34) along the main groove and timing groove lands is represented in figure 5.4.4, for a set of clearances and turning speeds.

At this point it is interesting to reflect on the work by Edge et al [3] who found that the most severe region for cavitation erosion was at the end of the inlet port and the start of the delivery port, as predicted by the new theoretical solution shown in figure 5.4.4. From figure 5.4.4a) and 5.4.4b) it is shown that cavitation at the entrance of the timing and main groove is more likely to appear as the clearance is reduced. The pressure asymmetry is accentuated as the central clearance decreases, and it is also accentuated as the inlet pressure decreases. It is important to notice that as the turning speed increases, figures 5.4.4c), 5.4.4d), the pressure distribution asymmetry increases as well and will tend to create a low pressure area at both sides of the groove entrance. Therefore it can be said that according to the theory developed, cavitation at the entrance of the timing and main grooves is more likely to appear when working under low pressures, small clearances, and high turning speeds.

In figure 5.4.5 some analytical and numerical pressure distributions are compared and it will be noticed that for the particular cases presented, the pressure distribution comparisons are very similar even when cavitation appears at the timing groove entrance. This gives further support to the analytical predictions being presented here.

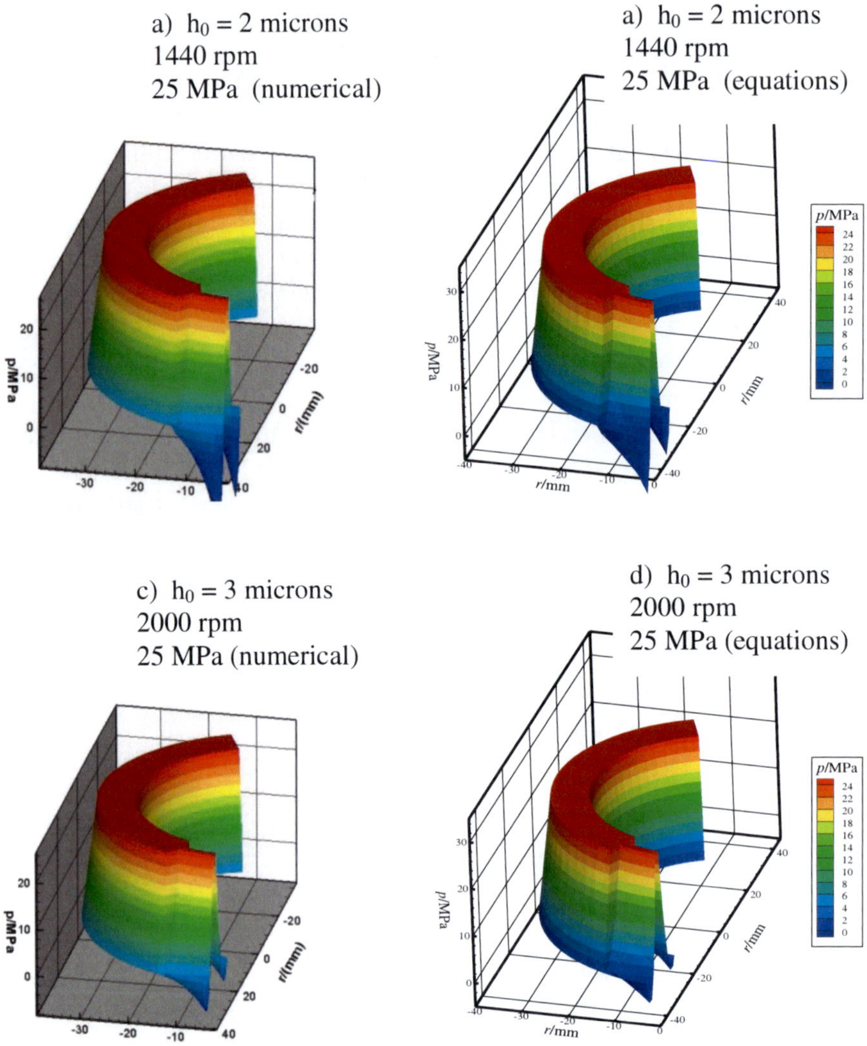

Figure 5.4.5. Pressure distribution along the main and small grooves for a set of different parameters. Maximum tilt, numerical and analytical solutions.

5.4.5.1.2. Leakage in the Main Groove and the Timing Groove

Figure 5.4.6 represents the leakage for different central clearances 'ho', where for each clearance the maximum tilt angle α possible has been used. A comparison between the numerical solution and the theoretical equations is also shown. Notice there exist an excellent agreement between the theoretical equations and the numerical solution.

Although not presented, the leakage variation with pressure differential is linear as expected for both grooves. From these graphs it is clearly noticed that leakage is much higher on the external land than on the internal one. Leakage also has the tendency to reach an asymptotic value as the barrel-plate distance increases. When comparing figures 5.4.6a) and

5.4.6b) it can be said that the leakage due to the timing groove is less than 5% of the main groove leakage. One of the reasons why the timing groove leakage is so small is because the land between the groove and the exterior is much bigger than in the main groove.

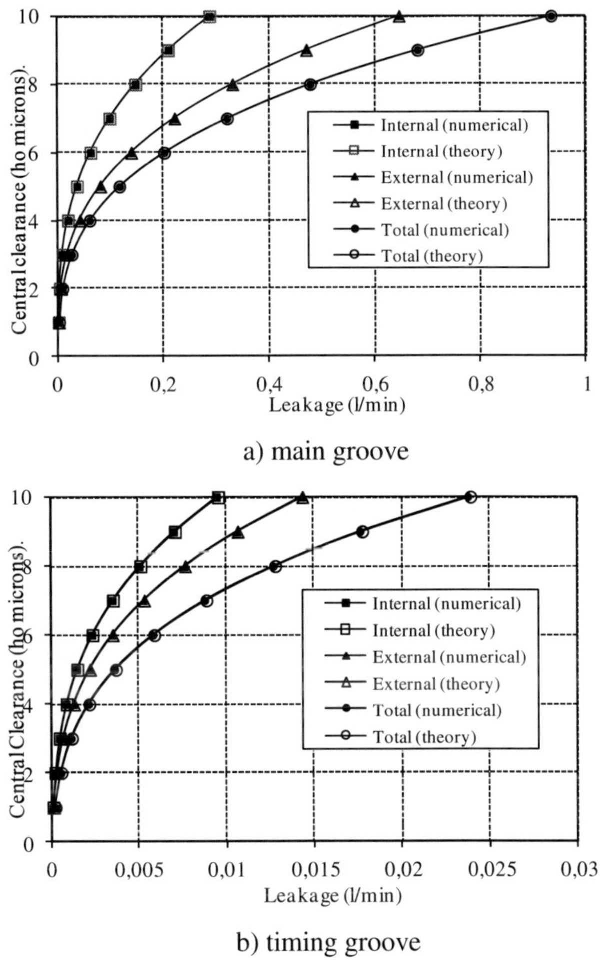

Figure 5.4.6. Leakage between the barrel and plate for different central clearances and for maximum α at each clearance. Internal pressure 25 MPa

5.4.5.1.3. Force Acting on the Barrel

When evaluating the force on the barrel due to the main groove, then according to equation (5.4.41) it is expected that a linear variation between the force and the internal pressure will exist. It is also noted from equation (5.4.41), that such a force will not depend on rotational speed or tilt, but just the remaining geometry.

This lack of significant force variation is due to the groove symmetry and when the symmetry disappears it is expected some further force variation will exist. In fact, the force variation with angular velocity and clearance exists when studying the force created by the timing groove. The force for the barrel-plate versus pressure differential, due to the timing groove, for a constant turning speed and a given clearance is represented by equation (5.4.49) from where it can be noticed that the force slightly increases with the central clearance.

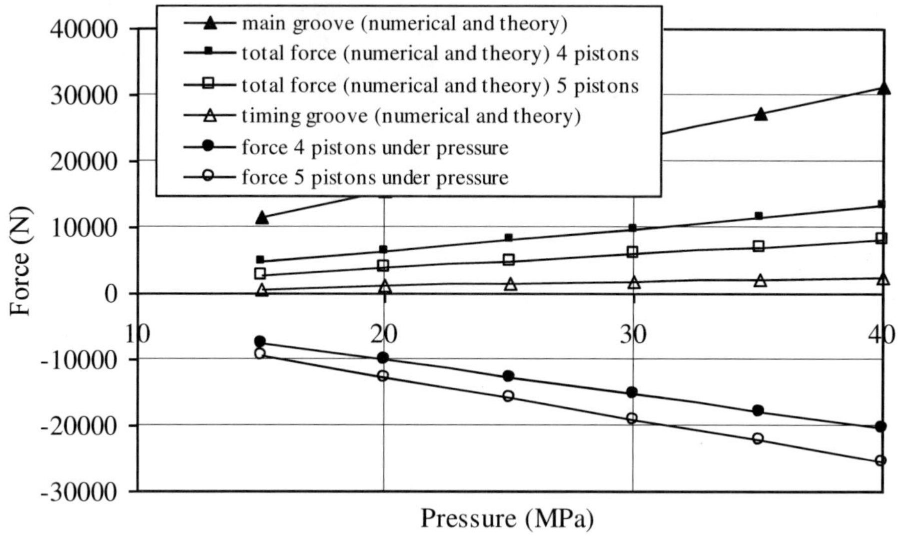

Figure 5.4.7. Mean force over the pump barrel for a set of pressure differentials, numerical solution and theory.

Although not presented, the force decreases with the increase of angular velocity. It must be taken into account that for pressures smaller than 15MPa cavitation is much likely to appear for this design with a 1μm clearance, but only in a very small zone around the small groove entrance. To find out the net force over the barrel, it is necessary to consider the opposing force due to the pressure inside each cylinder chamber, described in section 5.4.2.5. This force will depend on the number of pistons pressurised, therefore in figure 5.4.7 the overall force is presented when four or five pistons are under pressure. The reality is that values will fluctuate between the two total limits.

When computing the overall force due to the main groove, timing groove and cylinder pressure in figure 5.4.7, it is shown that the timing groove plays a very small role. Therefore, all changes in force due parameters other than pressure will be negligible. It can be said that the inclusion of the small groove would bring an increase of force of typically 10% for the case studied. Again, the numerical analysis results and analytical results show excellent agreement, the two approaches producing indiscernible predictions. It is very important to point out that when considering the force due to the cylinders, it is noticed that when the fifth piston is under pressure the total force over the barrel is much more balanced than when just 4 cylinders operate, in fact, the fluctuation of the total force when acting 4 or five pistons is of almost 40% of the total force acting over the barrel.

5.4.5.1.4. Mean Torques about the XX and YY Axes

Perhaps the most interesting aspect of this study regarding average steady state conditions is the fact that torque over the "X" and "Y" axes can now be clearly and explicitly defined.

Consider first the torque about the XX axis, equations (5.4.45) and (5.4.52), due to the main groove and timing groove effect is represented in figure 5.4.8 where it is noticed that the maximum torque over the "X" axis will be found for the highest turning speeds and smallest clearances. Notice that the torque found via the numerical approach is very close to the theoretical predictions.

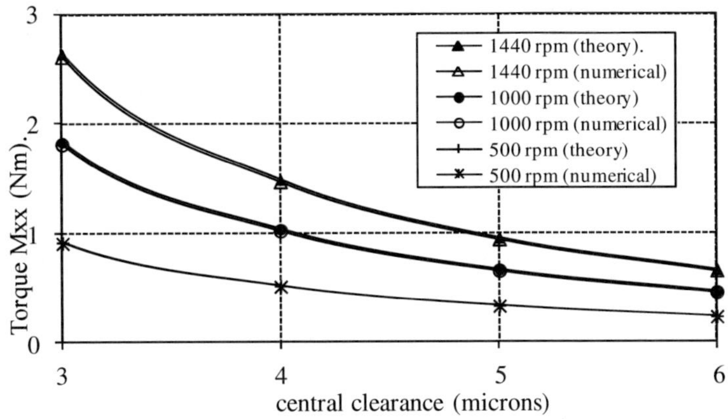

a) main groove

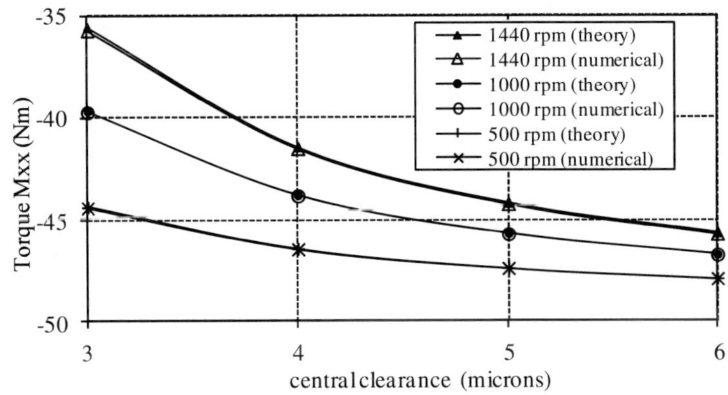

b) timing groove

Figure 5.4.8. Mxx Torque due to the main and timing groove effect, maximum tilt, 25MPa.

To be analytically appropriate, the minimum clearance has been set to 3μm since cavitation exists around the timing groove below this clearance, even though the area of cavitation is very small as previously discussed. It has to be pointed out that the torque increases with the central clearance for the timing groove and will be numerically greater than the effect due to the main groove. Note also that the torque due to the main groove is not zero due to the effect of angular velocity which distorts the pressure distribution around the groove. It is also noticed that the main groove torque acts in the opposite direction to the timing groove torque; the main groove torque is mainly created by the effect of rotational speed, but the timing groove torque is mainly created by the static pressure.

Consider next the torque about the YY axis, equations (5.4.47) and (5.4.53). It is important to point out that the torque due to the main port plate groove, equation (5.4.47), is independent of the central clearance, the turning speed and the tilt. Therefore it is expected that a linear behaviour versus the pressure differential will exist. On the other hand, the torque over the "Y" axis created by the timing groove makes it clear that it will depend on tilt angle, turning speed, central clearance, inlet pressure and geometrical dimensions, equation (5.4.53). Figure 5.4.9 shows the torque Myy for the main groove and the timing groove versus tilt angle and for a set of central clearances.

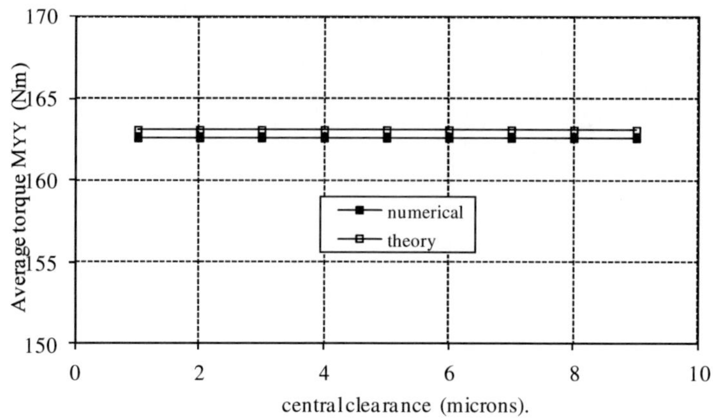

a) main groove

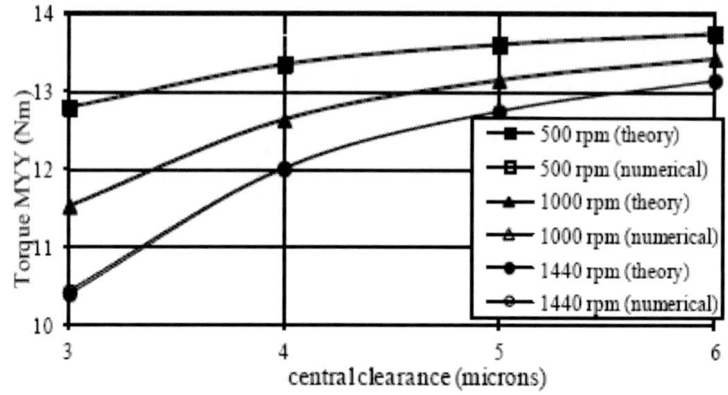

b) timing groove

Figure 5.4.9. Myy torque due to the main and timing groove effect, maximum tilt, 25MPa.

These have been represented in a consistent way to those for Mxx, but it is clear that the contribution from the main groove is substantially constant for a particular pump pressure. Theory also suggests that there is no speed effect and figure 5.4.9a) shows that the effect of central clearance change is negligible.

It is important to note when considering the main groove, that the contribution from the pistons varies with angular position. Therefore figure 5.4.9a) shows the average torque evaluated over a rotation of 40°. It will be demonstrated later that the cylinders torque fluctuation is about 7.5%. The timing groove effect is of course substantially lower than the main groove effect although Myy does increase with increasing central clearance and decreases with increasing speed. This effect is perfectly understandable once it is understood what happens with the pressure distribution on both lands of the timing groove when modifying the barrel turning speed and or the central clearance. As the rotational speed decreases and/or the clearance increases, the pressure distribution across and along the timing groove lands tends to be higher and more symmetrical, giving a higher value for the torques about both axis. Figures 5.4.8b) and 5.4.9b) clearly indicate that the maximum torques due to the timing groove occur at the lowest speed and highest central clearance. As in the previous cases, the results given by the equations presented in this paper and the numerical solution have a very good agreement.

5.4.5.2. Experimental Results

Measurements undertaken are divided into three parts; first some typical main sine waves measured by transducers T1, T2 and T3 will be presented, the test rig being presented in figures 5.4.2, 5.4.3, then, it will be introduced, the average distance between the port plate and the barrel aluminium disk, and finally analysis will be concentrated into understanding a second and much smaller amplitude fluctuation wave, which is sitting on the top of the main wave.

5.4.5.2.1. Position Transducers Direct Measurements

Figure 5.4.10, presents some measured results for the three transducers and for a pump outlet pressure of 10 MPa. This figure just presents the dynamic wave amplitude without considering the transducers zero positioning. The first think to notice is that there is a main sine wave and some fluctuations sitting on the top of it, which will be called the fluctuation wave. The main wave sine wave frequency, when considering the pump rotational speed of 1440 rpm is 24 Hz, the oscillation period being of 0.0416 seconds. This period is of course fixed for all the tests undertaken. The wave peak-to-peak amplitude is mostly due to the barrel run out, which was measured to be 7 microns. It has to be noticed nevertheless, that the peak-to-peak found during experimentation was not constant and varied between 7 and 12 microns, depending on the pressure and the position of the transducer. An analysis of these variations is necessary; both for the main wave and the fluctuation wave. Recall that transducer 1 always gives the maximum fluctuation wave amplitude.

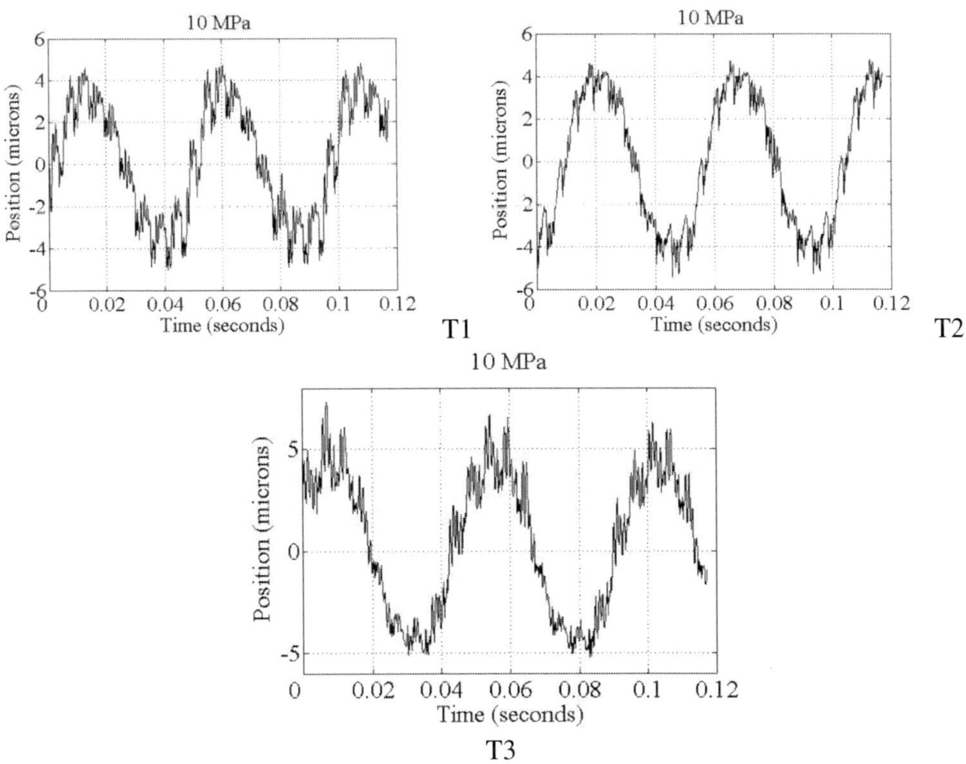

Figure 5.4.10. Position fluctuation measured from transducers T1, T2, T3. Pump pressure10 MPa. 20° swash plate angle and 45°C.

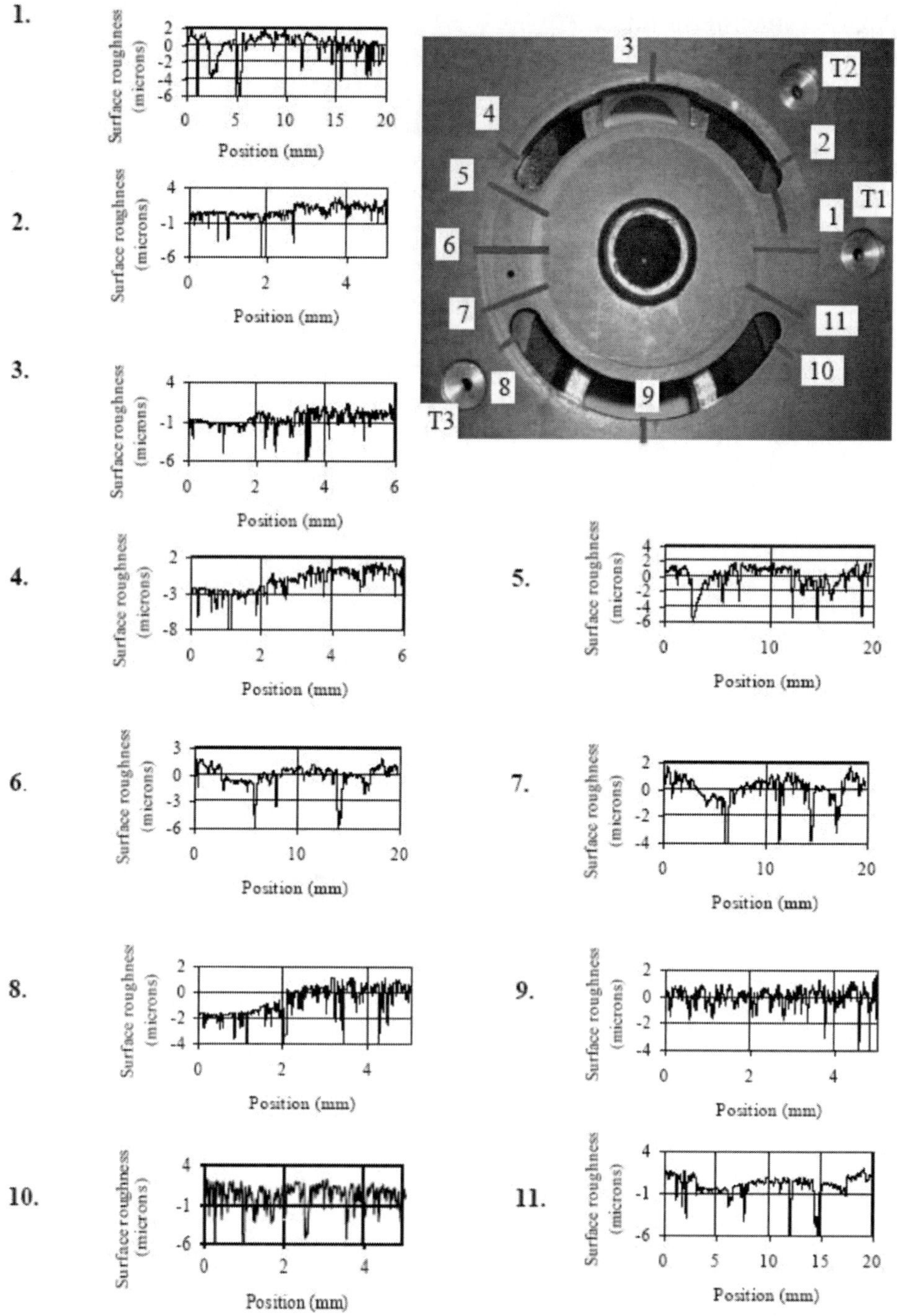

Figure 5.4.11. Port plate face surface direct measurements.

5.4.5.2.2. Average Distance between Port Plate and Barrel Aluminium Disc

Since the rotating aluminium measuring face has a run-out of 7 microns, the barrel portplate gap distance measured is changing over time. Therefore, the average distance is determined from a minimum of 20000 measurements for each position transducer location after taken into account each transducers initial position. It will be noticed that at some points, the average distances are smaller than the real average distance between the barrel face and the barrel aluminium disk, which is of 0.286 mm. This is due to the transducers being located outside the barrel sliding surface external diameter, see figure 5.4.2, where the tilt effect is more clearly noticeable. Also the pump plate has some small erosion due to metal-to-metal contact between the barrel and the plate and this erosion varies between 0 and 4 microns. A photograph of the plate inside face is presented in figure 5.4.11, showing several surface roughness graphs measured at different points located around port plate surface. Notice that erosion of around 4 microns is found in the area located between 45° and 225° anticlockwise from the position of transducer 1 shown in figure 5.4.11, in particular in positions 4 to 8. This clearly indicates that plastic metal to metal contact has occurred in this area. Similar erosion to the one presented in figure 5.4.11 was also found in the barrel sliding surface.

Figure 5.4.12 presents a typical graph showing the dynamic distance variation between the barrel aluminium disk and port plate, as measured by the three position transducers. The 7micron peak-to-peak fluctuation can clearly be seen, although this value is not constant and varies for each transducer and with pump output pressure and oil temperature.

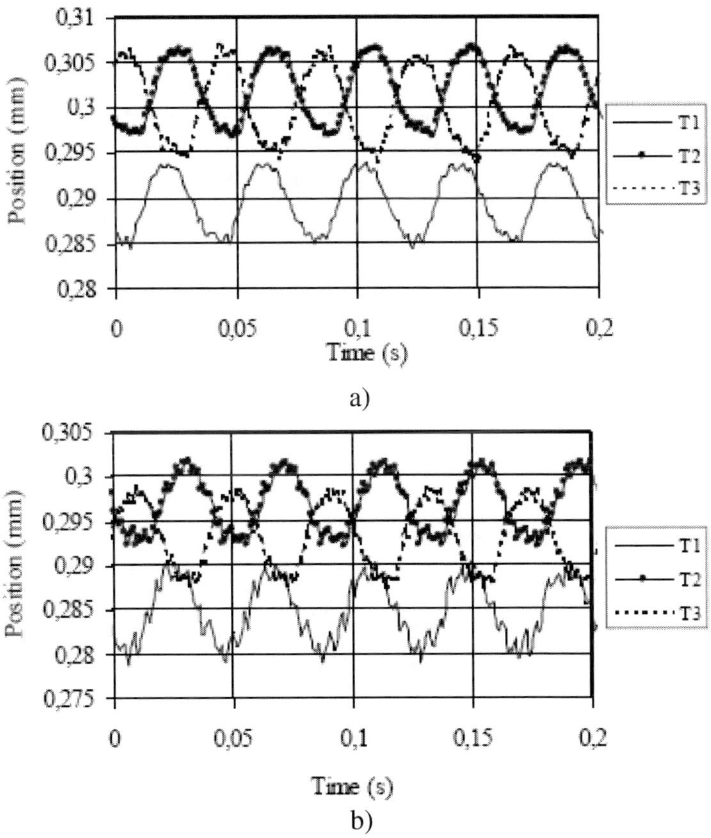

Figure 5.4.12. Typical measured distance between the barrel plate aluminium disk and the port plate. Swash plate angle 20 degrees, 28°C, a) 2.5 MPa; b)17.5 MPa.

It is necessary to consider that the barrel aluminium plate has a static run-out of 7 microns and therefore, the main wave amplitude is mostly due to the barrel dynamic run-out.

It is relevant to point out that transducer 1 always gives the minimum barrel/port plate distance, transducer 2 gives the maximum distance, and more clearly at high temperatures. Transducer 3 presents a distance which is smaller than transducer 2, but the phase given by transducer 3 is displaced about 180º with respect to the phase presented by transducer 2. Since transducers 2 and 3 are respectively on the upper and lower part of the plate, and separated nearly 180º, such phase difference had to be expected.

Transducers 1 and 2, have nearly the same phase, since they are close to each other. In most of the cases studied, the main wave amplitude from transducers 2 and 3, was bigger than the one given by transducer 1, indicating that a barrel oscillation exist about the X axis, defined in figure 5.4.2.

As a result of the previous explanation, the barrel position and film thickness need to be studied as a temporal average. From figure 5.4.12, it is also noticed, especially at high pressures, that a second fluctuation appears superimposed on the first one. This second fluctuation has been called in the previous section, the fluctuation wave, and it is directly related with the barrel vibration frequencies. Such a fluctuation is more relevant from transducer 1 and especially at high pressures and high temperatures, indicating that the torque fluctuation origin has to be along the X axis, see figure 5.4.2. Figure 5.4.12 shows that the average distance between the barrel aluminium plate and the port plate tends to slightly decrease as pressure increases.

In figure 5.4.13 the measured average distance between the barrel aluminium plate and port plate is shown as a function of pressure, temperature and swash plate angle. It is clearly noticed that as pressure increases, the average distance slightly decreases and therefore the average film thickness will be decreasing with pressure. It is also noticed that such a decrease is less than 12 microns and is higher at 20º swash plate angle than at 10º degrees swash plate angle. When comparing figures 5.4.13 a, b, it is noticeable that at high temperatures, the film thickness decrease with pressure increase, is slightly higher (about 6 microns) than at low temperatures, yet the film thickness is considerably higher at low temperatures than at high ones. These results indicate that the maximum average film thickness is in general to be found around the axis position Y+, see figure 5.4.2, then transducer 2 always gives the higher value.

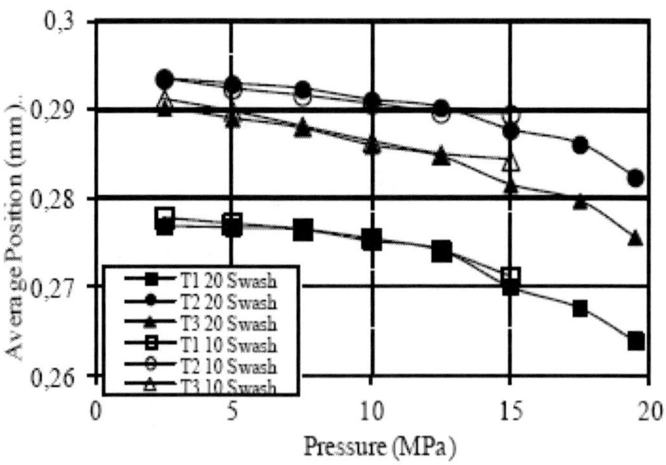

a)

Figure 5.4.13. (Continued).

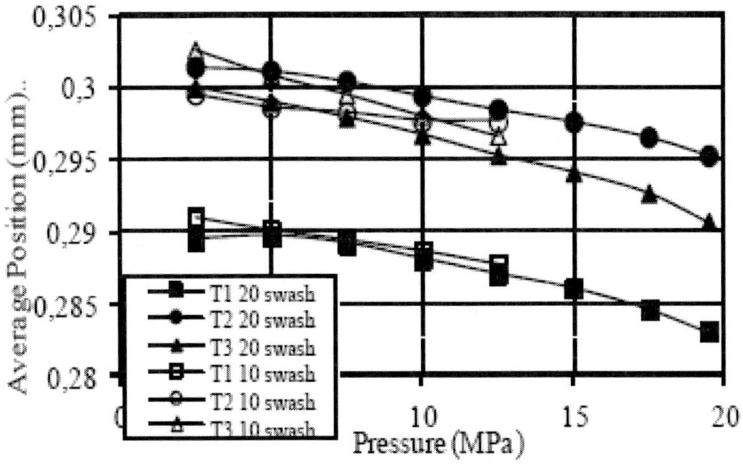

b)

Figure 5.4.13. Average measured transducers position for 10º and 20º swash plate angle. a) Oil temperature 45ºC. b) Oil temperature 28ºC.

Figure 5.4.11 demonstrates that around the area where transducer 2 is located, the metal-to-metal contact is not very significant since the surface is slightly croded. This indicates that the oil film thickness at this point is generally sufficient to prevent plastic metal-to-metal contact.

According to figure 5.4.13a) the average film thickness dependency on swash plate angle is negligible at high temperatures. Figure 5.4.13b) shows that at low pressures, low temperatures and low swash plate angles, transducer 3 is likely to give higher values than transducer 2, producing a shift in the position of the maximum film thickness, as it is reported in figure 5.4.14.

It can be concluded that in general the film thickness around the pump outlet kidney port is higher than the film thickness around the tank kidney port. When oil temperature, swash plate angles and pressure are small, the film thickness maximum position suffers a small displacement versus its generic position.

If the average values of the three transducers are understood as the average clearance measured between the aluminium plate and the port plate internal face, and assuming the three points as belonging to a plane, the rest of the average points around the barrel can be calculated.

The equation of a plane given three points takes the form, $aX+bY+cZ+d=0$. Where the constants a, b, c and d, have to be calculated from the X,Y,Z position of the three given points, using equation (5.4.56).

$$\begin{bmatrix} x-x_1 & y-y_1 & z-z_1 \\ x-x_2 & y-y_2 & z-z_2 \\ x-x_3 & y-y_3 & z-z_3 \end{bmatrix}=0; \qquad (5.4.56)$$

where (x_1, y_1); (x_2, y_2); (x_3, y_3); are the three position points where the transducers are located, values presented in section 5.4.4. The dimensions z_1, z_2, z_3, are the average distances between the port plate face and the barrel aluminium disk face, measured by the three transducers and presented in figure 5.4.13. The average position of any point in the same plane or in a parallel plane can easily be determined; therefore the location of the points where maximum and minimum average film thickness exists can be determined.

Figure 5.4.14 shows the general shape of the average film thickness at the barrel external land central diameter 2* $r_{m\ ext}$, and for two different swash plate angles, 10º and 20º, two different oil temperatures, 28ºC and 45ºC and two pump output pressures, 2.5 and 12.5MPa. It is necessary to consider that, port plate and barrel sliding surfaces are eroded, then, metal to metal contact will in fact appear for a film thickness of less than -4μm, therefore, the values in figure 5.4.14 smaller than -4μm, represents elastic and probably plastic metal to metal contact between the barrel face and the port plate surface.

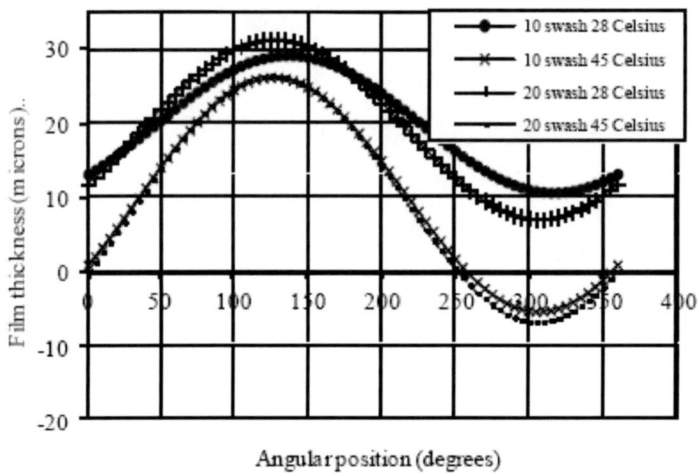

a)

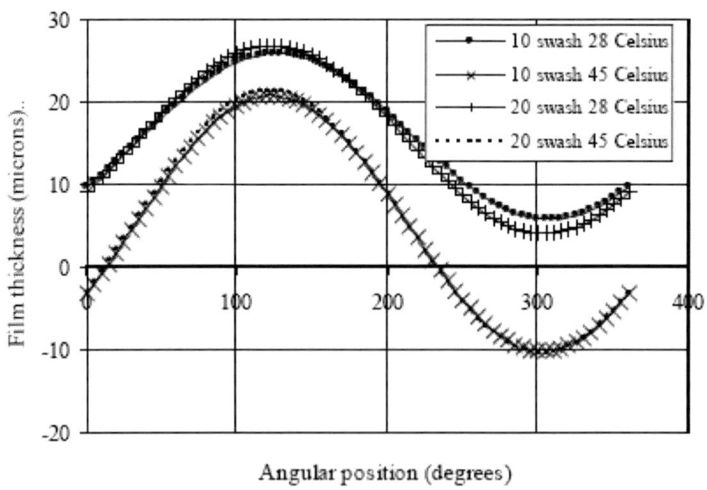

b)

Figure 5.4.14. Average calculated film thickness at two swash plate angles of 10º and 20º, oil temperatures, 28ºC, 45ºC , pressures, a) 2.5 MPa, b) 12.5 MPa.

According to figure 5.4.14, just for high temperatures and high pressures, a very narrow area of the port plate will be in contact with the barrel and as it will be seen in section 5.4.5.2.4, the elastic torque generated due to the contact is just a fraction of the pressure generated torque.

From figure 5.4.14, it is seen that at low pressures and low temperatures, the effect of the swash plate angle, slightly modifies the average film thickness. However as temperature increases, and/or pressure increases, the swash plate angle has no appreciable effect on the oil average film thickness. In both cases, at both low and high pressures, the predominant effect on the average film thickness variation is the oil temperature; as the fluid temperature increases, the film thickness decreases. Also when comparing figures 5.4.14a) with 5.4.14b) it is noticed that as pressure increases, the film thickness tends to slightly decrease. It must be pointed out that under some of the conditions studied, metal-to-metal contact exists and therefore mixed lubrication is often present. It needs to be considered that, according to figure 5.4.11 and due to the port plate erosion, the barrel can easily enter over 4 μm inside the port plate without having metal to metal contact. Mixed lubrication is more severe for high pressures and high oil temperatures. Figure 5.4.14 also shows, that the position of the maximum average film thickness for this particular pump is to be found at 120° clockwise versus the X+ axis, see figure 5.4.2d, and this maximum film thickness position remains constant for nearly all the test performed. At low pressures, low swash plate angles and low temperatures a shift in the average maximum film thickness location is being encountered. For such cases the maximum position is found at 140° clockwise versus the X+ axis, see figure 5.4.2d).

5.4.5.2.3. Barrel Dynamics, Fluctuation Wave

Figure 5.4.15 shows the fluctuation wave superimposed on the sinusoidal main wave. These graphs have been obtained via subtracting the sinusoidal main waves due mostly to the disk runout, from the overall waves measured for each case. An example of such overall waves is to be found in figures 5.4.10 and 5.4.12. As already explained, the fluctuation wave measured by transducer 1 is much higher than the one measured by other transducers. This happens under all conditions studied.

From figure 5.4.15 transducer 1, it can be seen that at 5 MPa, The fluctuation is sharp and irregular, it can not be seen any pattern, its peak to peak amplitude is about 1.75 microns and the frequency oscillates between 190 and 1000 Hz.

At 15 MPa, transducer 1, the fluctuation shape has a clear pattern, which is similar to the theoretical one presented in [45], also Yamaguchi [6] found a similar trend. In fact, as pressure increases the experimental signal increases its similarity with the theoretical wave defined in [45]. There are three peaks appearing between the small and main peaks, its origin will be studied in the next section. The maximum amplitude found is around 3 microns, the frequencies range between 200 and 1100 Hz.

Transducer 2 follows the same trend as transducer 1, since it is located nearby; transducer 3 produces a more random frequency wave, then its location is on the opposite site of the port plate. Although all graphs at different pressures are not presented in this chapter, when studying the evolution of the fluctuation peaks with pressure, it is found that as pressure increases the fluctuation wave amplitude increases, the decrease of oil temperature and or swash plate angle produces a decrease on the fluctuation wave amplitude, but the trend remains the same as the one presented in figure 5.4.15.

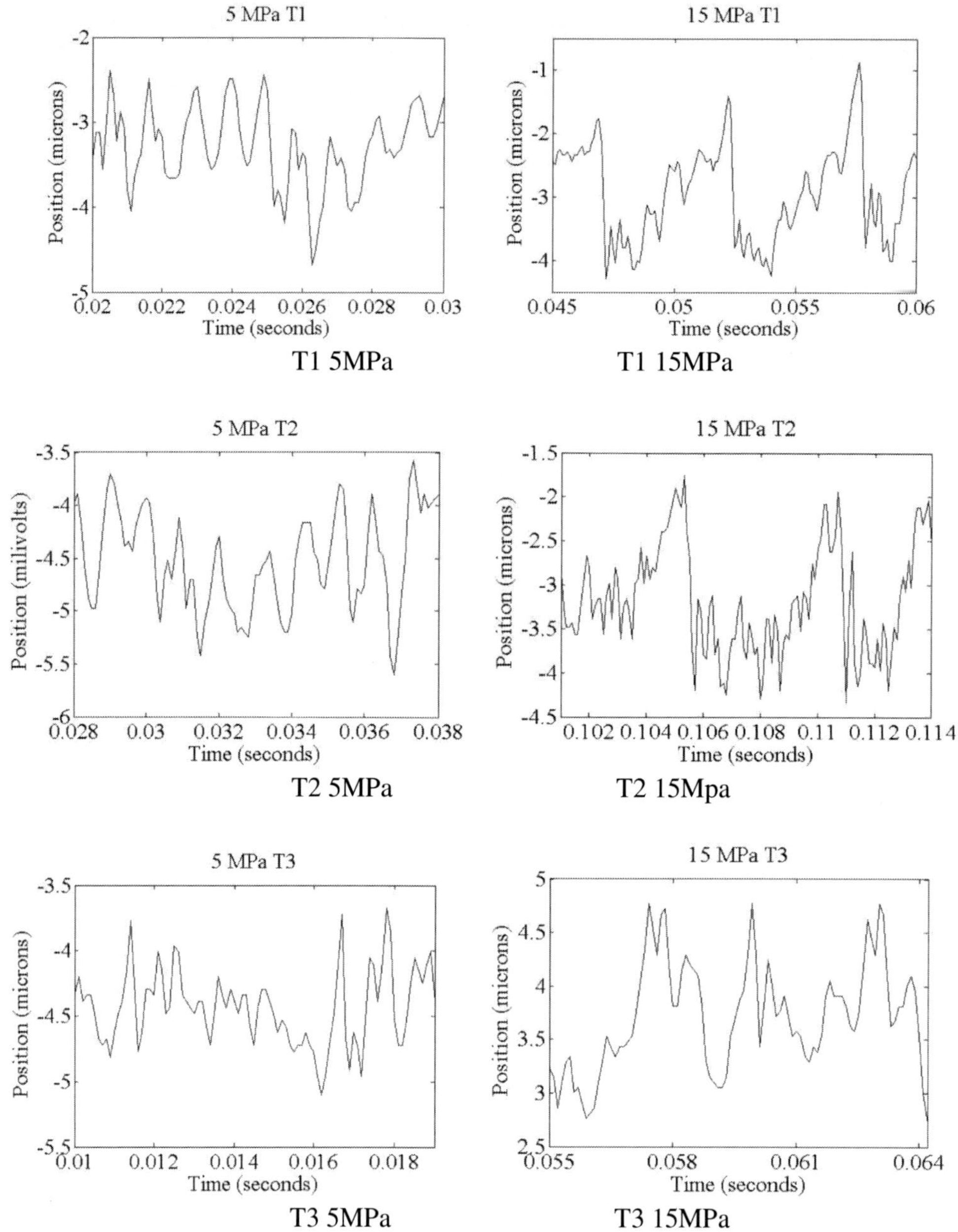

Figure 5.4.15. Position fluctuation from transducers 1, 2, 3 5, 15, MPa, oil temperature 45ºC, 20º swash plate angle.

Notice as well that in all cases studied, the minimum oscillation frequency appearing is about 200 Hz. A frequency of 216 Hz corresponds to the torque created by each piston when entering in contact with the timing groove and leaving the pressure kidney port, Bergada et al [45]. This explains why transducer 1 is always giving a clearer pattern than the other two transducers, then transducer 1 is located nearby the timing groove. Under most of the conditions studied, and more especially at high pressures, a high frequency of around 1100 Hz appears and is believed to be due to the metal-to-metal contact between the barrel face and the port plate. In fact, this frequency is believed to be the number of metal to metal contacts the barrel face and the swash plate are having per second, as it will be clarified in the next section.

5.4.5.2.4. Barrel Dynamics, Simulated Results

The fundamental independent dynamic equations over the two main barrel axes were derived in [45]. Such equations relate the input torques M_{XX}; M_{YY}; due to the pressure distribution acting over the barrel to the barrel inertia, damping coefficient, spring constant and barrel/port plate friction. However, as shown in figure 5.4.11, the port plate suffers elastic and plastic deformation. Therefore, in order to consider metal-to-metal elastic torque, which constant is M_E, a new term has been included in the previous barrel dynamic model, the barrel modified dynamic equations are summarised as equations (5.4.57; 5.4.58), see figure 5.4.1.

$$I_x\,\ddot{\delta}+B_x\,\dot{\delta}+K_x(\delta-\delta_0)+F_{forces_x}*\text{sign}(\dot{\delta})+\overbrace{\frac{E\,A_0}{T_{hi}}r^2}^{M_E}\,\max(|\delta|-\delta_L;0)=M_{XX} \quad (5.4.57)$$

$$I_y\,\ddot{\alpha}+B_y\,\dot{\alpha}+K_y(\alpha-\alpha_0)+F_{forces_Y}*\text{sign}(\dot{\alpha})+\overbrace{\frac{E\,A_0}{T_{hi}}r^2}^{M_E}\,\max(|\alpha|-\alpha_L;0)=M_{YY} \quad (5.4.58)$$

To develop the metal to metal elastic torque, the following procedure has been used:

The force F exerted by the material when compressed or stretched by ΔL and as a function of the original port plate thickness T_{hi}, contact area A_0 and the Young modulus E is:

$$F=\frac{E\,A_0\,\Delta L}{T_{hi}} \quad (5.4.59)$$

The distance the barrel enters into the port plate can be given as: $\Delta L = r\,\delta$. Taking into account that metal to metal contact appears just when the tilt is higher than the maximum angular value in which still does not exist metal to metal contact δ_L, then:

$$\Delta L = r\,(|\delta|-\delta_L). \quad (5.4.60)$$

$|\delta|$ is the angular value calculated with simulink, the absolute value considers that metal to metal contact can exist in both axis directions. If the angular distance displaced by the barrel is smaller than δ_L, there will not exist metal to metal contact, and this is expressed as:

$$\Delta L = r\;\max(|\delta|-\delta_L;0). \quad (5.4.61)$$

Finally, the torque created by the elastic forces will be:

$$M_{EF} = \frac{E\,A_0}{T_{hi}}\, r^2 \max\left(|\delta| - \delta_L ; 0\right) \tag{5.4.62}$$

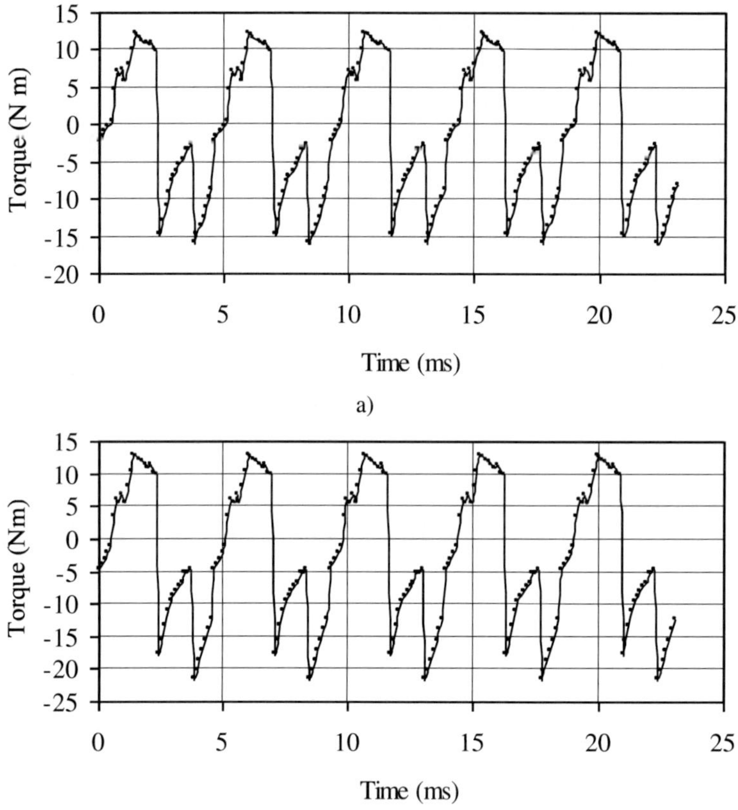

Figure 5.4.16. Calculated input torques Mx ,pump output pressures a) 15 MPa, b) 17.5 MPa.

The damping coefficient for the barrel piston assembly may be considered small, but it is highly dependent on oil temperature. Elastic metal to metal forces exist between the barrel face and the port plate; they act in opposite direction than the input torque and depend on the Young modulus, the area of contact and the distance the barrel penetrates into the plate. Although such elastic forces can in general be considered small when studying the barrel X, Y axis acceleration, they play a relevant role regarding the barrel dynamics. The torsional spring constant created by the pump axis and spring, although difficult to accurately determine, has to play an important role when studying the barrel dynamics, the spring located at the bottom of the barrel, for this particular pump is 22.5mm long, being its constant of 134.8 N/mm. Inertia effects will also need to be considered. Since the moment of inertia changes as the barrel rotates, the barrel plus pistons moment of inertia will be time dependent, although negligible in this example. The peak to peak variation is around 0.5% of the mean values Ix = Iy = 0.0127 Kg m^2. The input torques, Mx, M_Y, for the main and timing groove, were developed in [45] and presented here as equations 5.4.45, 5.4.47, 5.4.52 and 5.4.53. These torques act as inputs and are generated as data files to allow a dynamic solution via the MATLAB Simulink package.

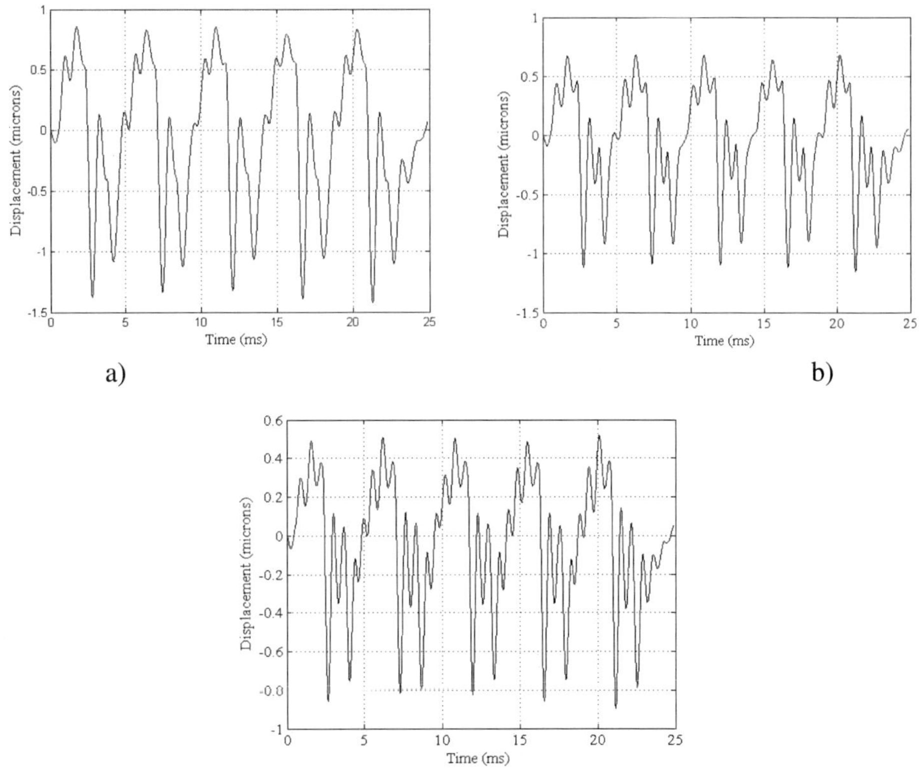

c)

a)15 MPa, K=6*10^5 Nm, B=35 Nm s/rad, M_E =0 Nm.
b)15 MPa, K=6*10^5 Nm, B=35 Nm s/rad, M_E =1.68*10^5 Nm.
c)15 MPa, K= 6*10^5 Nm, B=35 Nm s/rad, M_E =5.6*10^5 Nm.

Figure 5.4.17. Fluctuation wave simulation results.

Since this section is focused on understanding the origin of the barrel high frequency fluctuations, just the dominant fluctuation torque about the X axis (see figure 5.4.1), will be considered. Then as previously explained, the origin of the fluctuation wave is to be found along the Y axis (figure 5.4.1), (X axis from figure 5.4.2). Figure 5.4.16 presents three input perturbation torques using equations (5.4.45, 5.4.52) Mxx, generated versus the X axis, defined in figure 5.4.1, for two different pump output pressures, 15 and 17.5 MPa. The process of calculation of this torque was fully explained in [45] and due to its complexity shall not be repeated here. The graphs presented in figure 5.4.16 are also to be found in [46]. Introducing these torques as inputs, several dynamic barrel fluctuations as a function of the damping coefficient B, spring torsional constant K and the elastic metal-to-metal reaction torque constant M_E were found and presented in figure 5.4.17. Despite the fact that for the present simulation, the metal-to-metal friction torque has been considered negligible, since the objective was to study the effect of elastic forces, it has been found that the inclusion of such torque produces a similar effect than the elastic torque. Both enhance the fluctuation wave peaks amplitude. From these results it is noticed that if the spring torsional constant is big enough, the barrel will be able to follow the input torque generated by the pump. In fact, the metal-to-metal elastic forces play a decisive role when studding the barrel dynamics, as it is seen from figures 5.4.17a), b), c).

Notice that, the simulated results shown in figure 5.4.17, resembles very closely the experimental ones presented in figure 5.4.15. It is therefore demonstrated that the frequencies of around 1100 Hz, are in fact appearing when the pump barrel has a very high stiffness, and able to follow the small torque perturbations created by the pump versus the X axis. Such large stiffness appears at high pressures and high temperatures showing that under these conditions the barrel response is very fast

It is demonstrated that when the elastic torque is considered, the fluctuation peaks amplitude increase, figure 5.4.17b), c), clearly showing that elastic metal to metal forces need to be considered when aiming to understand the full barrel dynamics. In reality, elastic barrel/port plate forces can be quite different in time, since the metal-to-metal forces depend on the random contact between surfaces. This explains why the measured frequencies of the fluctuation peaks are not constant, although vary around 1100 Hz.

5.4.6. Conclusion

1. A set of new equations have been developed and are allowing progress to be made on the analytical solution for the pressure distribution, leakage, force and both torques between the barrel and port plate of an axial piston pump. Reynolds equation of lubrication has also been compared using a numerical method specifically applicable to the gap between the barrel and port plate. Results from the numerical model and the theoretical model were compared giving a very good agreement in all cases.
2. Leakage was found to be greater across the external land than the internal land, for the same operating pressure, and typically by a factor of 2 whether it be related to pressure or central clearance. As expected, the small timing groove produced a significantly lower leakage than the main groove and could probably be neglected from a total flow loss point of view. It was found that cavitation in pumps is more likely to appear for smaller clearances, smaller output pressures and bigger turning speeds.
3. The effect of precisely defining piston position was shown to be crucial when calculating the total force on the barrel, the total force reducing as the number of active pistons changes from 4 to 5 during one cycle. This was shown to be due to the balance between piston pressure effects and groove effects, the former becoming more dominant as the number of pistons active changes from 4 to 5.
4. Both dynamic torques acting over the barrel XX and YY axes were carefully studied and were introduced into a Simulink model to evaluate the barrel temporal position. Comparisons between numerically-analysed torques presented by another researcher and the ones presented in this paper show a good qualitative agreement, and clarify the role of the small grooves cut on the port plate regarding torque dynamics. In particular, the importance of precise modelling of the flow continuity mechanism was shown to be crucial in predicting the correct waveform shapes.
5. Temporal torque calculations showed a marked difference in shape for each axis considered, and due to piston pressure effects, the peak to peak values being much greater across the XX axis than the YY axis, although the average torque over the

YY axis is much higher than the one over the XX axis. This extends to the overall torque fluctuation when groove effects are also taken into account.

6. A test rig has been created to measure the barrel dynamic displacement, the results clearly show the torque increase as each piston groove is connected to the timing groove, as established by the theory presented. This was in spite of the requirement to extract the required data from a noisy signal due to barrel run-out. A good correlation between analytical and experimental results was found, and it can be concluded that barrel dynamic displacement above run-out displacement is created by torque dynamics. The torque increase, as the piston groove faces the timing groove, is responsible for the small peak fluctuation presented.
7. Barrel/port plate average film thickness and barrel dynamics have been experimentally evaluated for a range of operating conditions.
8. Average film thickness decreases mostly with the increase of the oil temperature, and has a smaller decrease with the increase of pump outlet pressure. Therefore at low pressures and low temperatures the average film thickness is at its maximum. As pressure and temperature increases, film thickness decreases, and mixed lubrication it is expected in most of the barrel/port plate surface. The barrel then undergoes a wobbling motion. The swash plate angle variation has a small effect on the average film thickness; nevertheless it has been found that at small swash plate angles, the average film thickness slightly increases.
9. The average film thickness around the pump outlet kidney port is thicker than the film thickness around the tank kidney port for nearly all conditions studied. The maximum film thickness for this particular pump is located around 120 degrees anticlockwise from the (X+) axis for most of the cases studied.
10. Due to the barrel aluminium measuring plate run-out, the position measured by the transducers presents a sinusoidal-type variation with a frequency of 24 Hz. In addition, a fluctuating component was found to be sitting onto the main wave. The displacement fluctuating wave has two main peaks, a small one related to the torque created when each piston enters in contact with the timing groove, and a main one created when each piston leaves the pressure kidney port. This occurs at a frequency of 216 Hz, that is, pump speed multiplied by the number of pistons. Along with the main fluctuation component, a second component occurs at a frequency around 1000-1100 Hz. This second component is more clearly seen when the system stiffness is high, system stiffness depending on the pump spring constant, pump central axis constant, and elastic/plastic metal-to-metal reaction forces.
11. As pressure increases, film thickness decreases, metal-to-metal contact increases and the fluctuating wave small perturbation peaks are more evident. The damping coefficient plays an important role regarding the barrel dynamics since as temperature increases, the damping coefficient decreases, allowing the barrel to move more freely. This is why at high temperatures and high pressures the fluctuation wave is more clearly seen. The elastic/plastic metal-to-metal forces, enhances the fluctuating wave peaks increasing their amplitude.

5.4.7. References

[1] Helgestad BO, Foster K, Bannister F K. (1974). Pressure Transients in an Axial Piston Hydraulic Pump. *Proceedings of the Institution of Mechanical Engineers*. 188:17 / 74. 189 – 199.

[2] Martin MJ and Taylor B. (1978). Optimised Port Plate Timing for an Axial Piston Pump. 5th Int Fluid Power Symposium. Cranfield, England. September 13 – 15, B5 – 51 66.

[3] Edge KA and Darling J. (1989). The Pumping Dynamics of Swash Plate Piston Pumps. ASME. *J Dynamic Systems, Measurement, and Control*. 111, 307-312.

[4] Jacazio G and Vatta F. (1981). The Block-lift in Axial piston Hydraulic Motors. The ASME/ASCE Bioengineering, *Fluids Engineering and Applied Mechanics Conference*. June 22-24, Boulder, Colorado USA, 1-7.

[5] Yamaguchi A. (1987). Formation of a Fluid Film between a valve plate and a Cylinder Block of piston Pumps and Motors. (2nd Report, A Valve Plate with Hydrostatic Pads). *Int Journal of the Japan Society of Mechanical Engineers*. 30:259, 87-92.

[6] Yamaguchi A. (1990). Bearing/seal characteristics of the film between a valve plate and a cylinder block of axial piston pumps: Effects of fluid types and theoretical discussion. *The Journal of Fluid Control*. 20:4, 7-29.

[7] Matsumoto K and Ikeya M. (1991). Friction and leakage characteristics between the valve plate and cylinder for starting and low speed conditions in a swashplate type axial piston motor. *Trans of the Japan Society of Mechanical Engineers,* part C. 57, 2023-2028.

[8] Matsumoto K and Ikeya M. (1991). Leakage characteristics between the valve plate and cylinder for low speed conditions in a swashplate-type Axial Piston Motor. *Trans of the Japan Society of Mechanical Engineers, part* C. 57, 3008-3012.

[9] Kobayashi S and Matsumoto K. (1993). Lubrication between the valve plate and cylinder block for low speed conditions in a swashplate-type axial piston motor. *Trans of the Japan Society of Mechanical Engineer*s, part C. 59, 182-187.

[10] Weidong G and Zhanlin W. Analysis for the real flow rate of a swash plate axial piston pump. *Journal of Beijing University of Aeronautics and Astronautics*. 22: 2, 223-227.

[11] Yamaguchi A. (1997). Tribology of Hydraulic Pumps. *ASTM Special technical publication*. No 1310, 49-61.

[12] Yamaguchi A, Sekine H, Shimizu S, Ishida S. (1987). Bearing/seal characteristics of the Oil film between a valve plate and a cylinderblock of axial pumps. *JHPS* 18:7, 543-550.

[13] Manring ND. (2000). Tipping the cylinder block of an axial-piston swash-plate type hydrostatic machine. *ASME Journal of Dynamic Systems Measurement and Control*. 122, 216-221.

[14] Manring ND. (2003). Valve-plate design for an Axial Piston pump operating at low displacements. *ASME J. of Mechanical Engineering*. 125, 200-205.

[15] Zeiger G and Akers A. (1985). Torque on the Swashplate of an Axial Piston Pump. Journal of Dynamic Systems, Measurement and Control. *ASME*. 107, 220-226.

[16] Zeiger G and Akers A. (1986). Dynamic analysis of an axial piston pump swashplate control. *Proc of the Institution of Mechanical Engineers*. 200:C1, 49-58.

[17] Inoue K and Nakasato M. (1994). Study of the operating moment of a swash plate type axial piston pump. First report, effects of dynamic characteristics of a swash plate angle supporting element on the operating moment. *The Journal of Fluid Control*, 22:1, 30-46.

[18] Inoue K and Nakasato M. (1994). Study of the operating moment of a swash plate type axial piston pump. Second report, effects of dynamic characteristics of a swash plate angle supporting element on the cylinder pressure. *Journal of Fluid Control*. 22:1, 7-29.

[19] Manring ND and Johnson RE. (1996). Modeling and designing a variable-displacement open-loop pump. *Journal of Dynamic Systems Measurement and Control,* 118, 267-271.

[20] Wicke V, Edge KA, Vaughan D. (1998). Investigation of the effects of swash plate angle and suction timing on the noise generation potential of an axial piston pump. FPST-5, Fluid Power Systems and Technology, *ASME*, 77-82.

[21] Manring ND. (1999). The control and containment forces on the swash plate of an axial piston pump. *Journal of Dynamic Systems Measurement and Control,* 121, 599-605.

[22] Gilardino L, Mancò S, Nervegna N, Viotto F. (1999). An experience in simulation the case of a variable displacement axial piston pump. *Forth JHPS International Symposium,* paper 109, 85-91.

[23] Ivantysynova M, Grabbel J, Ossyra JC. (2002). Prediction of Swash Plate Moment Using the Simulation tool CASPAR. Proceedings of IMECE - ASME International Mechanical Engineering Congress & Exposition. November 17-22, New Orleans, Louisiana USA. IMECE2002-39322, 1-9.

[24] Manring ND. (2002). The control and containment forces on the swash plate of an axial piston pump utilizing a secondary swash-plate angle. Proceedings of the American Control Conference. Anchorage, AK-USA May 8-10, 4837-4842.

[25] Manring ND. (2002). Designing a Control and Containment device for a Cradle-mounted, Axial-Actuated Swash Plates. *J. of Mechanical Design.* 124, 456-464.

[26] Bahr MK, Svoboda J, Bhat RB. (2003). Vibration analysis of constant power regulated swash plate axial piston pumps. *J. of Sound and Vibration*, 259:5, 1225-1236.

[27] Ivantysynova M. (2002). A New Approach to the Design of Sealing and Bearing Gaps of Displacement Machines. Fluid Power forth JHPS International Symposium. 45-50.

[28] Ivantysynova M; Huang C. (2002). Investigation of the Flow in Displacement Machines Considering Elastohydrodynamic Effect. *Proceedings of the 5th JFPS International Symposium on Fluid Power,* Nara, Japan. 1, 219-229.

[29] Jouini, N. and Ivantysynova, M. (2008). Valve Plate Surface Temperature Prediction in Axial Piston Machines. *Proc. of the 5th FPNI PhD Symposium*, Cracow, Poland, 95-110.

[30] Baker, J. and Ivantysynova, M. (2008). Investigation of Power Losses in the Lubricating Gap between the Cylinder Block and Valve Plate of Axial Piston Machines. *Proc. of the 5th FPNI PhD Symposium*, Cracow, Poland, 302-319. Best paper award.

[31] Ivantysynova, M. (2008). Innovations in Pump Design – What are future directions?. *Proceedings of the 7th JFPS International Symposium on Fluid Power*, Toyama, 59-64. Invited lecture organized session agriculture and mining machinery.

[32] Ivantysynova, M. and Baker, J. (2009). Power Loss in the Lubricating Gap Between Cylinder Block and Valve Plate of Swash Plate Type Axial Piston Machines. *International Journal of Fluid Power*, 10:2, 29-43.

[33] Pelosi, M, Zecchi, M. and Ivantysynova, M. (2010). A Fully Coupled Thermo-Elastic Model for the Rotating Kit of Axial Piston Machines. *ASME/ Bath International Symposium on Fluid Power and Motion Control* (FPMC2010), 217-234.

[34] Baker, J. and Ivantysynova, M. (2009). Advanced surface design for reducing power losses in axial piston machines.*Proceedings 11th Scandinavian International Conference on Fluid Power*, SICFP'09, June 2-4, Linköping, Sweden.

[35] Pelosi, M. and Ivantysynova M. (2009). A Novel Fluid-structure Interaction Model for Lubricating Gaps of Piston Machines. *Proceedings of the Fifth Fluid Structure Interaction Conference*, eds. C.A. Brebbia, WIT Press, Southampton, UK. 13-24.

[36] Ivantysynova M; Lasaar R. (2004). An Investigation into Micro – and Macrogeometric Design of Piston/Cylinder Assembly of Swash plate machines. *International Journal of Fluid Power,* 5:1, 23-36.

[37] Seeniraj, G. (2009). Model based optimization of axial piston machines focusing on noise and efficiency. PhD thesis, Purdue University.

[38] Johansson A. (2005). Design Principles for Noise Reduction in Hydraulic Piston Pumps Simulation, Optimisation and Experimental Verification. PhD thesis, Linkoping University.

[39] Metha V. (2006). Torque ripple attenuation for an axial piston swash plate type hydrostatic pump, noise considerations. PhD Thesis, Missouri University.

[40] Seeniraj G. K. and Ivantysynova M. (2006). Impact of valve plate design on noise, volumetric efficiency and control effort in an axial piston pump. Proceedings of ASME International Mechanical Engineering Congress and Exposition, IMECE2006, Chicago, Illinois, USA.

[41] Seeniraj G. K. and Ivantysynova M. (2008). Multi-obejctive optimization tool for noise reduction in axial piston machines. *SAE International Journal of Commercial Vehicles* 1:1, 544-552.

[42] Hong YS, Lee SY. (2008). A comparative study of Cr-X-N (X=Zr, Si) coatings for the improvement of the low speed torque efficiency of a hydraulic piston pump. *Metals and Materials International.* 14:1, 33-40.

[43] Hong YS, Lee SY, Kim SH, Lim HS. (2006). Improovement of the low-speed friction characteristics of a hydraulic piston pump by PVD-coating of TiN. *Journal of Mechanical Science and Technology*. 20:3, 358-365.

[44] Mandal NP, Saha R, Sanyal D. (2008). Theorethical simulation of ripples for different leading-side groove volumes on manifolds in fixed-displacement axial-piston pump. IMechE part I: *J. Systems and Control Engineering*. 222, 557-570.

[45] Bergada JM, Watton J, Kumar S. (2008). Pressure, Flow, Force and Torque Between the Barrel and port plate in an axial piston pump. *ASME Journal of Dynamic Systems, Measurement and control*. 130:1, 011011-1/16.

[46] Bergada JM, Davies D Ll, Kumar S, Watton J. (2012). The effect of oil pressure and temperature on barrel film thickness and barrel dynamics of an axial piston pump. *Meccanica*. 47, 639-654.

5.5. SPHERICAL JOURNAL BEARING

5.5.1. Introduction

A spherical bearing is a common device used in pumps to permit rotation about a central point in two orthogonal directions, offer an unequaled combination of high load capacity, high tolerance to shock load and self aligning ability and play an important role in pump performance. A great amount of research has been done on spherical bearings in piston pumps, although mainly the different researchers focused on studding the friction piston ball – slipper. Figure 5.5.1 represents a schematic diagram of spherical piston – slipper bearing under consideration in this work.

Spherical journal bearings in piston pumps have been studied in some depth, mainly the different researchers focused on studding the friction piston ball – slipper. Böinhoff [1] performed a first analytical study on the spherical bearing friction. Hooke et al [2] pointed out that friction on a piston ball, plays a major role in determining the behavior of the slipper. Later, Hooke at al [3] studied experimentally the couples acting on the slipper ball; they concluded that lubrication is under all conditions deficient, appearing metal to metal contact. Friction increases with pressure and small slipper plate tilt angles. Ball friction causes the piston to rotate. In [4, 5] Iboshi and Yamaguchi pointed out that friction on the spherical bearing affects significantly the slipper tilt angles, rotational speed affects the central clearance slipper-plate. It must be said that those results agree very well with Hooke`s considerations.

In [6] Hooke and Li analysed carefully the three different tilting couples acting on the slipper, finding that the tilting couple due to friction at the slipper running face, is much smaller than the ones created at the piston – cylinder, piston – slipper interfaces.

Kobayashi et al [7] studied experimentally the friction torque characteristics between the piston ball and the slipper. Different surface coatings, clearances and surface roughness were analyzed. They found that friction torque increased with pressure increase, tending to an asymptotic value, an increase on the swash plate angle created a decrease on the friction torque, the friction torque decreased as clearance increased; there was also a slight decrease on friction torque with increase of oil temperature. Regarding the materials, they found that the use of PST05 solid lubricant witch was coated with PTFE gave the lowest friction torque at all pressures. The use of different surface roughness did not prove any significant change on friction torque. The measurements on leakage showed a small leakage decrease when slipper spin increased.

Spherical bearing was also experimentally studied in a ball-piston pump by Abe at al [8, 9], the work was mainly focused on the friction coefficient, although piston velocity, and pressure contact between sphere and cam were also evaluated, they had a first attempt on explaining pressure distribution on the spherical bearing. Different shapes of pistons with restricting holes were evaluated.

Elastohydrostatic lubrication of piston balls and slipper bearings was studied by Kobayashi et al [10], in fact, most of the work was focused in presenting the slipper swash plate gap and the leakage as a function of the slipper main land length and the slipper central hole. It was found that minimum film thickness and leakage through slipper tended to reach an asymptotic value with the increase of the slipper land length.

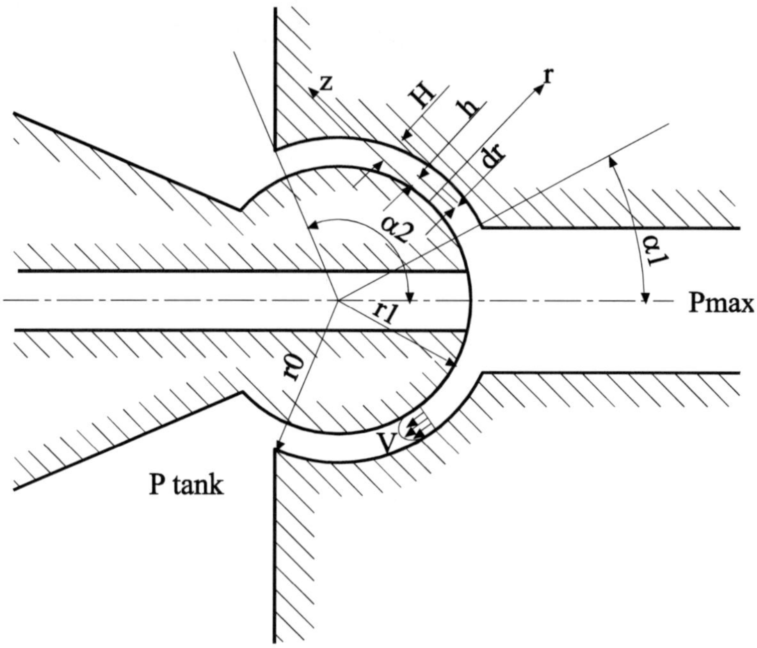

Figure 5.5.1. Spherical Journal main dimensions.

The transformed Reynolds equation of lubrication in spherical coordinates was accomplished by Meyer [11], the integration of such equation gives the pressure distribution along the sphere.

It is quite clear from the literature available that good research has been done on spherical bearing and the window for the further research is quite narrow. In what follows a very simple model of spherical bearing will be presented. Ball and bearing are considered as static. The model about to be presented, is to be found in Bergada [12] using two different approaches, one assuming the Poiseuille flow profile in the spherical bearing clearance and calculating the leakage from it and the second, by direct integration of Reynolds equation in spherical coordinates. In both cases an expression for leakage and pressure distribution was developed. Both techniques used by Bergada [12] are implemented in this chapter.

5.5.2. Mathematical Analysis

From all the studies presented, it can be stated that most of the work done on spherical bearings is related to friction torque, some attempts in evaluating pressure distribution and leakage were carried out experimentally by Abe et al [8], and recently Meyer [11] presented the transformed Reynolds equation of lubrication in spherical coordinates the integration giving the pressure distribution along the spherical bearing. Nevertheless, and due to the lack of information especially on the leakage thorough a spherical journal bearing it was decided to develop the following equations. Note that the leakage between the piston and slipper spherical bearing is expected to be small compared with the slipper-plate or the barrel-plate one. However it is interesting to develop the equations, then a thorough evaluation of the entire pump can later be performed.

The spherical journal main dimensions are:

$r_1 = 5.5$ mm
H = 2.54 microns
$\alpha_1 = 10.47$ degrees.
$\alpha_2 = 125.09$ degrees.

From figure 5.1 it can be stated:

$$\alpha_1 < \alpha < \alpha_2 \tag{5.5.1}$$

The surface differential in the gap between both spheres can be given as:

$$\int ds = \int_0^H 2\pi\,(r_1 + r)\sin\alpha\; dr \tag{5.5.2}$$

The volumetric flow rate between both spheres shall be presented as:

$$dQ = \int_0^H V\,2\pi\,(r_1 + r)\sin\alpha\; dr \tag{5.5.3}$$

The velocity distribution between two parallel plates, according to Poiseulle, shall be:

$$V = -\frac{1}{\mu}\frac{dP}{dz}\frac{r}{2}(H - r) \tag{5.5.4}$$

The volumetric flow will take the form:

$$dQ = \int_0^H -\frac{1}{\mu}\frac{dP}{dz}\frac{r}{2}\,(H - r)\,2\pi\,(r_1 + r)\sin\alpha\; dr \tag{5.5.5}$$

The relationship between angle differential and arc differential is:

$$dz = (r_1 + \frac{H}{2})\,d\alpha \tag{5.5.6}$$

Substituting equation (5.6) into (5.5) and integrating it is obtained:

$$Q = -\frac{1}{\mu}\frac{dP}{d\alpha}\frac{1}{(r_1 + \frac{H}{2})}\pi\sin\alpha\left[r_1\frac{H^3}{6} + \frac{H^4}{12}\right] \tag{5.5.7}$$

Pressure distribution along spherical journal angular position, will take the form:

$$dP = -Q\,\mu\,(r_1 + \frac{H}{2})\frac{1}{\pi\left[r_1\frac{H^3}{6} + \frac{H^4}{12}\right]\sin\alpha}d\alpha \tag{5.5.8}$$

The integration limits are:

$$\int_{P_{max}}^{P_{Tank}} dP = -Q\,\mu\,(r_1 + \frac{H}{2})\frac{1}{\pi\left[r_1\frac{H^3}{6} + \frac{H^4}{12}\right]}\int_{\alpha_1}^{\alpha_2}\frac{d\alpha}{\sin\alpha} \tag{5.5.9}$$

After integration it is obtained:

$$P_{max} - P_{tank} = Q\,\mu\,(r_1 + \frac{H}{2})\frac{1}{\pi\left[r_1\frac{H^3}{6} + \frac{H^4}{12}\right]}\ln\left[\frac{tg\frac{\alpha_2}{2}}{tg\frac{\alpha_1}{2}}\right] \tag{5.5.10}$$

The leakage between the two spheres will therefore be:

$$Q = \frac{(P_{max} - P_{tank})\;\pi\,(r_1\frac{H^3}{6} + \frac{H^4}{12})}{\mu(r_1 + \frac{H}{2})\;\ln\left[\frac{tg\frac{\alpha_2}{2}}{tg\frac{\alpha_1}{2}}\right]} \tag{5.5.11}$$

From equation (5.10) it can be obtained the pressure distribution as a function of the angular position, being:

$$P = P_{max} - \frac{Q\;\mu\;(r_1 + \frac{H}{2})\;\ln\left[\frac{tg\frac{\alpha}{2}}{tg\frac{\alpha_1}{2}}\right]}{\pi\;(r_1\frac{H^3}{6} + \frac{H^4}{12})} \tag{5.5.12}$$

Substituting equation (5.11) into (5.12) it is obtained:

$$P = P_{max} - \frac{(P_{max} - P_{tank})\;\ln\left[\frac{tg\frac{\alpha}{2}}{tg\frac{\alpha_1}{2}}\right]}{\ln\left[\frac{tg\frac{\alpha_2}{2}}{tg\frac{\alpha_1}{2}}\right]} \tag{5.5.13}$$

Equation (5.5.13) is the final equation which gives the pressure distribution as a function of the angular position and the inlet pressure.

Equations (5.5.11) and (5.5.13) can also be obtained via using the generalized Reynolds equations of lubrication in spherical coordinates. Such equation was found by Meyer [11] and have takes the form:

$$\frac{1}{\sin\alpha}\frac{\partial}{\partial\alpha}\left(h^3\sin\alpha\frac{\partial p}{\partial\alpha}\right)+\frac{1}{\sin^2\alpha}\frac{\partial}{\partial\phi}\left(h^3\frac{\partial p}{\partial\phi}\right)=6r^2\mu\left[\omega\cos\eta\sin\phi\frac{\partial h}{\partial\alpha}+\left(\omega\sin\eta+\omega\cos\eta\frac{\cos\alpha}{\sin\alpha}\cos\phi\right)\frac{\partial h}{\partial\phi}+2\frac{\partial h}{\partial t}\right] \quad (5.5.14)$$

The different parameters used in equation (5.5.14) are presented in figure 5.5.2.

For the case under study, it can be stated that:

$$\omega=0\,;\;\frac{\partial h}{\partial\phi}=0\,;\;\frac{\partial h}{\partial t}=0\,;\frac{\partial h}{\partial\alpha}=0\,;\;\frac{\partial p}{\partial\phi}=0\,; \quad (5.5.15)$$

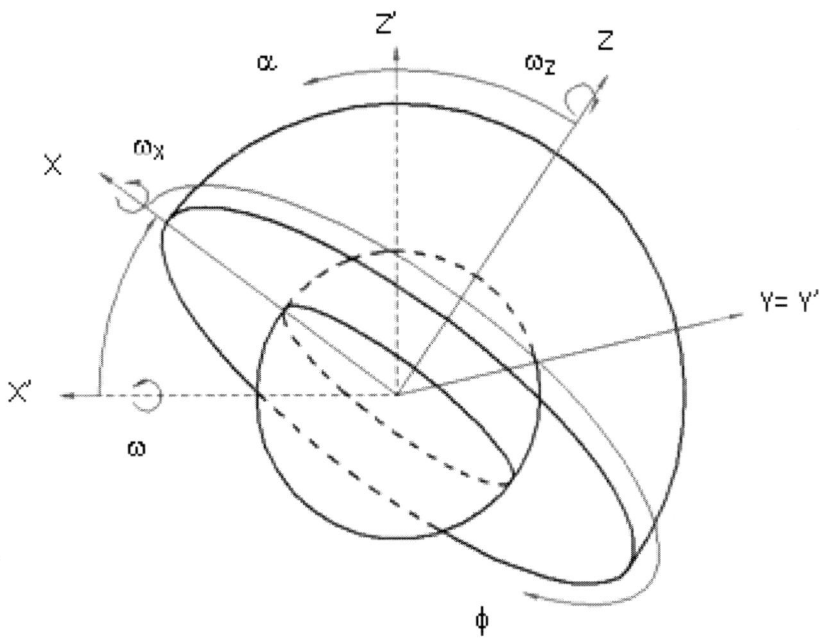

Figure 5.5.2. Parameters used in the Reynolds equation of lubrication in spherical coordinates. Equation (5.5.14).

Considering (5.15), equation (5.14) takes the form:

$$\frac{1}{\sin\alpha}\frac{\partial}{\partial\alpha}\left(h^3\sin\alpha\frac{\partial p}{\partial\alpha}\right)=0 \quad (5.5.16)$$

The integration of equation (5.16) gives birth to:

$$p=\frac{C_1}{h^3}\ln\left(tg\left(\frac{\alpha}{2}\right)\right)+C_2\,; \quad (5.5.17)$$

To find out the constants, the following boundary conditions are to be used:

$$\alpha=\alpha_1; \qquad p=p_{máx}$$
$$\alpha=\alpha_2;\ p=p_{\tan k} \tag{5.5.18}$$

Obtaining:

$$C_1=\left(p_{max}-p_{\tan k}\right)h^3\frac{1}{\ln\left(\dfrac{tg\left(\dfrac{\alpha_1}{2}\right)}{tg\left(\dfrac{\alpha_2}{2}\right)}\right)}; \tag{5.5.19}$$

$$C_2=p_{max}-\left(p_{max}-p_{\tan k}\right)\frac{\ln\left(tg\left(\dfrac{\alpha_1}{2}\right)\right)}{\ln\left(\dfrac{tg\left(\dfrac{\alpha_1}{2}\right)}{tg\left(\dfrac{\alpha_2}{2}\right)}\right)}; \tag{5.5.20}$$

Substituting the constants in equation (5.5.17) it is obtained:

$$p=p_{max}-\left(p_{max}-p_{\tan k}\right)\frac{\ln\left(\dfrac{tg\left(\dfrac{\alpha}{2}\right)}{tg\left(\dfrac{\alpha_1}{2}\right)}\right)}{\ln\left(\dfrac{tg\left(\dfrac{\alpha_2}{2}\right)}{tg\left(\dfrac{\alpha_1}{2}\right)}\right)}; \tag{5.5.21}$$

Notice that equation (5.5.21) and (5.5.13) are in reality the same equation.

To determine the leakage flow between the two spheres, it shall be used the same equation as the one used in the previous method, this is, equation (5.5.7), which has the form:

$$Q=-\frac{1}{\mu}\frac{dP}{d\alpha}\frac{1}{(r_1+\frac{H}{2})}\pi\sin\alpha\left[r_1\frac{H^3}{6}+\frac{H^4}{12}\right] \tag{5.5.22}$$

Now, the angular variation of pressure can be given as: $\frac{dp}{d\alpha}=\frac{C_1}{h^3\sin\alpha}$ being the constant C_1 already found as equation (5. 19), therefore:

$$\frac{dp}{d\alpha} = \frac{(p_{max} - p_{tan\,k})}{\sin\alpha} \frac{1}{\ln\left(\frac{tg\left(\frac{\alpha_1}{2}\right)}{tg\left(\frac{\alpha_2}{2}\right)}\right)} \tag{5.5.23}$$

Substituting equation (5.5.23) in (5.5.22) it is obtained:

$$Q = \frac{1}{\mu} \frac{\pi}{\left(r_1 + \frac{H}{2}\right)} \left[r_1 \frac{H^3}{6} + \frac{H^4}{12} \right] \frac{(p_{max} - p_{tan\,k})}{\ln\left(\frac{tg\left(\frac{\alpha_2}{2}\right)}{tg\left(\frac{\alpha_1}{2}\right)}\right)} \tag{5.5.24}$$

Notice that equation (5.5.24) is the same equation as (5.5.11), as a result, both methods produce the same results.

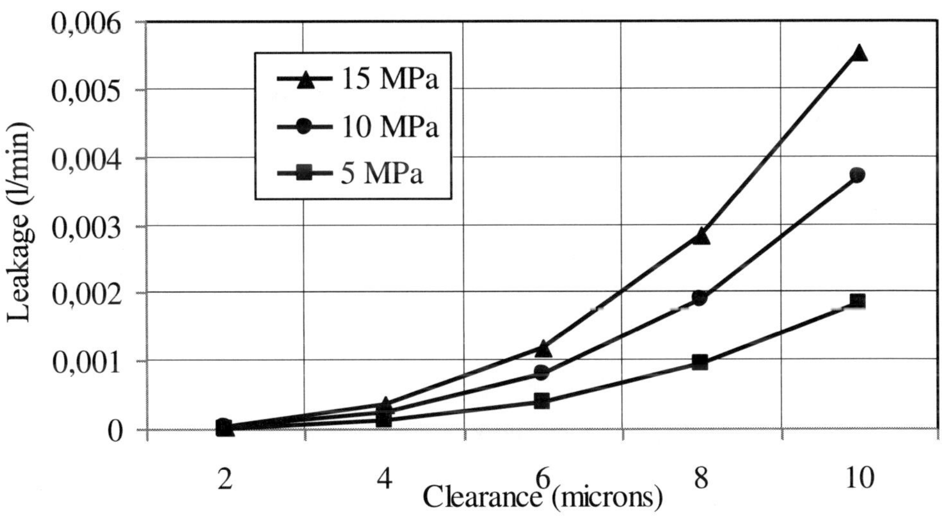

Figure 5.5.3. Spherical journal leakage as a function of the distance between the two spheres.

5.5.3. Results

Figure 5.5.3 presents the leakage across the original journal bearing as a function of the clearance between the two spheres and pressure differential. Equation (5.5.24) has been used to create this graph.

Figure 5.5.4 is the result of plotting equation (5.5.21) and presents the spherical journal bearing pressure decay as a function of the sphere angular position. Two different inlet pressures and maximum angles of the spherical journal are being employed.

5.5.4. Conclusion

A set of equations have been presented to calculate the leakage and pressure decay along a spherical journal bearing under laminar flow conditions. It can be concluded that spherical journal leakage is negligible when compared with other piston pump leakages already presented in the actual book chapter.

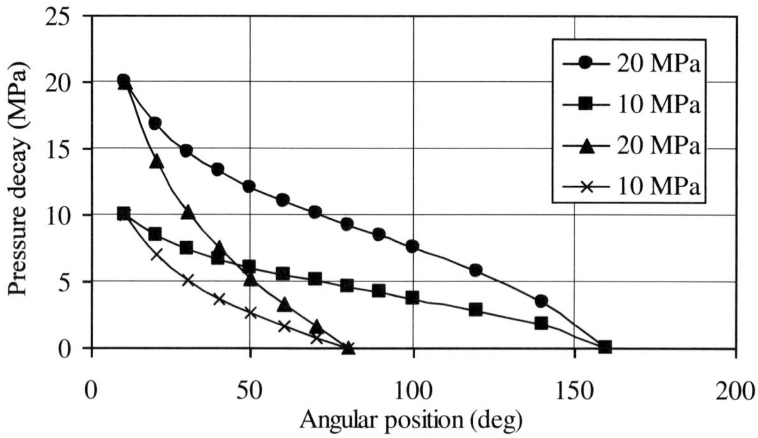

Figure 5.5.4. Spherical journal pressure decay as a function of the sphere angular position.

5.5.5. References

[1] Böinghoff, O. (1977). Untersuchen zum Reibungsverhalten der Gleitschuhe in Schrägscheiben-Axialkolbenmascinen. *VDI-Forschungsheft* 584. VDI-Verlag. 1-46.

[2] Hooke C.J. , Kakoullis Y.P. (1978). The lubrication of slippers on axial piston pumps. *5th International Fluid Power Symposium,* Durham, England. B2-13-26.

[3] Hooke C.J. , Kakoullis Y.P. (1981). The effects of centrifugal load and ball friction on the lubrication of slippers in axial piston pumps. *6th International Fluid Power Symposium,* Cambridge, England. 179-191.

[4] Iboshi N., Yamaguchi A. (1982). Characteristics of a slipper Bearing for swash plate type axial piston pumps and motors, theoretical analysis. *Bulletin of the JSME,* 25:210, 1921-1930.

[5] Iboshi N; Yamaguchi A. (1983). Characteristics of a slipper Bearing for swash plate type axial piston pumps and motors, experiment. *Bulletin of the JSME*, 26:219, 1583-1589.

[6] Hooke C.J., Li K.Y. (1989). The lubrication of slippers in axial piston pumps and motors. The effect of tilting couples. *Proceedings of the Institution of Mechanical Engineers*, part C, 203, 343-350.

[7] Kobayashi, S, Hirose M, Hatsue J, Ikeya M. (1988). Friction characteristics of a ball joint in the swashplate type axial piston motor. *Proc Eighth International Symposium on Fluid Power,* J2 Birmingham, England. 565-592.

[8] Abe K, Imai M, Kometani E. (1979). The performance of Ball-Piston Multi–Stroke Type Low Speed High Torque Motor. (Report N° 1, Experimental Study on Performance of Spherical Surface of Piston). *Bulletin of the JSME* 22:167, 700-706.
[9] Abe K, Ono K. (1979). The performance of Ball-Piston Multi–Stroke Type Low Speed High Torque Motor. (2nd report, practical analysis on multi-stroque type cam profile). *Trans. Japan Soc. Mech. Engrs. Part* B. 395, 974-981.
[10] Kobayashi S.; Ikeya M. (1993). Elastohydrostatic Lubrication of Piston Balls and Hydrostatic Slipper Bearings in Swashplate Type Axial Piston Motors. 10th International Conference on Fluid Power–the future for Hydraulics, Brugge, Belgium, 311-322.
[11] Meyer D. (2003). Reynolds Equation for Spherical bearings. *Journal of Tribology ASME.* 125, 203-206.
[12] Bergada J.M. (2012). Mecánica de fluidos. Breve introducción teórica con problemas resueltos. Barcelona, Iniciativa Digital Politècnica.

5.6. Piston Pump Full Dynamic Model

The equations about to be presented in this section have been developed in previous sections, nomenclature for each equation is to be found in each corresponding section.

5.6.1. Introduction

In theory the study of the piston-cylinder pressure ripple should be straightforward, since the differential equation involved is well known. The cylinder temporal pressure differential equation, depends on the leakage across the different piston pump clearances, but the dynamic equations linking the leakage across each piston pump gap and the pressure differential across the gap are not fully known.

To overcome this difficulty several researchers have used different approaches, Foster and Hannan [1], integrate the dynamic pressure differential equation of the cylinder and evaluate the leakage experimentally, an interesting point in this paper is that they take into account the effect of the oil volume at the inlet and delivery lines, the paper introduces a link between pressure transients into the cylinder and the noise generated by the pump.

Manring et al [2-4] assumes all leakage flows as laminar and uses a linear relation between pressure drop and flow, being necessary to find the leakage constant for every pump clearance.

Ivantysynova et al [5-9] integrate the Reynolds equation of lubrication, linked with the energy equation to evaluate dynamically every leakage, pressure distribution and temperature in all piston pump clearances. The implicit solution is performed via a numerical computer program, with all computer sub-routines linked to create a macro program called CASPAR which evaluates the entire pump behaviour.

Very recently Ji-en Ma et al [10], presented a single and multi-cylinder piston pump dynamics, in where they considered some typical equations to evaluate the leakage across the different piston/barrel gaps. They studied very carefully the leakage across the triangular timing grooves; they also considered the fluid inertia of the timing grooves, the simulated pressure ripple was compared with the experimental one clarifying the effect of the timing grooves regarding the piston dynamics, they also optimized the timing groove length for a particular application.

Among the models presented, the one based on numerical data it can be expected to produce slow results, the others define leakages using very simple equations. In the present section, an extensive set of explicit equations for every pump gap will be presented; all of the equations will be checked via performing a numerical analysis of the specified pump clearance, in fact, all leakage equations have been developed in previous sections, and they have been validated comparing them with numerical and experimental data. The equations will be joined together to study dynamically pump pressure ripple and leakages. The effect on the flow ripple when modifying the pump design will also be presented. Therefore in the present section, a simulation model based on analytical equations will be developed which produces very fast results and clarify very precisely the effect of different leakages through pump clearances.

5.6.2. Leakage Equation between Piston and Barrel

In section 5.2, has been developed an equation capable of giving the dynamic piston barrel leakage considering the grooves cut on the piston, equation (5.2.29). In figure 5.2.7 this equation was compared with a numerical simulation also presented in section 5.2, validating the equation developed. Since equation (5.2.29) shall be used in the present section, it is again presented next.

It must be recalled that equation (5.2.29) was developed using the following assumptions:

- Laminar flow is being considered in all cases.
- The flow is two-dimensional.
- Relative movement between piston and barrel exists.
- The gap piston cylinder is simulated as the gap between two flat plates.
- No eccentricity is considered.
- Each land and groove is modelled as a flat plate.

$$q_{piston-barrel} = \pi D_P \left[\frac{h_1 R_{sw} \tan\varepsilon \left(-\sin(\sigma t)\right) \sigma}{2} \right] - \frac{\pi D_P}{12} \left[\frac{P_{Tank} - P_{Piston} - 6 R_{sw} \tan\varepsilon \left(-\sin(\sigma t)\right) \sigma \mu \left[\frac{h_{10} - h_1}{h_{10}^3} \right] (l_2 + l_4 + l_6 + l_8 + l_{10})}{\frac{\mu}{h_{11}^3} \left(l_1 + l_2 + l_3 + \ldots + l_{11} - \frac{l_{11}}{2} - R_{sw} \tan\varepsilon \cos(\sigma t) \right) + \mu \left[\left(\frac{1}{h_2^3} - \frac{1}{h_1^3} \right) (l_2 + l_4 + l_6 + l_8 + l_{10}) \right]} \right]$$

(5.2.29) (5.6.1)

5.6.3. Leakage Equation in the Clearance Tilt Slipper and Swash Plate

In section 5.3, equations giving the leakage slipper swash plate when the slipper sits parallel or tilted versus the swash plate were presented, the equations were validated comparing them with experimental and numerical results, see figures 5.3.31 and 5.3.32 among others. In what follows, the equation capable of giving the leakage for the tilt slipper case, equation (5.3.68) is again introduced, notice that the remaining integral needs to be solved numerically. The constant K_1 for the slipper under study, was given as equation (5.3.52), being the constant equation for a slipper with a generic number of lands presented as equation (5.3.64).

$$Q_{leakage} = -\int_0^{2\pi} \frac{k_1}{12\,\mu}\, d\theta \qquad (5.3.68)\ (5.6.2)$$

$$k_1 = \frac{p_{tank} - p_{inlet} - \dfrac{3\mu\omega\alpha\sin\theta}{2}\left[\sum_{i=1}^{i=n} \frac{r_{mi}\left(r_{i-1}^2 - r_i^2\right)}{\left(h_{0i} + \alpha r_{mi}\cos\theta\right)^3}\right] - 3\mu\omega\alpha\sin\theta\left[\sum_{i=1}^{i=(n-1)} \frac{\ln\left(\dfrac{r_{i+1}}{r_i}\right)\left[\sum_{j=1}^{j=i} r_j^2\left(r_{m(j+1)} - r_{mj}\right)\right]}{\left(h_{0(i+1)} + \alpha r_{m(i+1)}\cos\theta\right)^3}\right]}{\sum_{i=1}^{i=n} \frac{\ln\left(\dfrac{r_i}{r_{(i-1)}}\right)}{\left(h_{0i} + \alpha r_{mi}\cos\theta\right)^3}}$$

(5.3.64) (5.6.3)

The assumptions used to generate equations (5.3.64) and (5.3.68) are:

- Flow will be considered laminar.
- The slipper plate clearance is not uniform; the slipper is tilted.
- Steady conditions are considered.
- Slipper spin is taken into account.
- Flow will be considered as radial.
- Slipper pocket, groove and slipper lands are flat.
- The only relative movement between slipper/swash plate is slipper spin.

5.6.4. Leakage Equations in the Clearance Barrel Port Plate

As defined in section 5.4, to determine the leakage flow in the clearance barrel and port plate, four leakage equations are needed, they are equations (5.4.27), (5.4.30), (5.4.36) and (5.4.37).

The first two equations define the leakage flow across the main groove external and internal lands, while the second two equations give the leakage across the timing groove external and internal lands. In section four it was demonstrated that leakage across the timing groove lands was negligible compared with the leakage across the main groove lands. This is why in the present section, just the leakage equations across the main groove shall be used, these equations are:

$$Q_{ext} = \frac{(p_{ext} - p_{int})}{12\,\mu \ln\left(\frac{r_{ext}}{r_{ext2}}\right)} \begin{bmatrix} h_0^3 \left[\theta\right]_{-\theta_i}^{\theta_j} + 3 h_0^2\, \alpha\, r_{m\,ext} \left[\sin\theta\right]_{-\theta_i}^{\theta_j} + \\ 3 h_0\, \alpha^2\, r_{m\,ext}^2 \left[\frac{1}{4}\sin(2\theta) + \frac{\theta}{2}\right]_{-\theta_i}^{\theta_j} + \alpha^3\, r_{m\,ext}^3 \left[\frac{1}{12}\sin(3\theta) + \frac{3}{4}\sin\theta\right]_{-\theta_i}^{\theta_j} \end{bmatrix} \quad (5.4.27)\ (5.6.4)$$

$$Q_{int} = -\frac{(p_{ext} - p_{int})}{12\,\mu \ln\left(\frac{r_{int}}{r_{int\,2}}\right)} \begin{bmatrix} h_0^3 \left[\theta\right]_{-\theta_i}^{\theta_j} + 3 h_0^2\, \alpha\, r_{m\,int} \left[\sin\theta\right]_{-\theta_i}^{\theta_j} + \\ 3 h_0\, \alpha^2\, r_{m\,int}^2 \left[\frac{1}{4}\sin(2\theta) + \frac{\theta}{2}\right]_{-\theta_i}^{\theta_j} + \alpha^3\, r_{m\,int}^3 \left[\frac{1}{12}\sin(3\theta) + \frac{3}{4}\sin\theta\right]_{-\theta_i}^{\theta_j} \end{bmatrix} \quad (5.4.30)\ (5.6.5)$$

The assumptions used to generate equations (5.4.27) and (5.4.30) are:

- Flow will be considered laminar.
- The barrel port plate clearance is not uniform; the barrel is tilted.
- Steady conditions are considered.
- Barrel turning speed is taken into account.
- Flow will be considered as radial.

5.6.5. Leakage Equation in the Piston Slipper Spherical Journal Bearing

The leakage across the spherical journal clearance has been obtained in section 5 and given as equation (5.5.11). Flow is considered as laminar and there is no relative movement between spheres.

$$Q = \frac{(P_{max} - P_{tank})\ \pi\ (r_l \frac{H^3}{6} + \frac{H^4}{12})}{\mu (r_l + \frac{H}{2})\ \ln\left[\frac{tg\frac{\alpha_2}{2}}{tg\frac{\alpha_1}{2}}\right]} \quad (5.5.11)\ (5.6.6)$$

5.6.6. Flow Leaving Each Piston-Barrel Chamber

Since the area through which the flow is leaving each piston-barrel chamber is much bigger than the rest, the flow is traditionally assumed as turbulent, for such cases the conventional equation used is:

$$Q_{out-piston} = \mathrm{sign}(p_{piston} - p_d) C_d\, A_c \sqrt{\frac{2}{\rho}\left|p_{piston} - p_d\right|} \quad (5.6.7)$$

The discharge coefficient is generally assumed as constant and equal to 0.6 although in reality depends on the cross sectional area and pressure differential. P_d is the pressure at the pump outlet, see figure 5.6.2. The temporal cross sectional area (A_c) has been calculated for each piston to its corresponding position, and it is the area, across which the output flow is leaving the piston-barrel chamber in direction pump outlet, it is represented in figure 5.6.1b. Notice that due to the entrance timing groove, the area increase has two different slopes.

It is noticed that the output area across which the flow towards the pump output will leave, goes from zero to a maximum every 165 degrees. When a piston enters in contact with the timing groove, the output area increase rather sharply, and once the piston enters in contact with the main groove, the area increases in a lower rate. While the piston is fully in contact with the main groove, the output area remains constant. When calculating the temporal timing groove area, it should also be considered the timing groove depth, since the input area is in reality the cross sectional area perpendicular to the fluid flow. For the present pump the timing groove depth is constant at all points and has a value of 1mm.

Figure 5.6.1a presents the top view schematic diagram of all nine pistons assembly, showing the barrel plate slots and its angular dimensions.

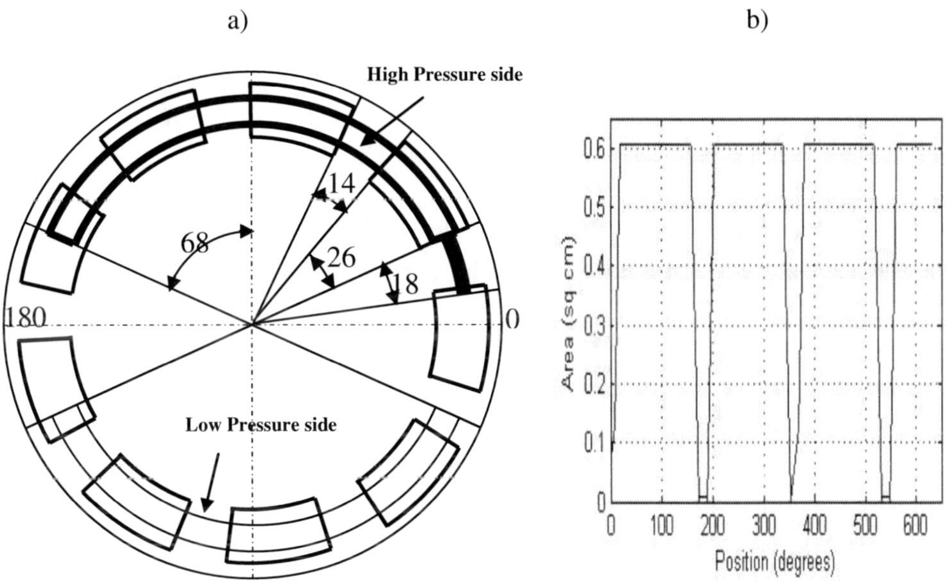

Figure 5.6.1. Temporal cross section area (A). a) Barrel/port plate view. b) Single piston/port plate temporal area.

5.6.7. Temporal Piston Cylinder Differential Equation

As the barrel turns around the swash plate, the volume of each piston-cylinder chamber and the area of its connecting side to the pump output changes with time. The connecting side area of each piston as a function of the angular position has been represented in figure 5.6.1. The temporal volume of each piston-cylinder chamber as a function of swash plate angular position is given by equation (5.6.8).

$$\forall = \forall_0 + \frac{\pi D_p^2}{4} R_{sw} \tan\varepsilon \left(\cos\lambda_{sw} - 1\right) \tag{5.6.8}$$

where, $\forall_0$ = is the cylinder volume when the piston is located at the bottom dead centre, D_p is the piston diameter, R_{sw} is the piston pith radius, ε is the swash plate tilt angle and λ_{sw} is the swash plate angular position.

The temporal pressure inside each piston can be found by applying continuity equation in integral form in the piston-cylinder chamber, as given in equation (5.6.9).

$$\frac{dp_{piston}}{dt} = \frac{\beta}{\forall}\left(-Q_{slip-plate} - Q_{barrel-plate} - Q_{sphere} - Q_{piston-barrel} - Q_{out-piston} - \frac{d\forall}{dt} \right) \quad (5.6.9)$$

Equation (5.6.9) gives the temporal pressure inside the piston-cylinder chamber as a function of the flow leaving the chamber, the fluid bulk modulus and the temporal volume of the chamber. According to Ma et al [10], the flow due to the fluid inertia when entering the timing groove, should also be included in equation (5.6.9), reminding that in the present case the timing groove has a constant depth, such small flow has been calculated, but under all conditions studied such flow was over 10 times smaller than the spherical leakage, which is by far the smallest leakage in the pump, as it will be presented in table 5.6.1. Therefore such leakage inertia effect has not been included in equation (5.6.9), since it is negligible.

Once all leakage equations are being substituted in equation (5.6.9) and after integration is performed, the time dependent pressure distribution inside one piston-cylinder chamber and the time dependant leakages through all pump gaps can be evaluated.

5.6.8. Temporal Outflow Ripple, Combination of Nine Pistons

In order to study the effect of the nine pistons regarding the entire pump dynamics, it was decided to apply the continuity equation in integral form. When combining the output flow from all the pistons connected to higher pressure side at any time instant, it results in equation (5.6.10), see figure 5.6.2.

$$\frac{dp_d}{d\theta} = \frac{\beta}{\sigma \forall_{valve+pipe}}\left(\sum_{i=1}^{n} Q_{out\text{-}piston} - Q_{outlet} \right) \quad (5.6.10)$$

n = number of pistons connected to the pump outlet at any time.

Notice that in equation (5.6.10) the volume considered is the one involving the output port of the pump, the volume of the tube connecting the pump and the relief valve, and the volume of the relief valve, volume submitted under pressure.

The flow leaving the pump, will have to pass though the pressure relief valve, the relation between this flow, the pressure differential between relief valve inlet and outlet and the valve cross section is given in equation (5.6.11).

$$Q_{outlet} = \mathrm{sign}(p_d - p_{tank}) C_d A_{valve} \sqrt{\frac{2}{\rho}\left| p_d - p_{tank} \right|} \quad (5.6.11)$$

The relief valve dimensions and its dynamics, do affect the pump output flow/pressure ripple, it is therefore necessary to consider the relief valve dynamics in order to include its behaviour when studding the overall pump behaviour. For the present study, the valve cross section was calculated for each working condition and substituted in equation (5.6.11).

The relief valve dimensions and its dynamics, do affect the pump output flow/pressure ripple, it is therefore necessary to consider the relief valve dynamics in order to include its behaviour when studding the overall pump behaviour. For the present study, the valve cross section was calculated for each working condition and substituted in equation (5.6.11).

5.6.9. Computational Technique

Figure 5.6.2 shows a combined flow assembly of all the pistons under pressure. Pistons numbered from 1 to 5 are connected to high pressure side and pistons numbered from 5 to 9 are connected to tank side. Piston 5 is shown twice, as it can be connected to high pressure side or tank side depending on barrel angular position. Notice that, the total pump leakage is the addition of barrel leakage, the slipper-plate, piston-barrel and spherical bearing leakages coming for each of the nine pistons at any given time. It can also be seen that pump outlet flow is the addition of the flow, coming from the pistons connected to the high pressure side. The rectangular box connecting pump outlet with the pressure relieve valve, represents the total volume of valve, pipe and pump outlet port.

A computer program has been written in MATLAB to combine all the equations (5.6.1-5.6.11) according to the flow chart shown in figure 5.6.2. First the contact area (Ac) for each piston corresponding to its position has been determined and then leakages through all four clearances have been calculated for all nine pistons. For the first time step, the pressure inside the piston chambers connected to high pressure side is assumed to be the given initial pressure condition. After that, all nine pistons have been combined using equation (5.6.10), finding the output pressure (P_d) when advancing in time. This calculated output pressure (P_d) provides the pressure just outside piston chamber (pump outlet) for the next time step. The numerical integrations have been performed by using fifth order Runge-Kutta method. It must be pointed out, from the equations presented in this chapter that the variables $P_{piston} = P_{int} = P_{inlet} = P_{max}$ will be seen as the same variable, which represents the temporal pressure at the piston chamber.

5.6.10. Experimental Test Rig

In order to find out the dynamic pressure ripple inside the piston-cylinder chamber, the test rig presented in figure 5.6.3 was created in Cardiff University.

Three very high response Kistler pressure transducers were located in cylinders 1, 4 and 7, figure 5.6.3a shows two of the transducers already in position. Notice that the connecting cables are coming out of the pump through the pump axis. Since the entire system, barrel, transducers and pump axis, turn, a slip ring assembly was needed outside the pump, see figure 5.6.3b, therefore the measurements taken by the different pressure transducers were send to the data acquisition system. Further details on the test rig can be found in Haynes PhD [11].

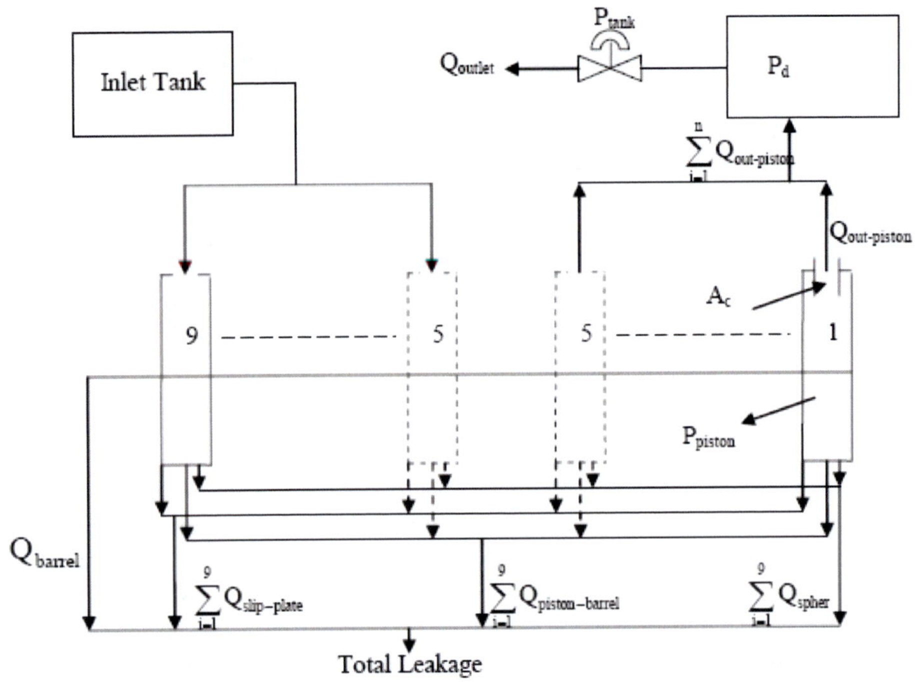

Figure 5.6.2. A combine assembly of all pistons with nomenclature in Axial Piston pump.

In figure 5.6.3b, can also be seen a fourth pressure transducer located outside the pump, just before the pressure relief valve. This fourth transducer was used to measure the pressure ripple just outside the pump outlet. The test rig used allowed to modify three parameters, turning speed, swash plate angle and output pressure. Test were performed for four different turning speeds, 200, 400, 700 and 1000 rpm, two swash plate angle tilts 10 and 20 degrees and output pressures ranging from 1 to 10 MPa every 1 MPa. Hydraulic oil temperature was kept constant at 37 Celsius for all tests performed.

All information was captured using a PC based data logger, which allowed data to be directly captured onto a Microsoft windows based workstation.

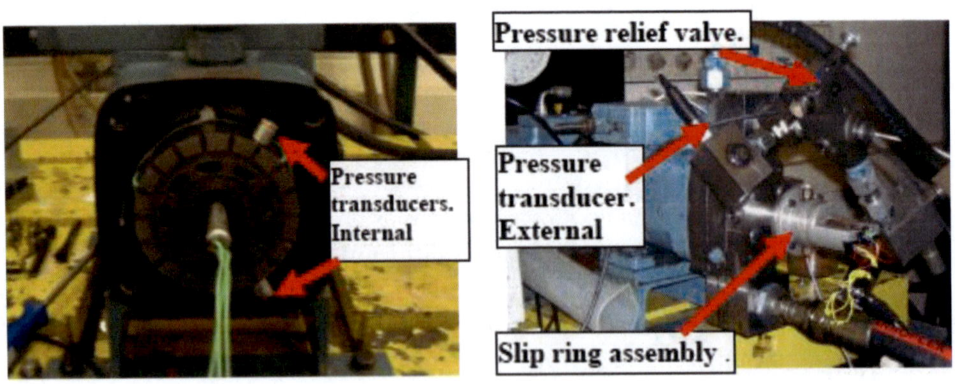

a) b)

Figure 5.6.3. Test rig used to measure directly the dynamic pressure inside a piston-cylinder chamber in an axial piston pump.

5.6.11. Results

5.6.11.1. Experimental Results

Figure 5.6.4 presents the cylinder pressure ripple for 20 degrees swash plate angle, 1000 rpm and for several output pressures, where it is noticed that the higher the output pressure will be, the higher is the pressure ripple inside the cylinder chamber. In fact, the pressure ripple in the cylinder chamber depends on the output pressure, the pump turning speed and swash plate angle. Pressure ripple is higher at high output pressures, high pump turning speeds and high swash plate angles. Figure 5.6.5 shows the variation of pressure ripple for 10 MPa output pressure and 20 degrees swash plate angle, as a function of different turning speeds, it is clearly seen that as turning speed decreases pressure ripple also decreases, at 1000 rpm the cylinder pressure ripple is of about 1 MPa, while at 400 rpm the peak to peak pressure ripple is less than 0.5 MPa. It is also noticed that the shape of pressure ripple changes with turning speed.

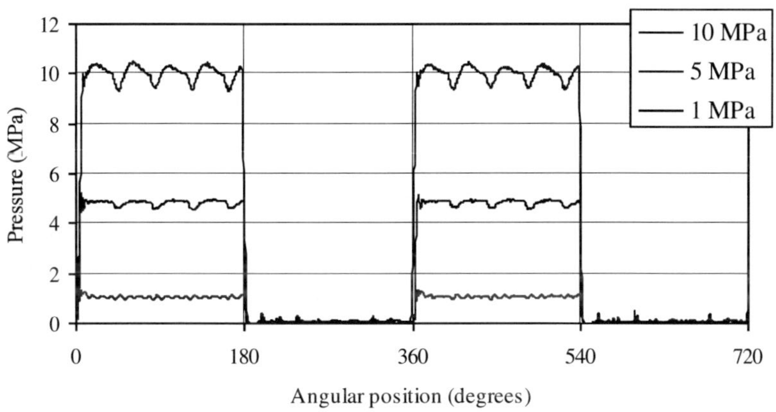

Figure 5.6.4. Cylinder pressure ripple for 20 degrees swash plate angle, 1000 rpm and several output pressures.

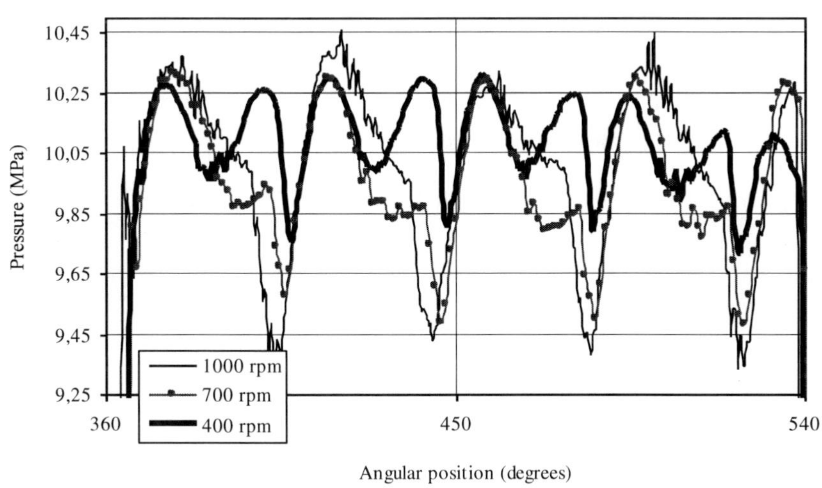

Figure 5.6.5. Cylinder pressure ripple for 20 degrees swash plate angle, 10 MPa and several pump turning speeds.

In figure 5.6.6 it is presented the cylinder pressure ripple variation as a function of swash plate angle, reducing the swash plate angle will bring a reduction of pressure ripple. It can be concluded that pressure ripple is primarily affected by output pressure being the effects of turning speed and swash plate angle, although important, less relevant than the output pressure ones.

Another interesting point which can be studied thanks to the experimentation undertaken is the relation between the pump output pressure, just before the pressure relief valve, see figure 5.6.3, and the pressure inside the cylinder chamber. For all the cases studied the output pressure was matching perfectly well the variations of cylinder pressure, output pressure was also under all conditions slightly lower than the pressure inside the cylinder chamber, such difference represents the pressure losses mostly inside the pump, since the pipe uniting the pump and the relief valve was very short. Pressure differential between the cylinder chamber and outside the pump, was found to be of about 0.25 MPa for 20 degrees swash plate angle, 1000 rpm and 10 MPa output pressure, such pressure differential tends to slightly decrease with the decrease of the swash plate angle and slightly increases, about 0.05 MPa, with the decrease of output pressure, such increase it is understood when noticing that as the output pressure decreases, leakage decreases, therefore a slightly higher flow leaves the pump through the pump output port. A decrease in pump turning speed brings a small decrease in pressure differential.

During experimentation, it was also noticed that cylinder pressure during the intake period, initially falls to a minimum and then increases a bit as the piston goes from the upper death centre to the lower death centre. The minimum pressure was found to be very near to 0 MPa, although just for a very short period of time.

The pressure increase is more relevant for small turning speeds and high swash plate angles. Also for small turning speeds, it is seen that pressure does not remain constant when the piston goes from the lower death centre to the upper death centre, but it decreases as the piston moves towards the upper death centre, the decrease is higher for low turning speeds and higher pressures. This phenomena is very well understandable when considering that as turning speed decreases, the piston needs a longer time to go from LDC to UDC, giving more time to the fluid to escape across the clearances, barrel-plate and slipper port plate towards tank. This phenomenon can also be seen in figure 5.6.5, notice that the curves at 700 and 400 rpm decrease with angular position, hence with time.

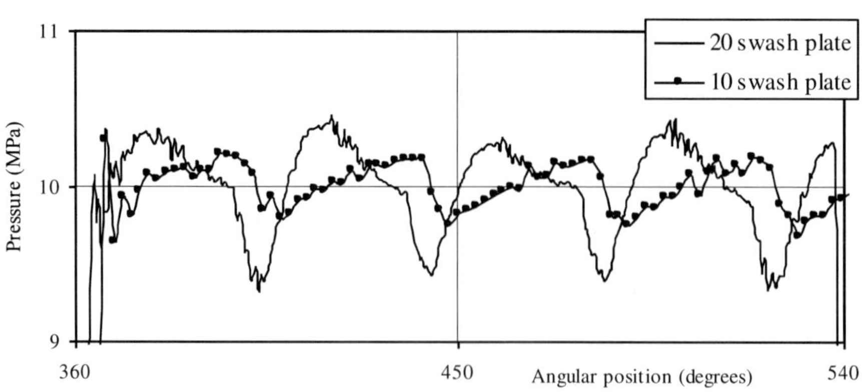

Figure 5.6.6. Cylinder pressure ripple for an output pressure of 10 MPa, 1000 rpm and two swash plate angles.

5.6.11.2. Numerical Results

Figure 5.6.7 presents the dynamic pressure inside a piston chamber at 1000 rpm pump turning speed when the pressure relieve valve is set at 5 and 10 Mpa outlet pressure. The barrel-port plate, slipper-swash plate, piston-cylinder and spherical bearing central clearances are assumed to be 5, 10, 5, 5 microns respectively. The choice of the clearances, barrel-port plate and slipper-swash plate, was based on literature [7, 12, 13], on the other hand the clearances for piston-cylinder and spherical bearing gap have been chosen based on manufacturers experience. Figure 5.6.7 shows a very good agreement between numerical and experimental results.

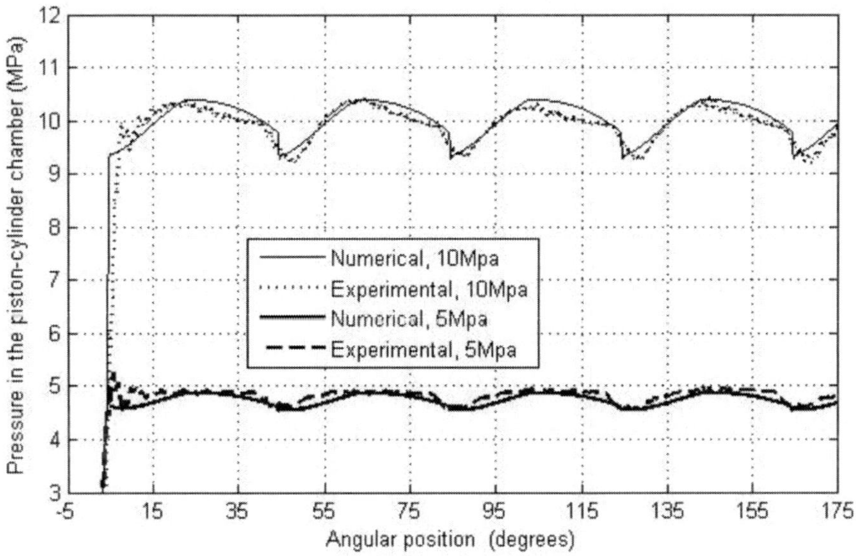

Figure 5.6.7. Pressure inside piston at 5 and 10 Mpa valve set pressure, 1000 rpm pump turning speed, Comparison between Numerical and Experiments.

Figure 5.6.8 presents the normalized temporal outflow at 1000 rpm pump turning speed, 10 Mpa outlet pressure, when being the new pump barrel, slipper, piston and spherical bearing clearances of 5, 10, 5, 5 microns respectively. In the same graph, it is presented the output flow ripple when the clearances increase in different percentages, simulating the pump erosion as it becomes old. It can be noticed, that regardless of the percentage increase in clearance, the shape of the temporal outflow ripple remains constant. Nevertheless the pump outflow decreases about 6% when the magnitude of the clearances doubles, which will result in a 6% decrease of volumetric efficiency.

Although not presented in the present chapter, for the different clearances studied, the shape of the simulated pump output pressure ripple remains also constant.

Table 5.6.1 presents, for three different clearance configurations, the average leakage across all four piston pump gaps over one pump rotation. It can be noticed that when clearances in all gaps have the same magnitude, the leakage through barrel port plate is dominant, giving about 70% of the total piston pump leakage. It can also be seen that when all clearances double from 5 to 10 microns, the total pump leakage given as a percentage to the total output flow, increase from 0.6 % to 4.6%. According to the literature [7, 12, 13], the slipper-port plate clearance is typically about 10 microns, and the barrel port plate clearance tends to fluctuate around 5 microns. The last two columns of table 5.6.1 specify that under these conditions, the main leakage source for the present piston pump is the slipper-swash plate, producing around the 70% of the total leakage. Further information is to be found in [14].

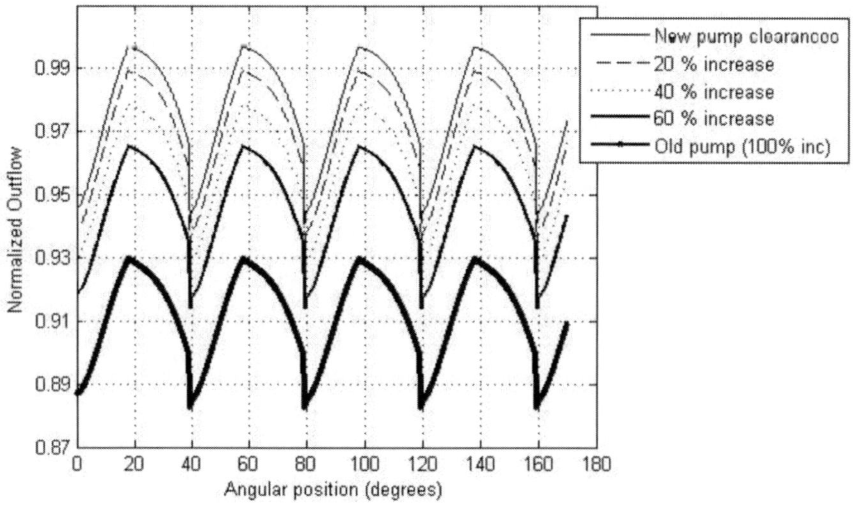

Figure 5.6.8. Normalized temporal out flow from the pump as the pump clearances increases, set valve pressure 10 Mpa, pump turning speed 1000rpm. New pump clearances: slipper clearance 10 μ, barrel clearance 5 μ, spherical bearing clearance 5 μ and piston cylinder clearance 5 μ.

Table 5.6.1. Leakages and out flow at different clearances, set valve pressure 10 Mpa

	Clearances (Microns)	%Flow	Clearances (Microns)	%Flow	Clearances (Microns)	%Flow
Barrel-plate	5 μ	71.4%	10 μ	70%	5 μ	26.04%
Slipper	5 μ	22.5%	10 μ	23.9%	10 μ	71.64%
Piston	5 μ	2.7%	10 μ	2.7%	5 μ	1.04%
Spherical bearing	5 μ	3.4%	10 μ	3.4%	5 μ	1.28%
Total Leakage	0.609 % of total output flow		4.57% of total output flow		1.63 % of total output flow	

5.6.12. Conclusion

1. A new piston pump model based on the new algebraic leakage equations is presented. The beauty of the model is its fast calculating speed and its good performance, since it is capable of simulating the pump output pressure ripple in

great detail. The model has been validated using directly experimental measurements of the dynamic pressure inside a piston-cylinder chamber.

2. A novel, state of the art test rig, designed in Cardiff University and able to measure the dynamic pressure inside the cylinder in a piston pump has been presented. Dynamic pressure measurements inside the cylinder are being undertaken as a function of pump turning speed, outlet pressure and swash plate angle. The results clearly show how pressure ripple is being affected by such parameters; the output pressure being the parameter which more directly affects pressure ripple. The pressure ripple outside the pump follows with great detail the dynamic pressure ripple in a given cylinder, although it is about 0.25 MPa lower than the pressure inside the cylinder, such pressure differential is mostly due to the pressure losses in the pump output channels.
3. It is demonstrated that the main source of the leakage in piston pumps is weather the slipper-swash plate or the barrel-port plate, producing over 94% of the total leakage. Depending on the magnitude of the clearances slipper-swash plate and barrel-port plate, the main source of the leakage would be weather one gap or the other.
4. The pump output flow reduction is over 6% when all the clearances double. The output flow and pressure ripple shape is found to be independent of clearances magnitude.

5.6.13. References

[1] Foster K, Hannan D.M. (1977). Fundamental Fluidborne and Airborne Noise Generation of Axial Piston Pumps. Seminar on Quiet Oil Hydraulic Systems. *Institution of Mechanical Engineers*. London, England. Paper C257/77.

[2] Manring N.D. (2000). The Discharge Flow Ripple of an Axial-Piston Swash-Plate Type Hydrostatic Pump. *Journal of Dynamic Systems, Measurement, and Control. ASME.* 122, 263-268.

[3] Manring N.D; Damtew F.A. (2001). The Control Torque on the Swash plate of an Axial Piston Pump utilizing Piston -Bore Springs. *Journal of Dynamic Systems, Measurement, and Control. ASME.* 123, 471-478.

[4] Manring N.D; Zhang Y. (2001). The Improved Volumetric Efficiency of an Axial - Piston Pump Utilizing a Trapped-Volume Design. *Journal of Dynamic Systems, Measurement, and Control. ASME.* 123, 479-487.

[5] Ivantysynova M; Huang C. (2002). Investigation of the Flow in Displacement Machines Considering Elastohydrodynamic Effect. *Proceedings of the* 5th *JFPS International Symposium on Fluid Power,* Nara, Japan. 1, 219-229.

[6] Ivantysynova M, Grabbel J, Ossyra JC. (2002). Prediction of a Swash Plate Moment Using The Simulation Tool CASPAR. ASME International Mechanical Engineering Congress & Exposition. November 17-22, New Orleans, Louisiana USA. *IMECE* 2002-39322, 1-9.

[7] Wieczorek U; Ivantysynova M. (2002). Computer Aided Optimization of Bearing and Sealing Gaps in Hydrostatic Machines–the Simulation Tool CASPAR. *International Journal of Fluid Power* 3:1, 7-20.

[8] Ivantysynova M. (1999). A New Approach to the Design of Sealing and Bearing Gaps of Displacement Machines. Fluid Power forth JHPS International Symposium, 45-50.

[9] Ivantysynova M; Lasaar R. (2004). An Investigation into Micro–and Macrogeometric Design of Piston/Cylinder Assembly of Swash plate machines. *International Journal of Fluid Power* 5:1, 23-36.

[10] Ma J, Fang B, Xu B, Yang H. (2010). Optimization of cross angle based on the pumping dynamics model. *Journal of Zhejiang University SCIENCE* A, 11:3, 181-190.

[11] Haynes JM. (2008). *Axial piston pump leakage modeling and measurement*. PhD Thesis, Cardiff University UK.

[12] Harris RM. Edge KA. And Tilley DG. (1993). Predicting the behaviour of slipper pads in swashplate-type axial piston pumps. *ASME Winter Annual Meeting*. New Orleans, Louisiana. 1-9.

[13] Bergada JM, Davies D. Ll, Kumar S, Watton J. (2012). The effect of oil pressure and temperature on barrel film thickness and barrel dynamics of an axial piston pump. *Meccanica* 47-3, 639-654.

[14] Bergada JM, Kumar S, Davies D. Ll, Watton J. (2012). A complete analysis of axial piston pump leakage and output flow ripples. *Applied Mathematical Modelling* 36, 1731-1751.

5.7. Some New Trends on Piston Pumps

According to the information gathered, leakage on piston pumps is especially high in the clearance slipper-swash plate and also in the barrel port plate clearance. To increase volumetric efficiency, slipper and barrel lands could be enlarged, the, a priory expected effect, would be a reduction of leakage flow, yet, mechanical efficiency would very likely decrease. As a result, the pump overall efficiency could decease. Friction slipper swash plate and barrel port plate could be diminished via using composite materials as defined for example in [42, 43] section 5.4. Even the noise due to the metal to metal friction, could suffer some reduction when using composite materials. At the moment the drawback of the use of such materials is their fragility, but it seams, the use of alternative materials will increase in future pumps design.

It has also been seen that slippers and even the barrel, wobble, which means they constantly fluctuate around a given tilted position. To stabilize slipper movement around the swash plate, grooves can be used, but at the moment is being studied the possibility of using spherical slippers, swash plate would also have a spherical shape. It is interesting to point out that the equations presented in section 5.5, could be used to calculate leakage and pressure distribution for spherical slippers. As a final conclusion, it seems that new materials and new piston pump configurations are likely to be used in the future.

5.8. Nomenclature, General

C_d	Discharge coefficient.
D_p	Piston diameter (m).

F	Generic force (N)
h_i	Generic clearance (m).
$P_i = p_i$	Generic pressure (Pa).
P_{tank}	Tank pressure (Pa).
P_{piston}	Pressure in the piston cylinder chamber (Pa).
Q_i	Generic volumetric flow rate (m^3/s).
R_{sw}	Piston pith radius (m).
$\forall$	Generic volume (m^3).
t	Time (s).
V_{SL}	Piston velocity, measured from the lower death centre (m/s).
β	Fluid bulk modulus (N/m^2).
ρ	Density of fluid (Kg/m^3).
μ	Fluid dynamic viscosity. (Kg/(m s))
$\lambda_{SW} = \sigma t$	Angular position around the swash plate (rad).
σ	Pump angular velocity (rad/s).
ε	Swash plate tilt angle (rad).

Nomenclature, Specific for Section 5.2.

A,C,E,G,I,K,M,O,Q,S,U.	Constants (m^2/s).
B,D,F,H,J,L,N,P,R,T,V.	Constants (N/m^2).
C_{A1}	Constants (Kg/s m^3).
C_{A2}	Constants (Kg/s^2m).
Dc	Cylinder diameter (m).
Ec1, Ec2	Minimum edge clearances (m).
h_l	Clearance piston cylinder (m).
$L_i = l_i$	Length of a given piston land (m).
L_t	Length of the piston inside cylinder at time t (m).
Lo	Length of the piston inside cylinder at t=0 (m).
L1	Piston cylinder axis intersection point position, figure 2.5 (m).
q	Leakage through piston-cylinder clearance (m^3/s).
T_x, T_y	Torque versus the x and y piston axis respectively (N m).
u	Piston velocity, measured from the upper death centre (m/s).
$\dot{\forall}$	Volumetric flow per unit depth (m^2/s).
$V_{S\theta}$	Angular surface velocity of piston (m/s).
x	Distance from the axis origin (m).
X1, Y_1	Coordinates of a generic point on cylinder surface (m; m).
X, Y	Generic coordinates axis (m; m).
α	Piston tilt to cylinder axis (rad).
θ, L	Generic piston coordinates axis (rad; m).

Nomenclature, Specific for Section 5.3

a	Coefficient of discrete momentum equation.
b	Source term in discrete momentum equation.
A_p	Coefficient in pressure correction equation.
B_p	Mass conservation term in continuity.
C	Constant (Nm)
C_1, C_3, C_5, C_7,	Constants (m^3/s).
C_2, C_4, C_6, C_8, k_2, k_4, k_6, k_8, k_M,	Constants (Pa).
dr, $d\theta$,dz	Grid size in r, θ and z direction.
h_1	Slipper pocket central clearance (m).
$h_2 = h_4$	Slipper first land central clearance (m).
h_3	Slipper groove central clearance (m).
h_{01}	Slipper pocket central clearance (m).
h_{02}	Slipper first land central clearance (m).
h_{03}	Slipper groove central clearance (m).
h_{min}	Slipper swash plate minimum clearance (m).
h_{max}	Slipper swash plate maximum clearance (m).
i, j, k	Grid coordinate in r, θ and z direction.
k_1, k_3, k_5, k_7, k_L,	Constants (N m).
M_{xx}; M_{yy}	Torque versus X and Y axis (N m).
P_{inlet}	Pressure at the slipper central pocket for a radius r_0 (Pa).
P_{outlet}	Pressure at the slipper external radius r_4 (Pa)
r	Slipper generic radius (m).
r_0	Slipper central pocket orifice radius (m).
r_1	Snner land inside radius (m).
r_2	Inner land outside radius (m).
r_3	Outer land inside radius (m).
r_4	Outer land outside radius (m).
r_m	Average radius between land borders (m).
r_{m1}	Average radius between slipper pocket borders (m).
r_{m2}	Average radius between inner land borders (m).
r_{m3}	Average radius between groove borders (m).
r_{m4}	Average radius between outer land borders (m).
r, θ , z	Cylindrical coordinates vector (m, rad, m).
S_ϕ	Source term in momentum equation for corresponding ϕ (Kg/m^2s^{-2}).
U	Slipper generic tangential velocity (m/s).
u	Fluid generic parabolic velocity (m/s).
V	Fluid velocity (m/s).
α	Slipper tilt angle (rad).
α_P	Under relaxation factor for pressure.
α_v	Under relaxation factor for velocity.
ϕ	Flux vector (m/s).

θ	Slipper angular position versus a coordinate axis (rad).
ω	Slipper spin (rad/s).
Γ	Computation domain boundary (m).
Ω	Computation domain (m^3).

Subscripts

r, θ, z	Component of vector in r, θ and z direction.
in, out	Corresponding to Inlet and outlet boundary.
p	Grid point under consideration.
nb	Neighbour grid point of point p.
E,W,N,S,T,B	East, west, north, south, top and bottom direction.

Superscripts

*	Imperfect computed field.
'	Correction in corresponding quantity.

Nomenclature, Specific for Section 5.4

A_0	Metal to metal contact area between the barrel and the port plate (m^2).
A_{cylin}	Cylinder area (m^2).
A_{flow}	Flow cross section at the end of the cylinder (m^2).
B	Barrel damping coefficient (Nm/rad s^{-1}).
c_1, c_3	Constants (Nm).
c_2, c_4	Constants (N/m^2).
E	Young modulus (N/m^2).
F	Force, due to the main groove (N).
F_{sg}	Force, created by the timing groove (N).
F_{forces}	Torque created due to friction (Nm).
h_o	Barrel port plate central clearance (m).
I	Barrel moment of inertia, versus a generic angle (kg m^2).
K	Spring torsional constant acting over the barrel (Nm).
M_E	Elastic metal to metal torque constant (Nm).
M_{EF}	Elastic metal to metal torque (Nm).
M_{XX}	Fluid generated torque versus the barrel X axis (Nm).
M_{YY}	Fluid generated torque versus the barrel Y axis (Nm).
$P_{int\ cyl}$	Pressure inside the cylinder (N/m^2).
p_{ext}	Pump inlet (tank) pressure (N/m^2).
p_{int}	Pump outlet pressure (N/m^2).
$p_{ext\ land}$	Pressure distribution across the external land, main port plate groove (N/m^2).
$p_{int\ land}$	Pressure distribution across the internal land, main port plate groove (N/m^2).

$p_{ext\ land\ sg}$	Pressure distribution across the internal land, timing groove (N/m^2).
$p_{int\ land\ sg}$	Pressure distribution across the internal land, timing groove (N/m^2).
Q_{ext}	Leakage across the external land, main port plate groove (m^3/s).
Q_{int}	Leakage across the internal land, main port plate groove (m^3/s).
$Q_{ext\ sg}$	Leakage across the external land, timing groove (m^3/s).
$Q_{int\ sg}$	Leakage across the internal land, timing groove (m^3/s).
r	Barrel generic radius (m).
r_{int}	Internal radius of the main groove (m).
r_{ext}	External radius of the main groove (m).
$r_{int\ 2}$	Internal land inner radius (m).
$r_{ext\ 2}$	External land outer radius (m).
$r_{m\ ext}$	External land average radius, between r_{ext} and r_{ext2} (m).
$r_{m\ int}$	Internal land average radius, between r_{int} and r_{int2} (m).
R_{int}	Internal radius of the timing groove (m).
R_{ext}	External radius of the timing groove (m).
$R_{m\ int}$	Timing groove internal land average radius, between r_{int2} and R_{int} (m).
$R_{m\ ext}$	Timing groove external land average radius, between r_{ext2} and R_{ext} (m).
r_m ; R_m	Average radius between land borders (m).
r_o	Groove central radius (m).
T_{hi}	Port plate thickness (m).
v_e	Flow generic velocity across a external land (m/s).
v_i	Flow generic velocity across a internal land (m/s).
α_{max}	Barrel maximum tilt angle (rad).
α	Barrel tilt angle perpendicular to X axis (rad).
α_0	Barrel initial angular position, perpendicular to X axis (rad).
α_L	Maximum barrel tilt angle perpendicular to X axis for a given working conditions (rad).
δ	Barrel tilt angle perpendicular to Y axis (rad).
δ_0	Barrel initial angular position, perpendicular to Y axis (rad).
δ_L	Maximum barrel tilt angle perpendicular to Y axis for a given working conditions (rad).
ΔL	Decrease in length due to compression forces (m).
γ	Timing groove angle (rad).
ν	Position transducers measured voltage (V).
θ	Barrel angular position, versus the maximum tilt axis. (rad).
θ_i, θ_j	Main groove angular dimension (rad).

Nomenclature, Specific for Section 5.5

C_1	Constant (N m).
C_2	Constant (Pa).
dr	Radial differential (m).
H	Spherical journal clearance (m).
P_{max}	Maximum pressure (Pa).

P_{tank}	Minimum pressure, tank pressure (Pa).
r	Generic radius (m).
r_1	Spherical journal internal radius (m).
r_0	Spherical journal external radius (m).
α	Generic angular position (rad).
α_1	Minimum angular position (rad).
α_2	Maximum angular position (rad).
η	Spherical journal angular position (rad).
ϕ	Spherical journal angular position (rad).

Chapter 6

ACCUMULATORS

6.1. INTRODUCTION TO ACCUMULATORS

Hydraulic systems are often rather inefficient, then for several time periods the relieve valve remains opened allowing high pressure fluid returning to tank. One way to avoid such waste of energy is via using pressure compensated variable displacement pumps, another possible way is via storing fluid energy in accumulators and releasing this energy when the circuit demand is high. Typically in fluid power systems, the power demand is cyclical, in such systems, the use of accumulators can substantially reduce the size of the pump, increasing system efficiency and reducing power requirements. According to Mordas [1], the use of accumulators can reduce electric power by 20 to 70%. As a conclusion, in a world where the cost of energy steadily increases, and considering the vast amount of hydraulic systems employed, it is difficult to understand why accumulators are still underused.

Despite the fact that energy storage is one of the most relevant accumulator applications, there exist some other important uses which might be interesting to consider. The use of accumulators as pulsation dampers is extremely important in many applications. Notice that all pumps, and very specially piston pumps, generate pressure pulsations during operation, this being undesirable and detrimental of both the smooth operation and operational life of components. The use of a blade type accumulator located downstream of the pump shall dampen the pulsation to an acceptable limit. Accumulators can also be used as emergency operators in the event of a power failure, as a volume compensators, in closed circuits where thermal expansion can cause an increase of fluid pressure, and among other applications, as shock absorbers or hydraulic dampers, whenever shock waves appear in hydraulic systems due to, for example, quick valves closing, or shocks in suspension systems, mobile cranes, agricultural machinery etc. appear, in all these cases, accumulators will be able to store the shock energy and avoid possible system failure.

6.2. TYPES OF ACCUMULATORS

There exist three main sorts of hydraulic accumulators, bladder, diaphragm and piston accumulators, see figure 6.1. Other sort of accumulators like weight loaded accumulators and spring loaded accumulators shall not be considered in this chapter then they are barely used.

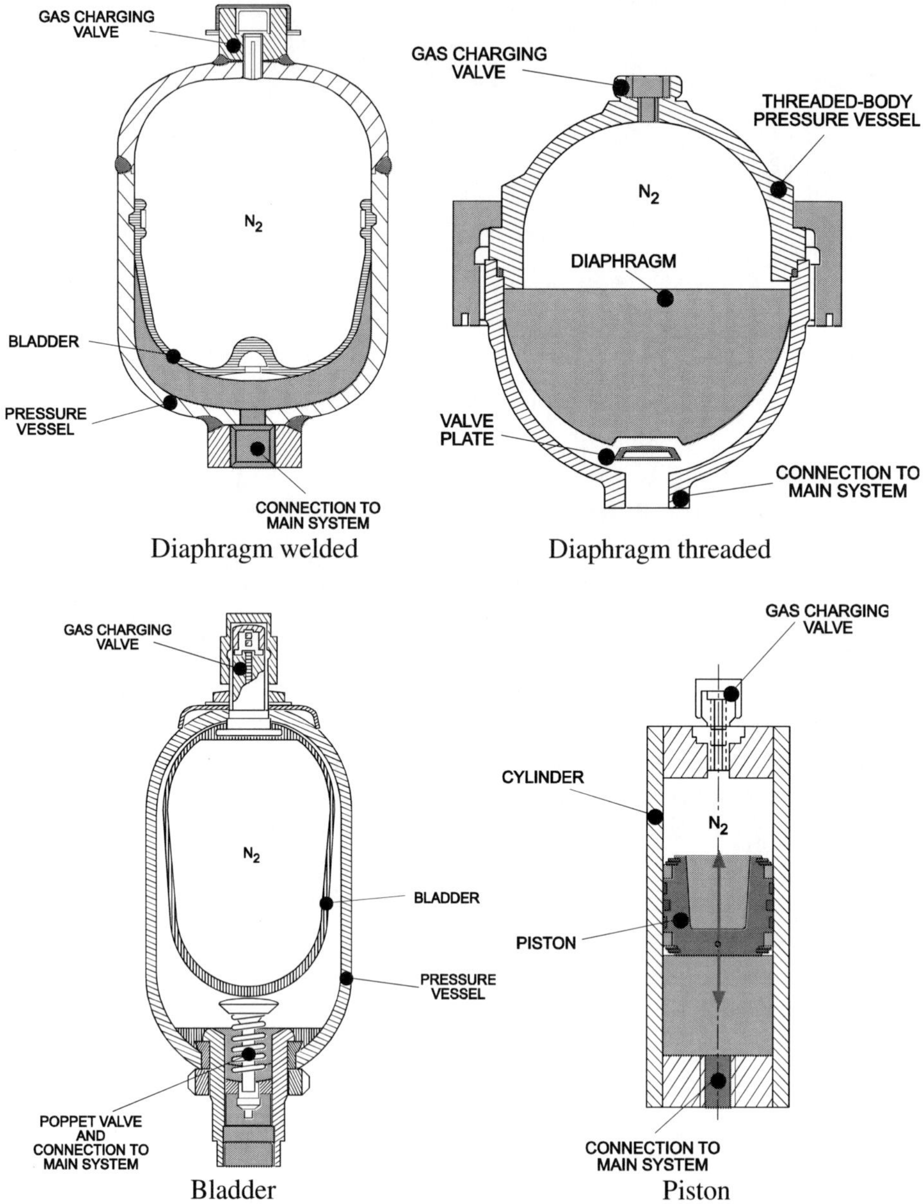

Figure 6.1. The three main types of accumulators.

Bladder, diaphragm and piston accumulators are pre-charged with gas, usually nitrogen, the fluid compartment and gas compartment are separated. The pre-charge pressure cause the bladder or diaphragm to completely fill the inside of the steel shell, and close the poppet or valve plate. Whenever the oil pressure in the hydraulic circuit matches the pressure inside the accumulator, the inlet valve will open and fluid will enter the accumulator, as fluid keeps entering, the volume of nitrogen will be further reduced, the gas is compressed and the pressure will increase. As soon as the pressure in the hydraulic circuit decreases, fluid will leave the accumulator tending to increase system pressure.

Bladder and diaphragm accumulators are extensively used; an advantage of the first sort is that a worn or damaged elastomeric bladder can be replaced. Diaphragm accumulators have two basic constructions, welded and threaded. Welded models are non reparable while threaded models can be disassembled to change the diaphragm, but they are more expensive. A very important difference between the different sorts of accumulators is the ratio between the maximum operating pressure and the pre-charge pressure. This ratio is about 4:1 for bladder accumulators, 8:1 for diaphragm welded accumulators and 10:1 for threaded ones. The higher the pressure ratio the greater the usable fluid volume will be, these accumulators shall be seen as more efficient. At this point it is interesting to highlight that piston accumulators, have in theory no restriction in pressure ratio.

Regarding the dynamic response, piston units have a slow one and therefore cannot be used to reduce pressure pulsations, as shock absorbers or hydraulic dampers. On the other hand bladder and diaphragm units are ideal for these fast response dynamic applications. Piston accumulators have also the disadvantage of being susceptible of leaking, which could lead to a failure with no previous clear indication. Piston accumulators are ideal for emergency service applications and can have large capacities.

6.3. Accumulators Design

In the present sub section, the methodology to be used to dimension accumulators for different applications will be presented.

6.3.1. Accumulator Used As Volume Accumulator/Energy Storage

Initially, the different parameters definition will be introduced.

P_0= Accumulator transportation pressure. Is the accumulator gas pressure used to maintain the bladder or diaphragm extended, also called gas pressure to transport the accumulator. (Usually 1 MPa).

P_1= Pre-charge pressure. Is the gas pressure required for the accumulator to be used in a given application.

P_2= Minimum operating pressure. As an approximate rule, the relation $P_1 \approx 0.9 * P_2$ can be stated.

P_3= Maximum operating pressure. The relation between the maximum operating pressure and the pre-charge pressure depends on the sort of accumulator and has been defined in the previous sub section.

$\forall_1$ Represents the maximum possible volume of the gas inside the accumulator.

$\forall_2$ Is the volume of the gas when the pressure is P_2.

$\forall_3$ Represents the volume of the gas when the pressure is P_3.

In figure 6.2 are represented the different parameters previously defined, notice that two different compression/expansion processes are defined, adiabatic and isothermal. The sub index a, is defining adiabatic compressions/expansions, while the sub index i, characterizes isothermal compressions/expansions. An adiabatic process is a quick one, the fluid has no time to transfer heat to the environment, on the other side, an isothermal process is a very slow one, the fluid has enough time to transfer heat to the environment and as a result maintaining the temperature as constant.

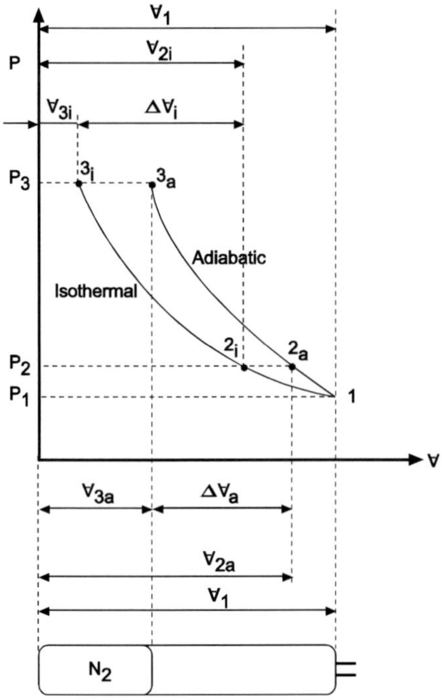

Figure 6.2. Adiabatic and isothermal compression process of a gas inside an accumulator.

The process to calculate the nominal volume of an accumulator used to store energy can be defined as follows.

Between points 1, 2 and 3 in figure 6.2 considering a generic, polytrophic compression process, can be established:

$$P_1 \forall_1^n = P_2 \forall_2^n \text{; from where: } \forall_2 = \left(\frac{P_1}{P_2}\right)^{\frac{1}{n}} \forall_1 \tag{6.1}$$

$$P_1 \forall_1^n = P_3 \forall_3^n \text{; from where: } \forall_3 = \left(\frac{P_1}{P_3}\right)^{\frac{1}{n}} \forall_1 \tag{6.2}$$

where n is the generic polytrophic parameter.

The accumulator useful volume will be $\forall_2 - \forall_3$.

$$\forall_2 - \forall_3 = \forall_1 \left[\left(\frac{P_1}{P_2}\right)^{\frac{1}{n}} - \left(\frac{P_1}{P_3}\right)^{\frac{1}{n}}\right] \tag{6.3}$$

Equation which can be rearranged to obtain:

$$\Delta\forall = \forall_2 - \forall_3 = \forall_1 \left(\frac{P_1}{P_2\,P_3}\right)^{\frac{1}{n}} \left[(P_3)^{\frac{1}{n}} - (P_2)^{\frac{1}{n}}\right] \tag{6.4}$$

The volume of the accumulator will be:

$$\forall_1 = \Delta\forall \left(\frac{P_2\,P_3}{P_1}\right)^{\frac{1}{n}} \left[\frac{1}{(P_3)^{1/n} - (P_2)^{1/n}}\right] \tag{6.5}$$

Equation (6.5) gives the volume of an accumulator as a function of the required volume, and the minimum and maximum operating pressures. This equation can be used for any polytrophic process, as a generic rule, the relation between the value of the parameter n and the time taken for a compression/expansion process is:

n = 1.4 adiabatic process, the compression/expansion process is done in few seconds or less.

n = 1.35 the compression/expansion process takes about 20-30 seconds.

n = 1.25 the process lasts 60-90 seconds.

n = 1.1 the time used for the compression/expansion is about 4 to 8 minutes.

n = 1 Isothermal process.

These values are not absolute and may change depending on the accumulation type its and possible insulation.

It is to be noticed that equation (6.5) does not consider the possibility of accumulators working at different temperatures, in order to include the effect of a minimum/maximum operating temperature, the following procedure is considered.

If the fluid temperature is considered to be T_2, between points 2 and 3 in figure 6.2 can be said:

$$P_2\,\forall_2^* = m\,R\,T_2\,; \tag{6.6}$$

$$P_3\,\forall_3^* = m\,R\,T_2\,; \tag{6.7}$$

And if the temperature was maintained at a value T_1, the relation between points 2 and 3 would be:

$$P_2\,\forall_2 = m\,R\,T_1\,; \tag{6.8}$$

$$P_3\,\forall_3 = m\,R\,T_1\,; \tag{6.9}$$

Dividing equation (6.6) by (6.8) and equation (6.7) by (6.9) it is obtained:

$$\frac{\forall_2^*}{\forall_2}=\frac{T_2}{T_1};\ \forall_2^*=\frac{T_2}{T_1}\forall_2 \tag{6.10}$$

$$\frac{\forall_3^*}{\forall_3}=\frac{T_2}{T_1};\ \forall_3^*=\frac{T_2}{T_1}\forall_3 \tag{6.11}$$

And subtracting (6.11) from (6.10) it is reached:

$$\Delta\forall^* = \forall_2^* - \forall_3^* = \frac{T_2}{T_1}(\forall_2 - \forall_3) = \frac{T_2}{T_1}\Delta\forall \tag{6.12}$$

When substituting equation (6.12) into (6.5) it is obtained the equation giving the volume of the accumulator as a function of the different operating temperatures.

At this point it is interesting to highlight that equation (6.5) estimates the volume of the required accumulator as a function of the pre-charge pressure, the minimum and maximum operating pressures, and the volume of fluid required. But in reality, the theoretical required volume needs to be corrected due to non idealities to obtain the real volume a given accumulator will give. Figure 6.3 introduces the correction factors to be used for isothermal and adiabatic processes. Notice that the correction factor K_i defined for an isothermal process and K_a characterizing an adiabatic one, are obtained in figure 6.3 as a function of the relation minimum/maximum operating pressures.

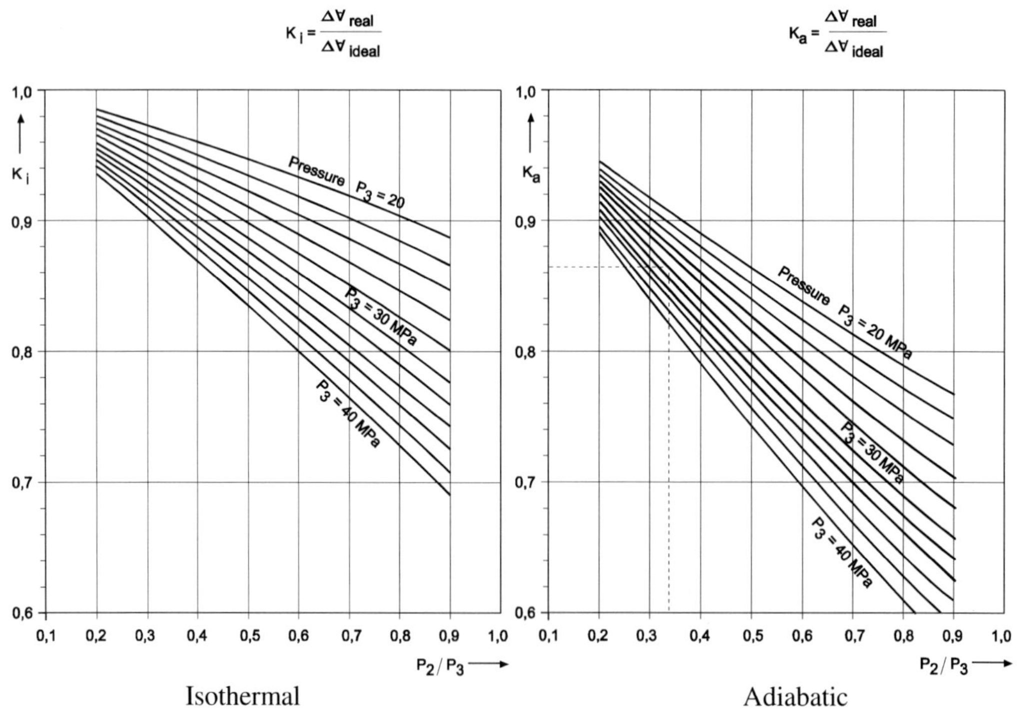

Figure 6.3. Correction factors for isothermal and adiabatic processes.

6.3.2. Accumulators Used As Pulsation Compensator

Another important use of accumulators is its application as pulsation compensators. For this particular application a small volume needs to come in and get out the accumulator in very short time periods, as a result, the process must be seen as adiabatic, n = 1.4. From equation 6.3 it is stated:

$$\forall_1 = \frac{\Delta\forall}{\left[\left(\frac{P_1^*}{P_2^*}\right)^{\frac{1}{n}} - \left(\frac{P_1^*}{P_3^*}\right)^{\frac{1}{n}}\right]} \tag{6.13}$$

The liquid volume $\Delta\forall$ to be considered for the present calculation is a function of the type and capacity of the pump. Since piston pumps are in general the ones producing the highest pressure ripple, the application to this sort of pumps is considered next.

The dynamic volume entering and leaving the accumulator is given as, $\Delta\forall = K^* * C_v$, where:

K^* = coefficient, function of the number of pistons of the piston pump and if pump is single or double acting. Table 6.1 presents some values of the coefficient K^*.

C_v = volumetric displacement of a single piston.

P^* = Average working pressure.

P_1^* = Pre-charge pressure. Gas pressure required for the accumulator to be used in a given application. As an approximate value, can be established $P_1^* \approx 0.6$ to $0.75 * P^*$; or $P_1^* \approx 0.8 * P_2^*$

P_2^* = Minimum system expected pressure, due to the pressure ripple, $P_2^* = P^* - B$.

P_3^* = Maximum system expected pressure, due to the pressure ripple, $P_3^* = P^* + B$.

A = Remaining desired pulsation $\pm(\%)$.

B = Deviation from the average pressure, $B = \frac{A * P^*}{100}$

Table 6.1. Values of coefficient K^* for several cases

Number of pistons.	K^* (single acting pump)	K^* (double acting pump)
1	0.69	0.29
2	0.29	0.17
3	0.12	0.07
4	0.13	0.07
5	0.07	0.023

6.3.3. Accumulator Used As a Shock Damper

In any hydraulic circuit water hammer might appear due to quick valves opening or closing, pump starting and stopping might also create the same phenomena. Water hammer causes a high increase in the circuit pressure due to the quick fluid acceleration or deceleration.

A very straightforward and easy way to determine the fluid pressure increase in the hydraulic circuit is defined by:

$$\Delta P = C\frac{\rho L V}{t} \tag{6.14}$$

where ρ is the fluid density, L is the pipe length, V is the fluid velocity and t is the valve shut down time. The constant C considers the pipe elasticity, its common value ranges between 1 and 2.

The volume of the accumulator required to reduce shock pressure within predetermined limits ΔP^*, can be obtained with the expression:

$$\Delta\forall = Q\left(\frac{C\rho L V}{\Delta P^*} - t\right)\frac{1}{2} \tag{6.15}$$

Substituting equation (6.15) into (6.3) it is obtained:

$$\forall_1 = \frac{Q\left(\frac{C\rho L V}{\Delta P^*} - t\right)\frac{1}{2}}{\left[\left(\frac{P_1^*}{P_2^*}\right)^{\frac{1}{n}} - \left(\frac{P_1^*}{P_3^*}\right)^{\frac{1}{n}}\right]} \tag{6.16}$$

Notice that for the present case the flow will enter the accumulator in a very short period of time, the process will need to be considered as adiabatic n = 1.4. For the present case, the variables are described as:

Q = Volumetric flow rate in the pipe.

P_1^* = Pre-charge pressure. Gas pressure required for the accumulator to be used in a given application. As an approximate value, can be established: $P_1^* \approx (0.6 \text{ to } 0.9) * P_2^*$

P_2^* = Operating pressure, when the valve is open.

P_3^* = Maximum allowable pressure. $P_3^* = P_2^* + \Delta P^*$.

6.4. Accumulators Application

6.4.1. Examples Accumulator Used as Energy Storage

Example 1.

The hydraulic circuit presented in figure 6.4 has a linear actuator which needs to perform cyclically. As a time function, the piston will need to displace according to the following steps.

During the first three seconds the piston is advancing , $T_1 = 3$ s. Piston advance velocity $V_{piston} = 0.02$ m/s. The force opposing to the movement will be F = 20000N.

The following second, $T_2 = 1$ s. Piston remains static $V_{piston} = 0$ m/s.

During the next 6 seconds, $T_3 = 6$ s, the piston returns to the initial position.

The last five seconds, $T_4 = 5$ s, the piston remains static and the accumulator is being charged.

Average pressure losses in the circuit, due to the directional valve and pipe can be considered to be of $18*10^5$ Pa. Pressure relieve valve is set to $200*10^5$ Pa. Piston dimensions are defined in figure 6.4.

The problem consists of determining the minimum volumetric flow the pump will need to provide and the required volume of the accumulator.

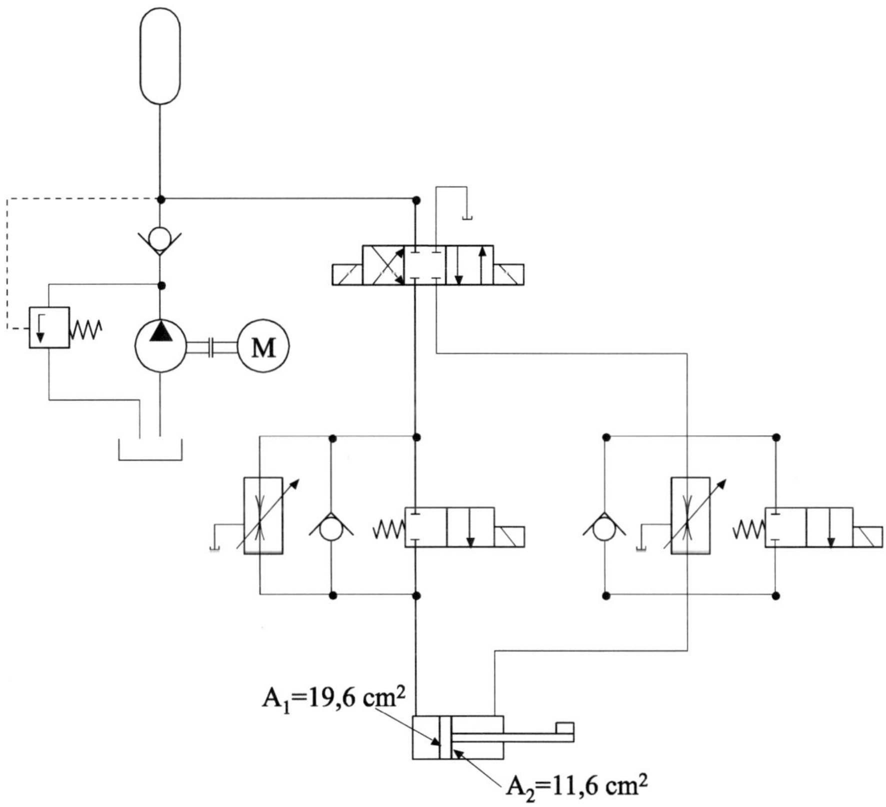

Figure 6.4. Hydraulic circuit used for the present example.

During the first 3 seconds the cylinder needs to move at a velocity of 0.02m/s, the cylinder surface is of $19.6*10^{-4}$ m^2, therefore the volumetric flow necessary for the displacement will be:

$$Q_{advance} = S_a * v_a = 19.6*10^{-4}*0.2 = 3.92*10^{-4}\ \frac{m^3}{s}$$

The cylinder length has to be: $L = v*T = 0.2*3 = 0.6m$

Since it is required for the cylinder to return in 6 seconds, the returning speed will be:

$$L = v*6 = 0.6m \; ; \; v = 0.1\frac{m}{s}$$

The volumetric flow required for the cylinder to return is to be found:

$$Q_{return} = S_r * v_r = 11.6*10^{-4}*0.1 = 1.16*10^{-4}\frac{m^3}{s}$$

The required minimum volumetric flow the pump will need to supply shall be:

$$Q_{pump} = \frac{Q_a T_1 + Q_r T_3}{T_1 + T_2 + T_3 + T_4} = \frac{3.92*10^{-4}*3 + 1.16*10^{-4}*6}{3+1+6+5} = 1.248*10^{-4}\frac{m^3}{s}$$

Now, assuming that the pump is supplying exactly this flow at the required pressure, the volumetric flow the accumulator will need to provide during the cylinder advance will be:

$$Q_{accumulator(a)} = Q_{advance} - Q_{pump} = 3.92*10^{-4} - 1.248*10^{-4} = 2.672*10^{-4}\frac{m^3}{s}$$

Therefore, the volume of fluid the accumulator will need to supply is to be:

$$\Delta\forall_{accumulator} = Q_{accumulator} * T_1 = 2.672*10^{-4}*3 = 8.016*10^{-4} m^3$$

Notice that this volume will enter the accumulator in three steps:

a. During the piston returning process, lasting 6 s, the fluid volume entering the accumulator will be: $\Delta\forall_r = \left(1.248*10^{-4} - 1.16*10^{-4}\right)*6 = 0.528*10^{-4} m^3$

b. During the period of 1 s. $\Delta\forall_1 = 1.248*10^{-4}*1 = 1.248*10^{-4} m^3$

c. During the final period of 5s. $\Delta\forall_4 = 1.248*10^{-4}*5 = 6.24*10^{-4} m^3$

$$\Delta\forall_{accumulator} = \Delta\forall_r + \Delta\forall_1 + \Delta\forall_4 = (6.24 + 1.248 + 0.528)*10^{-4} = 8.016*10^{-4} m^3$$

To determine the volume of the accumulator it is necessary to know the system operating pressure, the maximum operating pressure and the Nitrogen pre-charge pressure.

The required system fluid pressure needed for the cylinder to advance will be:

$$P_{cylinder} = \frac{F_{advance}}{S_{piston}} = \frac{20000}{19.6*10^{-4}} = 10204081.6 Pa$$

Considering the losses, the required pressure at the pump outlet and accumulator entrance will be:

$$P_{pump} = P_{cylinder} + P_{losses} = 10204081.6 + 18*10^5 = 12004081.6Pa$$

This pressure needs to be seen as the minimum operating pressure P_2. The maximum operating pressure, P_3, is to be the one limited by the pressure relieve valve, which according to the problem is $200*10^5$Pa.

The nitrogen pre-charge pressure shall be: $P_1 = 0.9*P_2 = 0.9*12004081.6 = 10803673.4Pa$

The ratio minimum to maximum operating pressure is:

$$\frac{P_2}{P_3} = \frac{12004081.6}{200*10^5} = 0.6;$$

According to figure 6.3, and considering the compression and expansion process as adiabatic, the correction factor is:

$$K_a^* = \frac{\Delta\forall_{real}}{\Delta\forall_{ideal}} = 0.84$$

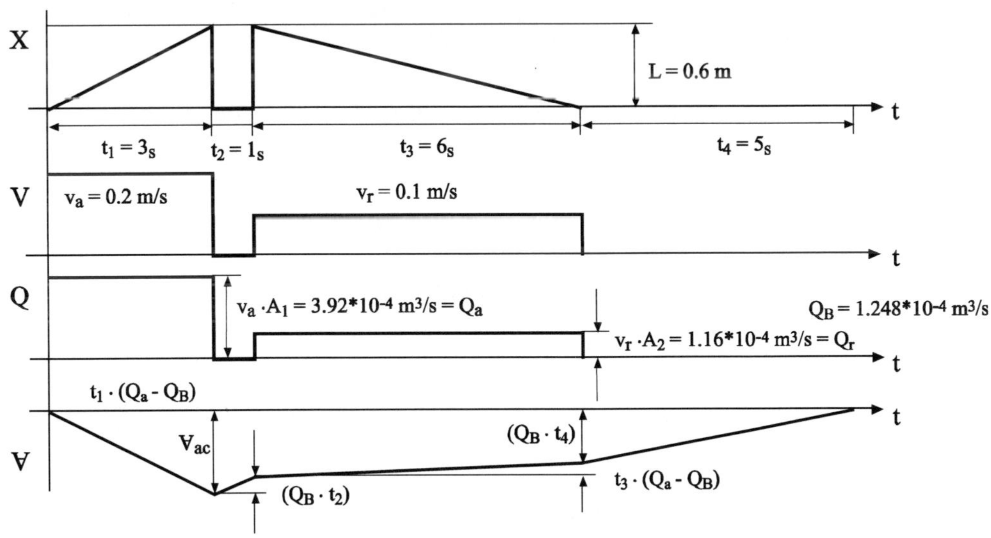

Figure 6.5. Time dependent actuator position, velocity and volumetric flow. The incoming and outgoing actuator fluid volume is also presented.

As a result, the volume of fluid the accumulator will need to store will be:

$$\Delta\forall_{ideal} = \frac{\Delta\forall_{real}}{0.84} = \frac{8.016*10^{-4}}{0.84} = 9.5428*10^{-4}m^3$$

And substituting this volume in equation 6.5 it is obtained:

$$\forall_1 = \Delta\forall\left(\frac{P_2 P_3}{P_1}\right)^{\frac{1}{n}}\left[\frac{1}{(P_3)^{1/n}-(P_2)^{1/n}}\right] = 9.5428*10^{-4}\left(\frac{12004081.6*200*10^5}{10803673.4}\right)^{\frac{1}{1.4}}\left[\frac{1}{\left(200*10^5\right)^{\frac{1}{1.4}}-(12004081.6)^{\frac{1}{1.4}}}\right]$$

$$\forall_1 = 3.3672*10^{-3}\,m^3$$

It is to be noticed that the pump needed has to be able to supply the ideal volume $\Delta\forall_{ideal}$, which means the volumetric flow supplied by the pump needs to be corrected. Figure 6.5 presents the actuator position, velocity, and volumetric flow required as a function of time, in the same graph is also presented the temporal fluid volume incoming and outgoing from the accumulator.

Example 2.

In this second example, the hydraulic circuit presented in figure 6.6 is quite similar to the circuit presented in the previous case, again a linear actuator needs to perform cyclically, but now the actuator is driven by a proportional valve. As a time function, the linear actuator will need to displace according to the following steps.

During the first three seconds the piston will accelerate from 0 to V_{piston} = 0.012 m/s. The force opposing to the movement will be F = 20000N.

The following 2 seconds, T_2 = 2 s. the piston moves as a constant velocity of V_{piston} = 12 m/s.

During the next 3 seconds, T_3 = 3 s, the piston decelerates from 12m/s to 0 m/s.

The following 1 second, T_4= 1s, the piston remains static.

During the next 8 seconds the piston returns to its initial position, during the first 4 = T_5 seconds the piston accelerates from 0m/s to a maximum velocity and during the remaining 4 = T_6 seconds the piston decelerates from the maximum velocity to 0m/s.

Average pressure losses in the circuit, due to the directional valve and pipe can be considered to be of $26*10^5$ Pa. Pressure relieve valve is set to $200*10^5$ Pa. Piston dimensions are defined in figure 6.6.

The problem will again consist of determining the minimum volumetric flow the pump will need to provide and the required volume of the accumulator.

The piston acceleration in the first 3 seconds will be:

$$a_1 = \frac{V}{T_1} = \frac{0.12}{3} = 0.04\frac{m}{s^2}$$

The piston displacement during this period is:

$$X_1 = \frac{a*T_1^2}{2} = \frac{0.04*3^2}{2} = 0.18m$$

In the second time period T_2=2s, the piston is moving at a constant velocity of 0.12m/s, the piston displacement will be: $X_2 = V_p * T_2 = 0.12*2 = 0.24m$

During the third period T_3=3s, the piston is decelerating, therefore:

$$a_3 = \frac{V}{T_3} = \frac{0-0.12}{3} = -0.04\frac{m}{s^2}$$

$$X_3 = \frac{-a_3 * T_3^2}{2} = \frac{0.04 * 3^2}{2} = 0.18m$$

The total distance displaced during the advancing process will be:

$$X_T = X_1 + X_2 + X_3 = 0.18 + 0.24 + 0.18 = 0.6m$$

The returning process is done in two periods of 4 seconds, the distance displaced in each period will be $X_4 = X_5 = 0.3m$, the piston acceleration and deceleration shall be:

$$a_{acce} = \frac{X_4 * 2}{T_5^2} = \frac{0.3*2}{4^2} = 3,75*10^{-2}\frac{m}{s^2}$$

$$a_{dece} = \frac{X_5 * 2}{T_6^2} = \frac{0.3*2}{4^2} = 3,75*10^{-2}\frac{m}{s^2}$$

And the maximum piston velocity is determined: $V_{max} = a_{acce} * T_5 = 0.0375 * 4 = 0.15\frac{m}{s}$

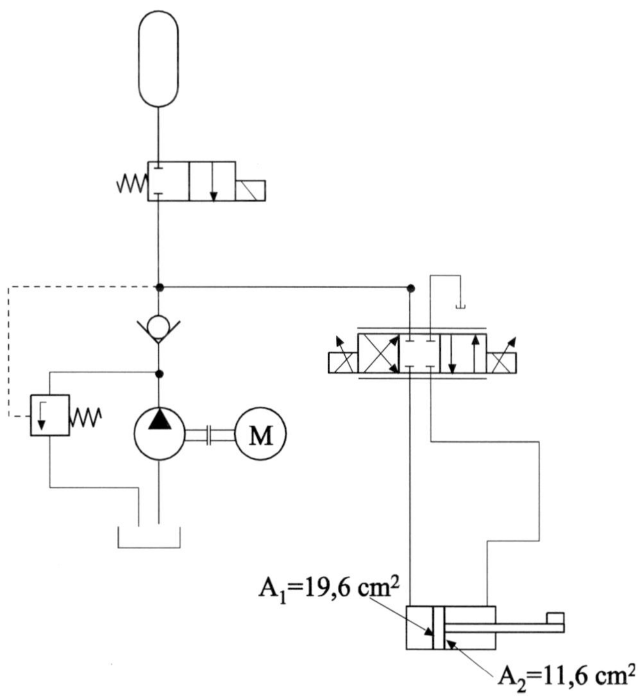

Figure 6.6. Hydraulic circuit used for the second example.

In figure 6.7 it is represented the time dependent position, velocity and acceleration of the piston.

The volumetric flow the pump will need to provide will be:

$$Q_{pump} = \frac{(X_1+X_2+X_3)*S_1+(X_4+X_5)*S_2}{T_1+T_2+T_3+T_4+T_5+T_6} = \frac{0.6*0.196*10^{-2}+0.6*0.116*10^{-2}}{17} = 1.1012*10^{-4}\frac{m^3}{s}$$

The fluid volume entering or leaving the accumulator at each time period is the difference between the volume needed to displace the accumulator minus the fluid volume supplied by the pump, it therefore can be calculated as:

For the first period of three seconds:

$$\forall_{T_1} = \int_0^{T_1=3}\left(a_1*S_1*t-Q_{pump}\right)dt = a_1*S_1*\frac{T_1^2}{2}-Q_{pump}*T_1 =$$
$$0.04*0.196*10^{-2}*\frac{3^2}{2}-1.1012*10^{-4}*3 = 0.2244*10^{-4}m^3$$

During the period of 2 seconds, when the cylinder is moving at a constant speed:

$$\forall_{T_2} = \int_0^{T_2=2}\left(V_p*S_1-Q_{pump}\right)dt = V_p*S_1*T_2-Q_{pump}*T_2 =$$
$$0.12*0.196*10^{-2}*2-1.1012*10^{-4}*2 = 0.25016*10^{-3}m^3$$

Notice that the second period integration starting time, is in reality the final time of the first period.

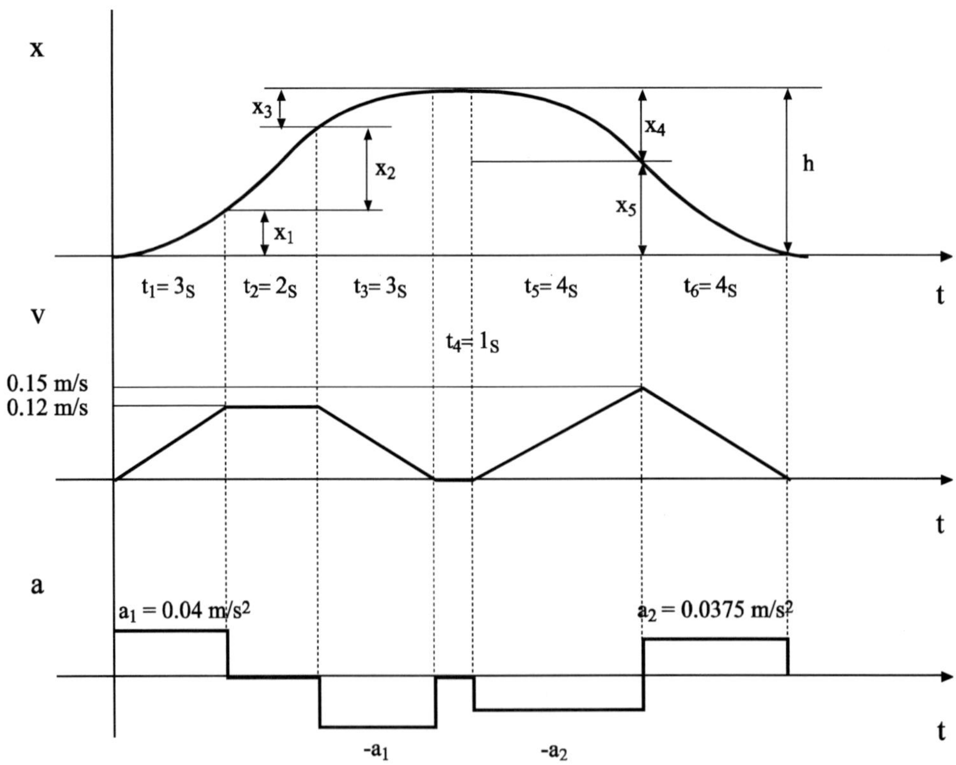

Figure 6.7. Time dependent actuator position, velocity and acceleration.

The next period of three seconds in which the piston decelerates, it is stated:

$$\forall_{T_3} = \int_0^{T_3=3} \left(a_3 * S_1 * t - Q_{pump}\right) dt = a_1 * S_1 * \frac{T_3^2}{2} - Q_{pump} * T_3 =$$

$$0.04 * 0.196 * 10^{-2} * \frac{3^2}{2} - 1.1012 * 10^{-4} * 3 = 0.2244 * 10^{-4} m^3$$

Again the initial time of this process is the final time of the previous one.

The next period last T_4=1 second, and the piston remains static:

$$\forall_{T_4} = Q_{pump} * T_4 = 1.1012 * 10^{-4} * 1 = 1.1012 * 10^{-4} m^3$$

The returning process is divided in two periods, an acceleration period and a deceleration one, for the acceleration period:

$$\forall_{T_5} = \int_0^{T_5=4} \left(a_{acce} * S_2 * t - Q_{pump}\right) dt = a_{acce} * S_2 * \frac{T_5^2}{2} - Q_{pump} * T_5 =$$

$$0.0375 * 0.116 * 10^{-2} * \frac{4^2}{2} - 1.1012 * 10^{-4} * 4 = -0.9244 * 10^{-4} m^3$$

And for the deceleration process:

$$\forall_{T_6} = \int_0^{T_6=4} \left(a_{dece} * S_2 * t - Q_{pump}\right) dt = a_{decele} * S_2 * \frac{T_6^2}{2} - Q_{pump} * T_6 =$$

$$0.0375 * 0.116 * 10^{-2} * \frac{4^2}{2} - 1.1012 * 10^{-4} * 4 = -0.9244 * 10^{-4} m^3$$

It is interesting to realize that the function $\forall_{T_1}$ has a maximum, the time at which this maximum is to be found is calculated as:

$$\frac{d\forall_{T_1}}{dt} = 0 = a_1 * S_1 * T_{1(max)} - Q_{pump}$$

$$T_{1(max)} = \frac{Q_{pump}}{a_1 * S_1} = \frac{1.1012 * 10^{-4}}{0.04 * 0.196 * 10^{-2}} = 1.4s$$

The maximum fluid volume entering the accumulator at this particular time is:

$$\forall_{T_1(max)} = a_1 * S_1 * \frac{T_{1(max)}^2}{2} - Q_{pump} * T_{1(max)} =$$

$$0.04 * 0.196 * 10^{-2} * \frac{1.4^2}{2} - 1.1012 * 10^{-4} * 1.4 = -0.7734 * 10^{-4} m^3$$

Performing the same process with the function $\forall_{T_6}$ it is found:

$$\frac{d\forall_{T_6}}{dt} = 0 = a_{dece} * S_2 * T_{6(max)} - Q_{pump}$$

$$T_{6(max)} = \frac{Q_{pump}}{a_{dece} * S_2} = \frac{1.1012*10^{-4}}{0.0375*0.116*10^{-2}} = 2.53s\,;$$

Time to be measured from the previous step final time.

$$\forall_{T_6(max)} = a_{dece} * S_2 * \frac{T_{6(max)}^2}{2} - Q_{pump} * T_{6(max)} =$$

$$0.0375*0.116*10^{-2} * \frac{2.53^2}{2} - 1.1012*10^{-4} * 2.53 = -1.3936*10^{-4}\,m^3$$

Notice that this volume entering the accumulator is to be measured versus the final volume obtained in the previous step. To clarify all those accumulator incoming and outgoing volumes, figure 6.8 is presented, in this figure the fluid entering and leaving the accumulator as a function of time is presented, all previous calculated volumes are clearly specified.

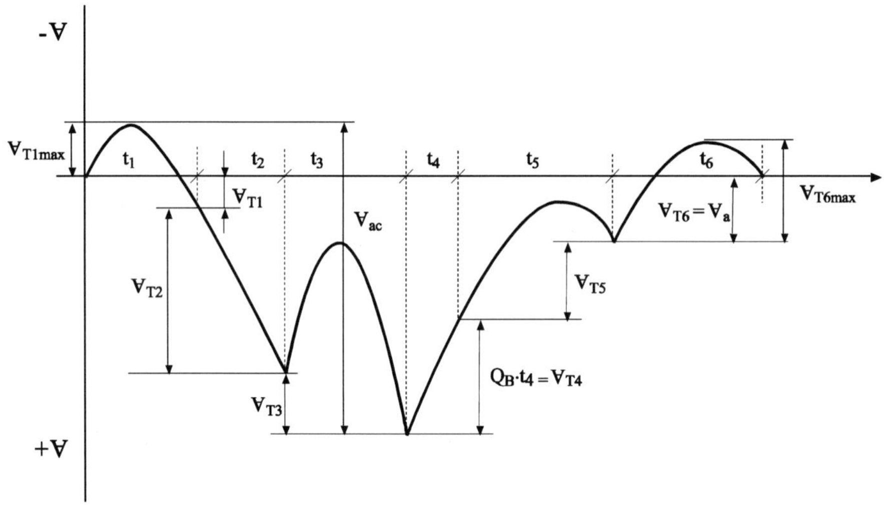

Figure 6.8. Accumulator fluid volume as a function of time.

To determine the required volume the accumulator will need to supply, it is just needed to add the required volumes at each time step, see figure 6.8, obtaining:

$$\Delta\forall_{real} = \Delta\forall_{T_1(max)} + \Delta\forall_{T_1} + \Delta\forall_{T_2} + \Delta\forall_{T_3} = (0.7734 + 0.2244 + 2.5016 + 0.2244)*10^{-4} = 3.7238*10^{-4}\,m^3$$

The required pressure at the cylinder entrance will be the same as in the previous example, since the opposing force and the cylinder surface are the same, therefore: $P_{cylinder} = 10204081.6Pa$

The pressure supplied by the pump will need to be:

$$P_{pump} = P_{cylinder} + P_{losses} = 10204081.6 + 26*10^5 = 12804081.6Pa$$

As established in the previous case, this pressure needs to be seen as the minimum operating pressure P_2. The maximum operating pressure, P_3, is to be the one limited by the pressure relieve valve, which according to the problem is $200*10^5$Pa.

The nitrogen pre-charge pressure shall be: $P_1 = 0.9*P_2 = 0.9*12804081.6 = 11523673.4$Pa

The ratio minimum to maximum operating pressure will now be:

$$\frac{P_2}{P_3} = \frac{12804081.6}{200*10^5} = 0.64\ ;$$

According to figure 6.3, and considering the compression and expansion process as adiabatic, the correction factor is approximately:

$$K_a^* = \frac{\Delta\forall_{real}}{\Delta\forall_{ideal}} = 0.83$$

As a result, the volume of fluid the accumulator will need to store will be:

$$\Delta\forall_{ideal} = \frac{\Delta\forall_{real}}{0.84} = \frac{3.7238*10^{-4}}{0.83} = 4.4865*10^{-4}\,m^3$$

Substituting the corresponding values into equation 6.5 it is obtained the required accumulator volume.

$$\forall_1 = \Delta\forall\left(\frac{P_2\,P_3}{P_1}\right)^{\frac{1}{n}}\left[\frac{1}{(P_3)^{1/n} - (P_2)^{1/n}}\right] = 4.4865*10^{-4}\left(\frac{12804081.6*200*10^5}{11523673.4}\right)^{\frac{1}{1.4}}\left[\frac{1}{\left(200*10^5\right)^{\frac{1}{1.4}} - (12804081.6)^{\frac{1}{1.4}}}\right]$$

$$\forall_1 = 1.7731*10^{-3}\,m^3$$

It is to be noticed that the pump needed has to be able to supply the ideal volume $\Delta\forall_{ideal}$, which means the volumetric flow supplied by the pump needs to be conveniently corrected.

6.4.2. Example Accumulator Used as Pulsation Compensator

In a given system, a piston pump having 5 single acting pistons and turning at 1450 rpm is supplying pressurized fluid at 10 MPa. The pump volumetric displacement is of $C_V = 1\,dm^3/rev.$ and the remaining pulsation is desired to be reduced to $A = 1\%$. Determine the required accumulator needed.

The volumetric displacement of a single piston will be:

$$C_V\Big|_{1piston} = \frac{1}{5} = 0.2\,dm^3/rev = 0.2*10^{-3}\,m^3/rev$$

According to table 6.1, the constant K, will have a value of: K = 0.07.
The allowed pressure deviation from the average working pressure shall be:

$$B = \frac{A*P^*}{100} = \frac{1*100*10^5}{100} = 1*10^5\,Pa$$

Pre-charge pressure shall take a value of: $P_1^* = 0.7*P^* = 0.7*100*10^5 = 70*10^5\,Pa$
The minimum and maximum expected system pressures will be:

$$P_2^* = P^* - B = 100*10^5 - 1*10^5 = 99*10^5\,Pa$$

$$P_3^* = P^* + B = 100*10^5 + 1*10^5 = 101*10^5\,Pa$$

Substituting all values in equation (6.13) it is obtained:

$$\forall_1 = \frac{\Delta\forall}{\left[\left(\frac{P_1^*}{P_2^*}\right)^{\frac{1}{n}} - \left(\frac{P_1^*}{P_3^*}\right)^{\frac{1}{n}}\right]} = \frac{0.07*0.2*10^{-3}}{\left[\left(\frac{70*10^5}{99*10^5}\right)^{\frac{1}{1.4}} - \left(\frac{70*10^5}{101*10^5}\right)^{\frac{1}{1.4}}\right]} = 1.264*10^{-3}\,m^3 = 1.264\,dm^3$$

Being this the accumulator volume required for the desired application.

6.4.3. Example Accumulator Used as a Shock Damper

The system considered consists of a pipe with an internal diameter of 0.05 m and a length of 100 m. A volumetric flow of 4.5 dm^3/s flows along the pipe, fluid density is 875 kg/m^3. The system operating pressure is $P_2^* = 8MPa$, the usual closing time the valve located at one end of the pipe is 0.5 seconds and for this particular application the allowable overpressure is of $\Delta P^* = 2*10^5 Pa$. The problem consists in determining the required accumulator volume to fulfill the established conditions.

The maximum overpressure the circuit will suffer, will be:

$$V = \frac{Q}{S} = \frac{4.5*10^{-3}}{\frac{\pi\,0.05^2}{4}} = 2.291\frac{m}{s}$$

$$\Delta P = C\frac{\rho\,L\,V}{t} = 2\frac{875*100*2.291}{0.5} = 801850\,Pa$$

The pre-charge pressure P_1^* and the maximum allowable pressure P_3^* will be:

$$P_1^* = 0.9 * P_2^* = 0.9 * 80 * 10^5 = 72 * 10^5 \, Pa$$

$$P_3^* = P_2^* + \Delta P^* = 80 * 10^5 + 2 * 10^5 = 82 * 10^5 \, Pa .$$

Substituting all values in equation (6.16) it is obtained:

$$\forall_1 = \frac{Q\left(\dfrac{C\,\rho\,L\,V}{\Delta P^*} - t\right)\dfrac{1}{2}}{\left[\left(\dfrac{P_1^*}{P_2^*}\right)^{\frac{1}{n}} - \left(\dfrac{P_1^*}{P_3^*}\right)^{\frac{1}{n}}\right]} = \frac{4.5 * 10^{-3}\left(\dfrac{2 * 875 * 100 * 2.2918}{2 * 10^5} - 0.5\right)\dfrac{1}{2}}{\left[\left(\dfrac{72 * 10^5}{80 * 10^5}\right)^{\frac{1}{1.4}} - \left(\dfrac{72 * 10^5}{82 * 10^5}\right)^{\frac{1}{1.4}}\right]} = 0.20887 m^3 = 208.87 dm^3$$

A normalized accumulator having a volume slightly superior to the calculated one shall be used.

6.5. NOMENCLATURE

a	Acceleration. (m/s^2).
C	Constant.
Ka	Adiabatic correction factor.
Ki	Isothermal correction factor.
K^*	Piston pump correction coefficient.
L	Length. Pipe length. (m).
m	Mass. (Kg).
n	Polytrophic coefficient.
P	Pressure. (Pa).
P^*	Average working pressure. (Pa).
Q	Volumetric flow. (m^3/s).
S	Surface. (m^2).
T	Temperature. Time. (K). (s).
t	Time. Valve shut down time. (s).
V	Velocity. (m/s).
X	Position. (m).
ρ	Density. (Kg/m^3).
$\forall$	Volume. (m^3).

6.6. REFERENCES

[1] Mordas JM. (1999). The accumulator a pump's best friend. *Hydraulics and Pneumatics.* 52:4. 59-75.

[2] Valencia E, Bergada JM, Ripoll M. (2006). *Oleohidráulica problemas resueltos.* Barcelona. Edicions UPC.

[3] Schneider RT. (2001). Don't forget to consider accumulators. *Hydraulics and Pneumatics*. 54:10. 43-44.

[4] Watton J, Xue Y. (1995). Identification of fluid power component behaviour using dynamic flowrate measurement. *Journal of mechanical engineering science*. 209. 179-191.

[5] Ho TH, Ahn KK. (2012). Design and control of a closed-loop hydraulic energy-regenerative system. *Automation in Construction*. 22. 444-458.

[6] Huayong Y, Wei S, Bing X. (2007). New Investigation in energy regeneration of hydraulic elevators. *ASME Transactions on mechatronics*. 12:5. 519-526.

[7] Yilei L, Zhencai Z, Guoan C, Guohua C. (2013). A novel energy regeneration system for emulsion pump test. *Journal of Mechanical Science and Technology*. 27:4. 1155-1163.

[8] Ho TH, Ahn KK. (2010). Modelling and simulation of hydrostatic transmission system with energy regeneration using hydraulic accumulator. *Journal of mechanical Science and Technology*. 24:5. 1163-1175.

[9] Li Py, Van de Ven JD, Sancken C. (2007). Open accumulator concept for compact fluid power energy storage. *ASME International Mechanical Engineering congress and R&D Exposition*. November 11-15, Seattle USA. 1-14.

[10] Kim J, Yoon GH, Noh J, Lee J, Kim K, Park H, Hwang J, Lee Y. (2013). Development of optimal diaphragm-based pulsation damper structure for high-pressure GDI pump systems through design of experiments. *Mechatronics 23*, 369-380.

[11] Pagilla PR, Garimella SS, Dreinhoefer LH, King EO. (2001). Dynamics and control of accumulators in continuous strip processing lines. *IEEE Transactions on industry applications*. 37:3, 934-940.

[12] Theron NJ, Els PS. (2007). Modeling of a semi-active hydropneumatic spring-damper unit. *Int. J Vehicle Design*. 45:4, 501-521.

[13] Minav TA, Virtanen A, Laurila L, Pyrhönen. (2012). Storage of energy recovered from an industrial forklift. *Automation in construction*. 22, 506-515.

[14] Yokota S, Somada H, Yamaguchi H. (1996). Study on an active accumulator. (active Control of high-frequency pulsation of flow rate in hydraulic systems). *JSME International Journal*. 29:1, 119-124.

[15] Rydberg KE. (2005). Hydraulic accumulators as key components in energy efficient mobile systems. *Proceedings of the sixth International Conference on Fluid Power Transmission and Control*. April 05-08 China. 124-129.

[16] Okoye CN, Jiang JH, Hu ZD. (2005). Application of Hydraulic power unit and accumulator charging circuit for electricity generation, storage and distribution. Proceedings of the sixth International *Conference on Fluid Power Transmission and Control*. April 05-08 China. 224-227.

[17] Godin EM, Korotkova V. (1981). Calculating the parameters of accumulators for pneumo-hydraulic cyclic pumps with an investigation of their most efficient spheres of application. *Vestnik Mashinostroeniya*. 61:10, 35-38.

[18] Lee J, Lee UY. (2012). Design optimization of an accumulator for reducing rotary compressor noise. *Proc. IMechE Part E: J Process Mechanical Engineering*. 226:4, 285-296.

[19] Lin T, Wang Q. (2012). Hydraulic accumulator-motor-generator energy regeneration system for a hybrid hydraulic excavator. *Chinese Journal of Mechanical Engineering*. 25:6, 121-129.

[20] Midgley WJB, Cathcart H, Cebon D. (2013). Modelling of hydraulic regenerative braking systems for heavy vehicles. *Proc. IMechE Part D: J Automobile Engineering*. 227:7, 1072-1084.

[21] Prupis LM. (1987). Selecting optimum parameters for a centralized hydraulic drive with gas-hydraulic accumulators and pumps. *Stanki I Instrument*. 58:10, 12-14.

[22] Prupis LM. (1985). Hydraulic drive with gasohydraulic accumulators. *Stanki i Instrument,* 5:12, 55-57.

Chapter 7

CONTAMINATION CONTROL IN FLUID POWER SYSTEMS

7.1. INTRODUCTION

Contamination control is of vital importance in any fluid power system, it is just necessary to consider that nearly 70% of all failures occurring in a circuit are due to contamination problems, the economical cost involved, according to MIT, may reach the 6% of a country GNP. Also American Navy estimated that contamination cost on hydraulic systems per hour was as high as 60% of the petrol cost required each hour. Minimizing contamination is therefore a key issue in the fluid power field.

7.2. SORTS OF CONTAMINATION

As a contaminant it is regarded all sort of solid particles and other fluids like air or water which might be mixed or just in contact with the hydraulic oil. Solid contaminants place themselves in the moving gaps like between valve spools and main body, piston and cylinder, poppet spool and relieve valve or pressure valve main body etc, enhancing erosion and often preventing a correct movement. Solid contaminants may also sediment and partially or fully block hydraulic fluid paths, this also affects the pump then it needs to increase its output pressure to maintain the required volumetric flow.

One of the most relevant problems associated with the existence of solid particles is linked with the sharp decrease of the lubrication capacity when small particles locate themselves in the gaps between moving parts, producing localized temperature increase, a quick erosion process and leakage increase.

Contamination level due to solid particles is defined according to the norm ISO 4406, three numbers indicate the amount of solid particles bigger than 2μm, 5μm, and 15μm, existing in 100ml of fluid. As an example, the code ISO 20/18/12 indicates that in 100ml of fluid exist between 0.5 and 1 million particles bigger than 2μm, between 130000 and 250000 particles bigger than 5μm, and between 2000 and 4000 particles bigger than 15μm. Table 7.1 presents the relation between the number associated to the ISO code and the amount of particles existing in the fluid.

Fluid solid particles contamination is due to three main sources:

Contamination incorporated: new hydraulic oil may have some solid contaminants immersed in it, therefore needs to be filtered before introduced in the system. New fluid power systems are likely to have contaminants left in the system during assembly or even in maintenance work, pipe threads, seal materials, grinding chips, welding beats, silicone tape shreds, are among the solid contaminants a new fluid power system might have. New systems need to run for about 50 hours, after that oil and filter cartridges must be changed. This process is called flushing.

Contamination generated: the most dangerous contamination in a fluid power system is the one generated by the system itself, since the particles generated by the system are very hard and aggressive, producing severe wear.

Environmental contamination: solid particles coming from the environment may enter the hydraulic system through the reservoir; water may also condense on the reservoir's interior. Appropriate air filters are required in tanks.

Solid particles are not the single contaminant appearing in hydraulic systems, fluids as air and water are also regarded as contaminants. Air can be found dissolved in hydraulic oil or in form of bubbles. Air bubbles, when compressed, originate heat and this localized temperature increase may destroy hydraulic oil additives. Air dissolved in fluid, originates a power loss transferred by the fluid, increases the noise level and working temperature associated to a hydraulic system, may also generate some non desirable chemical reactions and oil oxidation. The oxidized, contaminated, fluid particles increase wear and must be removed from the hydraulic circuit.

The presence of water in hydraulic circuits can be as destructive as air, but its elimination is more difficult than air. Water is commonly found due to air condensation inside reservoirs, as temperature decreases, air humidity condensates and forms small drops, these drops mixed with hydraulic oil. Water presence in a hydraulic circuit is often spotted thanks to the appearance of cavitation, usually in pumps although it can also happen in valves. Cavitation can easily be detected due to its high frequency noise associated, metal parts were cavitation bubbles implode suffer severe localized erosion. Whenever a hydraulic system may need to operate at temperatures below 0 C°, water particles may freeze and can easily block small gaps and orifices.

Water promotes metal oxidation and corrosion actuating as electrolyte and conducting electricity between two different materials, the internal surface of reservoirs is one of the first places where corrosion appears. Corrosion in aluminum and zinc alloys is to be seen as a white layer of oxide. Gears and steel journal bearings are among the first places where oxidation appears. Water reacts with hydraulic oil additives; it reacts with oxidation inhibitors producing acids which precipitate and increase wear, at high temperatures, above 60C°, water reacts with zinc based wear reduction additives and destroys them. Water helps joining together small contaminated sticky particles which may jeopardize free spool or poppet movements.

It is also interesting to consider that water enhances biologic contamination, microorganisms like bacteria, algae etc could also appear in a fluid power system, the presence of water and air speeds the process. The size of microorganisms, oscillate between 0.2 and 2 microns for unicellular ones and can reach 200 microns when cell colonies are being formed. Under favorable conditions, bacteria can reproduce each 20 minutes and therefore grow exponentially. Bacteria produce acid, which damage most metals, the bacteria itself, is also a huge problem in a hydraulic system, then it prevents the movement of components and may block fluid passages.

Table 7.1. Particles interval linked with ISO standards

ISO number	Interval number of particles to be found in 100 ml of fluid.	
	Minimum Number	Maximum number
1	1	2
2	2	4
3	4	8
4	8	16
5	16	32
6	32	64
7	64	130
8	130	250
9	250	500
10	500	1000
11	1000	2000
12	2000	4000
13	4000	8000
14	8000	16000
15	16000	32000
16	32000	64000
17	64000	130000
18	130000	250000
19	250000	500000
20	500000	1 Million
21	1 Million	2 Million
22	2 Million	4 Million
23	4 Million	8 million
24	8 Million	16 Million

A first evidence of biologic contamination is an unpleasant smell, produced by decomposing microorganisms. Oil viscosity and color might very well change. It has to be noticed that whenever these symptoms appear, the fluid as well as hydraulic system components could be severely damaged and a proper inspection is required. Anti-bacterial products and proper filtering shall prevent this problem, although the elimination of water and air from the hydraulic system has to be seen as the best way to prevent biological contamination.

7.2.1. Definitions Regarding Filtration

Nominal Filtration

It is a filtration value indicated by the filter manufacturer. For example, the specification MIL-F5504A defines a filter with nominal filtration of 10 microns, the one which retains 98% in weight of all contaminant particles bigger than 10 microns, being its concentration a specified one. Nevertheless, the majority of manufacturers consider nominal filtration as the particles dimension for which filtration efficiency is higher than 50%, the filter captures more than the incoming 50% of the particles bigger than a specified value.

Absolute Filtration

According to NFPA (National Fluid Power Association), absolute filtration specifies in microns the diameter of the biggest spherical hard particle able to cross the filter under determined conditions. Absolute filtration has to be seen as an indication of filter opening.

Beta rating β_x.

Beta rating is the international standard defining the efficiency of filter elements and it is based on fluid samples taken before and after the filter, it is also called the multi-pass method for evaluating filtration performance ISO 4572. The ratio β_x is the relation between the number of particles of a size equal or bigger than (x) in microns entering the filter, divided by the number of particles leaving the filter.

$$\beta_x = \frac{\text{Number of particles of size equal or bigger than (x) entering the filter}}{\text{Number of particles of size equal or bigger than (x) leaving the filter}}$$

A rating $\beta_5 = 75$ means that, for each 75 particles of a size equal or bigger than 5 microns entering the filter, 74 particles will be trapped and one particle will cross the filter, the filtration efficiency will be: 74/75=0,9866 (98.66%).

The relation between beta rating β_x and the filter efficiency η is given by: $\beta_x = \frac{100}{100-\eta}$

According to ISO standards, when the value $\beta_x \geq 75$ it is accepted that filtration is absolute. There are nevertheless many manufacturers which understand absolute filtration is defined when $\beta_x \geq 100$, (efficiency 99%). Since experience has demonstrated that when $\beta_x = 100$ fluid contamination is fully controlled even under the most rigorous applications.

7.2.2. Sort of Wear and Erosion in Hydraulic Systems

Abrasive wear. Hard particles place themselves between two surfaces having relative movement, one or both surfaces will be deteriorated and further abrasive particles may be generated. This is the most common sort of wear, representing between 20 and 50% of all wear appearing in fluid power systems.

Wear due to fatigue. Particles may obstruct clearances and generate micro cracks. This sort of wear represents between 10 and 20% of all wear appearing in a hydraulic circuit.

Adhesive wear. When lubricating oil film between two surfaces disappears, metal to metal contact and therefore wear is likely to happen. This represents between 7 and 15% of all wear.

Erosive wear. Very fine particles transported by the fluid at high speed erode surfaces, grooves etc. representing between 4 and 8% of overall wear.

Corrosive wear. Chemical contamination and water mixed in hydraulic oil, originate corrosion and chemical reactions which degrade surfaces. Represents between 3 and 13% of all cases.

Wear due to cavitation. The introduction of restrictions like a filter in the suction line may originate cavitation in the pump, pulling up solid particles of it.

Wear due to air in the system. Air bubbles mixed in the fluid can pull out particles from surfaces in the same way as cavitation.

7.3. Hydraulic Filters

In fluid power systems, filters are the components used to control contamination; they capture, contain and eliminate contaminant particles. Filters consist of two main parts, a body and a filtering material. In depth type or absorbent filters, fluid is forced through multiple layers of material and in surface type filters fluid flow though a single layer of woven mesh. Typical filter materials used for depth type filters are: metal fibers, glass fibers, in both cases, the fibers can be woven or matted and pressed, synthetic fibers, porous and permeable paper "resin coated" and sintered granular metals, are other possible materials to be used. Typical materials for surface type filters are: cellulose fiber, nylon cloth and steel wire cloth.

In depth type filters, are generally used to obtain a very fine filtration. Metal, glass fibers and synthetic fiber filters can be cleaned, nevertheless they are difficult to clean and sometimes have to be discarded, paper filters, regardless of the type, cannot be cleaned and must be discarded when filled with contaminants.

In fluid power systems, filters can be placed in four different locations, in the suction line, the pressure line, the return line and in an independent filtration circuit, see figure 7.1.

Suction line filters are low grade ones, they are usually metallic or of cellulose cloth, they protect mostly the pump. The main problem with suction filters is that they can cause a high pressure drop and cavitation, especially when using filters with ratings of 5 or 10 microns, which is usually not the case. Using magnetic separators reduce the risk of cavitation.

Pressure line filters are located upstream of contamination sensitive and expensive components like servovalves, proportional valves, motors etc. being its mission to protect such components. They are usually mounted directly onto the component to be protected. The usual filtration rate of these filters oscillates between 3 and 25 microns, they are often made of glass fibers and its type is an in depth one.

If the fluid power system has a pressure line filter located downstream of the main pump and other pressure filters located upstream of sensitive components, the grade of filtration of both filters should be the same, especially if the main pump flow passes though the sensitive components.

If the filters mounted directly upstream of the component have a filtration rate higher than the main pressure line filter, the component filters will in fact act as a main filter and will probably need to be replaced more often than expected.

Return line filters prevent contamination originated in the circuit to return to tank, these filters are low pressure elements although they must withstand possible pressure peaks, if required; it's filtration rate can be very small and therefore they are ideal to protect the main pump, then suction line filters of the same filtration rate would generate a pressure drop which could cause cavitation of the pump.

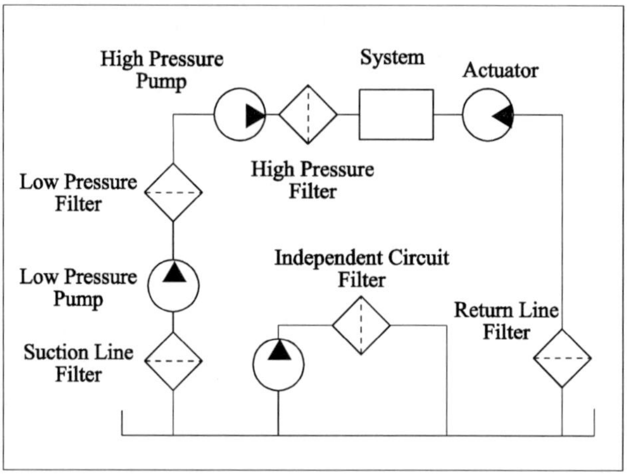

Figure 7.1. Different possible locations of filters.

Off line filtration or independent circuit filtration. Effectiveness of pressure line and return line filters is reduced when appearing pressure peaks, water hammer effects, fluid pulsations and or vibrations. Optimum filtration is achieved when continuous and pulsation free flow enters the filters. The best way to fulfill these requirements is via filtering using an independent circuit. Therefore, whenever system working conditions are severe and proper filtration via using pressure line or return line filters it is difficult to obtain, independent circuit filtration may be the best choice. In cases where the return line flow is very big, off line filtration may be the best solution. One of the advantages of using independent circuit filtration is that filter cartridges can be changed at any time, even the cartridges filtration rate may be modified without affecting the main system design. Another advantage is that the independent circuit can be placed in the most convenient location, providing space restrictions may exist. Another point to be considered is that these systems run independently of the main fluid power system, therefore can keep cleaning the oil even when the main system is at rest.

Magnetic filtration. Regardless of filter type and location, the filter central core may be made of a magnetic material, fluid entering the filter passes through the magnetic central core where ferromagnetic particles are trapped, remaining particles are to be collected on the conventional cartridge, which is the second line of defense the fluid must cross.

Regardless of filter location, the use of filters duplex, two filters in parallel, may be considered, especially when system needs to be continuously running.

It is also to be highlighted that nowadays, the majority of filters have an electronic/electric switch which measures the filter pressure differential and indicates the filter status, whenever a filter cartridge is to be replaced will clearly be indicated by the electronic/electric indicator. Notice as well that many filters have a bypass in parallel; filters with bypass are indicated for fluid power systems where the cost of stopping the system to replace the cartridge is too high. Cartridges shall in this case, be replaced at the end of the day, whenever the enterprise production stops.

7.3.1. Filtering Elements

Multiple Fabric Layers Filtering Elements

Experience gathered by researchers and technicians has brought to the industry to develop filters with multiple fabric layers. In these filtering elements, the external layers located on of both sides of the fabrics, internal and external, are usually metallic and hold the fabric layers between them. Contaminants are trapped by the fabric layers, usually made of glass fibers coated with resin. In order to maximize the filtering surface, fabric layers are folded in zigzag, which also increases the mechanical resistance of the filter cartridge.

This sort of filtering elements are characterized by a very fine filtration, they maintain the rate of filtration β_x for a wide range of pressure differentials, they have a high capacity to retain particles, have a large accumulation surface, a very good chemical resistance and they are able to handle high pressure differential peaks, they are able to work at high pressures. The presence of water mixed with the hydraulic oil, does not reduce the filtration capacity of these filters. Fluid usually flows from the external part towards the internal one.

Water Absorbent Filtering Elements

Water absorbent elements are designed to eliminate water mixed with the mineral hydraulic oils or synthetic fluids. One of the most known filtering elements is the Par-Gel from Parker, which is composed of layers of a highly absorbent inorganic copolymer, having no reaction with hydraulic oil and prevents the formation of bacteria and algae in the filter. Once the water is being trapped in the filtering element it will not go back to the system, even if the element is submitted to high pressures. It must be noticed that water absorbent elements have some capacity to retain solid particles from the system, but they must not be used to do so.

Coreless Filtering Elements

These elements constitute an ecologic alternative versus the conventional ones, since no metal parts are being used in its construction. The main characteristic is that the central tube, which is usually made of metal, is not a part of the replaceable filter cartridge, it is a separated part and therefore it is reusable.

In these sorts of filtering elements, the external layers located on of both sides of the fabrics, which hold the fabric layers between them, are non metallic; they are usually made of a polymeric cloth.

Due to this non metallic construction, used cartridges are between 50 and 60% lighter than the conventional ones, they can be more easily crushed reducing its volume in more than a 60%, and they can also be incinerated.

7.3.2. Pressure Losses in Filters

Pressure losses in filters are due to the pressure losses in the filter body plus the losses in the cartridge (filtering element). $\Delta P_{Total} = \Delta P_{Body} + \Delta P_{Cartridge}$

The first one is proportional to the flow crossing the filter to the power 2, the second one is usually proportional to the volumetric flow and it is time dependent, since increases as the contamination particles get trapped in the cartridge. Pressure losses in both cases increase with the increase of fluid density and or viscosity. Manufacturers provide filter pressure losses when the cartridge is clean and for a fluid with a given density and viscosity. Corrections must be undertaken if the working fluid is not the one considered by the manufacturer.

Typical correction expressions consider the pressure drop in the filter body proportional to the fluid density, pressure drop in the filtering element is considered to be proportional to the fluid viscosity.

Figure 7.2 presents typical pressure loss characteristics for a given filter body and a filter cartridge, the kinematic viscosity and density of the hydraulic oil used was respectively 32 cSt and 900Kg/m^3.

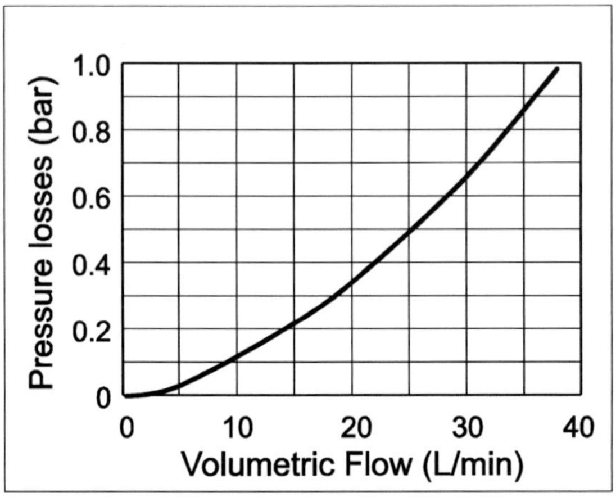

a

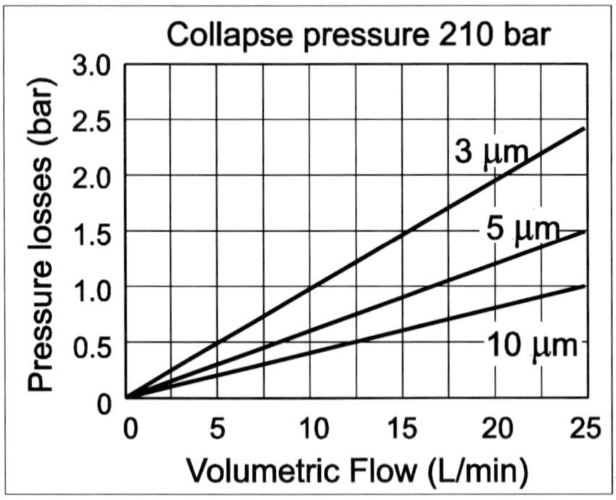

b

Figure 7.2. Pressure losses for a Hydraulic oil ISO 32 (32 cSt) and density 900Kg/m3. a) Pressure losses versus volumetric flow for the filter body. b) Pressure losses versus volumetric flow for the filter cartridge and for three different filtration dimensions 3, 5 and 10 microns. The cartridge is clean.

Notice that three straight lines are presented for the cartridge, each line represents pressure losses in a cartridge able to filter solid particles of a given dimension of 3, 5 and 10 microns respectively.

If for example, hydraulic oil with a kinematic viscosity of 35cSt and a density of 930Kg/m^3 was to be used, the correction expressions needed to modify each given point belonging to the pressure losses curves would be:

For the filter body $\Delta P_{Body} = \Delta P_{Curve\ from\ the\ graph} * \frac{930}{900}$

For the filtering element $\Delta P_{Cartridge} = \Delta P_{Curve\ from\ the\ graph} * \frac{35}{32}$

7.4. Strategies for Contamination Control

Strategies to control contamination in fluid power systems very much depend on the enterprise philosophy. There exist four different contamination control methodologies.

1. Passive contamination control. Filter cartridges are being changed periodically. The optimum cleanness level of the hydraulic system it is not being evaluated or measured at any point.
2. Reactive contamination control. Filter cartridges are being changed whenever possible, there is not a pre-established program, fluid power system cleanness level is not measured or evaluated at any point.
3. Active contamination control. Filter cartridges are being changed following a pre established program and according to the requirements of fluid power components manufacturers. Fluid samples are being taken one or two times a year, although often the results obtained are not considered regarding the modification of the pre established program.
4. Proactive contamination control. The optimum level of system cleanness is initially being determined, filters are being chosen according to the fluid power components manufacturers, fluid samples are being taken periodically and filter cartridges are being changed according to the results obtained from the samples.

In any case, the most recommended strategy is the proactive one, since it allows to precisely determine the contamination level of a hydraulic system, being this knowledge essential for predictive maintenance.

To implement proactive contamination control, three steps have to be followed:

a. Determine the optimum cleanness level of a fluid power system.
b. Install the appropriate filters.
c. Check periodically fluid samples to control system cleanness.

All hydraulic and lubrication systems need to have specified a required cleanness level according to ISO 4406. The most sensible component in a hydraulic system is the one which specifies the cleanness level in the system. Generally, manufactures specify which cleanness level each component requires. As a general rule, the higher the system working pressure the higher the required cleanness level.

It is important to realize that to achieve a certain level of filtration, two factors must be considered, the filter filtration rate and its position in the hydraulic system, notice as well that the higher the filtrated volumetric flow the higher the fluid cleanness will be.

To determine the filter size, the most relevant parameter is the maximum allowed pressure drop when cartridges are clean, as a general rule, this pressure drop shall be:

- For suction line: between 0.01 and 0.035 bar.
- For pressure line, when the filter has a bypass in parallel: 0.5 bar.
- For pressure line, and for filters without bypass: 1 bar.
- For return line: between 0.1 and 0.5 bar.

Table 7.2. Recommended fluid sampling interval. According to Vickers

System cleanness level	Working pressure	Daily working hours.	
		8 hours or less	More than 8 hours
17/15/12 Or more clean.	<14 MPa	4 months	3 months
	Between14 and 21 MPa	3 months	2 months
	>21 MPa	3 months	2 months
18/16/13 Or less clean	<14 MPa	6 months	4 months
	Between14 and 21 MPa	4 months	3 months
	>21 MPa	4 months	2 months

In order to assure a cartridge reasonable life, it is recommended a filter pressure drop to be less than half the pressure drop necessary to open the bypass valve. All filters need to have incorporated a device indicating filter status and when filter cartridge needs to be substituted. As a general rule, filter cartridges need to be substituted, when the filter indicator indicates so, or when filters have been operating more than 1000 hours, or each time hydraulic fluid is being replaced.

Once filters are being installed and the hydraulic system runs normally, fluid samples need to be collected periodically, table 7.2 indicates as a function of system pressure and level of cleanness required, how often samples should be taken, it is also recommendable for the samples to be taken from the return line.

Being the samples collected and analyzed, if the working fluid is not having the required level of cleanness, following actions should be considered:

a. Replace filter cartridge or cartridges, providing the filters indicators show that a particular cartridge needs to be replaced.
b. Check if any external source of contamination has affected the system, a tank being opened or contamination entering the system through cylinder seals etc, and correct the problem.
c. Check if filters are positioned appropriately and if they filter the maximum possible flow.
d. Consider the possibility of replacing filter cartridges and using cartridges with a higher filtration rate.
e. Add new filters to the system, the use of independent circuit filtration is quite often a good solution.

7.5. Nomenclature

P Pressure. (Pa).

$Q = \dot{\forall}$ Volumetric flow. (m^3/s).

β_x Filter beta rating.

η Filter efficiency.

ΔP Pressure differential. (Pa).

7.6. References

[1] Dapena C, Bergada JM. (1998). Control de la contaminación de sistemas oleohidráulicos. *Automatica e Instrumentación*. 287, 83-103.

[2] Radhakrishnan M. (2003). Hydraulic fluids, a guide to selection, test methods and use. New York. The American Society of Mechanical Engineers.

[3] Rohner P. (1991). *Industrial hydraulic control*. Honk Kong. Wiley.

[4] Vickers. (1993). Manual de oleohidráulica Industrial 5a edición. Vickers España.

[5] Mannesmann Rexroth. (1988). Proyecto y construcción de equipos hidráulicos 1ª edición. Mannesmann Rexroth España.

Chapter 8

INTRODUCTION TO CARTRIDGE VALVES

8.1. INTRODUCTION

During the sixties, fluid power systems faced the problem of working with large volumetric flows, initially conventional valves were used, but the forces acting on the spools and poppets forced to the manufacturers to use large springs, valves became, for these particular applications, too expensive and rather inefficient. To solve these problems cartridge valves were created. A cartridge valve is a logic element, it essentially consists of a spool inserted in a main block and fluid pressure acts on the spool both sides generating a hydrostatic equilibrium, see figure 8.1, a small spring assures the spool position at rest. Whenever the hydrostatic equilibrium is being modified, the spool displaces and opens the passage between ports A and B.

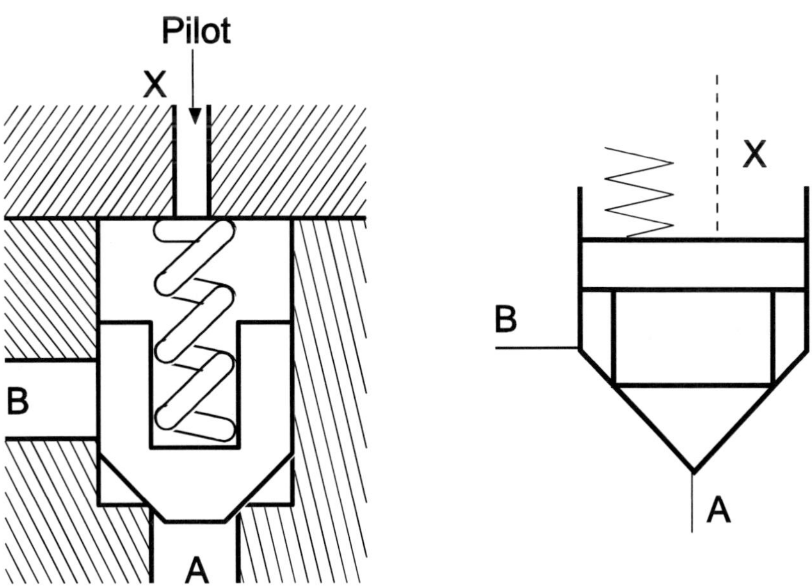

Figure 8.1. Basic logic element of a cartridge valve and its associated symbol.

Initially it was thought that cartridge valves could replace all existing valves, but soon it was seen that in a circuit the combination of cartridge and conventional valves was needed, in fact cartridge valves need to be driven by conventional valves of a much smaller size.

The use of cartridge valves is nowadays reduced to applications where large flow and or pressure is required, figure 8.2 presents a graph where approximately defines under which conditions these valves are to be used, notice that for flow higher than 150L/min and pressures higher than 21MPa, the use of cartridge valves is recommended, nevertheless each manufacturer decides which are the limiting conditions for which conventional or cartridge valves are to be used.

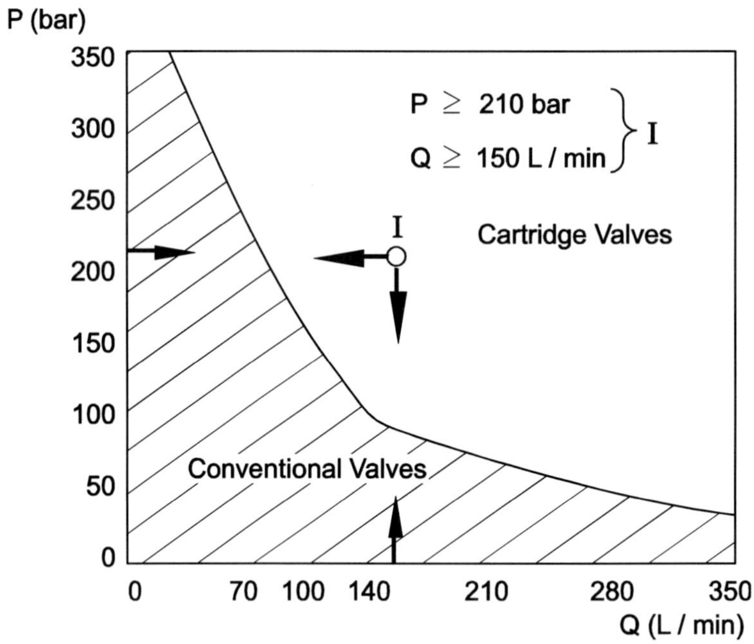

Figure 8.2. Generic limits for the use of conventional and cartridge valves.

It is important to realize that since cartridge valves are based on a logic element, all sort of valves existing in the conventional fluid power domain, can be implemented using cartridge valves, even if they need to be proportional. Table 8.1 presents a basic classification of the existing cartridge valves. In the present chapter a brief description of some of the most representative cartridge valve configurations will be presented.

8.2. Cartridge Valves, Main Parts and Classification

The main parts of a typical two way cartridge valve is presented in figure 8.3, it consist of three main parts, the valve main body, the spool which is being inserted in the main body and the control block. In reality the spool is the moving part of the logic element, which consists of a sleeve, a spring and the spool. Notice that generally this sort of valve does not operate by itself, since it usually needs a control block, where a piloting valve is usually being allocated and which regulates pressure onto the upper part of the spool.

Table 8.1. Cartridge valves generic classification

Cartridge valves generic classification.			
	Sort of valve	Actuation Method	Several existing versions.
Cartridge valves	Pressure relief valves	Manual	Standard
			With discharge via directional valve
			With shutting
			Adjusting 2 pressures
			Adjusting 3 pressures
		Proportional	All previous ones.
	Pressure reducing valves	Manual	Standard (manual adjusting)
			With pressure reducing piloting valve.
			With closing function
		Proportional	All previous three.
	Directional valves	Manual and proportional	Piloted through port A without piloting valve.
			Piloted through port A with piloting valve.
			Piloted through port B without piloting valve.
			Piloted through port B with piloting valve.
			Piloted through ports A and B
			Externally piloted
			Internally piloted
	Flow control valves	Manual and proportional	Simple restrictor
			Flow regulator 3 ports
			Flow regulator with pressure reducing valve
			Flow regulator with relief valve

In reality there exist three different ways to modify the pressure at the spool upper part, see figure 8.4. One way is via using a piloting valve directly allocated onto the valve main body, figure 8.4a. It is important to realize that the piloting valve will in reality characterize the functional operation of the cartridge valve, notice for example that figure 8.4a characterizes a valve which allows flow to go from port A to port B, whenever the pressure differential between A and B is high enough to displace the conical spool allocated in the piloting valve. Port X is to be for this example considered as blocked. Notice as well that whenever pressure at port B reaches a certain value, the valve will close, flow will never be allowed to go from port B to port A. Figure 8.4b, presents a second configuration, in which pressure at the spool upper part is being externally regulated though port X, spool will open whenever pressure differential between ports A and X will allow to do so, for this kind of valve flow is allowed to go from port A to port B and vice-versa. A third way which may be used to control a cartridge valve is presented in figure 8.4c, in this configuration, the port B and the upper part of the spool are directly connected, if pressure on port B is higher or equal to the one on port A, the valve will remain closed and no fluid will flow in any direction, if on the other hand pressure on port A is higher than the one on port B the valve will open allowing the flow to go from port A to port B, this valve acts as a check valve.

When introducing figure 8.4, nothing has been said regarding the areas where pressures on ports A, B and X are acting and clearly these areas play a decisive role. Figures 8.5 and 8.6 present the areas where fluid pressure acts, notice that there are three representative areas, A1, A2 and A3. Area A1 is usually taken as the reference one, cartridge valve dimensions are given as a function of the area A1, port A pressure acts onto this area.

The area A2 is an annular one, when the valve is at rest, the pressure of port B acts onto the area A2, it is important to realize that this area does not always exist, since in some cartridge configurations the areas A1 and A3 are the same. The area A3 characterizes the upper part of the spool, over this particular surface it may act the same pressure of port A, or port B or a different one, in fact a generic cartridge valve will behave as a different sort of valve depending of the pressure acting over the surface A3. Different relations between areas A1 and A3, are typically used for different sort of valves, a relation 1:1 is often used for pressure control valves, relations 1:1.1; 1:1.5 and 1:2, are normally used in flow directional valves.

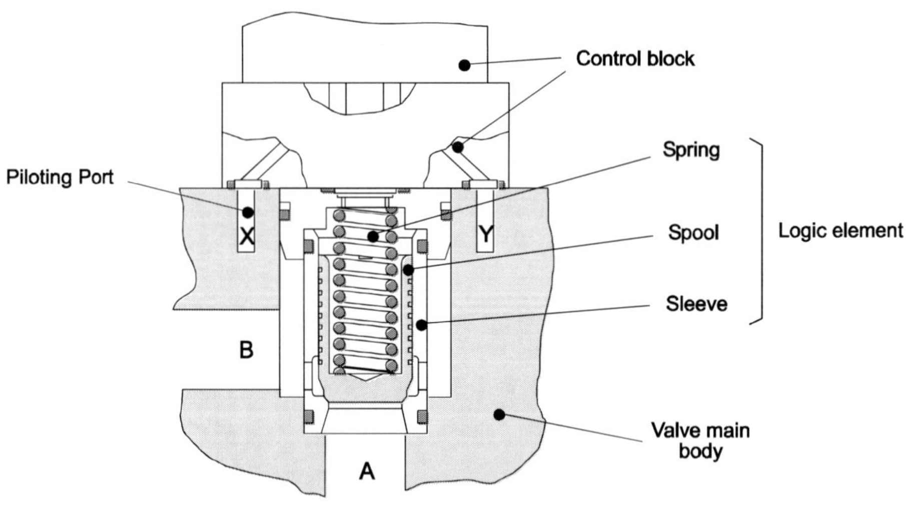

Figure 8.3. Main parts of a cartridge valve.

To understand why the area rate is of huge importance, it is just needed to consider the forces acting on each spool side, notice that under static conditions, forces on both spool sides, due to pressure and the spring, maintain equilibrium, but while valve opens or closes dynamic forces play an important role. Figure 8.5b presents a generic cartridge valve when the valve is slightly open.

Under these conditions, dynamic forces due to the fluid momentum interchange between the entrance area A1 and the minimum fluid passage between the spool and the valve main body, area A4, see figure 8.5b, shall be evaluated as:

$$F_{dynamic} = \dot{m}\left(V_1 - V_4 * \sin(\alpha)\right) \tag{8.1}$$

where $\dot{m}$ represents the mass flow crossing the valve, V_1 is the mean fluid velocity at the entrance, area A1, V_4 is the fluid velocity at the section A4 and (α) is the fluid inclination angle at the exit. It is important to realize that these forces, although often small compared with the static ones, destabilize the spool and may generate erratic behavior, notice as well that these forces often tend to close the spool since usually V_4 is much bigger than V_1.

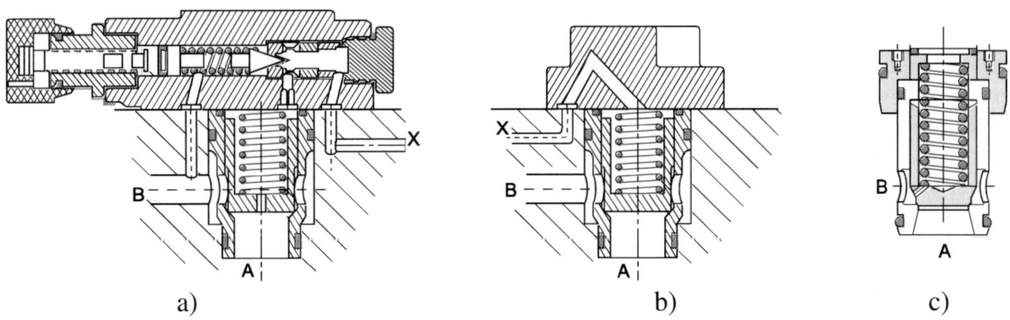

Figure 8.4. Different piloting systems used in cartridge valves.

8.3. Main Cartridge Valve Configurations

As already defined in table 8.1 cartridge valves can be used to create any sort of valve existing in the conventional fluid power field. In what follows a brief description of several main cartridge valve configurations will be presented. Initially pressure control valves will be introduced next several configurations of flow directional and flow control valves shall be presented.

8.3.1. Main Configurations of Pressure Control Cartridge Valves

There exist two types of valves which fall into this configuration, pressure relief valves and pressure reducing valves, the first ones are designed to limit the pressure in a fluid power system while the second ones maintain a constant pressure downstream of the valve.

8.3.1.1. Pressure Relief Cartridge Valves

Its function is limiting the pressure in a circuit via directing the fluid to the tank, the valve is normally closed and its area rate A1:A3 is usually 1:1. Figure 8.7a presents a typical configuration of a pressure relief cartridge valve, port A is connected to the main circuit while port B is connected to tank, pressure of port A acts on both sides of the spool maintaining the valve closed thanks to the spring force. Whenever the pressure on port A reaches a pre determined setting level, it displaces the conical seat valve located on top of the main cartridge valve, and allows the fluid to go from port A to port Y which is usually connected to tank. As a result, the pressure onto the main spool upper part decreases allowing the spool to move upwards and the main flow goes from port A to port B limiting the pressure in the circuit. Figure 8.7b presents the location of the pressure relief cartridge valve in a circuit. The sort of valve described in figure 8.7 is called manual adjusting pressure relief cartridge valve or Standard pressure relief valve, since the opening pressure is being set manually via adjusting the screw located on top of the valve.

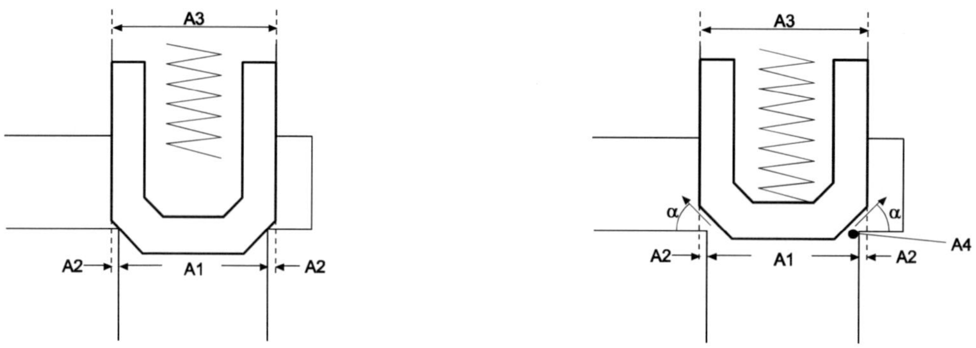

Figure 8.5. Different areas of a cartridge valve and forces acting on it.

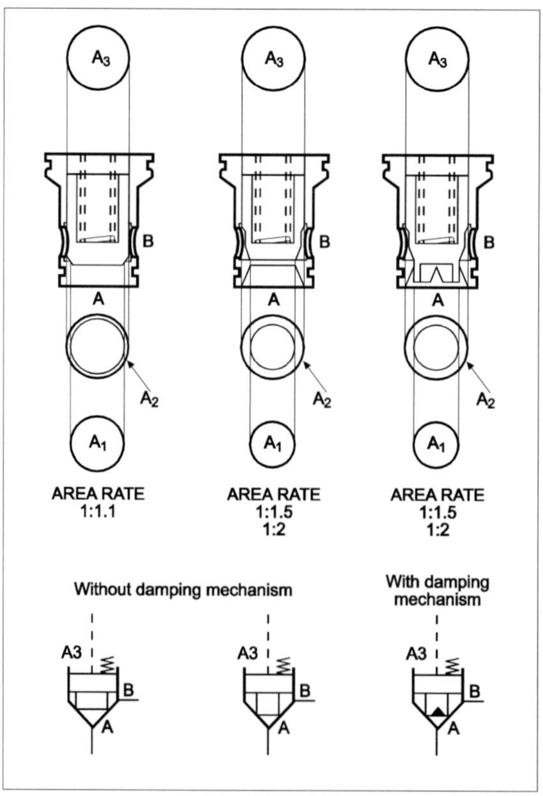

Figure 8.6. Different area rate in cartridge valve and its normalized symbol associated.

A second possible configuration of pressure relief cartridge valves is the one having manual adjusting and discharge via directional valve, figure 8.8. In fact this configuration is the same as the one presented in figure 8.7, but now when the directional valve is connected, the cartridge valve will open regardless of the pressure in port A allowing the flow to go from port A to port B (tank), and therefore depressurizing the system.

A third existing configuration is presented in figure 8.9, in this particular case, when the directional valve is at rest, the pressure relief cartridge valve will not open regardless of the pressure of port A or port B, the cartridge valve will act as a relief valve only when the directional valve will be activated. This configuration is called pressure relief cartridge valve with shutting function.

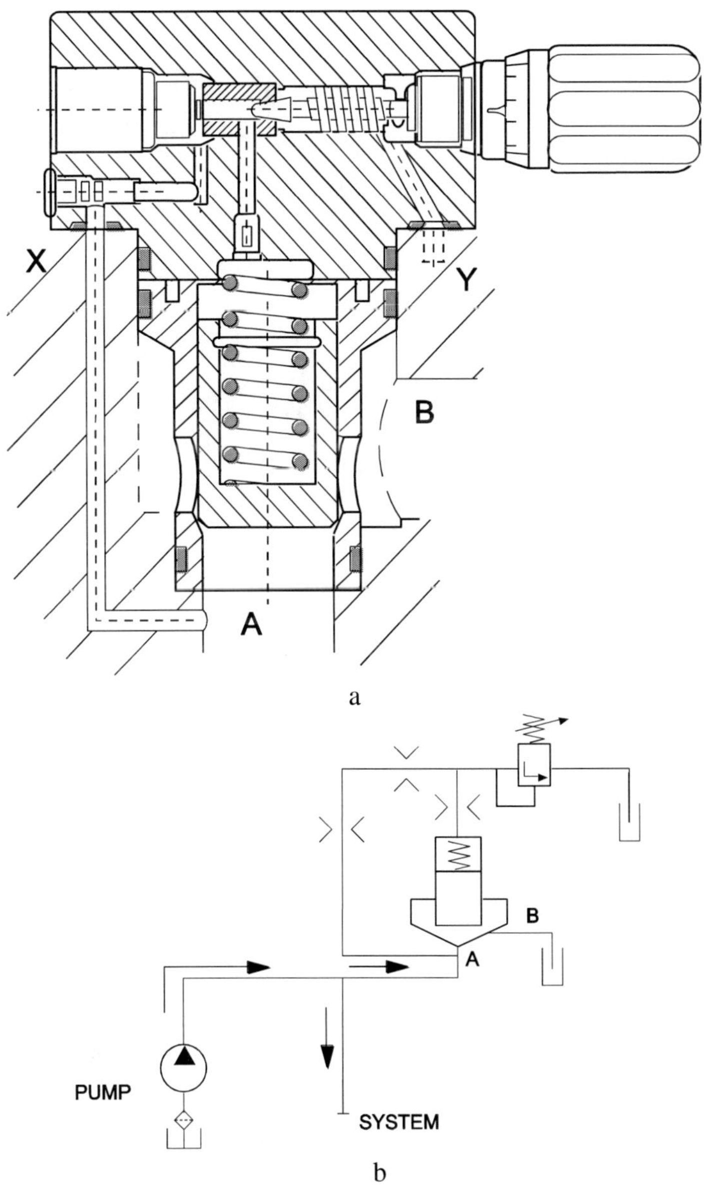

Figure 8.7. a, b. Manual adjusting pressure relief cartridge valve and its location in a circuit.

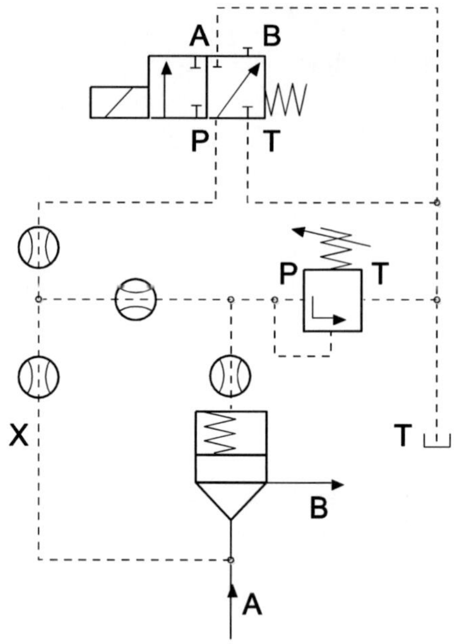

Figure 8.8. Pressure relief cartridge valve having manual adjusting and discharge via directional valve.

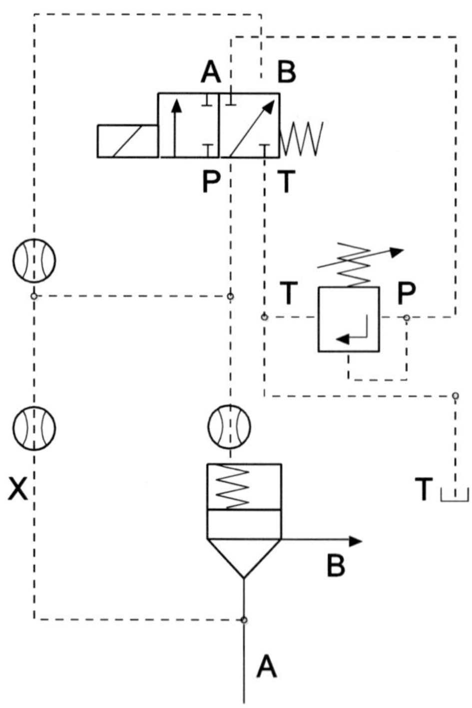

Figure 8.9. Pressure relief cartridge valve having manual adjusting and shutting function.

There exist as well several configurations which allow adjusting the limit pressure of a circuit in two and even three different levels, its typical circuit is defined in figure 8.10a, b. A valve with the possibility of adjusting two pressures, a minimum and a maximum one, is often used in circuits where pressure instabilities and or pressure jumps are high. Figure

8.10a, presents a circuit where two conventional pressure relief valves are connected in parallel, both acting onto the cartridge valve upper part, area A3. Providing the directional valve is not connected, circuit pressure is controlled by the pressure relief valve located nearby the cartridge valve, whenever the pilot directional valve will be connected, the relief valve having the minimum setting pressure will be the one controlling the maximum circuit pressure. Figure 8.10b, introduces the same configuration as figure 8.10a, but now with three possible adjusting pressures, when the directional valve is at rest, the diagram defines a similar configuration as the one defined in figure 8.10a, with two limiting pressures, and when the directional valve is connected then a third limiting pressure will be defined.

In any case, when a hydraulic system requires several different limiting pressure settings, the best option is the use of a pressure relief cartridge valve with proportional adjusting. In reality, the configuration of this sort of valve is very similar to the one used for standard pressure relief cartridge valves, the only difference is to be seen in the piloting relief valve which regulates the pressure onto the cartridge valve spool upper part, see figure 8.11. Notice that now this piloting relief valve is proportional, therefore pressure on the upper part of the cartridge valve spool is to be modified at will according to the current applied to the proportional piloting relief valve. This configuration is in fact the best one since allows modifying the maximum pressure of a fluid power circuit with high precision and at distance. Quite often, the proportional configuration has in parallel a standard pressure relief valve, which acts as a security valve in case of failure of the electric or electronic system, notice that this conventional pressure relief valve in parallel with the proportional one is already introduced in figure 8.11.

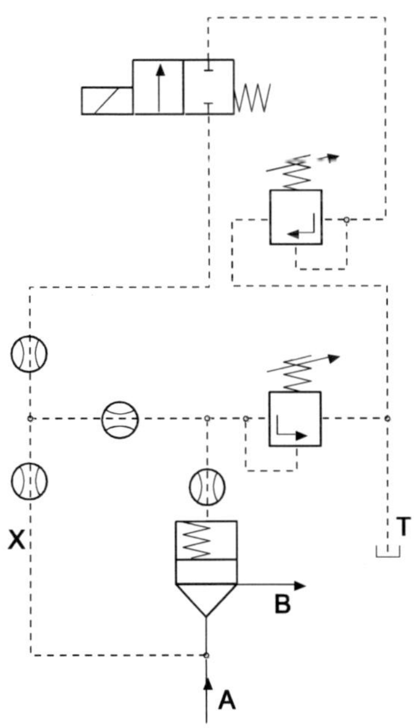

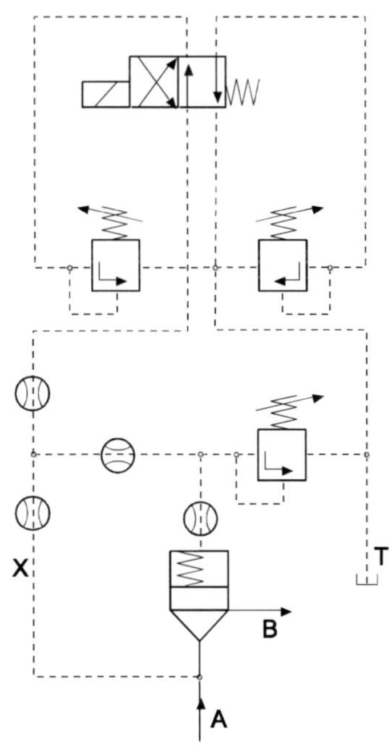

Figure 8.10. a, b. Pressure relief cartridge valve configurations with 2 and 3 different adjusting pressures. a) Two adjusting pressures. b) Three adjusting pressures.

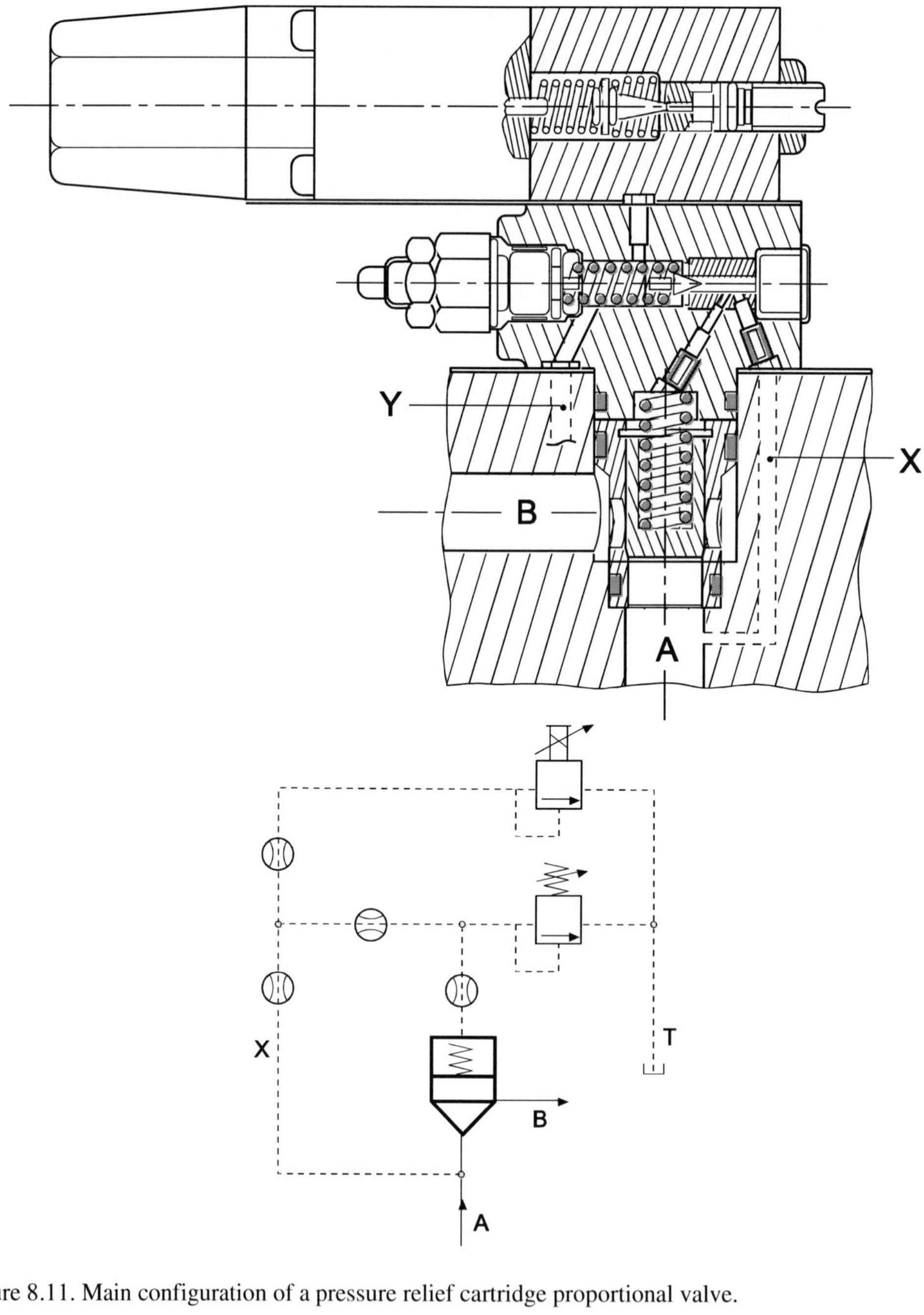

Figure 8.11. Main configuration of a pressure relief cartridge proportional valve.

Some typical characteristic curves of a given pressure relief cartridge valve are presented in figure 8.12, the graph presented on the left hand side characterizes the behavior of a conventional cartridge valve while the graph presented on the right hand side is characteristic of a proportional valve. It is interesting to realize that once the valve opens at a pre-determined pressure, the system pressure will remain rather constant and independent of the volumetric flow crossing the valve, in fact this phenomena is due to two main factors, the volumetric flow going though the piloting valve and the related overall forces, static and dynamic, acting onto the main cartridge/spool. Notice that once the main spool will open, a small pressure increase in port A will produce a flow increase through the piloting valve, which will generate a further pressure decrease onto the main spool upper part, forcing the spool to increase the opening passage, and therefore maintaining pressure on port A rather constant. Generally speaking, proportional valves allow a better pressure regulation, specially when having an electronic feedback.

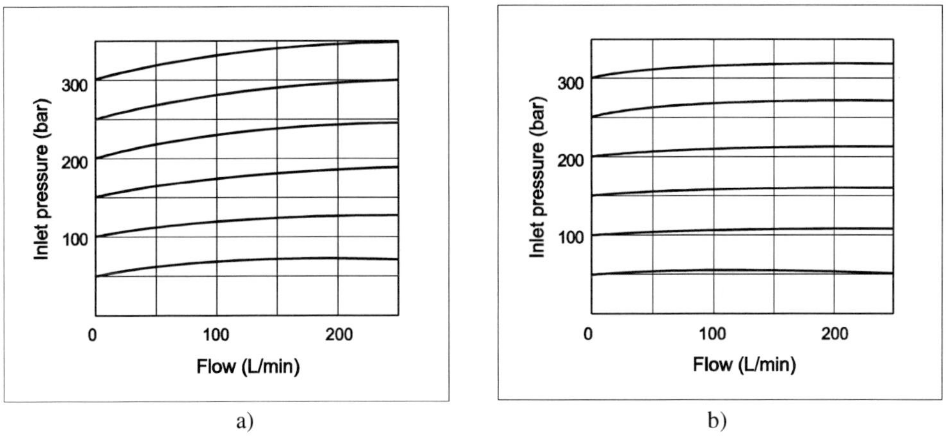

Figure 8.12. Generic pressure flow curves for pressure relief cartridge valve with manual and proportional adjusting. a) manual, b) proportional.

8.3.1.2. Pressure Reducing Cartridge Valves

Pressure reducing valves are designed to maintain a constant pressure downstream of the valve. The valve itself is normally open and regulates the opening section in order to maintain constant the downstream pressure. The two way pressure reducing valve, is in reality working together with the main pressure relief valve of a circuit, since in order to maintain constant the pressure downstream, the pressure reducing valve increases fluid pressure upstream forcing part of the incoming fluid from the pump to go to tank though the circuit main pressure relief valve. The standard configuration of a pressure reducing cartridge valve is presented in figure 8.13a, notice that now port B is the one connected to the piloting valve and to the upper part of the spool, notice as well that the valve is normally open and the passage section will tend to reduce as pressure on port B will increase, the piloting valve is to be adjusted manually, regulating in this way the downstream pressure. Figure 8.3b presents the location of a pressure reducing cartridge valve in a possible circuit.

A second existing configuration consist of a pressure reducing cartridge valve in which the piloting valve is a three port manual adjusting pressure reducing valve, see figure 8.14. In this configuration, the connection between ports A and B is maintained trough the piloting valve, which is a pressure reducing valve, normally open. It is to be noticed that initially, if port B is being set to have, for example, a pressure 80% of the one in port A, the main spool would be closed and the piloting valve would connect ports D with tank T. If pressure on port B decreases, the piloting valve would connect ports D and C, allowing fluid from port A to port B though the piloting valve, when doing so, the pressure onto the upper part of the cartridge valve spool will reduce and as a result the spool will displace upwards increasing the area passage between ports A and B and therefore increasing the pressure on port B.

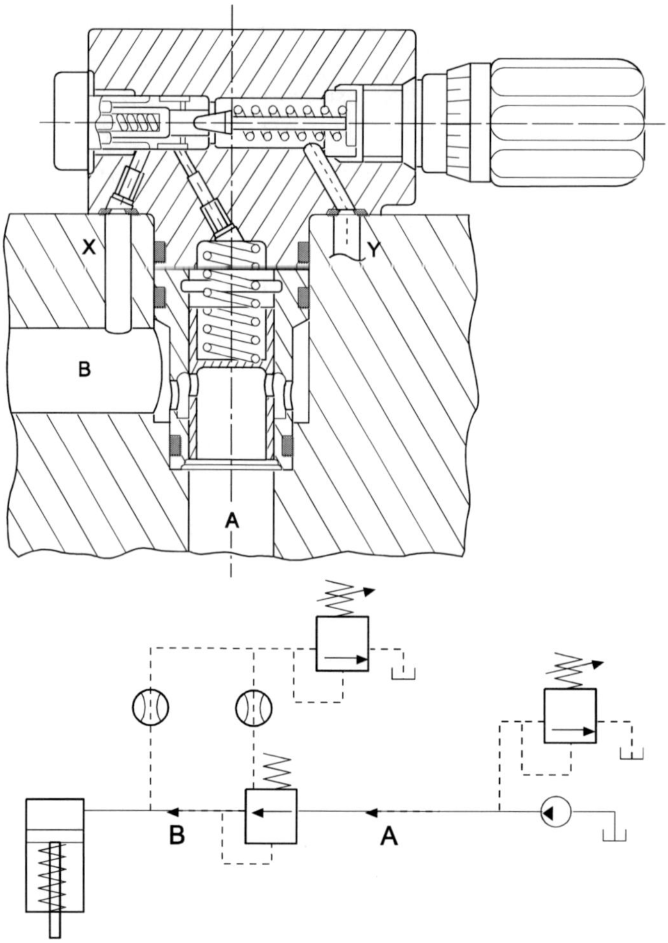

Figure 8.13. a, b. Standard pressure reducing cartridge valve and its position in a circuit.

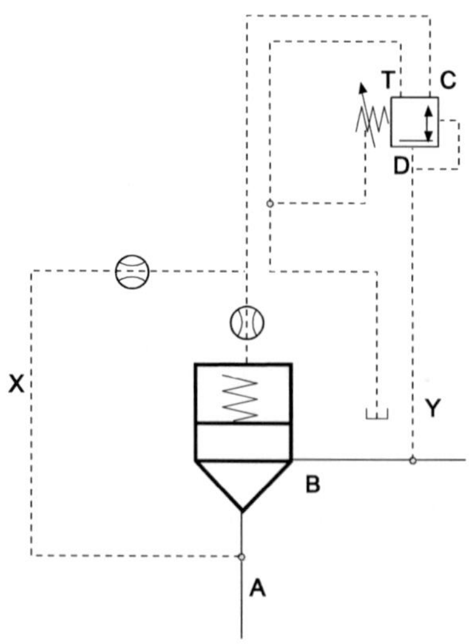

Figure 8.14. Pressure reducing cartridge valve with three port manual adjusting pressure reducing piloting valve.

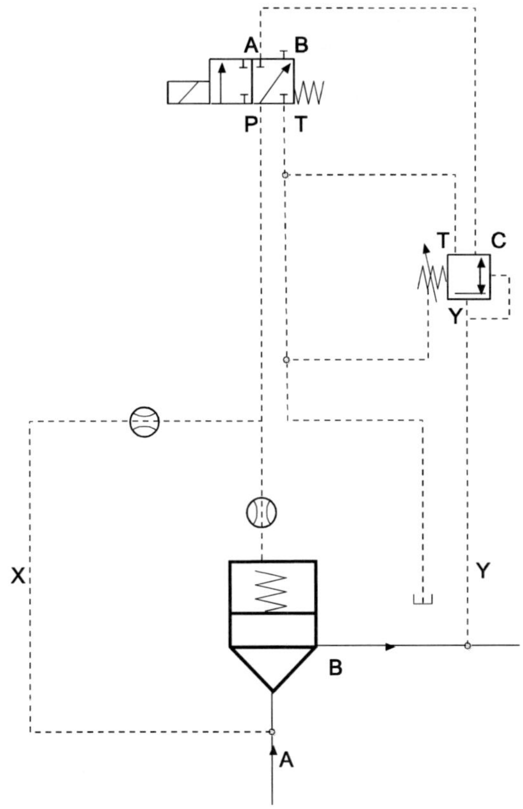

Figure 8.15. Pressure reducing cartridge valve with three port manual adjusting pressure reducing piloting valve and closing function.

A third possible configuration would be the same as the previous one, which was having a manual adjusting three port piloting pressure reducing valve. The difference now consists of an additional closing function regulated by a directional valve, see figure 8.15. It is to be seen that whenever the directional valve is at rest, regardless of the pressure in ports A or B, the cartridge valve will be closed, and fluid will not go from port A to port B. In this configuration, only when the directional valve is connected, pressurized fluid will be allowed to reach port C of the piloting valve and the system cartridge and piloting valve will operate as a pressure reducing cartridge valve, exactly in the same way as in the previous case presented.

As it has already seen with the pressure relief cartridge valves, the best way to regulate downstream circuit pressure using pressure reducing cartridge valves is via employing a proportional piloting valve. Proportional piloting valves have the advantage of very precise regulation and the possibility of doing it from a distance. In fact the three pressure reducing cartridge valve versions already presented are susceptible to be proportional, figure 8.16a, b, c presents the three previous configurations converted to the proportional field. Notice that in all three cases a proportional pressure relief valve is being employed to regulate whether the pressure onto the upper part of the cartridge valve spool figure 8.16a, or the counterbalance pressure on the pressure reducing piloting valve, figures 8.16b, c.

Regarding the characteristic curves of pressure reducing cartridge valves, the same two parameters as in relief valves are being evaluated, the difference resides in that the pressure considered is now the downstream pressure. It must at this point be remembered that, pressure reducing cartridge valves are designed to maintain a constant downstream pressure independent of the required volumetric flow. Figure 8.17 presents some typical curves of a cartridge pressure reducing valve, notice that in theory downstream pressure is to be maintained rather constant and independent of the flow crossing the valve, in reality nevertheless, downstream pressure slightly reduces as flow crossing the valve increases, this is due to the fact that pressure drop across the valve increases as a function of the crossing flow to the power two.

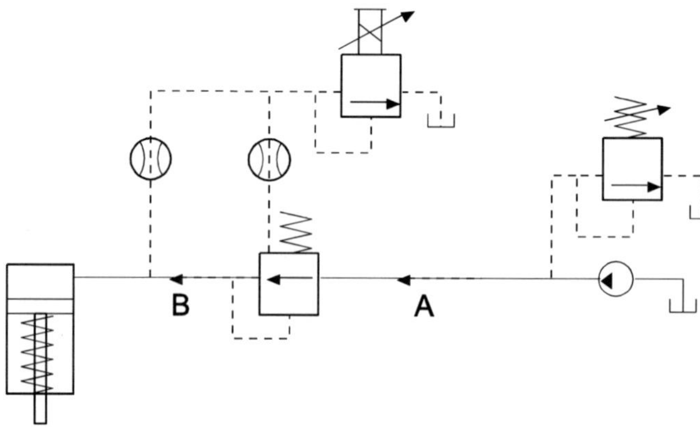

a)

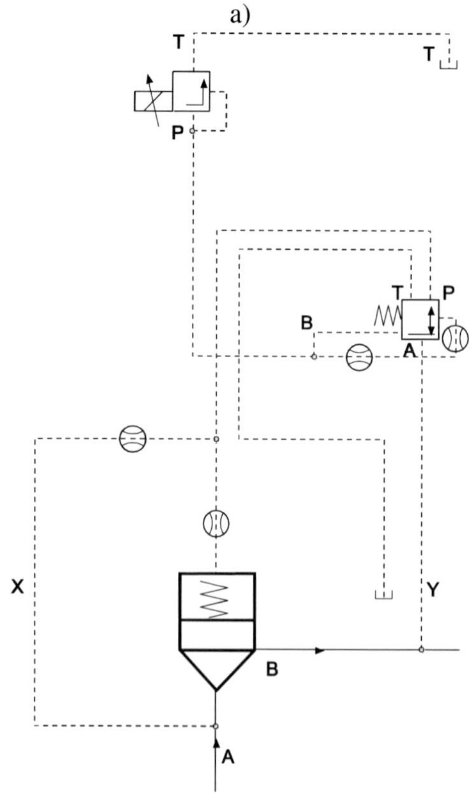

b)

Figure 8.16. a, b, c. (Continued).

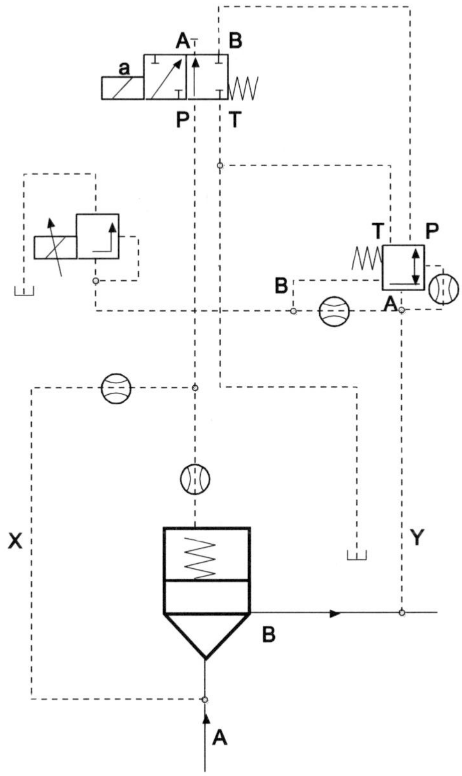

c)

Figure 8.16. a, b, c. Proportional configurations of pressure reducing cartridge valves.

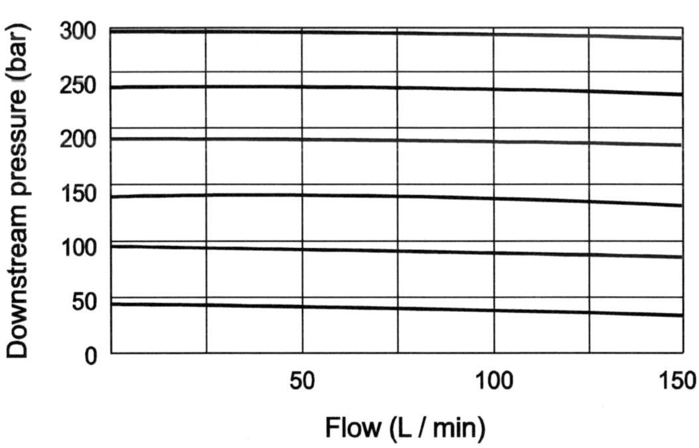

Figure 8.17. Typical characteristic curves of pressure reducing cartridge valves.

8.3.2. Directional Cartridge Valves

Directional cartridge valves are used to direct large fluid flows to several circuit branches. One of the main characteristics of directional cartridge valves is that the area ratio A1:A3, see figure 8.6, goes from 1:1.1 to 1:2. As the sort of cartridge valves presented in this chapter are two way valves, each cartridge valve shall be used to direct fluid from one particular port to another. Directional cartridge valves may use or not a piloting valve, figure 8.18a, b, presents both possible cases. The scheme presented in figure 8.18a, characterizes a directional cartridge valve without piloting valve, notice that when pressure on port A is higher than the pressure on port B, the valve will remain closed, flow will never go from port A to port B, on the other hand, whenever pressure on port B will overcome a pre determined value, flow will be allowed to go from port B towards port A. As it is, this valve characterizes a check valve. Figure 8.18b, presents the same configuration as in 8.18a, but now the valve is being controlled using a 4 way pilot directional valve. With this configuration it is to be seen that whenever the pilot directional valve is at rest, the directional control cartridge valve will operate in the same way as the configuration presented in figure 8.18a, but once the pilot directional valve will be connected, flow will be allowed to go from port A towards B or vice-versa.

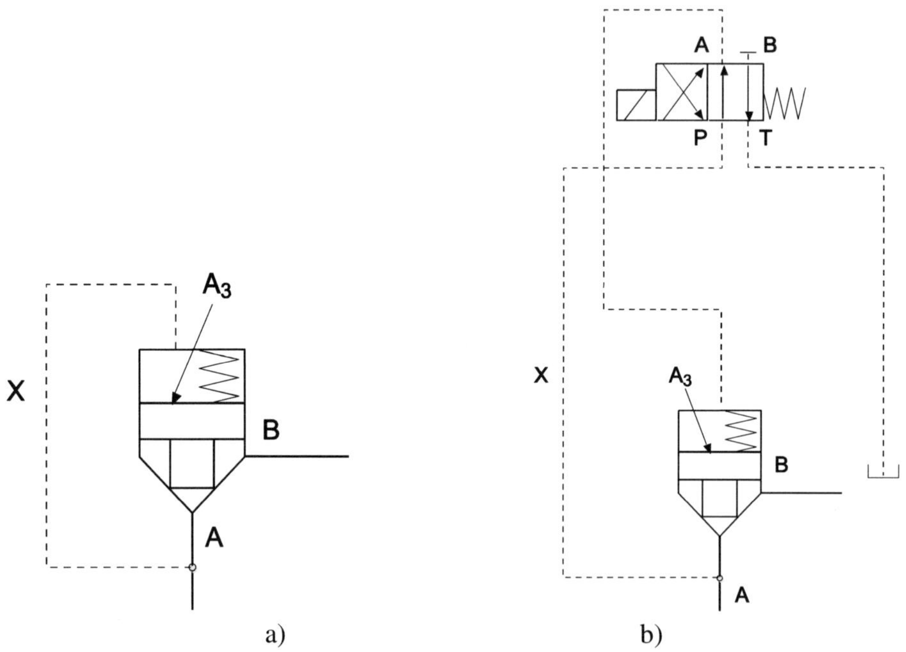

Figure 8.18. a, b. Scheme of a directional cartridge valve without and with piloting valve. Piloting through port A.

A second possible configuration, consist of the same kind of valve presented in figure 8.18 but now the feedback line is taken from port B instead of port A, see figure 8.19a, b. Figure 8.19a characterizes a check valve, flow will be never allowed to go from port B towards port A, and flow will be allowed to go from port A towards port B once pressure on port A will reach a pre determined value. Figure 8.19b, characterizes the same sort of valve providing the directional valve is at rest, once the directional valve will be connected, cartridge valve will open allowing fluid to go from port A to B and vice-versa.

A third possible configuration is a mixing of the previous two, in this one, see figure 8.20, both ports A and B are connected through the pilot valve to the spool upper side, therefore whenever the directional piloting valve is at rest, the valve will remain closed regardless of the pressure in ports A or B. Once the directional piloting valve will be connected, cartridge valve will open allowing fluid to go from ports A to B or vice-versa, flow will go from a high pressure port to a low pressure one.

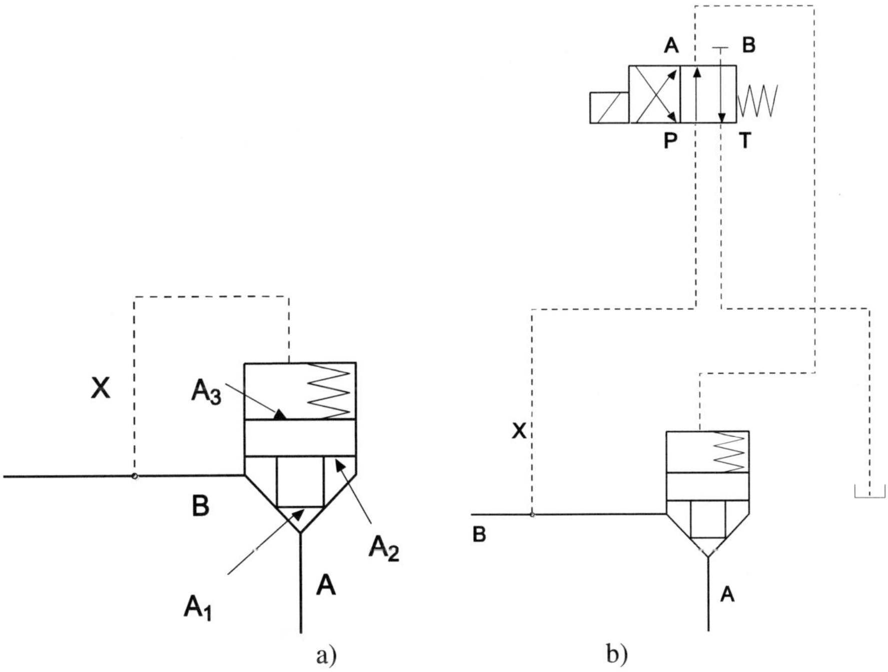

Figure 8.19. a, b. Scheme of a directional cartridge valve without and with piloting valve. Piloting through port B.

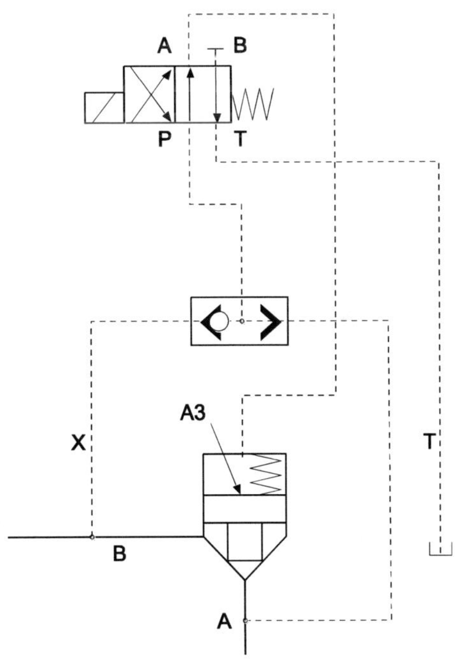

Figure 8.20. Scheme of a directional cartridge valve with piloting via ports A and B.

Having seen the previous configurations, it is easy to realize that another possible option would be connecting the port X to an external pressure, see figure 8.21a. In this configuration pressure onto the spool upper part can be modified at will, therefore opening a wide range of possibilities regarding the pressures at which flow is allowed to go from one port to another. At this point it is interesting to realize that if pressure on port X, via using a piloting proportional valve, could be modified at will, the cartridge directional valve would operate as a proportional one.

A case which could be considered as the opposite than the one just presented would be, when having an internal piloting, see figure 8.21b. Notice that port B and the cartridge spool upper side are now connected, clearly the flow will never be allowed to go from port B towards port A, and whenever pressure on port A will reach a value high enough to overcome the counterbalancing forces due to pressure on port B and the spring, the valve will open, allowing fluid going from port A to port B, again this configuration characterizes a check valve. Notice that a very similar sort of valve would be obtained if the internal piloting would connect port A and the spool upper side.

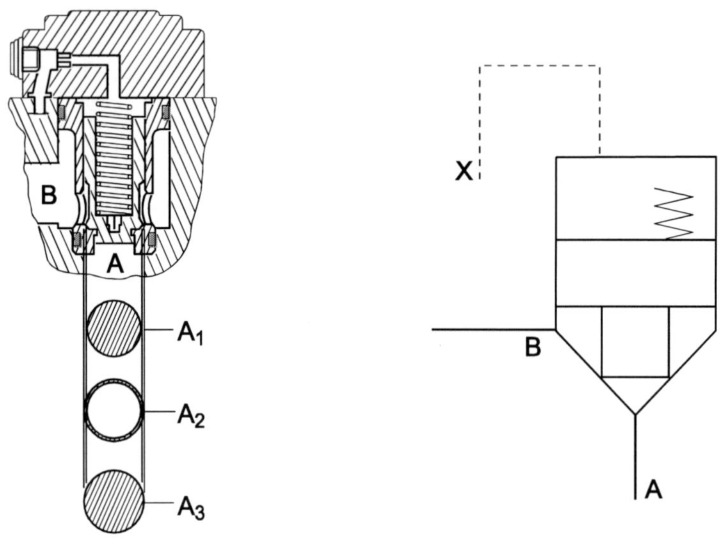

a)

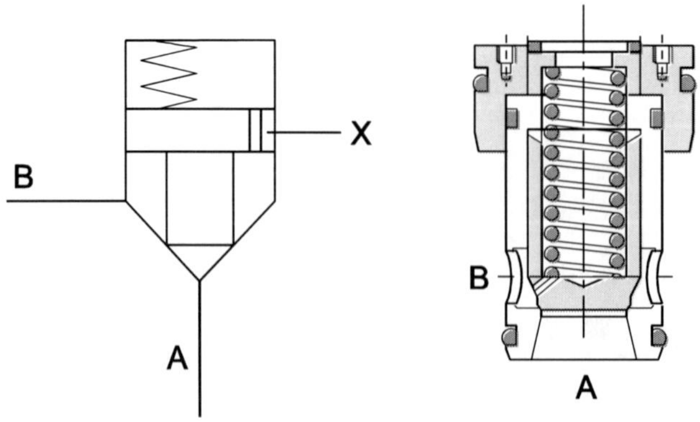

b)

Figure 21. a, b. Directional cartridge valves having external and internal piloting. a) external, b) internal.

Regarding the characteristic curves of a cartridge directional valve, the curves characterize the pressure differential between inlet and outlet as a function of the volumetric flow crossing the valve and for a given passage area. Usually the curves are defined for the maximum passage area, the valve is fully open, and therefore a single curve shall be provided for each directional cartridge valve nominal dimension. Notice as well that in the proportional field, a set of curves, for each passage area from fully closed to fully open should be defined. Figure 8.22a, presents the characteristic curves of a directional cartridge valve having a nominal dimension of 40. It is to be highlighted that any directional cartridge valve can be designed with and without damping mechanism; this is why figure 8.22a presents two curves for the same valve. A two way directional cartridge valve is modified to include a damping mechanism via extending the lower part of the spool, as presented in figure 8.6. The damping mechanism diminishes cartridge vibration and pressure fluctuations during the valve opening and closing, figure 8.22b, represents the estimated pressure fluctuations with and without damping mechanism when the valve opens and closes. Damping mechanism is used in circuits in which, for example, one or more actuators need to have a smooth starting or ending movement.

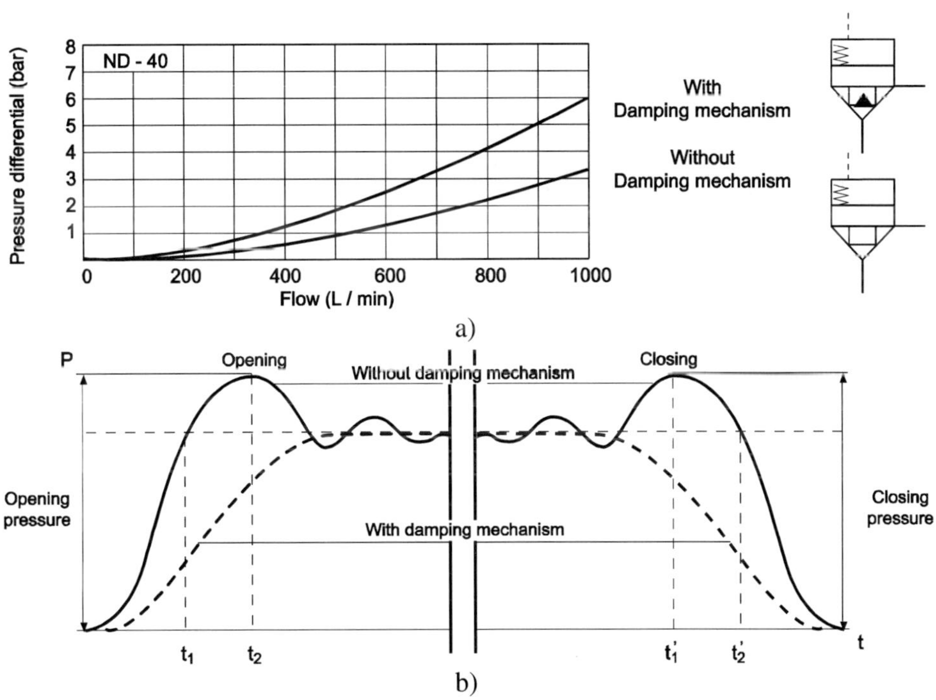

Figure 8.22a, b. Static and dynamic characteristic curves of a cartridge directional valve.

8.3.3. Flow Control Cartridge Valves

The mission of a flow control cartridge valve is to control the volumetric flow flowing to a particular circuit section. Usually the idea is to maintain a constant volumetric flow rather independent of the upstream/downstream pressure variation; these valves are called pressure compensated.

There exist therefore, two main groups of pressure control cartridge valves, the ones which are not pressure compensated and the ones which are pressure compensated. The first sort cannot maintain a constant downstream flow if there is an upstream or downstream pressure variation, downstream flow will therefore be affected by pressure fluctuations. The pressure compensated cartridge valves have the advantage of maintaining a constant downstream flow regardless of the system pressure fluctuations.

It is important to realize that, two way flow control no compensated cartridge valves are often able to regulate downstream flow because they interact with the main pressure relief valve of the system.

In reality, the simplest way to regulate downstream flow consists of increasing upstream pressure and direct part of the incoming flow from pump to tank via the main system pressure relief valve. Nevertheless in what follows several possible configurations shall be studied.

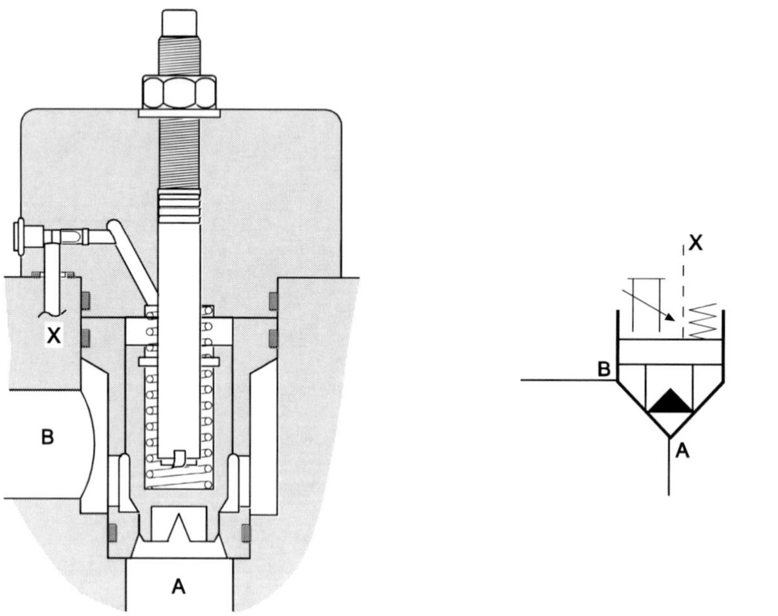

Figure 8.23. Two way flow control cartridge valve with damping mechanism and limiting spool displacement.

Figure 8.23 presents a flow control cartridge valve with damping mechanism and limiting spool displacement, notice that the spool displacement is being limited by a manual adjusting screw, which will delimitate the maximum opening section.

This is the simplest flow control cartridge valve configuration and it is called simple throttle. Considering figure 8.23, if for example downstream pressure, port B, increases, the flow across the valve would tend to decrease and in order to maintain the flow the opening section of the cartridge valve will tend to increase allowing a higher amount of flow crossing the valve, this regulation nevertheless is to be seen as partially effective.

A more effective way to maintain a constant flow is via using pressure compensated cartridge valves, an example of which is presented in figure 8.24. This particular configuration, consists of a flow restriction or a manual regulating throttle located below port B, and a pressure compensator which acts as a pressure reducing valve.

It is important to notice that the downstream pressure P_C, is used as a feedback and acts onto the upper side of the pressure reducing cartridge valve, notice as well that this particular valve is normally open, allowing fluid going from port A, pressure $P_{A,}$ to downstream of port B, pressure P_C.

If for example downstream pressure or load pressure, P_C, decreases, pressure acting onto the upper side of the pressure reducing cartridge valve will also be smaller and the cartridge valve will move upwards reducing the passage area, when doing so, pressure losses between port A and port B will increase, therefore decreasing the pressure on port B, P_B, as a result, pressure differential between downstream and upstream of the throttle, P_C - P_B will tend to remain constant and so will be the volumetric flow across it.

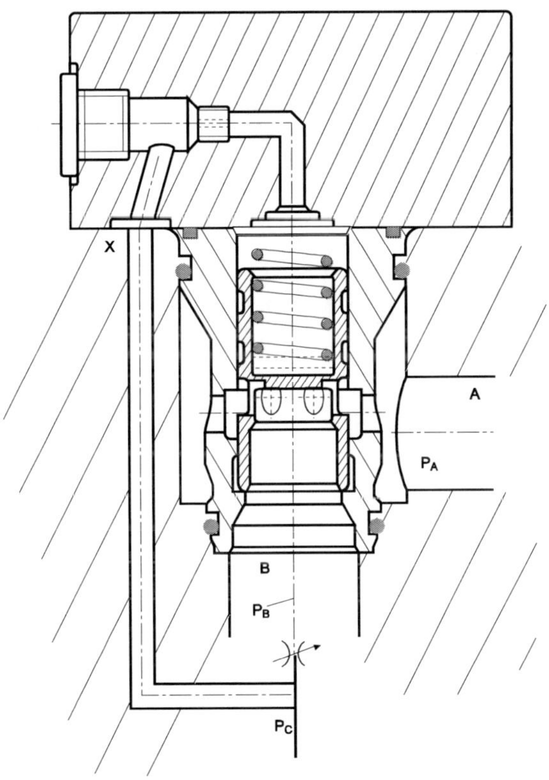

Figure 8.24. Pressure compensated cartridge valve consisting of a throttle and a pressure reducing valve.

Another possible configuration is the one shown in figure 8.25, consisting of a throttle and a cartridge relief valve connected in parallel. In this configuration, the cartridge relief valve spool is subjected to, on one side the throttle valve upstream pressure and on the other side the throttle valve downstream pressure. If downstream pressure decreases, the volumetric flow across the throttle valve will tend to increase, but as downstream pressure is acting onto the spool upper side, the cartridge relief valve opening will increase allowing part of the pump incoming fluid be directed to tank and tending to maintain a constant flow across the throttle valve.

A more sophisticated and very likely a highly efficient configuration than the ones presented until now, is the one introduced in figure 8.26. It consists of a downstream manual adjusting throttle valve, a cartridge relief valve acting as a main valve, a three way pressure reducing piloting valve and a proportional relief valve which controls the counterbalancing pressure on the pressure reducing piloting valve. Flow is mend to go from port A towards B and C, initially it appears as if the cartridge relief valve had to be closed since port A pressure acts on both sides of the main spool, but since a little amount of volumetric flow passes through the three way pressure reducing valve towards port B, in reality the pressure relief cartridge valve is partially open, therefore allowing the main flow going from port A towards B and C. Whenever downstream pressure, port C, for example decreases, the three way pressure reducing valve will increase the area passage between port B and tank, and at the same time reducing the area passage between ports A and B though the three way pressure reducing piloting valve, as a result the pressure onto the upper side of the cartridge valve spool will increase reducing the main flow area passage, as a conclusion the pressure on port B shall be reduced and therefore pressure differential between ports C and B shall be maintained, and so will be the volumetric flow.

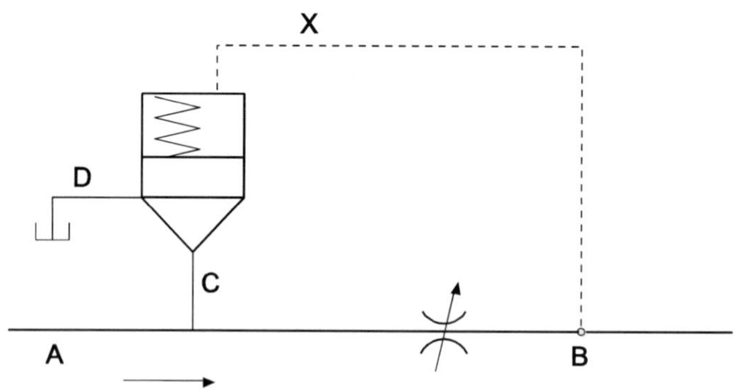

Figure 8.25. Pressure compensated cartridge valve consisting of a throttle and a pressure relief valve which directs flow to tank.

Notice as well that the proportional piloting pressure relief valve located onto the upper side of figure 8.26 will allow modifying at will the counterbalance pressure onto the three way pressure reducing valve, therefore allowing a higher degree of freedom regarding the downstream pressure at which the three way pressure reducing valve will switch.

As mentioned in previous sections, the maximum control in regulating any parameter is obtained via using proportional or servo valves. The key issue in proportional flow control cartridge valves consists in obtaining a full control of the opening section via using an electronic feedback of the main spool position. Figures 8.27 and 8.28 present, some example configurations, figure 8.27 consists of a main spool having its position controlled via a feedback electronic transducer and a proportional piloting valve which directs the downstream piloting flow whether to the spool upper side or towards tank. As the piloting valve can be proportionally controlled, the pressure onto the main cartridge spool upper side can be easily regulated and therefore the opening section shall be modified according to the user specifications.

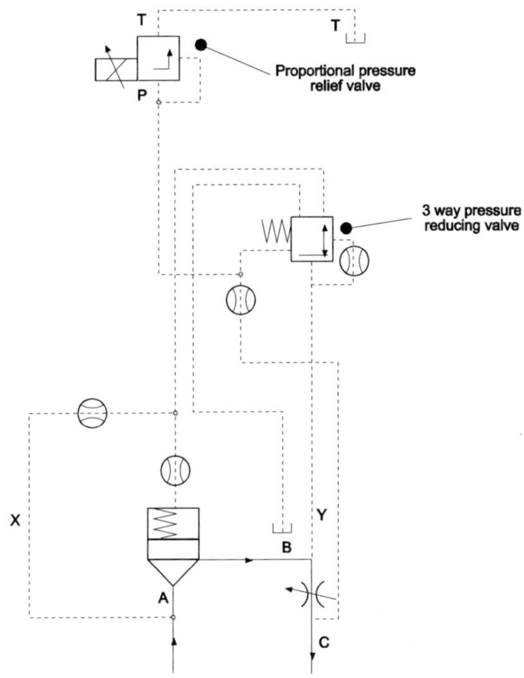

Figure 8.26. Flow control pressure compensated cartridge valve consisting of a throttle a three way pressure reducing valve and a proportional pressure relief valve.

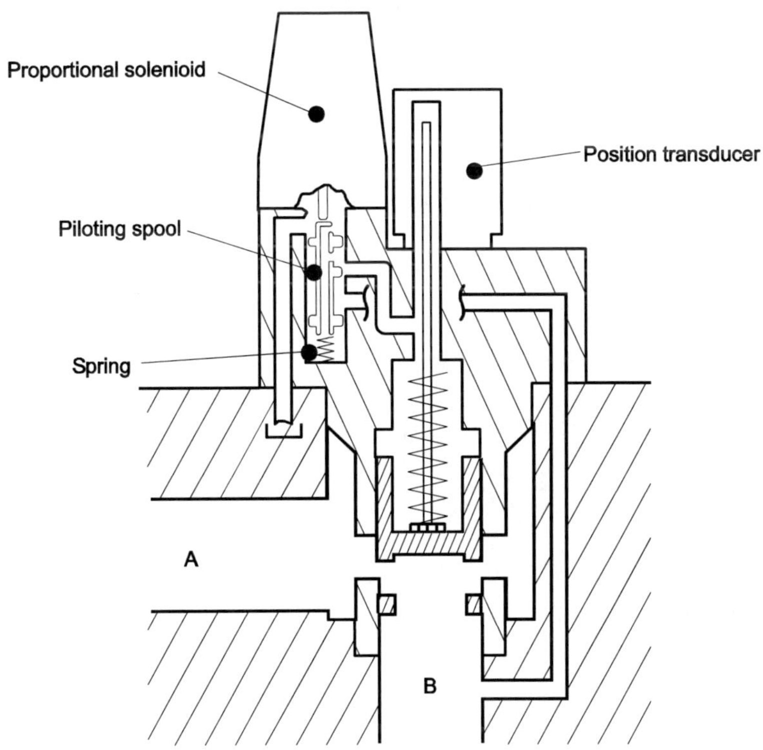

Figure 8.27. Example of a proportional flow control cartridge valve, the opening section is controlled via a piloting proportional directional valve.

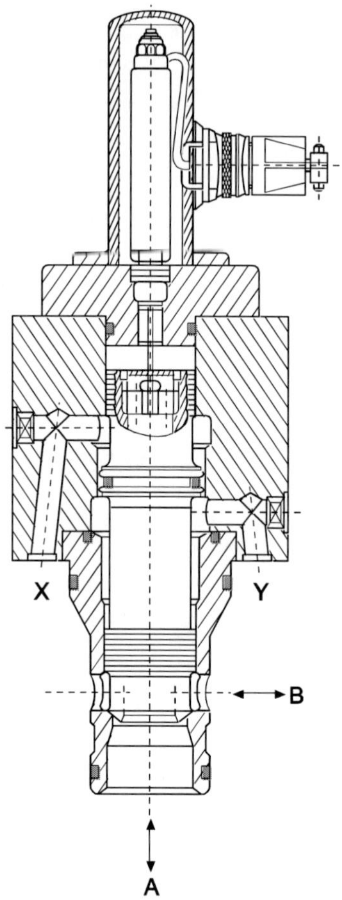

Figure 8.28. Example of a proportional flow control cartridge valve, the opening section is controlled via modifying the pressure on both sides of the main spool.

A similar configuration to the one just presented is introduced in figure 8.28, notice as well that the spool has a position feedback, but now there are two ports from which pressure at different spool sections can be supplied. Notice that port X acts onto the spool upper side and port Y acts onto the spool middle surface.

Pressure on ports X and Y has to be controlled by several external proportional piloting valves, not shown in the figure. If for example pressure is applied to port Y and port X is connected to tank, the spool will displace upwards, and the opposite will happen if pressure is applied to port X and port Y is connected to tank. Clearly, via regulating the pressure on both ports X and Y the spool position can be precisely defined, and so will be the opening section, the feedback transducer is used to compare and control the desired and real spool position.

Regarding the characteristic curves of flow control valves, there is no need to say that for a simple throttle and understanding that the area passage remains constant; the relation pressure differential versus flow crossing the valve is quadratic. For proportional valves, characteristic curves are often presented as a relation between the current applied to the proportional solenoid and the volumetric flow crossing the valve, curves obtained for each given pressure differential between valve inlet and outlet, figure 8.29 shows a generic plot where thanks to the proportional feedback the obtained relationship is linear.

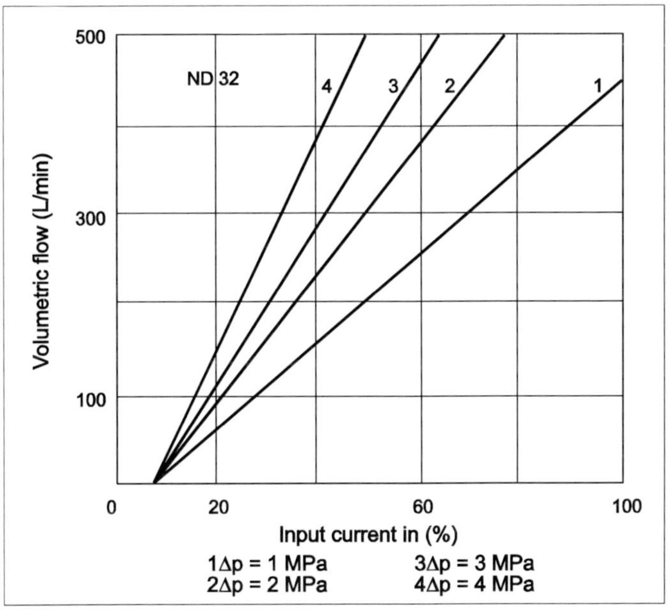

Figure 8.29. Example of characteristic curves for proportional flow control cartridge valves.

8.4. Example of Application

As a very simple example of a cartridge valves application, figure 8.30 presents a schematic representation of a circuit designed to move a hydraulic cylinder of very large dimensions. Notice that the circuit consists of four directional cartridge valves, a four way three positions piloting directional valve and the cylinder, the dash lines are the piloting ones. Figure 8.30 is showing the circuit connections which force the linear hydraulic cylinder to move forward. Notice that the piloting valve connects the pressure line with the upper side of the cartridge valves 2 and 4, therefore these two valves will remain closed. On the other hand the piloting valve connects as well the upper side (area A3) of cartridge valves 1 and 3 to tank, allowing the spool displacement. As a result, flow entering the circuit crosses the valve 3 and goes towards the cylinder right hand side, at the same time fluid from the cylinder left hand side is allowed to go towards valve 1 and it is directed to tank, see the arrows. It is important to realize that cartridge valve 1 allows a manual regulation of the opening section, and therefore it is possible to regulate the cylinder left hand side pressure. Although not presented in figure 8.30, it is important to see that the cylinder will pull back whenever the right hand side of the piloting valve will be connected, notice that under this situation, the piloting valve will pressurize the upper side of valve 1 preventing it from moving, but valves 2, 3 and 4 are allowed to displace whenever pressure on its lower spool side (area A1) will be applied. Under these conditions, incoming flow from the pump, is being directed through valves 2 and 3 towards both sides of the cylinder, but as valve 4 is allowing the fluid to go to tank the piston right hand side pressure will be smaller than the left hand side part, this can be regulated via using the throttle valve located before valve 3, as a result cylinder will pull back. Notice as well that whenever the piloting valve is at rest, central position, the cylinder will not move since the four cartridge valves will remain closed.

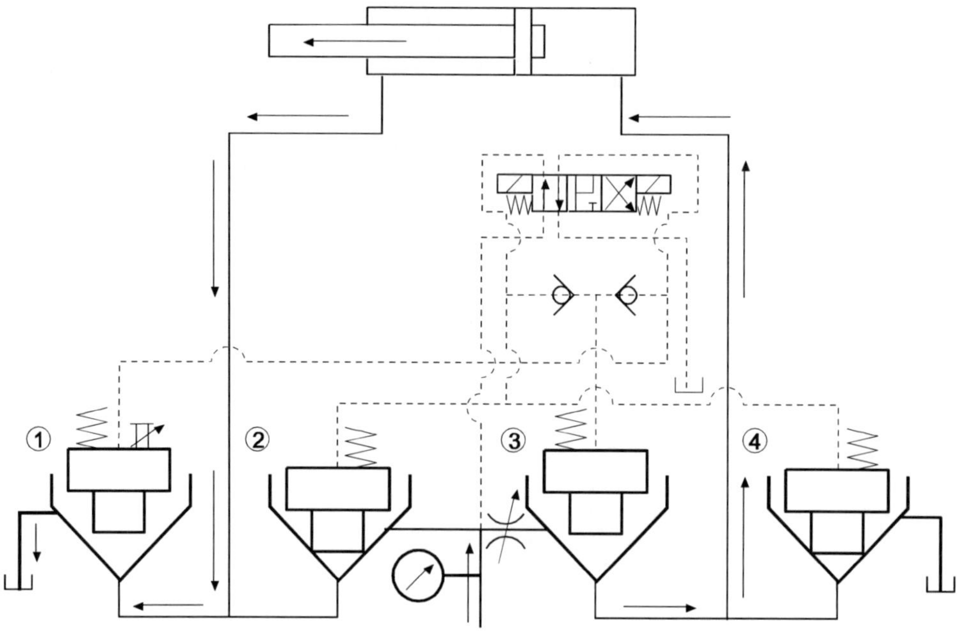

Figure 8.30. Example of application of directional cartridge valves.

Before finishing the present chapter I would like to point out that most of the figures and graphs here introduced are simple modifications of the graphs and figures found in the tutorials from Rexroth, Bosch and Vickers among others, where the reader will find extensive information from this particular field.

8.5. References

[1] Mannesmann Rexroth. (1989). Técnica de válvulas insertables de 2 vías. Main, Germany. Mannesmann Rexroth AG.

[2] Vickers. Valvulas de Cartucho para roscar. Germany. Vickers Systems SA.

[3] Garcia D, Bergada JM. (1997). Válvulas insertables de dos vías I. Barcelona Fluidos 560-561.

[4] Garcia D, Bergada JM. (1997). Válvulas insertables de dos vías II. Barcelona Fluidos 654-664.

INDEX

A

Absolute filtration, 344
Acceleration, 13, 14
Accumulators,
correction factors, 329
energy storage main expressions, 326-328, 331, 339
type of, 325-326
used as pulsation compensators, 329, 330, 339
used as shock damper, 330, 331, 339, 340
Angular momentum equation,
integral form, 39-41
application to turbomachinery, 41-43
non inertial coordinate systems, 43-44
Angular velocity, 18-20
Axial piston pump dynamic simulation, 309-319
Axial piston pump experimental test rig, 314, 315
Axial piston pumps main components, 183, 184
Axial piston pumps numerical results, 317, 319
Axial piston pump pressure ripple, 316, 317

B

Barrel port plate,
boundary conditions, 266
(CFD) results, 281
dynamics, experimental results, 286, 287
dynamics, fluctuation wave, 287, 289
dynamic simulation, 294, 295, 296
experimental test rig, 277, 279
film thickness, 292
fluctuation wave, 293
force, 271, 272, 283
force due to timing groove, 274, 275
leakage, 269
leakage as a function of the clearance, 282
main dimensions, 265
main dynamic equations, 294, 295
mathematical analysis, 264
pressure distribution, 267, 268
previous research, 261
surface roughness, 288
theoretical results, 280, 281
timing groove leakage, 271
torque, 273, 274, 284, 285
torque due to timing groove, 275, 276, 284, 285
transducers measured average position, 290

Bernoulli equation, 46
Beta rating, 344
Bulk modulus, 4, 9

C

Cartridge valves,
generic classification, 352
different area rates, 354, 355
dynamic forces, 354
logic element, 351
main parts, 352, 353
Cartridge valves working limits, 351
Circulation, 15
Conical seat relief valves, 131
(CFD) modeling, 143-145
Conical seat relief valves,
    forces, 139-143
    mathematical development, 136-143
    Contamination control strategies, 348, 349
Contamination ISO number, 343
    Contamination in hydraulic circuits, sorts of, 342-344
Continuity equation, 30
Cartesian coordinates, 30
    cylindrical coordinates, 31, 36
    spherical coordinates, 31, 36
    integral form, 29, 30
    differential form, 30, 31
Convergence criteria (CFD), 125, 126
Coreless filtering elements, 347
Couette flow, 54
Couette-Poiseulle flow, 53, 54
Continuum theory, 3, 4
Convective flow, 14
    Coupling pressure and velocity for (FV), (FD), 113, 115
Cylindrical journal bearings, 89-96
    force, 94
pressure distribution, 95, 96
torque, 94
volumetric flow, 92

D

Deborah number, 3
Deformation tensor, 22, 23
Directional cartridge valves,
    piloting through port A, 361
    piloting through ports A and B, 362
    piloting through port B, 362
    with external and internal piloting, 363
    static and dynamic characteristic curves, 363
Disk runout, 232, 233

E

Effect of groove position, 237-239
Effective bulk modulus, 7-9
Energy equation,
    integral form, 44, 45
    mechanical work, 45, 46
    application to turbomachinery, 47-49
    differential form, 49, 50

F

Filters location in hydraulic circuits, 346
Finite element method, spatial discretization, 115
Flow between two parallel plates, 50, 51
Flow control cartridge valves,
    application example, 368, 369
    proportional, 367, 368
    proportional curves, 368
Flow control pressure compensated, cartridge valves, 364, 365, 366
    with relief valve, 366
    with pressure reducing valve and proportional relief valve, 366
Flow in narrow gaps, 78-81
Flow tone generators, 171

Flow with negligible acceleration, 78-81
Fluid elastic oscillations, 173
Fluid kinematics, 20, 21
Fluid dynamic oscillations, 172
Fluid, mechanical point of view, 3
Fluid, reologic equations, 12
Fluid resonant oscillations, 172
Fluid sampling interval, 349

G

Galerkin finite element approximation, 117, 118
Grid independency (CFD), 126
Groove pressure, 234, 236

H

Hydraulic filters, 345

K

Kinematic viscosity, 13
Knudsen number, 4

L

Laminar flow between two concentric rotating tubes, 69-78
boundary conditions, 70
Laminar flow between two concentric pipes, 63-69
Laminar flow in annular tubes, 62
  boundary conditions, 63
Laminar flow inside circular ducts, 59-78
Leakage barrel-port plate, 311
Leakage piston-barrel, 310
Leakage slipper swash plate, 310
Leakage spherical journal bearing, 311
Lennard-Jones potential, 2
Linear deformation, 21, 22
Local thermodynamic equilibrium, 4

M

Matherial derivative, 13
Mechanical work, 45, 46
Mesh less method, 127
Mesh topology, 126, 127
Momentum equation, 106, 107
discretization, 107, 108
  discretization via finite differences, (FD), 111, 113
  discretization via finite elements, (FE), 115
  discretization via finite volumes, (FV), 108, 111
source term linearization, 110, 111
integral form, 31, 32
differential form, 32, 36
  integral form, non inertial coordinate systems, 37, 38
  differential form, non inertial coordinate systems, 39
Molecular attraction / repulsion, 1
Multiple fabric layers filtering elements, 347

N

Navier Stokes,
  Cartesian coordinates, compressible / incompressible, 35
  coordinate transformation, 118-120
  cylindrical coordinates, 36
  source terms linearization, 120
  spherical coordinates, 36
Navier Stokes equation, 35
  weak form, 115-117
Navier Stokes transformed equations,
  pressure and velocity coupling, 122, 125
spatial discretization, 120,122
Nominal filtration, 344

P

Pathlines, 15, 16
Petroff number, 95
Piston-barrel,
clearance under tilt conditions, 192
(CFD), 190
Piston-barrel cavitation, effect of grooves, 200, 201
Piston-barrel leakage, 187-190, 194, 202, 204
effect of grooves, 198, 201, 202, 204
Piston-barrel pressure distribution, 191, 195, 196, 197
previous research, 184-186
Piston-barrel torque, 202-204
Piston configurations studied, 195
Piston cylinder differential equation, 312, 313
Plane journal bearings, 81-86
Plane Poiseulle flow, 54,55
Poiseulle flow, 54, 55, 59-62
Pressure losses in filters, 347, 348
Pressure reducing cartridge valves,
characteristic curves, 360
standard, 358, 359
with three port manual adjusting piloting valve, 359
with three port manual adjusting piloting valve, and closing function, 359
Pressure relief cartridge valves,
proportional configurations, 360
characteristic curves, 358
manual adjusting, 355, 356
manual adjusting and discharge option, 355, 356
manual adjusting and shutting function, 355, 356
proportional, 357
two adjusting pressures, 356, 357
three adjusting pressures, 356, 357
Proportional directional control valves, steady state curves 150-152

R

Radial distribution function, 1
Rayleich flow, 55-59
Recommended fluid sampling interval, 349
Reynolds equation of lubrication, 187, 212, 218, 265, 305
Cartesian coordinates, 86, 89
cylindrical coordinates, 96-101
Reynolds transport equation, 25, 29
Rotational flow, 18-20

S

Selecting a grid, 103
Servovalves, 153
Servovalve,
discharge coefficients, 162-164
erratic performance, 165-175
flow instability, 164
forces acting on it, 154-162
instability zone, 167
static performance curves, 176, 177
vibration spectra, 167
vibration frequencies, 165
Sommerfeld number, 94, 95
Slipper,
dimensions, 231
experimental test rigs, 231, 234
Slippers flat,
boundary conditions, 214
(CFD), 226
dynamic, 239
dynamic, groove pressure, 240, 241
dynamic, leakage, 244
dynamic, torque, 240, 241
dynamic, swash plate, 242, 243
dynamic, vorticity inside the groove, 243-247, 252, 254
force, 215, 216

leakage, 215, 216, 234, 235
numerical model, 227-230
pressure distribution, 215, 216, 236, 237
Slippers,
previous research, 207-212
flat, static equations, 212,217
tilt, boundary conditions, 219
tilt, dynamic, leakage, 247-250, 256
tilt, force, 224
tilt, leakage, 223
tilt, dynamic, groove pressure, 255
tilt, dynamic, pressure distribution, 250, 251
tilt, dynamic, pressure differential inside the groove, 252, 256
tilt, pressure distribution, 221-223
tilt, static equations, 217-225
tilt, torque, 224, 225
Spherical journal bearing,
leakage, 304, 307
main parameters, 302
mathematical analysis, 303-307
pressure distribution, 305, 306
Streaklines, 16
Streamlines, 15, 17
Strouhal number, 168
Surface tension, 9, 10

T

Thermodynamic fluid properties, 2, 3

V

Valve classification, 130
Velocity distribution between two concentric cylinders, 67
Velocity gradient tensor, 22, 23
Viscosity, definition, 10-13
Volumetric pumps classification, 182
Vorticity, 18, 19, 20
Vorticity tensor, 22, 23

W

Water absorbent filtering elements, 347
Wear and erosion in hydraulic systems, 345